LAWS OF EXPONENTS

$$a^x a^y = a^{x+y}$$

$$a^{-x} = \frac{1}{a^x}$$

$$(a^x)^y = a^{xy}$$

$$(ab)^x = a^x b^x$$

$$a^0 = 1$$

$$\left(\frac{a}{b}\right)^x = \frac{a^x}{b^x}$$

$$a^1 = a$$

$$1^x = 1$$

$$\frac{a^x}{a^y} = a^{x-y}$$

LAWS OF LOGARITHMS

$$\log_a (xy) = \log_a x + \log_a y$$

$$\log_b x = \frac{\log_a x}{\log_a b}$$

$$\log_a \frac{1}{x} = -\log_a x$$

$$\log_a 1 = 0$$

$$\log_a a = 1$$

$$\log_a \frac{x}{y} = \log_a x - \log_a y$$

$$a^{\log_a x} = x$$

$$\log_a (x^c) = c \log_a x$$

$$\log_a (a^x) = x$$

COMPLEX NUMBERS

$$i^2 = -1 \qquad \overline{a + bi} = a - bi \qquad |a + bi| = \sqrt{a^2 + b^2}$$

$$(a + bi) + (c + di) = (a + c) + (b + d)i$$

$$(a + bi)(c + di) = (ac - bd) + (ad + bc)i$$

SEQUENCES AND SERIES

Arithmetic sequence: $a_j = a_1 + (j - 1)d$

$$\sum_{j=1}^{n} a_j = na_1 + \frac{n(n - 1)}{2} \cdot d$$

Geometric sequence: $a_j = ar^{j-1}$

$$\sum_{j=1}^{n} a_j = a \frac{1 - r^n}{1 - r}$$

Geometric series:

$$\sum_{j=1}^{\infty} ar^{j-1} = \frac{a}{1 - r} \quad \text{for } |r| < 1$$

PERMUTATIONS AND COMBINATIONS

$$n! = n(n - 1)(n - 2) \cdots 3 \cdot 2 \cdot 1 \qquad \binom{n}{k} = {}_nC_k = \frac{n!}{k!(n - k)!}$$

$$_nP_k = \frac{n!}{(n - k)!}$$

BINOMIAL THEOREM

$$(x + y)^n = \sum_{j=0}^{n} \binom{n}{j} x^{n-j} y^j$$

COLLEGE ALGEBRA

COLLEGE ALGEBRA

ROBERT ELLIS
University of Maryland
at College Park

DENNY GULICK
University of Maryland
at College Park

HARCOURT BRACE JOVANOVICH, INC.

New York   San Diego   Chicago   San Francisco   Atlanta
London   Sydney   Toronto

TO OUR FAMILIES

ISBN: 0-15-507905-0

Library of Congress Catalog Card Number: 80-85130

Printed in the United States of America

Technical Art by Mel Erikson Art Service

Photo Credits

Cover: Marian Griffith

P. 164: © Tiers, Monkmeyer Press Photo Service
   236: left, Official U.S. Navy Photograph
        right, © Jim Anderson
   286: UPI Photo
   293: The Metropolitan Museum of Art, New York
   430: Ken Karp
   449: The Oriental Institute, University of Chicago

PREFACE

This book provides a thorough treatment of the algebraic topics generally covered in a college course; it can be used by anyone with the equivalent of two years of high school algebra. The text is essentially identical to *College Algebra and Trigonometry*, by Ellis and Gulick, except for the omission of those subjects related to trigonometry. Thus there is ample material for a one-semester or two-quarter course, and instructors can choose topics in accordance with the background of the students and course objectives.

The initial chapter contains a review of basic algebraic concepts and can be covered quickly or more slowly, depending on the preparation of the reader. Chapters 2 through 5 present the main topics of college algebra and related topics from analytic geometry. These include equations and inequalities, functions and equations and their graphs, and the exponential and logarithmic functions. The remainder of the book examines a variety of other subjects related to algebra: systems of equations and matrices, complex numbers, roots of polynomials, and topics in discrete algebra (including mathematical induction, sequences, series, permutations and combinations, and the binomial theorem).

One major goal for us has been to provide an exceptionally readable book on college algebra. As a result, we have tried to present each new topic carefully, with plenty of explanation and illustration, supporting examples, and cautionary remarks (which are set off from the text to enhance their impact). The examples thoroughly reinforce new concepts and provide computational practice. Furthermore, we include those theorems and proofs of theorems that give additional insight into the fundamental ideas presented. The symbol □ signals the completion of an example, and ■ identifies the end of a proof.

The exercise sets have been designed to afford ample practice on the topics appearing in the text. The exercises are graded, from routine to more sophisticated. Generally they are paired, odd and even, in order to give additional practice and because answers to the odd-numbered exercises are included in the back of the book. Where appropriate we have provided calculator exercises; these are clearly identified with the symbol ▣. In keeping with current trends, both text and exercises include numerous applications to such diverse areas as chemistry, geology, acoustics, motion, demography, agriculture, investments, economics, archeology, architecture, and games. Both the metric and the English systems of measurement are employed in applied exercises. The most difficult exercises are accompanied by an asterisk for easy recognition. Each chapter concludes with a list of key terms and formulas and a set of review exercises that give practice on the major topics of the chapter. For convenient reference, the major formulas and laws from college algebra are also printed on the inside covers. An Answer Manual containing answers to the even-numbered exercises is also available to instructors.

We wish to thank the following reviewers for their thoughtful and constructive comments: Murray Cantor (University of Texas, Austin), Daniel Mosenkis (City College of the City University of New York), Nancy Jim Poxon (Sacramento State College), Helen Salzberg (Rhode Island College), Erik Schreiner (Western Michigan University), Mary Scott (University of Florida), and William Smith (University of North Carolina, Chapel Hill). Our gratitude also goes to the staff of Harcourt Brace Jovanovich, Inc., for their assistance on this, our second project with them. We would especially like to thank Marilyn Davis for the many ways in which she helped us in the preparation of the manuscript and for her patience as we completed it, Judy Burke for her meticulous editing of the manuscript, Harry Rinehart for his work in designing the book, and Tracy Cabanis for supervising its production.

Robert Ellis

Denny Gulick

CONTENTS

BASIC ALGEBRA

1

Chapter 1 is the first of four chapters that contain the core of our discussion of algebra. It is intended primarily as a review of the basic rules of algebra.

We begin the chapter with a description of the various kinds of real numbers and the basic properties of the arithmetic operations of addition, subtraction, multiplication, and division. Next we show how to associate the real numbers with the points on a line. Then we branch out and discuss exponents, radicals, polynomials, and rational expressions, all of which are derived in one way or another from the four basic arithmetic operations.

The results of this chapter not only serve as tools that will be used repeatedly throughout the remainder of the book but also form the foundation for the many subjects that will be developed in the remaining chapters.

1.1

REAL NUMBERS

Real numbers appear everywhere in our lives. Indeed, we encounter them anywhere from a grocery store to a doctor's office to a sports page. Let us begin by describing the various kinds of real numbers.

Kinds of Real Numbers

The most basic kind of real numbers are the **natural numbers** (sometimes called **counting numbers**), 1, 2, 3, ... Closely related to the natural numbers are the **integers**, $0, 1, -1, 2, -2, 3, -3, \ldots$. Those integers that are multiples of 2 (namely, 0, $2, -2, 4, -4, 6, -6, \ldots$) are **even** integers, and the remaining ones (namely, $1, -1, 3, -3, 5, -5, \ldots$) are **odd** integers. From the integers arise the **rational numbers,** those that can be written in the form a/b, where a and b are integers with $b \neq 0$. For example, $\frac{22}{7}$, $-\frac{3}{2}$, $\frac{123}{49,265}$, and 11.6 (which can be written as $\frac{116}{10}$) are rational numbers. Every integer a is a rational number, since a can be written as a/b with $b = 1$. Thus the rational numbers include the integers.

There are real numbers that are not rational, that is, cannot be written as a/b, where a and b are integers. Such numbers are called **irrational numbers.** Although we will not prove it, two examples of irrational numbers are π (which is the ratio of the circumference of a circle to its diameter) and $\sqrt{2}$ (which is the ratio of the length of the diagonal of a square to the length of a side).

Every real number has a decimal expansion. For example, $\frac{5}{4}$ has the decimal expansion 1.25. The decimal expansion of any rational number is either terminating, as is the decimal expansion 1.25 of $\frac{5}{4}$, or repeating, as are the decimal expansions 0.6666 . . . of $\frac{2}{3}$ and 3.545454 . . . of $\frac{39}{11}$. The decimal expansion of an irrational number may or may not be possible to determine. On the one hand, it is known that the decimal expansion of π begins

$$3.14159265358979 \ldots$$

but the expansion does not terminate, nor does it repeat, nor is there any known pattern to the digits in the expansion. On the other hand, the digits in the expansion of an irrational number *can* have an easily recognizable pattern (see Exercise 53).

There are other kinds of numbers besides real numbers, but the real numbers are the only ones we will use until Chapter 7.

Operations and Conventions

There are four fundamental operations of arithmetic. For two real numbers a and b, we denote the operations as follows:

ARITHMETIC OPERATIONS
Sum: $a + b$
Difference: $a - b$
Product: ab, or $a \cdot b$, or $(a)(b)$
Quotient: $\dfrac{a}{b}$, or a/b, or $a \div b$, with $b \neq 0$

The numbers a and b are ***factors*** of the product ab, and in the quotient a/b, a is the **numerator** and b the **denominator**. Frequently the quotient a/b of two real numbers is called a ***fraction***.

> **Caution:** You will notice that a/b is not defined for $b = 0$. Thus dividing by 0 is not permitted and the quotient $a/0$ is not defined. In particular, $0/0$ and $1/0$ are not defined.

It will be important to keep in mind two conventions that are universally applied in evaluating arithmetic expressions.

First, in evaluating an expression such as $a + bc$, multiplication and division take precedence over addition and subtraction. Thus

$$3 + 4 \cdot 5 = 3 + 20 = 23$$

and

$$3 - 2 \div 6 = 3 - \frac{1}{3} = \frac{8}{3}$$

Second, we perform operations inside parentheses before those outside the parentheses. As a result,

$$4(-5 + 1 \cdot 3) = 4(-2) = -8$$

More generally, computation progresses from the inside expressions outward. This implies that

$$5[-3 + (4 + 2 \cdot 6)] = 5(-3 + 16) = 5 \cdot 13 = 65$$

Properties of the Arithmetic Operations

We now list the most important properties of addition and multiplication.

PROPERTIES OF ADDITION AND MULTIPLICATION

Addition	*Multiplication*	
$a + b = b + a$	$ab = ba$	Commutative properties
$(a + b) + c =$ $a + (b + c)$	$(ab)c = a(bc)$	Associative properties
$a + 0 = a =$ $0 + a$	$a \cdot 1 = a = 1 \cdot a$	Identity properties
For any number a there is a number $-a$ (called the ***additive inverse*** of a) such that $a + (-a) =$ $0 = (-a) + a$	For any number $a \neq 0$ there is a number $1/a$, or a^{-1} (called the ***multiplicative inverse*** of a), such that $a \cdot \dfrac{1}{a} = 1 = \dfrac{1}{a} \cdot a$	Inverse properties
$a(b + c) = ab + ac$ $(a + b)c = ac + bc$		Distributive properties

The associative property says that $(a + b) + c = a + (b + c)$, and as a result we can eliminate the parentheses and write $a + b + c$ for either $(a + b) + c$ or $a + (b + c)$. Thus by definition,

$$a + b + c = (a + b) + c = a + (b + c)$$

In particular,

$$6 + 4 + 1 = (6 + 4) + 1 = 6 + (4 + 1)$$

all of which are equal to 11.

In the same vein, we define abc by

$$abc = (ab)c = a(bc)$$

If we apply the distributive law several times, we find that

$$(a + b)(c + d) = (a + b)c + (a + b)d = ac + bc + ad + bd$$

Consequently

$$(a + b)(c + d) = ac + bc + ad + bd \tag{1}$$

Thus to multiply $a + b$ and $c + d$, we add together the four products that we can form by multiplying one of the numbers a and b from the first sum $(a + b)$ by one of the numbers c and d from the second sum $(c + d)$. The four products are indicated in the following diagram:

$$
\begin{array}{c}
\overset{\displaystyle ac}{\overbrace{}} \\
\overset{\displaystyle bc}{\overbrace{}} \\
(a + b)(c + d) \\
\underbrace{}_{\displaystyle bd} \\
\underbrace{}_{\displaystyle ad}
\end{array}
$$

The use of formula (1) is illustrated in Example 1, in which we use the fact that the square a^2 of any number a is defined by

$$a^2 = a \cdot a$$

Example 1. Find the product $(3a + 2)(4a + 5)$.

Solution. By (1),

$$(3a + 2)(4a + 5) = (3a)(4a) + (2)(4a) + (3a)(5) + (2)(5)$$

$$= 12a^2 + 8a + 15a + 10$$

$$= 12a^2 + 23a + 10 \quad \square$$

A special case of (1) arises if $c = a$ and $d = b$:

$$(a + b)(a + b) = a \cdot a + b \cdot a + a \cdot b + b \cdot b$$

or more succinctly,

$$(a + b)^2 = a^2 + 2ab + b^2 \qquad (2)$$

Similarly, other special cases of (1) are

$$(a - b)^2 = a^2 - 2ab + b^2 \qquad (3)$$

and

$$(a + b)(a - b) = a^2 - b^2 \qquad (4)$$

Example 2. Using the fact that $(\sqrt{2})^2 = 2$, verify that $(3 + \sqrt{2})^2 = 11 + 6\sqrt{2}$.

Solution. By (2),

$$(3 + \sqrt{2})^2 = 3^2 + 2(3)(\sqrt{2}) + (\sqrt{2})^2$$
$$= 9 + 6\sqrt{2} + 2 = 11 + 6\sqrt{2} \quad \square$$

A further property of multiplication that we will use repeatedly in solving equations is the following.

ZERO PROPERTY If $ab = 0$, then either $a = 0$ or $b = 0$.

Properties of Subtraction and Division

The basic properties involving negatives of numbers and fractions are listed below:

$$a - b = a + (-b) \qquad\qquad -\left(\frac{a}{b}\right) = \frac{-a}{b} = \frac{a}{-b}$$

$$-(-a) = a \qquad\qquad \frac{a}{b} + \frac{c}{d} = \frac{ad + bc}{bd}$$

$$-a = (-1)a \qquad\qquad \frac{a}{b} + \frac{c}{b} = \frac{a + c}{b}$$

$$-(ab) = (-a)b = a(-b) \qquad\qquad \frac{a}{b} \cdot \frac{c}{d} = \frac{ac}{bd}$$

$$(-a)(-b) = ab \qquad\qquad \frac{ac}{bc} = \frac{a}{b}$$

$$a(b - c) = ab - ac \qquad\qquad \frac{1}{\frac{1}{a}} = a$$

$$\frac{a}{b} = \frac{c}{d} \text{ if and only if } ad = bc \qquad\qquad \frac{\frac{a}{b}}{\frac{c}{d}} = \frac{a}{b} \cdot \frac{d}{c} = \frac{ad}{bc}$$

$$\frac{a}{b} = a\left(\frac{1}{b}\right)$$

If a and b are integers with $b \neq 0$, then the fraction a/b is in *lowest terms* if a and b have no common integer factors besides 1 and -1. Thus $\frac{4}{7}$ is in lowest terms, whereas $\frac{6}{8}$ is not, because 2 is an integer factor of both 6 and 8. The law

$$\frac{ac}{bc} = \frac{a}{b}$$

assists us in reducing a fraction to lowest terms. The idea is to determine integer factors common to numerator and denominator and then to cancel like factors in pairs, one from the numerator and one from the denominator.

Example 3. Reduce the following fractions to lowest terms.

a. $\dfrac{8}{20}$ b. $\dfrac{198}{54}$

Solution.

a. Since $8 = 4 \cdot 2$ and $20 = 4 \cdot 5$, it follows that

$$\frac{8}{20} = \frac{4 \cdot 2}{4 \cdot 5} = \frac{2}{5}$$

Because 2 and 5 have no common integer factors besides 1 and -1, $\frac{2}{5}$ is reduced to lowest terms.

b. Since $198 = 9 \cdot 22 = 9 \cdot 11 \cdot 2$ and $54 = 9 \cdot 3 \cdot 2$, it follows that

$$\frac{198}{54} = \frac{9 \cdot 11 \cdot 2}{9 \cdot 3 \cdot 2} = \frac{11}{3}$$

Because 11 and 3 have no common integer factors besides 1 and -1, $\frac{11}{3}$ is reduced to lowest terms.   □

Example 4. Write the sum $\dfrac{b}{a - b} + \dfrac{a}{a + b}$ as one fraction.

Solution. Using the properties of fractions, we have

$$\frac{b}{a - b} + \frac{a}{a + b} = \frac{b(a + b) + (a - b)a}{(a - b)(a + b)}$$

$$= \frac{ba + b^2 + a^2 - ba}{a^2 - b^2}$$

$$= \frac{a^2 + b^2}{a^2 - b^2} \quad □$$

If we add $\frac{3}{8}$ and $\frac{7}{12}$ by using the formula

$$\frac{a}{b} + \frac{c}{d} = \frac{ad + bc}{bd}$$

we obtain

$$\frac{3}{8} + \frac{7}{12} = \frac{3 \cdot 12 + 8 \cdot 7}{8 \cdot 12} = \frac{92}{96}$$

An equivalent way of obtaining the same result is to write $\frac{3}{8}$ and $\frac{7}{12}$ with the same denominator, called a *common denominator:*

$$\frac{3}{8} = \frac{3 \cdot 12}{8 \cdot 12} = \frac{36}{96} \quad \text{and} \quad \frac{7}{12} = \frac{7 \cdot 8}{12 \cdot 8} = \frac{56}{96}$$

Then

$$\frac{3}{8} + \frac{7}{12} = \frac{36}{96} + \frac{56}{96} = \frac{36 + 56}{96} = \frac{92}{96} \tag{5}$$

The common denominator 96 was obtained by taking the product of the denominators 8 and 12 of the fractions $\frac{3}{8}$ and $\frac{7}{12}$. However, 24 is also a common denominator of $\frac{3}{8}$ and $\frac{7}{12}$, because

$$\frac{3}{8} = \frac{3 \cdot 3}{8 \cdot 3} = \frac{9}{24} \quad \text{and} \quad \frac{7}{12} = \frac{7 \cdot 2}{12 \cdot 2} = \frac{14}{24}$$

Since the common denominator 24 is less than 96, the calculations required to add $\frac{3}{8}$ and $\frac{7}{12}$ are simpler if we use 24 instead of 96:

$$\frac{3}{8} + \frac{7}{12} = \frac{9}{24} + \frac{14}{24} = \frac{23}{24} \tag{6}$$

Of course $\frac{23}{24} = \frac{92}{96}$, so that the results of (5) and (6) are the same. In general it is simplest to use the smallest possible positive common denominator when adding two fractions. That common denominator is called the ***least common denominator*** (often abbreviated l.c.d.) of the two fractions. For instance, the least common denominator of $\frac{3}{8}$ and $\frac{7}{12}$ is 24.

Care in Using Rules of Algebra

When the rules of algebra are used correctly, they lead to valid conclusions. However, it is easy to be careless in using the rules or to manufacture rules that may look plausible but unfortunately are not valid. Below we list several formulas that are false for almost all values of a, b, and c; beside each we state the corresponding correct formula:

False	Correct
$(a + b)^2 \overset{?}{=} a^2 + b^2$	$(a + b)^2 = a^2 + 2ab + b^2$
$a(b + c) \overset{?}{=} ab + c$	$a(b + c) = ab + ac$
$(ab)(ac) \overset{?}{=} abc$	$(ab)(ac) = a^2bc$
$a - (b - c) \overset{?}{=} a - b - c$	$a - (b - c) = a - b + c$
$\dfrac{-a}{-b} \overset{?}{=} -\dfrac{a}{b}$	$\dfrac{-a}{-b} = \dfrac{a}{b}$
$\dfrac{a}{b + c} \overset{?}{=} \dfrac{a}{b} + \dfrac{a}{c}$	$\dfrac{a}{b + c}$ remains $\dfrac{a}{b + c}$
$a/(b + c) \overset{?}{=} a/b + c$	$a/(b + c)$ remains $a/(b + c)$
$\sqrt{a + b} \overset{?}{=} \sqrt{a} + \sqrt{b}$	$\sqrt{a + b}$ remains $\sqrt{a + b}$
$\dfrac{a}{c} \cdot \dfrac{b}{c} \overset{?}{=} \dfrac{ab}{c}$	$\dfrac{a}{c} \cdot \dfrac{b}{c} = \dfrac{ab}{c^2}$

EXERCISES 1.1

In Exercises 1–6, evaluate the given expression.

1. $-5 + (5 \cdot 2 + 1)$
2. $3 \cdot 6 - 6$
3. $\frac{2}{3} - (\frac{1}{2} - \frac{5}{6})$
4. $-2[(4 - 2 \cdot 3) - (2 - 6) + 6]$
5. $[(2 - 3)5 + (-1 + 4)(-2)][4 - 2 \cdot 3]$
6. $2[(\frac{1}{2} - \frac{1}{3}) + \frac{1}{3} \cdot 4] - \frac{1}{2}$

In Exercises 7–16, find the given product or power.

7. $(a + 1)(a + 2)$
8. $(2a - 1)(3a + 4)$
9. $(-a + 1)(a + 1)$
10. $(\frac{1}{2}a + 4)(a - \frac{1}{2})$
11. $(1 - \sqrt{3})(1 + \sqrt{3})$
12. $(2 + \sqrt{2})(5 - 3\sqrt{2})$
13. $(5 + \sqrt{2})^2$
14. $(\frac{1}{2} - \frac{2}{3}\sqrt{3})^2$
15. $(-3 + \frac{1}{3})^2$
16. $(-4 - 2\sqrt{2})^2$

17. Write $(a + b + 2)(c + d - 3)$ without parentheses.
18. Write $(a + bc)(d + ef)$ without parentheses.

In Exercises 19–24, reduce the given fraction to lowest terms.

19. $\dfrac{15}{20}$
20. $\dfrac{60}{42}$
21. $\dfrac{63}{210}$

22. $-\dfrac{286}{520}$
23. $-\dfrac{105}{147}$
24. $\dfrac{480}{612}$

In Exercises 25–30, carry out the indicated operation. Then reduce your answer to lowest terms.

25. $\dfrac{3}{4} \cdot \dfrac{5}{6}$
26. $\dfrac{21}{25} \cdot \dfrac{5}{9}$
27. $\left(-\dfrac{4}{9}\right) \cdot \left(-\dfrac{27}{10}\right)$

28. $\dfrac{-\dfrac{1}{6}}{\dfrac{1}{12}}$
29. $\dfrac{\dfrac{2}{3}}{\dfrac{8}{9}}$
30. $\dfrac{-\dfrac{2}{15}}{-\dfrac{7}{75}}$

C In Exercises 31–34, use a calculator to approximate the given expression.

31. $\dfrac{3.487 - 2.3496}{48.63 + 3.012}$
32. $\dfrac{(3.0107)(16.38)^2}{49.07}$

33. $\dfrac{23}{247} - \dfrac{59}{1001}$
34. $\dfrac{\pi - \sqrt{2}}{\pi + \sqrt{2}}$

In Exercises 35–40, find the least common denominator of the two fractions and use it to perform the indicated operation on the two fractions.

35. $\dfrac{1}{6} + \dfrac{1}{8}$
36. $\dfrac{5}{12} - \dfrac{7}{9}$
37. $\dfrac{5}{9} + \dfrac{7}{15}$

38. $-\dfrac{11}{24} + \dfrac{13}{36}$

39. $\dfrac{23}{30} - \dfrac{29}{36}$

40. $\dfrac{2}{35} + \dfrac{3}{49}$

In Exercises 41–44, write the given expression as one fraction.

41. $\dfrac{a-1}{a+1} - \dfrac{a+1}{a-1}$

42. $\dfrac{1}{a+b} + \dfrac{b}{a^2 - b^2}$

43. $\dfrac{2}{a} - \dfrac{3}{b} + \dfrac{4}{ab}$

44. $\dfrac{1}{a+2} + \dfrac{4}{a-2} - \dfrac{2a}{a^2 - 4}$

In Exercises 45–52, correct the given incorrect formula.

45. $(a+1)(b+1) \overset{?}{=} ab + 1$

46. $\dfrac{\frac{1}{a}}{\frac{1}{b}} \overset{?}{=} \dfrac{1}{ab}$

47. $a - (b+c) \overset{?}{=} a - b + c$

48. $(-a)(-b) \overset{?}{=} -ab$

49. $(a+b)^3 \overset{?}{=} a^3 + b^3$

50. $(-a)^2 \overset{?}{=} -a^2$

51. $\dfrac{1}{a+b} \overset{?}{=} \dfrac{1}{a} + \dfrac{1}{b}$

52. $\dfrac{a}{b} + \dfrac{c}{d} \overset{?}{=} \dfrac{a+c}{b+d}$

53. Explain why the number 0.10110111011110 . . . , whose decimal expansion has increasingly long strings of 1's, is irrational.

54. Use formula (1) to prove formulas (3) and (4).

55. Show that $(2a)^2 + (a^2 - 1)^2 = (a^2 + 1)^2$ for any real number a.

56. Show that

$$n = \left(\frac{n+1}{2}\right)^2 - \left(\frac{n-1}{2}\right)^2$$

for any real number n. Conclude that every odd integer is the difference of the squares of two integers.

57. a. Show that subtraction is neither commutative nor associative by finding numbers a, b, and c such that $a - b \neq b - a$ and $a - (b - c) \neq (a - b) - c$.

 b. Show that division is neither commutative nor associative by finding numbers a, b, and c such that $a/b \neq b/a$ and $a/(b/c) \neq (a/b)/c$.

58. Show that $a \cdot 0 = 0$ for any real number a. (*Hint:* First use the fact that $a \cdot 0 = a(0 + 0)$. Then use the distributive property. Finally, subtract $a \cdot 0$ from both sides of the resulting equation.)

59. Prove the Zero Property, that is, show that if a and b are real numbers with $ab = 0$, then either $a = 0$ or $b = 0$. (*Hint:* If $a \neq 0$, multiply both sides of the equation $ab = 0$ by a^{-1}.)

60. Show that if a is a real number with $a^2 = a$, then $a = 0$ or $a = 1$. (*Hint:* Rewrite the equation $a^2 = a$ first as $a^2 - a = 0$ and then as $a(a - 1) = 0$, and use Exercise 59.)

61. Let a be a real number. Show that $a = -a$ if and only if $a = 0$.

62. Show that if a and b are nonzero real numbers, then $ab(a^{-1} + b^{-1}) = a + b$.

63. Let a, b, and c be real numbers. Show that if $a + b = a + c$, then $b = c$.

64. Let a, b, and c be real numbers. Show that if $ab = ac$ and $a \neq 0$, then $b = c$.

65. Show that $(a + b + c)^2 = a^2 + b^2 + c^2 + 2ab + 2bc + 2ca$ for any real numbers a, b, and c.

1.2
THE REAL LINE

It is possible to associate the real numbers with the points on a given line l so that every real number is associated with one and only one point on l, and so that, conversely, every point on l is associated with one and only one real number. To establish such an association we first select a particular line l, usually drawn horizontally as in Figure 1.1, and then choose an arbitrary point 0 on l to associate with the number 0. The point 0 is called the **origin.** Next we select the points to associate with the positive integers 1, 2, 3, . . . , by marking off line segments of equal length to the right of 0 as in Figure 1.1, and points to associate with the negative integers $-1, -2, -3, . . .$, by marking off similar line segments to the left of 0. To determine the points on l associated with the rational numbers, we subdivide appropriate portions of l into smaller line segments of equal length. For example, to determine the points to associate with $\frac{1}{3}$ and $\frac{2}{3}$ we subdivide the line segment determined by 0 and 1 into three segments of equal length (Figure 1.1). All points on l that are not associated with rational numbers are associated with irrational numbers. The points associated with the irrational numbers π and $\sqrt{2}$ are exhibited in Figure 1.1.

The **positive direction** on l, pointing from left to right, is indicated by an arrow on l (Figure 1.1). Those numbers corresponding to points to the right of 0 are positive numbers, and those numbers corresponding to points to the left of 0 are the negative numbers.

The number that is associated with an arbitrary point A on l is called the **coordinate** of A, and the association of the points on l with real numbers is frequently called a **coordinate system** for l. The line l with a coordinate system is often referred to as a **real number line**, or **real line**.

Inequalities

Let a and b be real numbers. If $a - b$ is positive, we say that a is **greater than** b and write $a > b$; or alternatively, we say that b is **less than** a and write $b < a$. Geometrically, $a > b$ means that the point corresponding to a on the real line in Figure 1.2 lies to the right of

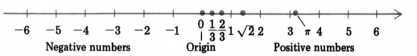

FIGURE 1.1

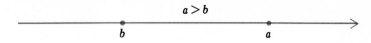

FIGURE 1.2

the point corresponding to b. Using this geometric interpretation and the fact that one of two distinct points on the real line must lie to the right of the other, we conclude that if a and b are real numbers, then exactly one of the following three possibilities is true:

$$a < b, \quad a = b, \quad \text{or} \quad a > b$$

This result is called the **trichotomy law.*** Taking $b = 0$ in the trichotomy law, we see that any real number a satisfies exactly one of the following:

$$a < 0, \quad a = 0, \quad \text{or} \quad a > 0$$

The relations $a < b$ and $a > b$ are called **inequalities,** and the symbols $<$ and $>$ are **inequality signs.** Although we will return to the general laws governing inequalities in Sections 2.6 through 2.8, we will now present a few basic properties of inequalities. To simplify the statements of the properties, we say that two nonzero numbers a and b have the **same sign** if $a > 0$ and $b > 0$, or if $a < 0$ and $b < 0$. If $a > 0$ and $b < 0$, or if $a < 0$ and $b > 0$, then we say that a and b have **opposite signs.** We are thus ready to give the properties:

$ab > 0$ if and only if a and b have the same sign.   (1)

$ab < 0$ if and only if a and b have opposite signs.   (2)

$a > 0$ if and only if $-a < 0$, and $a < 0$ if and only if   (3)
$-a > 0$.

Caution: It is sometimes tempting to regard $-a$ as a negative number simply because the expression $-a$ contains a minus sign. However, as the second half of (3) indicates, if a is negative, then $-a$ is actually positive. For instance, if $a = -5$, then $-a = -(-5) = 5$, a positive number.

Example 1. Let $a \neq 0$. Show that a and $1/a$ have the same sign.

Solution. Recall that

$$a \cdot \frac{1}{a} = 1 > 0$$

---

* The word *trichotomy* comes from a Greek word meaning "threefold division."

Since the product of a and $1/a$ is positive, a and $1/a$ must have the same sign by (1). $\square$

We write $a \geq b$ to mean that either $a > b$ or $a = b$, and express this by saying that a is **greater than or equal to** b. Alternatively, we can write $b \leq a$ and say that b is **less than or equal to** a. The symbols $\geq$ and $\leq$ are also called **inequality signs.** If $a \geq 0$, then a is not negative, so we say that a is **nonnegative.** For example, the number of miles a person travels during a given day is a nonnegative number.

If a, b, and c are three numbers, then the compound inequality $a < c < b$ means that $a < c$ and $c < b$, and we say that c **is between a and b.** For example, $2 < 2.13 < \frac{5}{2}$. Thus 2.13 is between 2 and $\frac{5}{2}$. Other compound inequalities, such as $a < c \leq b$ and $a \leq c \leq b$, are defined analogously.

> **Caution:** In any compound inequality, the inequality signs must all point in the same direction. We never write a compound inequality such as $3 > x \leq 5$, in which the inequality signs point in opposite directions.

Absolute Value

One consequence of endowing a line l with a coordinate system is that we can measure the distance between two points on l by using their coordinates. To be precise, let A and B be points on l having coordinates a and b, respectively. We define the **distance** $d(A, B)$ between A and B to be either $a - b$ or $b - a$, whichever is nonnegative (Figure 1.3). Thus the distance $d(2, -5)$ between 2 and -5 is $2 - (-5) = 7$, and the distance $d(3, 9)$ between 3 and 9 is $9 - 3 = 6$ (Figure 1.4).

By our definition, the distance $d(0, A)$ between 0 and A is either a or $-a$, whichever is nonnegative. Of the two numbers a and $-a$, the one that is nonnegative is very important in mathematics and is called the **absolute value** $|a|$ of a:

$$|a| = \begin{cases} a & \text{if} \quad a \geq 0 \\ -a & \text{if} \quad a < 0 \end{cases}$$

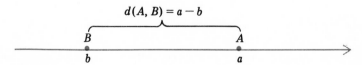

FIGURE 1.3

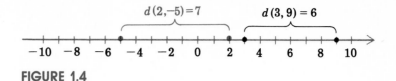

FIGURE 1.4

In the absolute value notation, the distance between 0 and A is given by

$$d(0, A) = |a|$$

Since $b - a = -(a - b)$, it follows from the definition of absolute value that $|a - b|$ is the nonnegative number of the two numbers $a - b$ and $b - a$, so that the distance between two points A and B can be written in the succinct form

$$d(A, B) = |b - a|$$

Because we have identified points on the line with their coordinates, we also speak of the distance between two numbers a and b, and we write $d(a, b)$ for that distance, which of course is the distance between the corresponding points on the line. Thus

$$d(a, b) = |b - a|$$

Example 2. Find $|5|$, $|-5|$, $|0|$, $|\pi - 3|$, and $|\sqrt{2} - 7|$.

Solution. Since 5, 0, and $\pi - 3$ are nonnegative, whereas -5 and $\sqrt{2} - 7$ are negative, we have

$$|5| = 5 \qquad\qquad |-5| = -(-5) = 5$$
$$|0| = 0 \qquad\qquad |\sqrt{2} - 7| = -(\sqrt{2} - 7) = 7 - \sqrt{2}$$
$$|\pi - 3| = \pi - 3 \qquad\qquad\qquad\qquad\qquad \square$$

Example 3. Find the distances between the following pairs of points on the real line.

    a. $A = -9, B = 0$ c. $A = 7, B = -1$
    b. $A = 2, B = 6$ d. $A = -2, B = -5$

Solution.
    a. $d(A, B) = d(-9, 0) = |0 - (-9)| = |9| = 9$
    b. $d(A, B) = d(2, 6) = |6 - 2| = |4| = 4$
    c. $d(A, B) = d(7, -1) = |-1 - 7| = |-8| = 8$
    d. $d(A, B) = d(-2, -5) = |-5 - (-2)| = |-3| = 3$ $\square$

We complete the section with a list of several special properties of absolute values:

$$|a| = 0 \text{ if and only if } a = 0 \qquad |a + b| \le |a| + |b|$$

$$|a| \ge 0 \qquad\qquad\qquad\qquad |ab| = |a||b|$$

$$|a| = |-a| \qquad\qquad\qquad \left|\frac{a}{b}\right| = \frac{|a|}{|b|}$$

$$|a - b| = |b - a|$$

From the property $|a + b| \le |a| + |b|$ it follows that either $|a + b| = |a| + |b|$ or $|a + b| < |a| + |b|$. See if you can find values of a and b for which $|a + b| = |a| + |b|$, and values for which $|a + b| < |a| + |b|$. Exercise 57 discusses exactly when equality holds.

EXERCISES 1.2

1. Draw a line, set up a coordinate system on the line, and locate the points corresponding to $-1, 2, -2, 3, -3, \frac{5}{2}, -\frac{5}{2}, \frac{7}{4},$ and $-\frac{7}{4}$.

In Exercises 2–12, write out the given statement using the symbols for inequalities.

2. x is less than 0.

3. x is greater than or equal to $\sqrt{2}$.

4. a is between 1 and 2.

5. $6 - r$ is between -1 and 1.

6. y is positive.

7. z is negative.

8. $x + 1$ is nonnegative.

9. $|x - 2|$ is less than 0.01.

10. $|x - 2|$ is greater than or equal to d.

11. c is less than or equal to $\frac{1}{10}$.

12. $4x$ is greater than or equal to 8.

In Exercises 13–20, write out the pairs of numbers in the form $a = b$, $a < b$, or $a > b$, whichever is correct.

13. $\sqrt{2}, 1$ 14. $|-4|, 4$ 15. $(-2)^2, 3$

16. $(-5)^2, 25$ 17. $0, \pi - 3$ 18. $\frac{22}{7}, \pi$

19. $\sqrt{16}, 4$ 20. $\frac{5}{7}, 0.7$

In Exercises 21–28, find the distance between A and B.

21. $A = 0, B = -1$ 22. $A = 5, B = 5$

23. $A = 6, B = 0$ 24. $A = -1, B = 3$

25. $A = -2, B = -1$ 26. $A = \frac{1}{2}, B = \frac{1}{4}$

27. $A = 9.6, B = 1.1$ 28. $A = \pi, B = \sqrt{2}$

In Exercises 29–42, write the numbers without using absolute values.

29. $|4 - 9|$ 30. $|5 + 1|$

31. $3 - |2 - 4|$

32. $|6 - 4| + |-2 - 5|$

33. $-2 - |-2|$

34. $|4 - \sqrt{2}| - 5$

35. $|3 - \pi| + 3$

36. $\frac{1}{2}|6 - 4|$

37. $\frac{1}{3}|4 - 10|$

38. $\dfrac{6}{|-4|}$

39. $|-7| + |-9|$

40. $\dfrac{-|-4|}{|12|}$

41. $|-5| + |5|$

42. $|-5| - |5|$

In Exercises 43–52, write the expression without absolute values.

43. $|x^2|$

44. $|(-4 - x)^2|$

45. $|x^2 + 1|$

46. $|-2 - y^2|$

47. $|a - 4|$ if $a \geq 4$

48. $|a - 4|$ if $a < 4$

49. $|a - b|$ if $a \geq b$

50. $|a - b|$ if $a < b$

51. $|a - b| - |b - a|$

52. $\dfrac{|a - b|}{|b - a|}$ if $a \neq b$

53. Show that $|a| = |-a|$.
54. Show that $|a^2| = |a|^2$.
55. Show that $-|a| \leq a \leq |a|$.
56. Show that $a^2 \geq 0$.
*57. Show that $|a + b| = |a| + |b|$ if and only if one of the following conditions is satisfied:
  a. $a = 0$ or $b = 0$
  b. a and b have the same sign

1.3

INTEGRAL EXPONENTS

In Section 1.1 we noted that

$$a^2 = a \cdot a$$

Similarly, the cube a^3 is defined by the formula

$$a^3 = a \cdot a \cdot a$$

In general, if n is any positive integer, the expression a^n stands for the product of n factors of a:

$$a^n = \overbrace{a \cdot a \cdot a \cdots a}^{n \text{ factors}}$$

with the understanding that $a^1 = a$. The expression a^n is read "a to the nth power," or "the nth power of a." Thus

$$2^5 = 2 \cdot 2 \cdot 2 \cdot 2 \cdot 2 = 32, \qquad \left(\frac{1}{3}\right)^2 = \frac{1}{3} \cdot \frac{1}{3} = \frac{1}{9}$$

and

$$(1.07)^4 = (1.07)(1.07)(1.07)(1.07) = 1.31079601$$

The expression $(1.07)^4$ might actually appear in a formula involving compound interest.

Next we define negative powers of numbers. Let n be a positive integer. We define a^{-n} by the formula

$$a^{-n} = \frac{1}{a^n}, \quad \text{for } a \neq 0 \tag{1}$$

Example 1. Compute

a. 2^{-3}

b. $\left(\frac{1}{3}\right)^{-4}$

Solution.

a. By (1),

$$2^{-3} = \frac{1}{2^3} = \frac{1}{8}$$

b. By (1),

$$\left(\tfrac{1}{3}\right)^{-4} = \frac{1}{\left(\tfrac{1}{3}\right)^4} = \frac{1}{\frac{1}{81}} = 81 \quad \square$$

To complete the definition of integral powers of numbers, we define

$$a^0 = 1, \quad \text{for } a \neq 0 \tag{2}$$

For example,

$$(1.7)^0 = 1 \quad \text{and} \quad \left(-\frac{1}{6}\right)^0 = 1$$

Caution: In (1) we did not define a^{-n} for $a = 0$, since the expression $\frac{1}{0}$ is meaningless. In (2) we did not define a^0 for $a = 0$ because no definition would be reasonable for all the various ways one might wish to assign a value to 0^0. Thus 0^n is not defined for any integer $n \leq 0$.

As a result of (1) and (2), a^n is defined for any integer n, with the restriction that $a \neq 0$ if $n \leq 0$. In the expression a^n, a is called the **base** and n the **exponent** or **power**. Thus a^n has the form

$$(\text{Base})^{\text{Exponent}}$$

Laws of Exponents

There are many ways of combining powers of numbers. First of all, let us consider the product $2^4 \cdot 2^3$, rearranged as follows:

$$2^4 \cdot 2^3 = (2 \cdot 2 \cdot 2 \cdot 2)(2 \cdot 2 \cdot 2) = (2 \cdot 2 \cdot 2 \cdot 2 \cdot 2 \cdot 2 \cdot 2)$$

$$= 2^7 = 2^{4+3}$$

Thus

$$2^4 \cdot 2^3 = 2^{4+3}$$

More generally, if a is any number and m and n are positive integers, then

$$a^m a^n = \overbrace{(a \cdot a \cdot a \cdots a)}^{m \text{ factors}} \overbrace{(a \cdot a \cdot a \cdots a)}^{n \text{ factors}} = \overbrace{a \cdot a \cdot a \cdots a}^{(m+n) \text{ factors}} = a^{m+n}$$

Therefore

$$a^m a^n = a^{m+n}$$

Actually, this formula is valid even if we remove the restriction that m and n be positive. This and several other formulas involving exponents are listed together below.

> **LAWS OF INTEGRAL EXPONENTS** Let a and b be real numbers and m and n integers. Each of the following formulas is valid for all values of a and b for which both sides of the equation are defined.
>
> i. $a^m a^n = a^{m+n}$ iv. $\left(\dfrac{a}{b}\right)^n = \dfrac{a^n}{b^n} = a^n b^{-n}$
>
> ii. $(a^m)^n = a^{mn}$ v. $a^{-n} = \dfrac{1}{a^n}$
>
> iii. $(ab)^n = a^n b^n$ vi. $\dfrac{a^m}{a^n} = a^{m-n}$

Recall that division by 0 is meaningless, as is raising 0 to a power that is 0 or negative. This implies, for example, that (iv) does not hold if $b = 0$, or if $a = 0$ and $n \le 0$.

Example 2. Simplify the following expressions.

    a. $5^7 5^{-4}$ b. $[(-3)^3]^2$ c. $(\frac{2}{5})^3(\frac{5}{4})^3$ d. $\dfrac{4^{17}}{4^{13}}$

Solution.

    a. By (i),

$$5^7 5^{-4} = 5^{7-4} = 5^3 = 125$$

    b. By (ii),

$$[(-3)^3]^2 = (-3)^{3 \cdot 2} = (-3)^6 = 729$$

    c. By (iii),

$$\left(\frac{2}{5}\right)^3 \left(\frac{5}{4}\right)^3 = \left(\frac{2}{5} \cdot \frac{5}{4}\right)^3 = \left(\frac{1}{2}\right)^3 = \frac{1}{8}$$

d. By (vi),

$$\frac{4^{17}}{4^{13}} = 4^{17-13} = 4^4 = 256 \quad \square$$

Example 3. Simplify the following expressions.

a. $\dfrac{(ab)^{-2}}{a^{-3}b^4}$ b. $\dfrac{x - y}{x^{-1} - y^{-1}}$

Solution.

a. Using the Laws of Integral Exponents, we have

$$\frac{(ab)^{-2}}{a^{-3}b^4} = \frac{a^{-2}b^{-2}}{a^{-3}b^4} = \frac{a^{-2}}{a^{-3}} \cdot \frac{b^{-2}}{b^4} = a^{-2-(-3)}b^{-2-4} = ab^{-6}$$

b. Using the Laws of Integral Exponents, we have

$$\frac{x - y}{x^{-1} - y^{-1}} = \frac{x - y}{\dfrac{1}{x} - \dfrac{1}{y}} = \frac{x - y}{\dfrac{y - x}{xy}} = (x - y)\left(\frac{xy}{y - x}\right)$$

$$= \frac{-(y - x)}{y - x}(xy) = -xy \quad \square$$

Caution: We have not included a law for evaluating $(a + b)^n$. It might be tempting to equate $(a + b)^n$ with $a^n + b^n$, but in general the two expressions are not equal. In fact, $(a + b)^n \neq a^n + b^n$ if a, b, and $a + b$ are different from 0. A correct formula for evaluating $(a + b)^n$, which is rather involved, appears in the last section of the book (Section 9.6).

The exponential laws can be extended to include products of more than two numbers. For example,

$$a^m a^n a^p = a^{m+n+p}$$

$$(abc)^n = a^n b^n c^n$$

Example 4. Simplify the following expressions and write them with only positive exponents.

a. $(xy^2)^3(x^2z^3)^{-2}(xyz)^2$ b. $\dfrac{(2r^3s^{-2})^4}{(rs^{-2}t^{-3})^2}$

Solution.

a. $(xy^2)^3(x^2z^3)^{-2}(xyz)^2 = (x^3y^6)(x^{-4}z^{-6})(x^2y^2z^2)$

$$= (x^3x^{-4}x^2)(y^6y^2)(z^{-6}z^2)$$

$$= x^{3-4+2}y^{6+2}z^{-6+2}$$

$$= x^1y^8z^{-4}$$

$$= \frac{xy^8}{z^4}$$

b. $\dfrac{(2r^3s^{-2})^4}{(rs^{-2}t^{-3})^2} = \dfrac{16r^{12}s^{-8}}{r^2s^{-4}t^{-6}}$

$= 16r^{12-2}s^{-8-(-4)}t^{-(-6)}$

$= 16r^{10}s^{-4}t^6 = \dfrac{16r^{10}t^6}{s^4}$ □

Let a be any real number. It follows from (1) in Section 1.2 that

$$a^2 \geq 0 \qquad\qquad (3)$$

In contrast,

$$a^3 \quad \text{has the same sign as} \quad a \qquad\qquad (4)$$

The statements in (3) and (4) are special cases of the following general results:

> a^n is nonnegative if n is an even integer.
>
> a^n has the same sign as a if n is an odd integer.

Example 5. Determine whether the given number is positive or negative.
  a. $(-2.17)^3(-4.63)^{-2}$ b. $(-1)^5(-2)^{-3}(\tfrac{3}{4})^{-14}$

Solution.
  a. Notice that

  $(-2.17)^3$ is negative because 3 is odd and -2.17 is negative

  $(-4.63)^{-2}$ is positive because -2 is even

  Therefore the product $(-2.17)^3(-4.63)^{-2}$, being the product of a negative and a positive number, is negative.
  b. Observe that

  $(-1)^5$ is negative because 5 is odd and -1 is negative

  $(-2)^{-3}$ is negative because -3 is odd and -2 is negative

  $(\tfrac{3}{4})^{-14}$ is positive because -14 is even

  Consequently the number in (b) is the product of two negative numbers and a positive number and is therefore positive. □

Scientific Notation

Quantities in the physical world come in all sizes, from the microscopic to the astronomical. For instance, the mass of an electron is approximately 0.000000000000000000000000000009 kilograms, and the mass of the sun is approximately 1,987,000,000,000,000,

000,000,000,000,000 kilograms. Many, perhaps most, quantities that arise in the physical sciences are either very small or very large. To make writing such numbers more convenient, scientists have adopted the standard practice of writing any quantity, regardless of its size, as a product $b \times 10^n$, where $1 \leq b < 10$ and n is an integer. This notation for the number is called the **scientific notation** for the number. In scientific notation the mass of the electron mentioned above is approximately 9×10^{-31} kilograms, and the mass of the sun is approximately 1.987×10^{30} kilograms. These numbers are obviously easier to write and remember when given in scientific notation.

Let a given positive number be equal to $b \times 10^n$ in scientific notation. On the one hand, if the number is greater than or equal to 1, then n is the number of places the decimal point must be moved to the *left* in order to make the decimal expansion of the resulting number lie between 1 and 9.9999 . . . . For example, $n = 3$ for the number 2341, since 2.341 lies between 1 and 9.9999 . . . . Thus

$$2341 = 2.341 \times 10^3$$

On the other hand, if the number is less than 1, then n is the negative of the number of places the decimal point must be moved to the right in order to make the decimal expansion of the resulting number lie between 1 and 9.9999 . . . . For example, $n = -4$ for the number 0.00073, since 7.3 lies between 1 and 9.9999 . . . . Thus

$$0.00073 = 7.3 \times 10^{-4}$$

Example 6. Write the following numbers in scientific notation.

    a. 14,753   d. 0.00000912
    b. 0.00632   e. 1,000,000
    c. 0.23

Solution.

    a. $14{,}753 = 1.4753 \times 10^4$
    b. $0.00632 = 6.23 \times 10^{-3}$
    c. $0.23 = 2.3 \times 10^{-1}$
    d. $0.00000912 = 9.12 \times 10^{-6}$
    e. $1{,}000{,}000 = 1 \times 10^6$ (or simply 10^6) □

Since most calculators display no more than ten digits, they normally use scientific notation in displaying very large or very small numbers. For example, in scientific notation, 2^{50} is approximately

$$1.125899907 \times 10^{15}$$

A calculator might display this as

$$1.125899907 \qquad 15$$

In contrast, 3^{-26} is approximately

$$3.934117957 \times 10^{-13}$$

and a calculator might display this as

$$3.934117957 \qquad -13$$

EXERCISES 1.3

In Exercises 1–8, compute the given number.

1. 2^0 2. 2^{-2} 3. $\left(\dfrac{1}{2}\right)^3$ 4. $\left(\dfrac{1}{2}\right)^{-3}$

5. 4^{-1} 6. 10^3 7. 10^{-3} 8. $(0.03)^2$

In Exercises 9–22, simplify the given expression.

9. $4^2 \cdot 4^3$ 10. $(-7)^4(-7)^8$ 11. $2^4(-2)^5$

12. $\dfrac{2^3}{2^5}$ 13. $\dfrac{(-7)^5}{7^6}$ 14. $2^5 \div 2^3$

15. $2^5 \div 2^{-3}$ 16. $3^{-6} \cdot 3^4$ 17. $10^9 \cdot 10^{-11}$

18. $(3 \cdot 3^3)^2$ 19. $4^4 \cdot 4^2 \cdot 4$ 20. $\dfrac{\pi^2 \pi^5}{\pi^3}$

21. $[(-4)^5]^6$ 22. $[(-3)^{-3}]^{-3}$

In Exercises 23–28, write the expression as a quotient $\dfrac{a}{b}$, where a and b are integers.

23. $2^3 \cdot 3^{-2}$ 24. $\dfrac{4^2}{5^3}$ 25. $\dfrac{3^4 2^{-3}}{3^2 2^{-1}}$

26. $\left(\dfrac{9}{5}\right)^3 \left(\dfrac{5}{9}\right)^4$ 27. $(147)^5 \div (147)^6$ 28. $(2^4 \cdot 5^3) \div 30$

In Exercises 29–44, simplify the expression.

29. $a^5 a^7$ 30. $a^4 a^{-2}$ 31. $y^{-2} y^{-6}$

32. $r^{-8} r^8$ 33. $b^4 \dfrac{1}{b^2} b^3$ 34. $\dfrac{b^3}{b^{-5}}$

35. $(-c^2)^4$ 36. $(-c^2)^5$ 37. $(xy^2)^3$

38. $(x^2 y^3 z)^4$ 39. $(\tfrac{2}{3} x^4)^{-2}$ 40. $(z^2/x^3)^5$

41. $rs^2(r^5 s^4)^3$ 42. $(t^{-1})^{-1}$ 43. $\dfrac{(st^{-1})^{-1}}{s^{-1} t^{-1}}$

44. $\dfrac{\dfrac{1}{s^{-1}} + \dfrac{1}{t^{-1}}}{s^{-1} t^{-1}}$

In Exercises 45–50, write the expression as a quotient involving only positive exponents.

45. $a^{-1} b^{-1}$ 46. $a^{-1} + b^{-2}$

47. $(a^{-1} + b^{-1})^{-3}$ 48. $(a + b)^{-1}(a^{-1} + b^{-1})$

49. $\dfrac{(a^{-1} + b^{-1})^{-1}}{(ab)^{-1}}$

50. $\dfrac{a^{-2}}{a^{-2} + b^{-2}}$

In Exercises 51–54, determine whether the given number is positive or negative.

51. $(3.1)^{-2}(2)^{-3}$

52. $(-3.2)^2 4^3$

53. $(-1)^3(5.2)^3$

54. $(-1.3)^{-2}7^{-2}$

In Exercises 55–60, write the number in scientific notation.

55. 483.2

56. 0.791

57. 1.009

58. 891,134

59. 0.9999

60. 0.0000134

c In Exercises 61–72, use a calculator to approximate the given value and write the answer in scientific notation.

61. $(9876)(2751)$

62. $(3715)(0.0015)$

63. $(0.4646)(0.3801)$

64. $(55.55)(6.148)$

65. $(0.0012)(0.00025)$

66. $(0.00000007)(0.000009)$

67. $(7.9 \times 10^3)(2.3 \times 10^5)$

68. $(4.791 \times 10^7)(9.31 \times 10^{-8})$

69. $9824 \div 112,344$

70. $83.74 \div 0.012$

71. $(3.246 \times 10^{-6}) \div (4.158 \times 10^{-4})$

72. $(1.111 \times 10^8) \div (5.876 \times 10^{-9})$

In Exercises 73–79, write the statement as an equation.

73. The circumference C of a circle is 2π times the radius r.

74. The area A of a circle is π times the square of the radius r.

75. The area A of a square is the square of a side s.

76. The area A of a triangle is half the product of the base b and the height h.

77. The volume V of a sphere is $\frac{4}{3}\pi$ times the cube of the radius r.

78. The volume V of a cylinder is π times the product of the height h and the square of the radius r of the base.

79. The volume V of a cone is one third the product of the height h and the square of the radius r of the base.

80. The mass of the earth is approximately 5.98×10^{24} kilograms. Write this number in decimal form.

81. The mass of a proton is approximately 1.67×10^{-27} kilograms. Write this number in decimal form.

82. a. Show that if a and b have the same sign, then $(ab)^n > 0$ for any integer n.
    b. Show that if a and b have different signs, then $(ab)^n > 0$ if and only if the integer n is even.

83. Show that $(a^2 + b^2)(c^2 + d^2) = (ac + bd)^2 + (ad - bc)^2$.

84. Show that $3(2ab)^2 + (a^2 - 3b^2)^2 = (a^2 + 3b^2)^2$.

*85. a. For what values of a, m, and n does part (ii) of the Laws of Integral Exponents not hold?
    b. For what values of a, b, and n does part (iii) of the Laws of Integral Exponents not hold?

© 86. A *light year* is the distance light in a vacuum travels in one year (approximately $365\tfrac{1}{4}$ days). Assuming that light in a vacuum travels 186,000 miles per second, use a calculator to compute the number of miles in one light year. Write your answer in scientific notation.

© 87. The average distance between the earth and the sun is approximately 149,000,000 kilometers. Assuming that 1 kilometer is equal to 0.621 miles, use a calculator to compute the average distance in miles between the earth and the sun. Write your answer in scientific notation.

1.4

RADICALS

If the area of a square is a square units, then we denote the length of one side of the square by $\sqrt{a}$. This leads us to define $\sqrt{a}$ for any nonnegative number a as follows:

$$\sqrt{a} = b \quad \text{if and only if} \quad b \geq 0 \text{ and } b^2 = a$$

For example,

$$\sqrt{9} = 3 \quad \text{because} \quad 3 \geq 0 \text{ and } 3^2 = 9$$

$$\sqrt{1.21} = 1.1 \quad \text{because} \quad 1.1 \geq 0 \text{ and } (1.1)^2 = 1.21$$

$$\sqrt{\frac{1}{4}} = \frac{1}{2} \quad \text{because} \quad \frac{1}{2} \geq 0 \text{ and } \left(\frac{1}{2}\right)^2 = \frac{1}{4}$$

Since $b^2 \geq 0$ for any real number b, $\sqrt{a}$ is defined only for nonnegative numbers a. That such a number b exists for any nonnegative number a is proved in more advanced books. We call $\sqrt{a}$ the *square root* of a. The expression $\sqrt{a}$ is also called a *radical*; a is the *radicand* of $\sqrt{a}$, and $\sqrt{}$ is the *radical sign.*
    If $a \geq 0$, then

$$(-\sqrt{a})^2 = (\sqrt{a})^2 = a$$

so the squares of the two numbers $-\sqrt{a}$ and $\sqrt{a}$ are the same, namely a. For that reason, $\sqrt{a}$ is occasionally called the *principal square root* of a, in order to emphasize that $\sqrt{a}$ cannot be negative.
    Two basic rules for combining radicals pertain to products and quotients:

$$\sqrt{ab} = \sqrt{a}\,\sqrt{b} \tag{1}$$

and

$$\sqrt{\frac{a}{b}} = \frac{\sqrt{a}}{\sqrt{b}} \tag{2}$$

Frequently we use (1) and (2) to simplify radicals and combinations of radicals.

Example 1. Simplify the following expressions.

a. $\sqrt{128}$ b. $\sqrt{8}\sqrt{18}$ c. $\dfrac{\sqrt{54}}{\sqrt{24}}$

Solution.

a. By (1),

$$\sqrt{128} = \sqrt{64 \cdot 2} = \sqrt{64}\sqrt{2} = 8\sqrt{2}$$

b. By (1),

$$\sqrt{8}\sqrt{18} = \sqrt{8 \cdot 18} = \sqrt{144} = 12$$

c. By (2),

$$\frac{\sqrt{54}}{\sqrt{24}} = \sqrt{\frac{54}{24}} = \sqrt{\frac{9 \cdot 6}{4 \cdot 6}} = \sqrt{\frac{9}{4}} = \frac{3}{2} \quad \square$$

Example 2. Assume that all quantities appearing in the following radicals are positive. Simplify the expressions.

a. $\sqrt{a^3 b^2}$ b. $\sqrt{6r^2 s}\sqrt{30r^4 s^3}$ c. $\dfrac{\sqrt{x^4 y^{-6} z^3}}{\sqrt{x^2 y^3 z}}$

Solution.

a. By (1),

$$\sqrt{a^3 b^2} = \sqrt{a^2 a b^2} = \sqrt{(ab)^2 a} = \sqrt{(ab)^2}\sqrt{a} = ab\sqrt{a}$$

b. By (1),

$$\sqrt{6r^2 s}\sqrt{30r^4 s^3} = \sqrt{(6r^2 s)(6 \cdot 5r^4 s^3)} = \sqrt{6^2 \cdot 5r^6 s^4}$$
$$= \sqrt{(6r^3 s^2)^2 \cdot 5} = \sqrt{(6r^3 s^2)^2}\sqrt{5} = 6r^3 s^2\sqrt{5}$$

c. By (2),

$$\frac{\sqrt{x^4 y^{-6} z^3}}{\sqrt{x^2 y^3 z}} = \sqrt{\frac{x^4 y^{-6} z^3}{x^2 y^3 z}} = \sqrt{\frac{x^2 z^2}{y^9}} = \sqrt{\frac{x^2 z^2}{y^8}\frac{1}{y}}$$
$$= \sqrt{\left(\frac{xz}{y^4}\right)^2 \frac{1}{y}} = \sqrt{\left(\frac{xz}{y^4}\right)^2}\sqrt{\frac{1}{y}} = \frac{xz}{y^4}\sqrt{\frac{1}{y}}$$
$$= \frac{xz}{y^4}\frac{1}{\sqrt{y}} = \frac{xz}{y^4\sqrt{y}} \quad \square$$

Caution: Despite the product and quotient rules for radicals given in (1) and (2), a corresponding sum rule fails. In fact, $\sqrt{a + b} \neq \sqrt{a} + \sqrt{b}$ (unless $a = 0$ or $b = 0$; see Exercise 80). To support this claim, we notice that

$$\sqrt{16 + 9} = \sqrt{25} = 5 \quad \text{but} \quad \sqrt{16} + \sqrt{9} = 4 + 3 = 7$$

so that

$$\sqrt{16 + 9} \neq \sqrt{16} + \sqrt{9}$$

It is common to alter a fraction such as $1/\sqrt{a}$ so that the denominator has no radical. The procedure is to multiply the fraction by 1, written in the form

$$\frac{\sqrt{a}}{\sqrt{a}}$$

Then we have

$$\frac{1}{\sqrt{a}} = \frac{1}{\sqrt{a}} \cdot \frac{\sqrt{a}}{\sqrt{a}} = \frac{\sqrt{a}}{\sqrt{a}\sqrt{a}} = \frac{\sqrt{a}}{a}$$

and the denominator of the last fraction contains no radicals. The process is called *rationalizing the denominator* of the fraction.

Example 3. Simplify the following fractions by rationalizing the denominator.

a. $\dfrac{3}{\sqrt{2}}$ b. $\dfrac{\sqrt{6}}{\sqrt{5}}$ c. $\dfrac{\sqrt{2}}{\sqrt{2} - 1}$ d. $\dfrac{\sqrt{a} - \sqrt{b}}{\sqrt{a} + \sqrt{b}}$

Solution.

a. Using the strategy discussed above, we multiply the fraction by $\sqrt{2}/\sqrt{2}$, which gives us

$$\frac{3}{\sqrt{2}} = \frac{3}{\sqrt{2}} \cdot \frac{\sqrt{2}}{\sqrt{2}} = \frac{3\sqrt{2}}{2} = \frac{3}{2}\sqrt{2}$$

b. Here we multiply the fraction by $\sqrt{5}/\sqrt{5}$, which yields

$$\frac{\sqrt{6}}{\sqrt{5}} = \frac{\sqrt{6}}{\sqrt{5}} \cdot \frac{\sqrt{5}}{\sqrt{5}} = \frac{\sqrt{30}}{5}$$

c. This time we multiply the fraction by $(\sqrt{2} + 1)/(\sqrt{2} + 1)$, obtaining

$$\frac{\sqrt{2}}{\sqrt{2} - 1} = \frac{\sqrt{2}}{\sqrt{2} - 1} \cdot \frac{\sqrt{2} + 1}{\sqrt{2} + 1}$$

$$= \frac{\sqrt{2}(\sqrt{2} + 1)}{2 - 1} = 2 + \sqrt{2}$$

d. We multiply the fraction by $(\sqrt{a} - \sqrt{b})/(\sqrt{a} - \sqrt{b})$, which yields

$$\frac{\sqrt{a} - \sqrt{b}}{\sqrt{a} + \sqrt{b}} = \frac{\sqrt{a} - \sqrt{b}}{\sqrt{a} + \sqrt{b}} \cdot \frac{\sqrt{a} - \sqrt{b}}{\sqrt{a} - \sqrt{b}}$$

$$= \frac{a - 2\sqrt{a}\sqrt{b} + b}{a - b} = \frac{a - 2\sqrt{ab} + b}{a - b} \quad \square$$

Let us observe that the square root and the absolute value (defined in Section 1.2) are intimately related by the formula

$$\sqrt{a^2} = |a| \quad \text{for any real number } a \tag{3}$$

After all, if $a \geq 0$, then $\sqrt{a^2} = a = |a|$. But if $a < 0$, then since $-a > 0$ and $(-a)^2 = a^2$, we conclude that

$$\sqrt{a^2} = -a = |a|$$

More generally, one can prove that for any real number a and any positive integer m,

$$\sqrt{a^{2m}} = |a|^m$$

(See Exercise 79.) Thus

$$\sqrt{\pi^{12}} = \pi^6 \quad \text{and} \quad \sqrt{(-11)^{10}} = 11^5$$

Also

$$\sqrt{a^6} = |a|^3$$

nth Roots   If the volume of a cube is a cubic units, then we denote the length of one side of the cube by $\sqrt[3]{a}$. This leads us to define $\sqrt[3]{a}$ for any real number a as follows:

$$\sqrt[3]{a} = b \quad \text{if and only if} \quad b^3 = a$$

For example,

$$\sqrt[3]{64} = 4 \quad \text{because} \quad 4^3 = 64$$

and

$$\sqrt[3]{-27} = -3 \quad \text{because} \quad (-3)^3 = -27$$

The number $\sqrt[3]{a}$ is called the *cube root* (or *third root*) of a. Notice that unlike the square root, the cube root is defined for *all* real numbers.

More generally, for any integer $n \geq 2$ we define the *nth root* $\sqrt[n]{a}$ of a as follows:

$$\sqrt[n]{a} = b \quad \text{if and only if}$$

$$b^n = a \quad \begin{cases} \text{for } a \geq 0 \text{ if } n \text{ is even} \\ \text{for any real number } a \text{ if } n \text{ is odd} \end{cases}$$

If $n = 2$, then $\sqrt[n]{a}$ becomes $\sqrt[2]{a}$, which is normally written $\sqrt{a}$. The number n in $\sqrt[n]{a}$ is called the *index* of the root; the index of $\sqrt{a}$ is 2.

Caution: Notice carefully that when n is an even integer, $\sqrt[n]{a}$ is defined only for nonnegative values of a.

Example 4. Simplify the following expressions.

     a. $\sqrt[4]{81}$ b. $\sqrt[5]{-32}$ c. $\sqrt[6]{\dfrac{64}{729}}$

Solution.

     a. $\sqrt[4]{81} = 3$, since $3^4 = 81$.
     b. $\sqrt[5]{-32} = -2$, since $(-2)^5 = -32$.
     c. $\sqrt[6]{\frac{64}{729}} = \frac{2}{3}$, since $(\frac{2}{3})^6 = \frac{64}{729}$. $\square$

The same kinds of laws hold for general nth roots as for square roots.

> **LAWS OF NTH ROOTS** Let a and b be real numbers and m and n positive integers. Each of the following formulas is valid for all values of a and b for which both sides of the equation are defined.
>
>   i. $(\sqrt[n]{a})^n = a$ iv. $\sqrt[n]{\dfrac{a}{b}} = \dfrac{\sqrt[n]{a}}{\sqrt[n]{b}}$
>
>   ii. $\sqrt[n]{a^n} = \begin{cases} |a| & \text{if } n \text{ is even} \\ a & \text{if } n \text{ is odd} \end{cases}$ v. $\sqrt[m]{\sqrt[n]{a}} = \sqrt[mn]{a}$
>
>   iii. $\sqrt[n]{ab} = \sqrt[n]{a}\,\sqrt[n]{b}$

Simplifying nth roots is similar to simplifying square roots, but with nth roots we use the laws listed above and factor out nth powers of numbers or variables where possible.

Example 5. Simplify the following expressions.

     a. $\sqrt[3]{-81}$ c. $\sqrt{\sqrt[3]{729}}$

     b. $\sqrt[4]{8}\,\sqrt[4]{162}$ d. $\dfrac{\sqrt[4]{32x^8y^6}}{\sqrt[4]{x^2y^2}}$

Solution.

     a. By (iii) and (ii),

$$\sqrt[3]{-81} = \sqrt[3]{(-27)3} = \sqrt[3]{(-3)^3 3} = \sqrt[3]{(-3)^3}\,\sqrt[3]{3} = -3\sqrt[3]{3}$$

     b. By (iii) and (ii),

$$\sqrt[4]{8}\,\sqrt[4]{162} = \sqrt[4]{(2^3)(2 \cdot 3^4)} = \sqrt[4]{2^4 3^4} = \sqrt[4]{(2 \cdot 3)^4} = 6$$

     c. By (v) and (ii),

$$\sqrt{\sqrt[3]{729}} = \sqrt[6]{729} = \sqrt[6]{3^6} = 3$$

d. By (iv) and (ii),

$$\frac{\sqrt[4]{32x^8y^6}}{\sqrt[4]{x^2y^2}} = \sqrt[4]{\frac{32x^8y^6}{x^2y^2}} = \sqrt[4]{32x^6y^4} = \sqrt[4]{(2^4x^4y^4)(2x^2)}$$

$$= \sqrt[4]{2^4x^4y^4}\ \sqrt[4]{2x^2} = 2|xy|\sqrt[4]{2x^2}\quad\square$$

EXERCISES 1.4

In Exercises 1–8, simplify the expression.

1. $\sqrt{16}$ 2. $\sqrt{81}$ 3. $\sqrt{\dfrac{1}{25}}$ 4. $\sqrt{\dfrac{4}{49}}$

5. $\sqrt{1.69}$ 6. $\sqrt{2.25}$ 7. $\sqrt{0.04}$ 8. $\sqrt{0.0036}$

In Exercises 9–22, simplify the expression.

9. $\sqrt{48}$
10. $\sqrt{9 \cdot 16}$

11. $\sqrt{\dfrac{1}{4} \cdot \dfrac{1}{36}}$
12. $\sqrt{3 \times 10^{12}}$

13. $\sqrt{4 \times 10^{11}}$
14. $\sqrt{0.5}\,\sqrt{4.5}$

15. $\sqrt{6}\,\sqrt{12}$
16. $\sqrt{32}\,\sqrt{72}$

17. $\sqrt{5 \times 10^5}\,\sqrt{20 \times 10^7}$
18. $\dfrac{1}{\sqrt{18}}$

19. $\dfrac{4}{\sqrt{96}}$
20. $\dfrac{\sqrt{63}}{\sqrt{21}}$

21. $\dfrac{\sqrt{75}}{\sqrt{147}}$
22. $\dfrac{\sqrt{6 \times 10^5}}{\sqrt{2 \times 10^{11}}}$

In Exercises 23–40, assume that all letters denote positive numbers. Simplify the expressions.

23. $\sqrt{a^2b^3}$
24. $\sqrt{24x^6y^{-4}}$

25. $\sqrt{(-32)x^3(-y)^5z^2}$
26. $\sqrt{\dfrac{12}{p^4q^7}}$

27. $\sqrt{\dfrac{p^5}{27q^8}}$
28. $\sqrt{3rs^{-3}t^2}\,\sqrt{27r^3s^5t^6}$

29. $\sqrt{8r/s^2}\,\sqrt{16s^2/r^4}$
30. $\dfrac{\sqrt{c^2d^6}}{\sqrt{4c^3d^{-4}}}$

31. $\dfrac{\sqrt{c^{-10}d^{-12}}}{\sqrt{c^{14}d^{-3}}}$
32. $\sqrt{(x + y)^2}$

33. $\sqrt{x^2 + 2xy + y^2}$
34. $\sqrt{x^2 - 2xy + y^2}$

35. $\sqrt{(4a - 7b)^4}$
36. $\sqrt{(9a^2 - 11b^3)^4}$

37. $\sqrt{\sqrt{16a^4b^8}}$
38. $\sqrt{\sqrt{9a^8/b^{10}}}$

39. $\sqrt{ab}\left(\dfrac{1}{\sqrt{b}} + \dfrac{1}{\sqrt{a}}\right)$
40. $\left(\sqrt{x} - \dfrac{1}{\sqrt{x}}\right)^2$

In Exercises 41–52, simplify the expression.

41. $\sqrt[3]{27}$ 　　　　　　42. $\sqrt[3]{-1/64}$ 　　　　　43. $\sqrt[3]{0.125}$

44. $\sqrt[3]{0.000125}$ 　　45. $\sqrt[4]{256}$ 　　　　　　46. $\sqrt[4]{0.0016}$

47. $\sqrt[5]{243}$ 　　　　　48. $\sqrt[7]{-1/128}$ 　　　49. $\sqrt[3]{8 \cdot 2}$

50. $\sqrt[3]{32}$ 　　　　　　51. $\sqrt[4]{16 \cdot 8}$ 　　　52. $\sqrt[4]{64}$

In Exercises 53–62, assume that all letters denote positive numbers. Simplify the expressions.

53. $\sqrt[3]{a^3 b^6}$ 　　　　　　　　　　54. $\sqrt[4]{16 a^4 b^{12}}$

55. $\dfrac{1}{\sqrt[3]{-125 x^3 y^6 z}}$ 　　　　56. $\sqrt[3]{\dfrac{16}{x^2 y^5}}$

57. $\dfrac{\sqrt[4]{(x+y)^4}}{\sqrt[4]{81 x^{12} y^8}}$ 　　　　　58. $\dfrac{\sqrt[5]{64 r^8 s^9 t^{13}}}{\sqrt[5]{2 r^3 s^{-6} t^{-7}}}$

59. $\sqrt[3]{t \sqrt[3]{t^9}}$ 　　　　　　　60. $\sqrt[3]{(3x - 5)^3}$

61. $\sqrt[4]{(4x + 2y)^8}$ 　　　　62. $\sqrt[3]{a + b} \, \sqrt[3]{a^2 - ab + b^2}$

In Exercises 63–72, simplify the given expression by rationalizing the denominator.

63. $\dfrac{1}{\sqrt{3}}$ 　　　　64. $\dfrac{\sqrt{9}}{\sqrt{7}}$ 　　　　65. $\dfrac{1}{1 + \sqrt{2}}$

66. $\dfrac{1}{1 - \sqrt{3}}$ 　　67. $\dfrac{\sqrt{6}}{3 + \sqrt{6}}$ 　　68. $\dfrac{2}{\sqrt{3} - \sqrt{2}}$

69. $\dfrac{4 + \sqrt{3}}{4 - \sqrt{3}}$ 　　70. $\dfrac{6 - \sqrt{2}}{6 + \sqrt{2}}$ 　　71. $\dfrac{\sqrt{a} + \sqrt{b}}{\sqrt{a} - \sqrt{b}}$

72. $\dfrac{\sqrt{a} - 2\sqrt{b}}{\sqrt{a} + 2\sqrt{b}}$

In Exercises 73–75, write the statement as an equation.

73. The length s of a side of a cube is the cube root of the volume V.

74. The radius r of a sphere is equal to the square root of the quantity obtained by dividing the surface area S by 4π.

75. The radius r of a sphere is the cube root of the quantity obtained by dividing 3 times the volume V by 4π.

C　In Exercises 76–77, use a calculator to see whether it gives the same value for each of the two numbers.

76. $\dfrac{\sqrt{(2.34)^4 (1.79)^3}}{\sqrt{(5.21)^3 (4.08)^2}}$ and $\sqrt{\dfrac{(2.34)^4 (1.79)^3}{(5.21)^3 (4.08)^2}}$

77. $\dfrac{\sqrt[3]{(3.29)^4 (-1.136)^5}}{\sqrt[3]{671.209}}$ and $\sqrt[3]{\dfrac{(3.29)^4 (-1.136)^5}{671.209}}$

78. In the Middle Ages, the formula*

$$\sqrt{a^2 + b} \approx a + \frac{b}{2a + 1}$$

was sometimes used to approximate square roots.

* The symbol $\approx$ is read "is approximately equal to."

a. Taking $a = 3$ and $b = 1$, use this formula to approximate $\sqrt{10}$.
b. From modern mathematics we know that a better approximation is given by

$$\sqrt{a^2 + b} \approx a + \frac{b}{2a}$$

Use this formula with $a = 3$ and $b = 1$ to approximate $\sqrt{10}$. (Observe that the approximation is better than that found in (a).)
c. An even better approximation is given by

$$\sqrt{a^2 + b} \approx a + \frac{b}{2a} - \frac{b^2}{8a^3}$$

Use this formula with $a = 3$ and $b = 1$ to approximate $\sqrt{10}$.

79. Show that for any real number a and any positive integer m we have $\sqrt{a^{2m}} = |a|^m$.
80. Show that if $a > 0$ and $b > 0$, then $\sqrt{a + b} \neq \sqrt{a} + \sqrt{b}$. (*Hint:* Square both $\sqrt{a + b}$ and $(\sqrt{a} + \sqrt{b})$.)
81. a. Show that $\sqrt{a^2 b^2} = |a||b|$ for any real numbers a and b.
    b. Use (1), (3), and part (a) to show that $|ab| = |a||b|$.

1.5

FRACTIONAL EXPONENTS

In the preceding two sections we discussed integral powers and roots of real numbers. Now we define a^r, the number a raised to an arbitrary rational number r.

To begin, we introduce the alternative notation $a^{1/2}$ for $\sqrt{a}$, valid whenever $a \geq 0$. Thus

$$a^{1/2} = \sqrt{a} \quad \text{for } a \geq 0$$

For example,

$$4^{1/2} = 2, \quad \left(\frac{16}{9}\right)^{1/2} = \frac{4}{3}, \quad \text{and} \quad 1.69^{1/2} = 1.3$$

By the definition of $\sqrt{a}$, we have

$$(a^{1/2})^2 = (\sqrt{a})^2 = a$$

Since

$$a^{(1/2)(2)} = a^1 = a$$

it follows that

$$(a^{1/2})^2 = a^{(1/2)(2)}$$

Thus Exponent Law (ii) of Section 1.3,

$$(a^m)^n = a^{mn}$$

also holds when $m = \tfrac{1}{2}$ and $n = 2$.

Similarly, we use the alternative notation $a^{1/3}$ for $\sqrt[3]{a}$, so that

$$a^{1/3} = \sqrt[3]{a}$$

More generally, for any integer $n \geq 2$, we adopt the alternative notation $a^{1/n}$ for $\sqrt[n]{a}$, defined for any real number a if n is odd and any nonnegative number a if n is even. Thus

$$a^{1/n} = \sqrt[n]{a} \quad \begin{cases} \text{for any } a \text{ if } n \text{ is odd} \\ \text{for } a \geq 0 \text{ if } n \text{ is even} \end{cases}$$

For example,

$$27^{1/3} = 3, \quad (-32)^{1/5} = -2, \quad \text{and} \quad 1^{1/6} = 1$$

By the definition of $\sqrt[n]{a}$, we have

$$(a^{1/n})^n = (\sqrt[n]{a})^n = a$$

Since

$$a^{(1/n)(n)} = a^1 = a$$

we conclude that

$$(a^{1/n})^n = a^{(1/n)(n)}$$

Therefore the exponential law

$$(a^m)^n = a^{mn} \tag{1}$$

given in Section 1.3 holds even when $m = 1/n$.

Recall that any rational number r can be put into the form m/n, where m and n are integers with $n > 0$ and m/n in lowest terms. Thus to define a^r for the rational number r it suffices to define $a^{m/n}$, where m and n are integers and $n > 0$. Using our new notation for the nth root, and noticing that

$$\frac{m}{n} = \frac{1}{n} \cdot m$$

we define $a^{m/n}$ by the formula

$$a^{m/n} = (a^{1/n})^m \tag{2}$$

However, there is an alternative formula for $a^{m/n}$. To obtain that formula, we observe that

$$(a^{m/n})^n = [(a^{1/n})^m]^n = (a^{1/n})^{mn} = (a^{1/n})^{nm} = [(a^{1/n})^n]^m = a^m$$

Taking the *n*th root of the far left and far right members yields

$$a^{m/n} = (a^m)^{1/n} \tag{3}$$

which gives us a second way of representing $a^{m/n}$. In computing $a^{m/n}$, we may use either (2) or (3). Which formula we use depends on the expression to be evaluated. For example, let us evaluate $8^{5/3}$ with both formulas:

$$\text{by (2): } 8^{5/3} = (8^{1/3})^5 = 2^5 = 32$$

$$\text{by (3): } 8^{5/3} = (8^5)^{1/3} = (32{,}768)^{1/3} = 32$$

Clearly, in this case it is easier to use (2) than (3), because large numbers are avoided by the use of (2). In contrast, it is easier to use (3) to compute $(\sqrt{32})^{-2/5}$:

$$\text{by (2): } (\sqrt{32})^{-2/5} = [(\sqrt{32})^{1/5}]^{-2} = [(\sqrt{2^5})^{1/5}]^{-2}$$

$$= \{[(\sqrt{2})^5]^{1/5}\}^{-2} = (\sqrt{2})^{-2} = \frac{1}{2}$$

$$\text{by (3): } (\sqrt{32})^{-2/5} = [(\sqrt{32})^{-2}]^{1/5} = \left[\frac{1}{(\sqrt{32})^2}\right]^{1/5} = \left(\frac{1}{32}\right)^{1/5} = \frac{1}{2}$$

Caution: If *n* is even and *m* is odd, the number $a^{m/n}$ is defined only for $a \geq 0$ (so that the right side of (2) is meaningful). Similarly, if $m \leq 0$, then *a* must be different from 0 in order for $a^{m/n}$ to be defined.

After studying logarithms in Chapter 5, we will be able to define a^r when *r* is *any real number*, irrational as well as rational. But in the meantime we will assume that *r* is rational.

All the previous laws of exponents remain valid for rational exponents. They are given next.

LAWS OF RATIONAL EXPONENTS Let *a* and *b* be real numbers and *r* and *s* rational. Each of the following formulas is valid for all values of *a* and *b* for which both sides of the equation are defined.

i. $a^r a^s = a^{r+s}$

ii. $(a^r)^s = a^{rs}$

iii. $(ab)^r = a^r b^r$

iv. $\left(\dfrac{a}{b}\right)^r = \dfrac{a^r}{b^r}$

v. $a^{-r} = \dfrac{1}{a^r}$

vi. $\dfrac{a^r}{a^s} = a^{r-s}$

Example 1. Simplify the following expressions.

a. $x^{2/5}y^{-2/3}(x^6y^3)^{4/3}$

b. $\left(\dfrac{x^{2/3}}{y^{1/5}}\right)^{15/7}$

Solution.

a. Using first (iii), then (ii), and finally (i), we have

$$x^{2/5}y^{-2/3}(x^6y^3)^{4/3} = x^{2/5}y^{-2/3}(x^6)^{4/3}(y^3)^{4/3}$$

$$= x^{2/5}y^{-2/3}x^8y^4 = x^{2/5\,+8}y^{-2/3\,+\,4} = x^{42/5}y^{10/3}$$

b. Using (iv) and (ii), we have

$$\left(\frac{x^{2/3}}{y^{1/5}}\right)^{15/7} = \frac{(x^{2/3})^{15/7}}{(y^{1/5})^{15/7}} = \frac{x^{10/7}}{y^{3/7}} \quad \square$$

As was the case with integral exponents, the Laws of Rational Exponents can be extended. For example,

$$(abc)^r = a^r b^r c^r$$

and

$$\left(\frac{ab}{c}\right)^r = \frac{a^r b^r}{c^r}$$

Example 2. Simplify $\left(\dfrac{9x^4y^7}{z^2}\right)^{2/5}\left(\dfrac{3x^{1/2}}{yz}\right)^{4/5}$.

Solution. Using the Laws of Rational Exponents and their extensions, we find that

$$\left(\frac{9x^4y^7}{z^2}\right)^{2/5}\left(\frac{3x^{1/2}}{yz}\right)^{4/5} = \left(\frac{9^{2/5}x^{8/5}y^{14/5}}{z^{4/5}}\right)\left(\frac{3^{4/5}x^{2/5}}{y^{4/5}z^{4/5}}\right)$$

$$= \frac{9^{2/5}3^{4/5}x^{8/5}x^{2/5}y^{14/5}}{y^{4/5}z^{4/5}z^{4/5}}$$

$$= \frac{(3^{4/5}3^{4/5})x^{10/5}y^{10/5}}{z^{8/5}}$$

$$= \frac{3^{8/5}x^2y^2}{z^{8/5}} \quad \square$$

Sometimes radicals and exponents appear in the same expression. In order to simplify such an expression it is usually convenient to convert all radicals to exponents by using the relation

$$\sqrt[n]{a} = a^{1/n}$$

Example 3. Simplify $\left(\dfrac{\sqrt{x}\ \sqrt[3]{y^2}}{z}\right)^{12}\left(\dfrac{\sqrt[4]{z}}{xy}\right)^2$.

Solution.

$$\left(\frac{\sqrt{x}\ \sqrt[3]{y^2}}{z}\right)^{12}\left(\frac{\sqrt[4]{z}}{xy}\right)^2 = \left(\frac{x^{1/2}(y^2)^{1/3}}{z}\right)^{12}\left(\frac{z^{1/4}}{xy}\right)^2$$

$$= \frac{x^6(y^2)^4}{z^{12}}\cdot\frac{z^{1/2}}{x^2y^2} = \frac{x^6y^8}{z^{12}}\cdot\frac{z^{1/2}}{x^2y^2} = \frac{x^4y^6}{z^{23/2}} \quad \square$$

EXERCISES 1.5

In Exercises 1–18, simplify the given expression.

1. $8^{5/3}$ 2. $16^{3/2}$ 3. $125^{-2/3}$

4. $64^{3/2}$ 5. $(-64)^{-4/3}$ 6. $64^{5/6}$

7. $64^{-5/6}$ 8. $(-1000)^{5/3}$ 9. $\left(\frac{9}{25}\right)^{-3/2}$

10. $(10^{-5})^{2/5}$ 11. $(16)^{0.25}$ 12. $\left(\frac{1}{4}\right)^{-1.5}$

13. $50^{3/2}$ 14. $81^{4/3}$ 15. $(2.25)^{-3/2}$

16. $\left(\frac{9}{8}\right)^{3/2}$ 17. $2^{-1.5}$ 18. $5^{2.5}$

In Exercises 19–38, assume that all letters denote positive numbers. Simplify the given expression.

19. $x^{1/3}x^{2/5}x^{4/15}$ 20. $x^{-4/3}x^{7/5}x^{-13/9}$

21. $(36a^3b^4)^{3/2}$ 22. $(x^3y^6z^9)^{2/3}$

23. $(x^3y^6z^8)^{2/3}$ 24. $\left(\dfrac{8a^4}{27b^2}\right)^{2/3}$

25. $\left(\dfrac{16}{z^3}\right)^{-3/4}$ 26. $\left(\dfrac{x^4y^7}{z^5}\right)^{-3/7}$

27. $\left(\dfrac{x^3y^7z^{-3}}{x^2y^{-5}z^{-10}}\right)^{2/5}$ 28. $\left(\dfrac{x^3y^7z^{-3}}{x^2y^{-5}z^{-10}}\right)^0$

29. $(32a\sqrt{a^3})^{2/5}$ 30. $[(a^{-2}b^3)^3]^{-2/3}$

31. $(p^2q)^{2/3}(pq^2)^{-2/3}$ 32. $(\sqrt{pq})^{2/3}$

33. $(\sqrt[3]{p^2+q^2})^{3/2}$ 34. $\left(\dfrac{\sqrt[3]{x}\,y^2}{\sqrt[4]{z}}\right)^6$

35. $\left(\dfrac{z^{1/3}\sqrt{x-y}}{2(x-y)}\right)^6$ 36. $\sqrt[3]{\dfrac{b}{27a^3}}$

37. $\sqrt{b^3}\ \sqrt[3]{b^2}$ 38. $(a^2+b^2)^{2/3} - \dfrac{a^2}{\sqrt[3]{a^2+b^2}}$

C In Exercises 39–44, use a calculator to approximate the given expression.

39. $4^{2/5}$ 40. $10^{-1/4}$ 41. $\pi^{1/3}$

42. $(\sqrt{2})^{1/5}$ 43. $(1.27)^{-4/9}$ 44. $\dfrac{(2.45)^{1.7}}{(3.97)^{2.13}}$

1.6

POLYNOMIALS

We have already used letters such as a, b, c, x, y, and z to represent real numbers. In the remaining sections of this chapter we will focus on combinations of expressions involving such letters.

The basic building block of the expressions we will consider is the *monomial* ax^k, where a is a *constant* (that is, a specific, preassigned number), x is a *variable* (that is, any real number, neither specified nor preassigned), and k is a nonnegative integer. The constant a is called the *coefficient* of the monomial. Examples of monomials are

$$16x^2, \quad -\sqrt{2}x^6, \quad \text{and} \quad -\frac{3}{5}x^{58}$$

Since $x^1 = x$ and $x^0 = 1$ for $x \neq 0$, we normally write

$$ax \quad \text{for} \quad ax^1 \quad \text{and} \quad a \quad \text{for} \quad ax^0$$

When we add the two monomials ax^k and bx^m, we obtain the *binomial*

$$ax^k + bx^m$$

It follows that

$$x - 2, \quad 2x^2 + x, \quad \text{and} \quad x^8 - 3x^3$$

are binomials.

More generally, the sum of a finite number of monomials is a *polynomial* (or *polynomial expression*). The general form of a polynomial is

$$a_n x^n + a_{n-1} x^{n-1} + \cdots + a_1 x + a_0$$

where n is a nonnegative integer, x is a variable, and a_n, a_{n-1}, $\ldots$, a_1, a_0 are constants. Examples of polynomials are

$$5, \quad x - \frac{4}{3}, \quad x^2 + 6x + 9, \quad x^3 - 27, \quad \text{and} \quad -\sqrt{7}x^6 - x^4$$

Notice that whereas the number n in x^n is an exponent, the number n in a_n simply helps to identify the coefficient.

The monomials that make up a polynomial are called *terms* of the polynomial. Thus the terms of $x^2 - 6x + 1$ are x^2, $-6x$, and 1. If $a_n \neq 0$, then the coefficient a_n of $a_n x^n$ is the *leading coefficient* of the polynomial and n is the *degree* of the polynomial. Thus the degree of $x^3 - 27$ is 3 and the leading coefficient is 1; likewise, the degree of $-\sqrt{7}x^6 - x^4$ is 6 and the leading coefficient is $-\sqrt{7}$. Finally, the number a_0 is called the *constant term* of the polynomial. The constant term may be 0, as in the polynomial $2x^3 - x^2 + x$.

Polynomials are classified according to their degrees. The polynomial 0 has no degree attached to it. For the other polynomials we have:

Polynomial	Degree	Form	Example
constant	0	$a_0 \ (a_0 \neq 0)$	$-\dfrac{4}{7}$
linear	1	$a_1 x + a_0 \ (a_1 \neq 0)$	$2x - 8$
quadratic	2	$a_2 x^2 + a_1 x + a_0 \ (a_2 \neq 0)$	$4x^2 - 6x + \sqrt{3}$
cubic	3	$a_3 x^3 + a_2 x^2 + a_1 x + a_0 \ (a_3 \neq 0)$	$x^3 - \dfrac{5}{2} x^2 + \pi$
nth-degree	n	$a_n x^n + a_{n-1} x^{n-1} + \cdots + a_1 x + a_0 \ (a_n \neq 0)$	$x^n + 1$

Caution: In a polynomial, the exponents of the powers of x must be integers; rational powers of x are not allowed. Thus $4x^{3/2} + 3x - 7$ is not a polynomial.

We say that two polynomials are **equal** if they have the same degree and if the corresponding powers of x have the same coefficients. In particular, the two polynomials

$$x^2 - 5x + 4 \quad \text{and} \quad ax^2 + bx + c$$

are equal if and only if $a = 1$, $b = -5$, and $c = 4$.

In a given polynomial, if we replace the variable x by a particular real number, the resulting expression is meaningful and represents a real number. Thus if we replace x by 2 in the polynomial $x^2 + 6x + 9$, then the polynomial becomes $2^2 + 6(2) + 9$, which simplifies to $4 + 12 + 9$, or 25. The process of replacing x by a particular real number is called **substitution.** The number obtained when a real number c is substituted into a polynomial is called the **value** of the polynomial for $x = c$.

Example 1. Find the values of $x^2 + 6x + 9$ for $x = 0$, $x = -\frac{4}{3}$, and $x = \sqrt{3}$.

Solution. We find in turn that

$$\text{for} \quad x = 0, \quad x^2 + 6x + 9 = 0^2 + 6(0) + 9$$
$$= 0 + 0 + 9 = 9$$

$$\text{for} \quad x = -\frac{4}{3}, \quad x^2 + 6x + 9 = \left(-\frac{4}{3}\right)^2 + 6\left(-\frac{4}{3}\right) + 9$$
$$= \frac{16}{9} - 8 + 9 = \frac{25}{9}$$

$$\text{for} \quad x = \sqrt{3}, \quad x^2 + 6x + 9 = (\sqrt{3})^2 + 6(\sqrt{3}) + 9$$
$$= 3 + 6\sqrt{3} + 9 = 12 + 6\sqrt{3} \quad \square$$

So far we have used the letter x for the variable of a polynomial. However, it is possible to use other letters as well. Thus $y^2 + 6y + 9$ is a polynomial, as is $z^2 + 6z + 9$. Occasionally we refer to a polynomial whose variable is y as a *polynomial in y.* In this vein, $y^2 + 6y + 9$ is a polynomial in y. Next we observe that if we substitute a number such as $-\frac{4}{3}$ for y in $y^2 + 6y + 9$, we obtain

$$\left(-\frac{4}{3}\right)^2 + 6\left(-\frac{4}{3}\right) + 9$$

which has the value $\frac{25}{9}$. This is the same value we found in Example 1. The reason it is the same is that $x^2 + 6x + 9$ and $y^2 + 6y + 9$ have the same degree and the same coefficients for corresponding powers. The fact that different letters have been used for the variable in the two polynomials is immaterial to the value of the polynomial when a real number is substituted.

Addition and Subtraction of Polynomials

We can combine numbers by adding, subtracting, multiplying, and dividing them. We can do the same with polynomials because they represent numbers. All the rules (including the commutative, associative, and distributive laws) that apply to real numbers may be applied to polynomials as well. In particular, the sum of any two polynomials in x can be obtained by adding the coefficients of like powers of x, which frequently can be done most easily by lining up terms with like powers of x vertically.

Example 2. Find the sum $(x^3 - 2x^2 + 4x - 1) + (-2x^5 - 3x^3 + \sqrt{5}\,5x^2 + 3)$.

Solution. By lining up terms with like powers of x vertically and using the laws mentioned above, we find that

$$
\begin{array}{l}
\phantom{-2x^5 -{}} x^3 - \phantom{+\sqrt{5}\;} 2x^2 + 4x - 1 \\
\underline{-2x^5 - 3x^3 + \sqrt{5}\ x^2 + 3} \\
-2x^5 - 2x^3 + (-2 + \sqrt{5})x^2 + 4x + 2 \quad \square
\end{array}
$$

Adding polynomials can also be accomplished in a horizontal format. In doing so we combine all terms with like powers of x.

$$(x^3 - 2x^2 + 4x - 1) + (-2x^5 - 3x^3 + \sqrt{5}\ x^2 + 3)$$
$$= -2x^5 + (x^3 - 3x^3) + (-2x^2 + \sqrt{5}\ x^2) + 4x + (-1 + 3)$$
$$= -2x^5 - 2x^3 + (-2 + \sqrt{5})x^2 + 4x + 2$$

Which format we use, vertical or horizontal, depends on personal preference. It is probably easier to keep coefficients of like powers of the variable straight with the vertical format, but it takes more space on the page of a book. Therefore, except in this section, we will generally adopt the horizontal format.

We subtract polynomials by subtracting terms with like powers.

Example 3. Find the difference $(x^4 - 6x^3 - 2x + 1) - (x^3 - 2x^2 - 3x - 5)$.

Solution. Subtracting terms with like powers, we find that

$$
\begin{array}{r}
x^4 - 6x^3 \qquad\quad - 2x + 1 \\
x^3 - 2x^2 - 3x - 5 \\
\hline
x^4 - 7x^3 + 2x^2 + \quad x + 6 \quad \square
\end{array}
$$

Multiplication of Polynomials

To multiply two polynomials we use the law of exponents

$$x^m x^n = x^{m+n}$$

in conjunction with the distributive law.

Example 4. Find the product $(3x - 2)(4x + 5)$.

Solution. We multiply in a vertical fashion, as we would multiply a pair of two-digit numbers:

$$
\begin{array}{r}
3x - 2 \\
4x + 5 \\
\hline
15x - 10 \\
12x^2 - \quad 8x \\
\hline
12x^2 + \quad 7x - 10 \quad \square
\end{array}
$$

Example 5. Find the product $(2x^2 - 3x - 4)(x^3 - 6x^2 + 1)$.

Solution. Using the same format, we have

$$
\begin{array}{r}
2x^2 - 3x \quad - 4 \\
x^3 - 6x^2 + 1 \\
\hline
2x^2 - 3x \quad - 4 \\
- 12x^4 + 18x^3 + 24x^2 \\
2x^5 - \quad 3x^4 - \quad 4x^3 \\
\hline
2x^5 - 15x^4 + 14x^3 + 26x^2 - 3x \quad - 4 \quad \square
\end{array}
$$

Two observations are in order. First, as with numbers, it does not matter in which order we add or multiply polynomials; the answer is the same. But again as with numbers, it *does* matter in which order we subtract polynomials. Second, the degree of the sum or difference of two polynomials is normally the higher degree of the two given polynomials (see Example 2 or 3). In contrast, the degree of the product of polynomials of degrees n and m is always $n + m$, which means that the degree of the product of polynomials is always the sum of the respective degrees. For example, the degrees of the polynomials to be multiplied in Example 5 are 2 and 3, and consequently the degree of the product is $2 + 3 = 5$.

Polynomials in Two Variables

A polynomial in the two variables x and y is a sum of monomials (or terms) of the form $cx^k y^m$, where c is a constant, x and y are variables, and k and m are nonnegative integers. Examples are

$$2x + y^2, \qquad x^2 + 2xy + y^2, \quad \text{and} \quad x^4 + 4x^3y^2 + 9xy^3 + y^6$$

One can define a polynomial in the three variables x, y, and z, or in more than three variables, in the same way. A polynomial in more than one variable is usually referred to as a *polynomial in several variables.*

Adding and multiplying polynomials of several variables proceeds as with ordinary polynomials. To compute a sum we add the terms with like powers of all the variables.

Example 6. Find the sum $(2x^2 - 6x^2y + 3xy^2 - 6x + y^2) + (-x^2 + 3x^2y + xy^2 + y^3 - y^2)$.

Solution. We add vertically, lining up terms with like powers of x and like powers of y:

$$\begin{array}{l} 2x^2 - 6x^2y + 3xy^2 - 6x + y^2 \\ \underline{-x^2 + 3x^2y + \ xy^2 \qquad\quad -y^2 + y^3} \\ \ \ x^2 - 3x^2y + 4xy^2 - 6x \qquad + y^3 \ \ \square \end{array}$$

To compute a product we again use the distributive law and then add terms with like powers of all the variables.

Example 7. Find the product $(x^2 - xy + y)(3x^2 - 4y)$.

Solution. We find that

$$\begin{array}{l} \qquad\qquad x^2 - \ xy + \ y \\ \qquad\qquad\ \ 3x^2 \quad - 4y \\ \hline \qquad -4x^2y + 4xy^2 - 4y^2 \\ \underline{3x^4 - 3x^3y + 3x^2y} \\ 3x^4 - 3x^3y - \ x^2y + 4xy^2 - 4y^2 \ \ \square \end{array}$$

Notice that in Examples 6 and 7 we have systematically written the answers with the x's before the y's in each term. It is very helpful to make such a choice about order and then stick to it.

Of the infinite collection of products of polynomials, several recur frequently. We list a few of these here. The first three essentially appeared as (2), (3), and (4), respectively, in Section 1.1.

$$(x + y)^2 = x^2 + 2xy + y^2 \tag{1}$$

$$(x - y)^2 = x^2 - 2xy + y^2 \tag{2}$$

$$(x + y)(x - y) = x^2 - y^2 \tag{3}$$

$$(x + y)^3 = x^3 + 3x^2y + 3xy^2 + y^3 \tag{4}$$

$$(x - y)^3 = x^3 - 3x^2y + 3xy^2 - y^3 \tag{5}$$

In each of these formulas, x or y may be replaced by another letter, by a specific number, or even by an expression of its own.

Example 8. Find the following products.

    a. $(x - 3y)^2$ b. $(x^2 + 2y^5)^3$

Solution.

    a. By (2), with $3y$ substituted for y, we have

$$(x - 3y)^2 = x^2 - 2x(3y) + (3y)^2 = x^2 - 6xy + 9y^2$$

    b. By (4), with x^2 substituted for x and $2y^5$ substituted for y,

$$(x^2 + 2y^5)^3 = (x^2)^3 + 3(x^2)^2(2y^5) + 3(x^2)(2y^5)^2 + (2y^5)^3$$

$$= x^6 + 6x^4y^5 + 12x^2y^{10} + 8y^{15} \quad \square$$

EXERCISES 1.6

In Exercises 1–4, find the value of the given polynomial for $x = 0$, $x = 2$, and $x = -1$.

1. $2x^2 - 5x + 3$
2. $x^3 - 3x^2 + 3x - 1$
3. $x^4 - 5x^2 + \sqrt{5}$
4. $x(x^4 - 5x^2 + \sqrt{5})$

In Exercises 5–26, perform the indicated operations and then simplify.

5. $(3x^2 - 2x - 1) + (4x^2 + 2x - 5)$
6. $(5x^4 + 3x^2 - 7) + (6x^3 + 5x^2 - 4x + 9)$
7. $(2x^3 + \frac{1}{2}x^2 + 4x) - (3x^3 - \frac{1}{2}x^2 + 2x - 4)$
8. $(\sqrt{3}x^3 - \sqrt{2}x^2) - (2x^3 - \sqrt{2}x^2 - x - 1)$
9. $(2x - 3)(4x - 5)$
10. $(-2x + 1)(7x - 3)$
11. $(\frac{1}{2}x + 3)(\frac{1}{3}x + 4)$
12. $(-4x - 2)(-3x + 6)$
13. $(3x - 2)(-x + 5) + (2x - 1)(5x - 7)$
14. $(x + 3)^2 - (x - 3)^2$
15. $(2x - 1)^2 - 4(x - 2)^2$
16. $3y(y^2 - 1)$
17. $\sqrt{2}y(y^2 - \sqrt{2})$
18. $y^2(4y^3 - y^2) + (y^2 - 1)^2$
19. $(r + 3)^2 + (r - 4)^2$
20. $(2 - s)(s - 4)^2$
21. $(2 - 3x)^2(3x - 1)^2$

22. $(x^2 + 2x + 4)(x - 2)$
23. $(2y^4 - 4y^2 + 8)(y^2 + 2)$
24. $(y^3 + 3y - 1)(2y^3 - y^2 - 2)$
25. $(x^{15} + 3x^{10} - 2x^5)(4x^{10} - 5x^5)$
26. $(-2x^{34} - 7x^{17} + 1)(x^{34} - 4x^{17} - 1)$

In Exercises 27–48, perform the indicated operations and then simplify.

27. $(x^2 + 2xy + y^2) + (3x^2 - xy + y^2)$
28. $(-x^2y^2 + xy + y^2) + (2x^2y^2 - x^2y + x + 2y^2)$
29. $(x^2 + 2xy) - (xy + y^2)$
30. $-(-7x^3y + 2xy + 3y^2)$
31. $-2(x^2y^2 - 3xy + 6y^2)$
32. $(2x^2 - y^3)^2$
33. $(\frac{1}{2}x + \frac{1}{4}y)^2$
34. $(2x - 3y)^3$
35. $(4x - y^2)^3$
36. $(p + 2q^2)^3$
37. $(x + h)^2 - x^2$
38. $(x + h)^3 - x^3$
39. $(x - y)(x^2 + xy + y^2)$
40. $(x + y)(x^2 - xy + y^2)$
41. $(r^2 - s^2)(r^3s + rs^3)$
42. $(x + y + z)(x + y - z)$
43. $(x + 2y - 3z)^2$
44. $(-2x - 5y + z)(4x - 3y - z)$
45. $(u^{1/2} + v^{1/2})^2$
46. $(u^{1/3} + v^{1/3})^3$
47. $\left(\dfrac{1}{r} - \dfrac{1}{s}\right)^2$
48. $\left(\dfrac{1}{r} + \dfrac{1}{s}\right)^2$

49. Show that $x(x + y)(x + 2y)(x + 3y) = (x^2 + 3xy + y^2)^2 - y^4$.
50. Show that $x^3 - y^3 = (x - y)^3 + 3xy(x - y)$.
51. Show that if x is replaced by $y - b/3a$, then the cubic polynomial $ax^3 + bx^2 + cx + d$ is transformed into a cubic polynomial of the form $ay^3 + ey + f$ with 0 as the coefficient of y^2.

1.7
FACTORING POLYNOMIALS

In the preceding section we multiplied polynomials. Now we consider the opposite procedure: writing a polynomial as a product of other polynomials, called *factors.* For example, since

$$4x^2 + 12x + 8 = 4(x + 2)(x + 1)$$

it follows that 4, $x + 2$, and $x + 1$ are factors of $4x^2 + 12x + 8$. Similarly,

$$x^4 - 6x^3 + 9x^2 = x^2(x - 3)^2$$

so that x^2 and $(x - 3)^2$ are factors of $x^4 - 6x^3 + 9x^2$. Of course, since $x^2 = x \cdot x$ and $(x - 3)^2 = (x - 3)(x - 3)$, we have

$$x^4 - 6x^3 + 9x^2 = x \cdot x(x - 3)(x - 3)$$

so that x and $x - 3$ are also factors of $x^4 - 6x^3 + 9x^2$. The process of rewriting a polynomial as the product of factors is called *factoring,* and factoring is important in the analysis of properties of polynomials and quotients of polynomials. Our interest lies in finding factors of degree 1 or higher, which are called ***nontrivial factors.***

Factors of Certain Quadratic Polynomials

To determine which polynomials have nontrivial factors is a difficult problem, and actually finding them can range from easy to impossible (and is likely to be hard if the degree of the polynomial is large). Even when the degree of the polynomial is 2, finding factors can be involved; the general discussion in this case must wait until Section 2.3. Since at this point we wish to become familiar with factors of polynomials, we will now look at polynomials that have the following special features:

(a) The polynomial is of degree 2.
(b) The coefficients of the polynomial are integers.
(c) Two (possibly identical) nontrivial factors exist and have integer coefficients.

Now let $x^2 + dx + e$ be a polynomial satisfying (a)–(c) so that there are factors $x + a$ and $x + b$ with a and b integers such that

$$x^2 + dx + e = (x + a)(x + b) \tag{1}$$

Multiplying out the right side of (1) gives us

$$x^2 + dx + e = x^2 + ax + bx + ab = x^2 + (a + b)x + ab$$

Since $x^2 + dx + e$ and $x^2 + (a + b)x + ab$ are equal, like powers of x must have the same coefficients. Therefore, equating the coefficients of x, we find that

$$a + b = d \tag{2}$$

and equating the constant terms, we find that

$$ab = e \tag{3}$$

It follows that we need only look among the factors of the constant term e to find a and b. We will work through an example to see how this can be accomplished.

Let us factor $x^2 + 9x + 8$ so that

$$x^2 + 9x + 8 = (x + a)(x + b)$$

where a and b are integers. Then (3) becomes $ab = 8$. Thus if a and b are to be integers as (c) insists, then the choices for a and b are 1, -1, 2, -2, 4, -4, 8, and -8. Since $ab = 8$, a and b have the same sign. But by (2), $a + b = 9$, so a and b are positive, and one is even, the other odd. We conclude that $a = 1$ and $b = 8$ (or $a = 8$ and $b = 1$). Either possibility yields the factors $x + 1$ and $x + 8$, so that

$$x^2 + 9x + 8 = (x + 1)(x + 8)$$

Notice that at first many choices for a and b seemed possible, but by studying the situation carefully we were able to reduce the number of possibilities dramatically. Below we list three observations that help in identifying possible integer values of a and b so that

$$x^2 + dx + e = (x + a)(x + b) \qquad (4)$$

    i. If $e < 0$, then a and b have different signs.
    ii. If $e > 0$ and $d > 0$, then $a > 0$ and $b > 0$.
    iii. If $e > 0$ and $d < 0$, then $a < 0$ and $b < 0$.

Example 1. Factor the following polynomials.
    a. $x^2 - 7x + 12$ b. $x^2 - 4x - 5$

Solution.
    a. If a and b are integers such that

$$x^2 - 7x + 12 = (x + a)(x + b)$$

then $ab = 12$, and by (iii), $a < 0$ and $b < 0$. The integer choices for a and b as a pair are thus -1 and -12, -2 and -6, and -3 and -4. Since $a + b = -7$, we conclude that $a = -3$ and $b = -4$ (or $a = -4$ and $b = -3$). Either possibility yields the factors $x - 3$ and $x - 4$, so that

$$x^2 - 7x + 12 = (x - 3)(x - 4)$$

    b. If integers a and b satisfy

$$x^2 - 4x - 5 = (x + a)(x + b)$$

then $ab = -5$. By (i), a and b have different signs. Choices for a and b as a pair are 1 and -5, and -1 and 5. Since $a + b = -4$, we find that $a = 1$ and $b = -5$ (or $a = -5$ and $b = 1$). Consequently $x + 1$ and $x - 5$ are factors, so that

$$x^2 - 4x - 5 = (x + 1)(x - 5) \quad \square$$

Two special forms of $x^2 + dx + e$ are

$$x^2 + 2ax + a^2 \quad \text{(where } d = 2a \text{ and } e = a^2)$$

and

$$x^2 - a^2 \quad \text{(where } d = 0 \text{ and } e = -a^2)$$

Using (1) and (3) of Section 1.6, with a instead of y, we can find factors for these special polynomials:

$$x^2 + 2ax + a^2 = (x + a)^2 \qquad (5)$$

and

$$x^2 - a^2 = (x + a)(x - a) \qquad (6)$$

Example 2. Factor the following polynomials.

  a. $x^2 + 6x + 9$ b. $x^2 - 16$ c. $49 - y^2$

Solution.

  a. By (5), with $a = 3$, we have

$$x^2 + 6x + 9 = x^2 + 2(3x) + 3^2 = (x + 3)^2$$

  b. By (6), with $a = 4$, we have

$$x^2 - 16 = x^2 - 4^2 = (x + 4)(x - 4)$$

  c. By (6), with $x = 7$ and y substituted for a, we find that

$$49 - y^2 = 7^2 - y^2 = (7 + y)(7 - y) \quad \square$$

In spite of the results of Examples 1 and 2, there are polynomials of degree 2 that do not have any nontrivial factors with real coefficients. To prove this we consider the polynomial $x^2 + c^2$, with $c \neq 0$. If $x - a$ were a factor of $x^2 + c^2$, then we would have

$$x^2 + c^2 = (x - a)(\text{polynomial})$$

But the right side is 0 for $x = a$, whereas the left side is positive for every value of x (because $c \neq 0$). Consequently $x^2 + c^2$ has no nontrivial factors. Therefore none of the polynomials

$$x^2 + 1 \ (=x^2 + 1^2), \qquad x^2 + 3 \ (=x^2 + (\sqrt{3})^2),$$
$$\text{and} \quad x^4 + 2 \ (=(x^2)^2 + (\sqrt{2})^2)$$

has a nontrivial factor.

A two-variable version of (4) is

$$x^2 + (a + b)xy + aby^2 = (x + ay)(x + by)$$

as you can verify by multiplying out the right side. Thus if we wish to factor $x^2 + dxy + ey^2$ so as to have

$$x^2 + dxy + ey^2 = (x + ay)(x + by)$$

we must once again have

$$a + b = d \quad \text{and} \quad ab = e$$

which appeared in (2) and (3). Moreover, (i)–(iii) remain valid in this context.

Example 3. Factor $x^2 - 2xy - 8y^2$.

Solution. If a and b are integers such that

$$x^2 - 2xy - 8y^2 = (x + ay)(x + by)$$

then $ab = -8$, so by (i), a and b have different signs. It follows that the choices for a and b as a pair are 1 and -8, -1 and 8, 2 and -4, and -2 and 4. Since $a + b = -2$, we conclude that $a = 2$ and $b = -4$ (or $a = -4$ and $b = 2$). Either possibility yields the factors $x + 2y$ and $x - 4y$, so that

$$x^2 - 2xy - 8y^2 = (x + 2y)(x - 4y) \quad \square$$

If the leading coefficient of the polynomial is not 1, then similar techniques can be used.

Example 4. Factor $3x^2 + 2x - 1$.

Solution. If the factors are to have integer coefficients, then the choices for the coefficients of x as a pair must be 3 and 1 or -3 and -1, and the constant terms must be 1 or -1. By trial and error we find that

$$3x^2 + 2x - 1 = (3x - 1)(x + 1) \quad \square$$

Factoring General Polynomials

Although it is impossible to give instructions for factoring polynomials in general, we can give guidelines in certain cases. In one type of polynomial there is a nontrivial factor of every term. Such a factor is called a *common factor.*

Example 5. Factor the following polynomials.
  a. $x^3 - 7x^2 + 12x$ b. $x^4 + 3x^2$

Solution.
  a. The factor common to each term is x, so we factor it out using the distributive law:

$$x^3 - 7x^2 + 12x = x(x^2 - 7x + 12)$$

By part (a) of Example 1, we know that

$$x^2 - 7x + 12 = (x - 3)(x - 4)$$

Consequently

$$x^3 - 7x^2 + 12x = x(x - 3)(x - 4)$$

b. Here x^2 is a common factor, and we factor it out, obtaining

$$x^4 + 3x^2 = x^2(x^2 + 3) \quad \square$$

In another type of polynomial there is a common factor not of every term but of each of several groups of terms together comprising the polynomial. Such common factors can then be factored out by grouping them together. The method is known as *factoring by grouping*.

Example 6. Factor the following polynomials by grouping.
a. $(x^4 + 2)x^2 - 3x(x^4 + 2)$ c. $2x^3 + 5x^2 - 6x - 15$
b. $x^3 - 3x^2 + 2x - 6$ d. $ax^2 + bxy - axy - by^2$

Solution.

a. The quadratic polynomial $x^4 + 2$ is a common factor, and we factor it out:

$$(x^4 + 2)x^2 - 3x(x^4 + 2) = (x^4 + 2)(x^2 - 3x)$$

Next we notice that x is a common factor of $x^2 - 3x$, so it can also be factored out, giving us

$$(x^4 + 2)x^2 - 3x(x^4 + 2) = x(x^4 + 2)(x - 3)$$

b. We observe that $x - 3$ is a factor of $x^3 - 3x^2$ and of $2x - 6$, and consequently

$$
\begin{aligned}
x^3 - 3x^2 + 2x - 6 &= (x^3 - 3x^2) + (2x - 6) \\
&= x^2(x - 3) + 2(x - 3) \\
&= (x^2 + 2)(x - 3)
\end{aligned}
$$

c. Since

$$2x^3 + 5x^2 = x^2(2x + 5) \quad \text{and} \quad -6x - 15 = -3(2x + 5)$$

it follows that $2x + 5$ is a common factor and that

$$
\begin{aligned}
2x^3 + 5x^2 - 6x - 15 &= (2x^3 + 5x^2) + (-6x - 15) \\
&= x^2(2x + 5) - 3(2x + 5) \\
&= (x^2 - 3)(2x + 5)
\end{aligned}
$$

d. We find that

$$ax^2 + bxy = x(ax + by) \quad \text{and} \quad -axy - by^2 = -y(ax + by)$$

Consequently

$$
\begin{aligned}
ax^2 + bxy - axy - by^2 &= x(ax + by) - y(ax + by) \\
&= (x - y)(ax + by) \quad \square
\end{aligned}
$$

Observe that in the solution of (c) we could further factor $x^2 - 3$ and obtain

$$x^2 - 3 = (x - \sqrt{3})(x + \sqrt{3})$$

but since we are only interested in factors with integer coefficients at this time, we left the answer as it was.

Finally, we give formulas for higher-degree polynomials that appear frequently in mathematics:

$$x^n - a^n = (x - a)(x^{n-1} + ax^{n-2} + a^2 x^{n-3} + \cdots$$
$$+ a^{n-2}x + a^{n-1}) \quad \text{for } n \text{ positive} \qquad (7)$$
$$x^n + a^n = (x + a)(x^{n-1} - ax^{n-2} + a^2 x^{n-3} - \cdots$$
$$- a^{n-2}x + a^{n-1}) \quad \text{for } n \text{ positive and odd} \qquad (8)$$

For $n = 3$, these formulas become

$$x^3 - a^3 = (x - a)(x^2 + ax + a^2) \qquad (9)$$

and

$$x^3 + a^3 = (x + a)(x^2 - ax + a^2) \qquad (10)$$

Example 7. Factor
    a. $x^3 - 27$ b. $x^5 + 32$ c. $x^6 - 2^6$

Solution.
  a. Using (9) with $a = 3$, we obtain

$$x^3 - 27 = x^3 - 3^3 = (x - 3)(x^2 + 3x + 9)$$

  b. Using (8) with $n = 5$ and $a = 2$, we have

$$x^5 + 32 = x^5 + 2^5 = (x + 2)(x^4 - 2x^3 + 4x^2 - 8x + 16)$$

  c. There are several ways to factor $x^6 - 2^6$:
     i. Use (7) with $n = 6$ and $a = 2$. This yields

$$x^6 - 2^6 = (x - 2)(x^5 + 2x^4 + 4x^3 + 8x^2 + 16x + 32)$$

     ii. Use (9) with $a = 2^2$ and x replaced by x^2, and then (6) with $a = 2$. This yields

$$x^6 - 2^6 = (x^2)^3 - (2^2)^3 = (x^2)^3 - 4^3 = (x^2 - 4)(x^4 + 4x^2 + 16)$$
$$= (x - 2)(x + 2)(x^4 + 4x^2 + 16)$$

     iii. First we use (6) with $a = 2^3$ and x replaced by x^3:

$$x^6 - 2^6 = (x^3)^2 - (2^3)^2 = (x^3)^2 - 8^2 = (x^3 - 8)(x^3 + 8)$$

Then we apply (9) to $x^3 - 8$ and (10) to $x^3 + 8$ and conclude that

$$x^6 - 2^6 = (x^3 - 8)(x^3 + 8)$$
$$= (x - 2)(x^2 + 2x + 4)(x + 2)(x^2 - 2x + 4) \quad \square$$

As part (c) of Example 7 illustrates, there are sometimes several ways of factoring a polynomial.

Caution: Notice that (8) is valid only if n is an odd integer. If n is even, then $x + a$ is *not* a factor of $x^n + a^n$, although there may be nontrivial factors of $x^n + a^n$ (see Exercise 72).

EXERCISES 1.7

In Exercises 1–22, factor the given polynomial.

1. $x^2 + 8x + 12$
2. $x^2 - 2x - 3$
3. $x^2 - 7x + 6$
4. $t^2 - t - 12$
5. $t^2 - 3t - 4$
6. $x^2 - 5x + 6$
7. $y^2 + 13y + 36$
8. $4y^2 - 1$
9. $x^2 - 19x + 60$
10. $x^2 + 10x + 9$
11. $21 - 10b + b^2$
12. $x^2 - 4x + 4$
13. $x^2 - 4$
14. $25 - x^2$
15. $a^2 - 14a + 49$
16. $a^2 - 5a + \frac{25}{4}$
17. $x^2 + 11x + \frac{121}{4}$
18. $49z^2 - 36$
19. $16 - 9z^2$
20. $x^2 - 20x - 800$
21. $x^2 - 2\sqrt{2}\,x + 2$
22. $x^2 + 2\sqrt{6}\,x + 6$

In Exercises 23–30, factor the given polynomial.

23. $2x^2 + 7x + 3$
24. $5x^2 + 4x - 1$
25. $-x^2 + 5x - 6$
26. $7x^2 + 8x - 12$
27. $7x^2 + 17x - 12$
28. $6t^2 - 5t - 6$
29. $6t^2 + 16t - 6$
30. $40x^2 + 14x - 45$

In Exercises 31–40, factor the given polynomial.

31. $x^2 + 2xy + y^2$
32. $x^2 + 2xy - 8y^2$
33. $x^2 - 4y^2$
34. $a^2 - 3ab - 4b^2$
35. $144a^2 - b^2$
36. $12x^2 + xy - y^2$
37. $x^4 - 2x^2y^2 + y^4$
38. $x^4 + 5x^2y^2 + 6y^2$
39. $5x^2 - 14xy - 3y^2$
40. $2x^2 - 5xy + 3y^2$

In Exercises 41–56, factor the given polynomial by finding a common factor or by grouping.

41. $x^2 + 2x$

42. $x^3 - 5x^2 + 4x$

43. $x^7 - x^6$

44. $15x^4 + 3x^3$

45. $(3x + 5)x^2 + (3x + 5)x$

46. $x^2 - 1 + 3(x - 1)$

47. $(x^2 - 5)^2 - 8(x^2 - 5) + 16$

48. $x^2 - 16 - 2(x + 4)$

49. $x^3 - 2x^2 - x + 2$

50. $x^3 + 2x^2 + 4x + 8$

51. $x^3 + x^2 - x - 1$

52. $x^5 + 2x^4 - x^3 - 2x^2$

53. $(3x + 2y)^2 + (3x + 2y) - 12$

54. $4x^2 + 4xy + y^2 - 16$

55. $x^2 - 10x + 25 - 4y^2$

56. $x^2 - 12x + 36 - 16y^2$

In Exercises 57–70, factor the given polynomial.

57. $x^3 + 1$

58. $x^3 - 1$

59. $x^3 - 8$

60. $8x^3 - 27$

61. $8x^3 - 1$

62. $y^4 - 16$

63. $32x^5 - 1$

64. $t^5 + 1$

65. $x^6 - y^6$ (*Hint:* Use (6), (9), and (10).)

*66. $x^6 + y^6$

67. $x^{10} - 1$

68. $(x + y)^3 - z^3$

69. $(x - 2y)^4 - 1$

70. $(x + y + z)^2 - (x + y - z)^2$

71. Show that $x^4 + 1 = (x^2 - \sqrt{2}\, x + 1)(x^2 + \sqrt{2}\, x + 1)$.

72. Show that $x^4 + a^4 = (x^2 - \sqrt{2}\, ax + a^2)(x^2 + \sqrt{2}\, ax + a^2)$.

*73. Show that $x^2 + x + 1$ cannot be factored as $(x + c)(x + d)$, where c and d are real numbers, even if c and d are not required to be integers. (*Hint:* Show that if $x^2 + x + 1 = (x + c)(x + d)$, then $c + d = 1$ and $cd = 1$. Then show from these equations that $0 < c < 1$ and $0 < d < 1$. But then $cd < 1$, which contradicts the equation $cd = 1$.)

1.8

RATIONAL EXPRESSIONS

In Section 1.6 we added, subtracted, and multiplied polynomials, and in each case the result was a polynomial. When we divide one polynomial by another, the result in general is not a polynomial. Quotients of polynomials are *rational expressions.* Some examples are

$$\frac{1}{x + 1}, \quad \frac{x^2 - 5x + 4}{x - 3}, \quad \text{and} \quad \frac{xy}{3x^2 + 4y^2}$$

A rational expression bears the same relationship to a polynomial as a rational number does to an integer. As with rational numbers, the polynomial appearing on top of a rational expression is its *numerator*, and the polynomial on the bottom is its *denominator.*

We have already discussed substituting real numbers into polynomials. We can also substitute real numbers into rational expressions, but we must use caution, since division by 0 is undefined. For example, it is not possible to substitute the number -1 into

$$\frac{1}{x + 1}$$

since the result would be $\frac{1}{0}$, which is meaningless. Similarly, since $x^3 + 3x^2 + 2x = x(x + 1)(x + 2)$, it is not possible to substitute 0, -1, or -2 into

$$\frac{-2x + 1}{x^3 + 3x^2 + 2x}$$

In general we can substitute into a rational expression any number for which the denominator is not 0. For example, we can substitute 2 for x in the preceding rational expression and obtain

$$\frac{-2(2) + 1}{2^3 + 3(2)^2 + 2(2)} = -\frac{3}{24} = -\frac{1}{8}$$

The set of all numbers that can be substituted for the variable in a rational expression is called the **domain** of the rational expression. Thus the domain of

$$\frac{1}{x + 1}$$

consists of all real numbers except -1, and the domain of

$$\frac{-2x + 1}{x^3 + 3x^2 + 2x}$$

consists of all real numbers except 0, -1, and -2. In contrast, the domain of

$$\frac{1}{x^2 + 1}$$

consists of all real numbers, since $x^2 + 1 \neq 0$ for all x.

A polynomial can be represented as a rational expression whose denominator is the constant polynomial 1. For example, the polynomial $2x^4 - 3x - 7$ can be represented as the rational expression

$$\frac{2x^4 - 3x - 7}{1}$$

The domain of any polynomial consists of all real numbers.

In the same way that we normally reduce a fraction to lowest terms, we can also reduce a rational expression to lowest terms by canceling factors common to numerator and denominator. In reducing a rational expression we first identify factors common to numerator and denominator and then use the law

$$\frac{ac}{bc} = \frac{a}{b} \tag{1}$$

to cancel the common factors.

Example 1. Reduce the following rational expressions to lowest terms.

a. $\dfrac{x^2 - 4}{x - 2}$ b. $\dfrac{x^2 - x - 6}{x^3 + 3x^2 + 2x}$

c. $\dfrac{x^3 + 3x^2 + 3x + 1}{(x + 1)(x^2 - 1)}$

Solution.

a. By (6) in Section 1.7,

$$x^2 - 4 = (x + 2)(x - 2)$$

Therefore we use (1) to cancel, obtaining

$$\frac{x^2 - 4}{x - 2} = \frac{(x + 2)(x - 2)}{x - 2} = x + 2$$

b. We notice that

$$x^2 - x - 6 = (x + 2)(x - 3)$$

and

$$x^3 + 3x^2 + 2x = x(x^2 + 3x + 2) = x(x + 2)(x + 1)$$

Thus by (1),

$$\frac{x^2 - x - 6}{x^3 + 3x^2 + 2x} = \frac{(x + 2)(x - 3)}{x(x + 2)(x + 1)} = \frac{x - 3}{x(x + 1)}$$

c. We observe that

$$x^3 + 3x^2 + 3x + 1 = (x + 1)^3$$

and

$$(x + 1)(x^2 - 1) = (x + 1)(x + 1)(x - 1) = (x + 1)^2(x - 1)$$

Consequently by (1),

$$\frac{x^3 + 3x^2 + 3x + 1}{(x + 1)(x^2 - 1)} = \frac{(x + 1)^3}{(x + 1)^2(x - 1)} = \frac{x + 1}{x - 1} \quad \square$$

Combinations of Rational Expressions

All the rules for adding, subtracting, multiplying, and dividing rational numbers hold for rational expressions. Multiplying and dividing rational expressions are usually less complicated than adding and subtracting them, because we do not need to employ common denominators. For multiplication of rational expressions we apply the formula

$$\frac{a}{b} \cdot \frac{c}{d} = \frac{ac}{bd} \tag{2}$$

and then cancel all common factors in order to reduce the expression to lowest terms.

Example 2. Find the product $\dfrac{x^3 - x^2}{x - 2} \cdot \dfrac{x + 3}{x^3 - 2x^2 + x}$ and reduce it to lowest terms.

Solution. Notice that

$$x^3 - x^2 = x^2(x - 1)$$

and

$$x^3 - 2x^2 + x = x(x^2 - 2x + 1) = x(x - 1)^2$$

Using (2) and then canceling common factors yields

$$\frac{x^3 - x^2}{x - 2} \cdot \frac{x + 3}{x^3 - 2x^2 + x} = \frac{(x^3 - x^2)(x + 3)}{(x - 2)(x^3 - 2x^2 + x)}$$

$$= \frac{x^2(x - 1)(x + 3)}{(x - 2)x(x - 1)^2}$$

$$= \frac{x(x + 3)}{(x - 2)(x - 1)} \quad \square$$

When we divide rational expressions we use the rule

$$\frac{a}{b} \div \frac{c}{d} = \frac{ad}{bc} \tag{3}$$

and as before we cancel all common factors to reduce the expression to lowest terms.

Example 3. Find the quotient $\dfrac{x + 1}{x - 1} \div \dfrac{x^2 + 2x + 1}{x^3 - 1}$ and reduce it to lowest terms.

Solution. First we notice that

$$x^2 + 2x + 1 = (x + 1)^2 \quad \text{and} \quad x^3 - 1 = (x - 1)(x^2 + x + 1)$$

Using (3), we find that

$$\frac{x + 1}{x - 1} \div \frac{x^2 + 2x + 1}{x^3 - 1} = \frac{(x + 1)(x^3 - 1)}{(x - 1)(x^2 + 2x + 1)}$$

$$= \frac{(x + 1)(x - 1)(x^2 + x + 1)}{(x - 1)(x + 1)^2}$$

Finally, we cancel common factors in the last expression, obtaining

$$\frac{x + 1}{x - 1} \div \frac{x^2 + 2x + 1}{x^3 - 1} = \frac{x^2 + x + 1}{x + 1} \quad \square$$

When we add or subtract rational expressions whose denominators have no nontrivial common factors, we first use the rule

$$\frac{a}{b} + \frac{c}{d} = \frac{ad + bc}{bd} \tag{4}$$

or

$$\frac{a}{b} - \frac{c}{d} = \frac{ad - bc}{bd} \tag{5}$$

and then simplify if possible.

Example 4. Find the sum $\dfrac{x + 1}{x - 2} + \dfrac{x + 2}{x + 3}$, and simplify.

Solution. Using (4), we have

$$\frac{x + 1}{x - 2} + \frac{x + 2}{x + 3} = \frac{(x + 1)(x + 3) + (x - 2)(x + 2)}{(x - 2)(x + 3)}$$

$$= \frac{(x^2 + 4x + 3) + (x^2 - 4)}{(x - 2)(x + 3)}$$

$$= \frac{2x^2 + 4x - 1}{(x - 2)(x + 3)} \quad \square$$

When adding rational expressions with common factors in the denominators, we may simplify the algebraic computations by using a *least common denominator* of the denominators in the original rational expressions—that is, a polynomial of lowest degree that is a multiple of the denominators of both rational expressions.

(This is reminiscent of the procedure applied when we added fractions by utilizing the least common denominator of two numbers in Section 1.1.)

Example 5. Express $\dfrac{2}{5x} + \dfrac{3}{x^2}$ as a rational expression in lowest terms.

Solution. A least common denominator is $5x^2$, so we rewrite each fraction so its denominator will be $5x^2$:

$$\frac{2}{5x} + \frac{3}{x^2} = \frac{2x}{5x^2} + \frac{15}{5x^2} = \frac{2x + 15}{5x^2} \quad \square$$

Example 6. Express $\dfrac{1}{x^2 + 5x + 4} - \dfrac{1}{x^2 + 8x + 16}$ as a rational expression in lowest terms.

Solution. Since

$$x^2 + 5x + 4 = (x + 4)(x + 1)$$

and

$$x^2 + 8x + 16 = (x + 4)^2$$

a least common denominator is $(x + 4)^2(x + 1)$. Consequently

$$\frac{1}{x^2 + 5x + 4} - \frac{1}{x^2 + 8x + 16} = \frac{1}{(x + 4)(x + 1)} - \frac{1}{(x + 4)^2}$$

$$= \frac{x + 4}{(x + 4)^2(x + 1)} - \frac{x + 1}{(x + 4)^2(x + 1)}$$

$$= \frac{(x + 4) - (x + 1)}{(x + 4)^2(x + 1)}$$

$$= \frac{3}{(x + 4)^2(x + 1)} \quad \square$$

For the sum of three or more rational expressions the technique is similar.

Example 7. Express $\dfrac{2x - 1}{x^2 + 4x - 5} + \dfrac{3}{x^2} - \dfrac{x - 3}{x^2 + 5x}$ as a rational expression in lowest terms.

Solution. First notice that the denominators are

$$x^2 + 4x - 5 = (x + 5)(x - 1), \quad x^2, \quad x^2 + 5x = x(x + 5)$$

Therefore a least common denominator is $x^2(x + 5)(x - 1)$, so we proceed as follows:

$$\frac{2x - 1}{x^2 + 4x - 5} + \frac{3}{x^2} - \frac{x - 3}{x^2 + 5x}$$

$$= \frac{2x - 1}{(x + 5)(x - 1)} + \frac{3}{x^2} - \frac{x - 3}{x(x + 5)}$$

$$= \frac{(2x - 1)x^2}{x^2(x + 5)(x - 1)} + \frac{3(x + 5)(x - 1)}{x^2(x + 5)(x - 1)} - \frac{x(x - 3)(x - 1)}{x^2(x + 5)(x - 1)}$$

$$= \frac{(2x - 1)x^2 + 3(x + 5)(x - 1) - x(x - 3)(x - 1)}{x^2(x + 5)(x - 1)}$$

$$= \frac{2x^3 - x^2 + 3x^2 + 12x - 15 - x^3 + 4x^2 - 3x}{x^2(x + 5)(x - 1)}$$

$$= \frac{x^3 + 6x^2 + 9x - 15}{x^2(x + 5)(x - 1)} \quad \square$$

Similar techniques apply if there are more than one variable.

Example 8. Express $\dfrac{1}{x^2 y} + \dfrac{1}{xy^2}$ as a rational expression in lowest terms.

Solution. The common denominator we use is $x^2 y^2$. As a result,

$$\frac{1}{x^2 y} + \frac{1}{xy^2} = \frac{y}{x^2 y^2} + \frac{x}{x^2 y^2} = \frac{y + x}{x^2 y^2} \quad \square$$

Combinations of expressions that are not necessarily polynomials but are nevertheless obtained by adding, subtracting, multiplying, dividing, or taking roots of polynomials are called **algebraic expressions.** The same general rules apply when combining algebraic expressions. In the following example, we will reduce the numerator and the denominator separately and then combine, noting that

$$\frac{\dfrac{a}{b}}{\dfrac{c}{d}} = \frac{a}{b} \div \frac{c}{d} = \frac{a}{b} \cdot \frac{d}{c} \qquad (6)$$

Example 9. Simplify $\dfrac{1 - \dfrac{1}{x + 1}}{1 + \dfrac{1}{x - 1}}$.

Solution. First we rearrange the numerator:

$$1 - \frac{1}{x + 1} = \frac{x + 1}{x + 1} - \frac{1}{x + 1} = \frac{(x + 1) - 1}{x + 1} = \frac{x}{x + 1}$$

The denominator is altered similarly:

$$1 + \frac{1}{x - 1} = \frac{x - 1}{x - 1} + \frac{1}{x - 1} = \frac{(x - 1) + 1}{x - 1} = \frac{x}{x - 1}$$

Therefore by (6),

$$\frac{1 - \dfrac{1}{x + 1}}{1 + \dfrac{1}{x - 1}} = \frac{\dfrac{x}{x + 1}}{\dfrac{x}{x - 1}} = \frac{x}{x + 1} \cdot \frac{x - 1}{x} = \frac{x - 1}{x + 1} \quad \square$$

If the denominator of an algebraic expression contains radicals, then one can sometimes rationalize the denominator by multiplying by 1 written in a special way. For example, if the denominator had the form $\sqrt{x} - \sqrt{a}$, we would multiply by

$$\frac{\sqrt{x} + \sqrt{a}}{\sqrt{x} + \sqrt{a}}$$

Example 10. Rationalize the denominator of $\dfrac{1}{\sqrt{x} - \sqrt{2}}$.

Solution. We multiply by $\dfrac{\sqrt{x} + \sqrt{2}}{\sqrt{x} + \sqrt{2}}$ and obtain

$$\frac{1}{\sqrt{x} - \sqrt{2}} = \frac{1}{\sqrt{x} - \sqrt{2}} \cdot \frac{\sqrt{x} + \sqrt{2}}{\sqrt{x} + \sqrt{2}} = \frac{\sqrt{x} + \sqrt{2}}{(\sqrt{x} - \sqrt{2})(\sqrt{x} + \sqrt{2})}$$

$$= \frac{\sqrt{x} + \sqrt{2}}{x - 2} \quad \square$$

EXERCISES 1.8

In Exercises 1–4, determine the domain of the rational expression, and find the value of the rational expression at the given number.

1. $\dfrac{-4}{x^2 + 5x + 6}$; 0

2. $\dfrac{x^2 - 3x}{x^3 + 2x^2 - 15x}$; -1

3. $\dfrac{x^2 + 1}{x^3 - 1}$; $\dfrac{1}{2}$

4. $\dfrac{x^2 + 7x + 12}{x^2 + 3}$; $\sqrt{3}$

In Exercises 5–14, reduce the rational expression to lowest terms.

5. $\dfrac{x + 1}{x^2 + 5x + 4}$

6. $\dfrac{x^2 + 3x - 28}{x^2 - 3x + 4}$

7. $\dfrac{y^2 - 5y - 24}{y^2 - 9}$

8. $\dfrac{9y^2 - 4}{3y^2 - 3y + \frac{2}{3}}$

9. $\dfrac{2b^2 + b - 3}{8b^2 + 2b - 15}$

10. $\dfrac{s^4 - 8s^2 + 16}{s^2 - s - 2}$

11. $\dfrac{s^4 - 13s^2 + 36}{s^2 - 7s + 12}$ 12. $\dfrac{s^4 - 5s^2 - 36}{s^2 - 7s + 12}$ 13. $\dfrac{x^2 - y^2}{x^3 - y^3}$

14. $\dfrac{x^2 + y^2}{x^4 - y^4}$

In Exercises 15–26, write as one rational expression in simplified form.

15. $\dfrac{x^2 - 1}{x^2 + x - 2} \cdot (x^2 - 4)$

16. $\dfrac{x^2 - 16}{(x^2 + 3x - 28)^2} \cdot (x^2 - 49)$

17. $\dfrac{y^2}{y + 3} \cdot \dfrac{y^2 + y - 6}{y}$

18. $\dfrac{y + 3}{y - 2} \cdot \dfrac{y^2 - 4}{y + 3}$

19. $\dfrac{z^2 - 9}{z^2 - 4} \cdot \dfrac{z^2 + 6z + 9}{z^2 - 4z + 3}$

20. $\dfrac{x - 2}{x - 4} \div \dfrac{x}{x - 4}$

21. $\dfrac{x + 3}{x - 1} \div \dfrac{x - 1}{x + 3}$

22. $\dfrac{a^2 + 5a - 50}{a^2 - 1} \div \dfrac{a^2 - 7a + 10}{a^2 + 6a + 5}$

23. $\dfrac{b^2 + 3b + 2}{b^2 - 2} \div \dfrac{b^2 - 7b - 18}{b^3 - b^2 - 2b + 2}$

24. $\dfrac{x^2 + y^2}{x^3 + y^3} \div \dfrac{x^2 + y^2}{x + y}$

25. $\dfrac{x^2 - 2xy + y^2}{x^2 + xy + y^2} \div \dfrac{x^2 - y^2}{x^3 - y^3}$

26. $\dfrac{x^2 + 2xy + y^2}{x^2 + xy + y^2} \div \dfrac{x^2 + y^2}{x^3 - y^3}$

In Exercises 27–52, write as one rational expression in simplified form.

27. $\dfrac{1}{x} + \dfrac{2}{x - 1}$

28. $\dfrac{x}{x - 3} + \dfrac{4}{x + 2}$

29. $\dfrac{5}{y^2 - 9} + \dfrac{3}{y + 3}$

30. $\dfrac{1}{y^2 - 7y + 6} + \dfrac{1}{y^2 - 2y - 24}$

31. $\dfrac{y}{y^2 + 5y - 24} + \dfrac{1}{3 - y}$

32. $\dfrac{z + 1}{z - 1} + \dfrac{z - 1}{z + 1}$

33. $\dfrac{2z - 3}{6z + 1} + \dfrac{3z - 1}{z + 4}$

34. $\dfrac{t^2 + 1}{t^2 - 1} + \dfrac{1}{1 - t}$

35. $\dfrac{4}{t - 1} + \dfrac{t + 7}{t^2 - 4t + 3}$

36. $\dfrac{3u}{u - 2} - \dfrac{4u}{u + 5}$

37. $\dfrac{2u + 1}{3u - 1} - \dfrac{3 - u}{2u - 3}$

38. $\dfrac{v - 1}{v + 1} - \dfrac{1 - 7v}{v^2 - 2v - 3}$

39. $\dfrac{v + 5}{v - 4} - \dfrac{12v + 6}{v^2 - 2v - 8}$

40. $\dfrac{1}{v + 2} - \dfrac{1}{v^2 + 3v} - \dfrac{1}{v^2 + 5v + 6}$

41. $\dfrac{a}{b} + \dfrac{b}{a}$

42. $\dfrac{a}{a - b} + \dfrac{b}{b - a}$

43. $1 - \dfrac{a^2}{a^2 + b^2}$

44. $1 + \dfrac{a^2}{a^2 + b^2}$

45. $\dfrac{1}{b} - \dfrac{2}{ab^2} + \dfrac{4}{ab^3}$

46. $\dfrac{1}{p - q} + \dfrac{1}{p + q} - \dfrac{1}{p^2 - q^2}$

47. $\dfrac{1}{p} + \dfrac{1}{q} + \dfrac{1}{r}$

48. $\dfrac{q}{p} + \dfrac{r}{q} + \dfrac{p}{r}$

49. $\dfrac{1}{h}\left(\dfrac{1}{x + h} - \dfrac{1}{x}\right)$

50. $\dfrac{1}{h}\left[\dfrac{1}{(x + h)^2} - \dfrac{1}{x^2}\right]$

51. $\left(\dfrac{1}{y} - \dfrac{1}{x}\right) \div \left(\dfrac{1}{y} + \dfrac{1}{x}\right)$

52. $\left(\dfrac{x}{y} + \dfrac{y}{x}\right) \div \dfrac{x^2 + y^2}{xy}$

In Exercises 53–62, simplify the expression.

53. $\dfrac{\frac{1}{x} + x}{\frac{2}{x} + 1}$

54. $\dfrac{\frac{1}{x} - 1}{x^2 + 2}$

55. $\dfrac{x^2 + \frac{4}{x}}{x + \frac{4}{x^2}}$

56. $\dfrac{\frac{x + 4}{x} - 1}{3 - \frac{4 - x}{x}}$

57. $\dfrac{\frac{z - 2}{z + 5} - \frac{z - 1}{z + 1}}{\frac{z - 3}{z + 1} - \frac{z}{z + 4}}$

58. $\dfrac{\frac{x - y}{x}}{\frac{x^2 - y^2}{xy}}$

59. $\dfrac{\frac{x}{y} - \frac{y}{x}}{\frac{1}{x} + \frac{1}{y}}$

60. $\dfrac{\frac{x^2}{y} - \frac{y^2}{x}}{\frac{1}{x} + \frac{1}{y}}$

61. $\dfrac{1}{\sqrt{x}} - \dfrac{1}{\sqrt{y}}$

62. $\dfrac{x}{\sqrt{y}} + \dfrac{y}{\sqrt{x}}$

In Exercises 63–69, rationalize the denominator.

63. $\dfrac{1}{\sqrt{x} - \sqrt{3}}$

64. $\dfrac{1}{\sqrt{x} - \sqrt{5}}$

65. $\dfrac{1}{\sqrt{x} - 9}$

66. $\dfrac{1}{\sqrt{x} + 6}$

67. $\dfrac{x - 3}{\sqrt{x} + \sqrt{3}}$

68. $\dfrac{1}{\sqrt{x} - \sqrt{y}}$

69. $\dfrac{\sqrt{x} - \sqrt{y}}{\sqrt{x} + \sqrt{y}}$

KEY TERMS

real number
  natural number
  rational number
  irrational number
factor
real line
  origin
  positive direction
  coordinate

distance
absolute value
base
exponent
scientific notation
square root
radical
cube root
nth root

polynomial	rational expression
monomial	numerator
coefficient	denominator
binomial	domain
leading coefficient	rational expression in lowest
degree	terms
constant term	least common denominator
polynomial in several vari-	rationalizing the denominator
ables	algebraic expression

KEY FORMULAS

$$(a + b)(c + d) = ac + bc + ad + bd$$
$$(a + b)^2 = a^2 + 2ab + b^2$$
$$(a - b)^2 = a^2 - 2ab + b^2$$

$$\frac{a}{b} + \frac{c}{d} = \frac{ad + bc}{bd}$$

$$\frac{a}{b} \cdot \frac{c}{d} = \frac{ac}{bd}$$

$$\frac{a}{b} \div \frac{c}{d} = \frac{ad}{bc}$$

$$a^r a^s = a^{r+s}$$
$$(a^r)^s = a^{rs}$$
$$(ab)^r = a^r b^r$$

$$a^{-r} = \frac{1}{a^r}$$

$$x^2 - a^2 = (x + a)(x - a)$$
$$x^3 - a^3 = (x - a)(x^2 + ax + a^2)$$
$$x^3 + a^3 = (x + a)(x^2 - ax + a^2)$$

$$x^n - a^n = (x - a)(x^{n-1} + ax^{n-2} + a^2 x^{n-3} + \cdots + a^{n-2}x + a^{n-1})$$
$$x^n + a^n = (x + a)(x^{n-1} - ax^{n-2} + a^2 x^{n-3} - \cdots - a^{n-2}x + a^{n-1})$$
$$(n \text{ odd})$$

REVIEW EXERCISES

In Exercises 1–12, calculate the value of the given expression.

1. $\dfrac{|7 - 9|}{|6 - 3|} - \dfrac{|-1|}{|4 - 8|}$

2. $\dfrac{3}{10} - \dfrac{4}{35} \cdot \dfrac{7}{2}$

3. $9^{-3/2}$

4. $\left(-\dfrac{1}{8}\right)^{5/3}$

5. $\dfrac{2^{-3}}{2^{-7}}$

6. $(\sqrt[3]{-3})^6$

7. $\sqrt{0.09}$

8. $9^{2.5}$

9. $3^4 \cdot 3^{-2}$

10. $\dfrac{(\sqrt{2})^3(\sqrt{2})^{-4}}{(\sqrt{2})^5}$ 11. $\sqrt[3]{\dfrac{-8}{27}}$ 12. $\sqrt[5]{\dfrac{1}{32}}$

In Exercises 13–14, write out the given statement using the symbols for inequalities.

13. $a - \sqrt{5}$ is nonnegative.

14. 0.2 is greater than $|3x + 1|$.

In Exercises 15–16, find the distance between A and B.

15. $A = 2.7$, $B = -1.6$ 16. $A = -\frac{4}{5}$, $B = -\frac{2}{3}$

In Exercises 17–20, write the given number in scientific notation.

17. 159,000   18. 0.00314

19. $(16)^{3/2} \times 10^{-5}$ 20. $\dfrac{231 \times 10^9}{11 \times 10^{-3}}$

In Exercises 21–36, simplify the given expression.

21. $\left(\dfrac{1}{2}\, a^{-2}\right)^3 a^4$ 22. $(\sqrt{3} + \sqrt{2})^2$

23. $\sqrt{2} + \sqrt{50}$ 24. $\dfrac{1}{2}(\sqrt{a + b} + \sqrt{a - b})^2$

25. $\dfrac{(15a^3)^2}{(15a^2)^3}$ 26. $\dfrac{(a^3b^{-3}c^{-1})^6}{(b^{-5}c^7)^{-2}}$

27. $\dfrac{(x - 5)^3}{|x - 5|}$ 28. $\dfrac{\sqrt{x}}{\sqrt[5]{x}}$

29. $(\sqrt[3]{2x})^9$ 30. $\sqrt{3x}\,\sqrt[4]{9x^2}$

31. $\dfrac{\sqrt[3]{32a^5b^{-5}}}{\sqrt[3]{4a^2b^{-4}}}$ 32. $(x^5y^{-2}z^6)^{3/2}$

33. $\dfrac{(a + b)^2 - (a - b)^2}{(a + b)^2 + (a - b)^2}$ 34. $[(a - b)^{-1} - (a + b)^{-1}]^{-1}$

35. $\dfrac{a - \dfrac{a^2}{a + b}}{b - \dfrac{b^2}{a + b}}$ 36. $\dfrac{\dfrac{x}{x + y} + \dfrac{y}{x - y}}{x^2 + y^2}$

In Exercises 37–44, perform the indicated operation and simplify.

37. $(2a - 4)(5a - 1)$ 38. $(\frac{1}{2}a - \frac{2}{3})^2$

39. $(2x - 9)(5x + 2) - 3(-2x + 7)(4x - 2)$

40. $(3x + 4)^2 - (3x - 5)^2$

41. $(x^{1/3} - x^{-1/3})^3$ 42. $(y/x - 2x/y)^2$

43. $(2x^2 + 5y)^3$ 44. $(x^2 + y^2 - 2z^2)^2$

In Exercises 45–46, simplify the given expression by rationalizing the denominator.

45. $\dfrac{1}{\sqrt{5} - 2}$ 46. $\dfrac{\sqrt{2a} - \sqrt{8b}}{\sqrt{2a} + \sqrt{8b}}$

In Exercises 47–48, assume that $a > 0$, and determine whether the expression is positive or negative.

47. $a - \sqrt{a^2 + 4}$

48. $a - \sqrt{a^2 - 1}$

In Exercises 49–64, factor the given polynomial.

49. $x^2 - 6x - 27$

50. $y^2 + 8y - 105$

51. $t^2 - 20t + 100$

52. $4u^2 + 4u + 1$

53. $12x^2 - 11x + 2$

54. $x^3 - 2x^2 + 2x - 4$

55. $(y - 1)^5 - 4(y - 1)^3$

56. $x^3 - x^2 - 90x$

57. $z^4 + z^2 - 2$

58. $z^6 - z^4$

59. $x^3 + x^2 + x + 1$

60. $x^3y + xy^3$

61. $x^2 + 2xy - 35y^2$

62. $9x^4 - 6x^2y^2 + y^4$

63. $16x^4 - 1$

64. $x^8 - 256$

In Exercises 65–66, find the domain of the given expression.

65. $\dfrac{-7x^2 + 2}{x^2 + 9x + 20}$

66. $\dfrac{2x - 3}{x(4x - 3)^3}$

In Exercises 67–70, find the value of the given expression at the given number.

67. $4x^3 - 2x^2 + 3x - 19$; 1

68. $\dfrac{3x - 5}{2x + 1}$; 2

69. $\dfrac{3x^2 + 1}{-4x^3 - 11}$; -3

70. $\dfrac{x^2}{x^6 - 2}$; $\sqrt{2}$

In Exercises 71–86, write as one rational expression in simplified form.

71. $\dfrac{x^2 - 6x + 5}{x^2 - 3x - 10}$

72. $\dfrac{x^2 + 2xy + y^2}{x^3 - x^2y - y^2x + y^3}$

73. $\dfrac{4x^2 - 1}{x^2 + x - 12} \cdot \dfrac{x^2 - 5x + 6}{6x^2 + x - 1}$

74. $\left(\dfrac{1}{x^2} - \dfrac{1}{x^3}\right)\left(\dfrac{1}{x} + \dfrac{1}{x^2}\right)$

75. $\dfrac{x^2 - 4}{x^2 - x - 6} \div \dfrac{x^2 + 2x - 8}{x^2 + x - 12}$

76. $\dfrac{3}{2x - 1} + \dfrac{2}{3x - 2}$

77. $\dfrac{2x - 3}{x - 2} + \dfrac{x}{x + 2}$

78. $\dfrac{x}{(x - 1)^2} - \dfrac{1}{x^2 - 1}$

79. $\dfrac{3x - 4}{x^2 + x} - \dfrac{5x - 2}{x^2 - x}$

80. $\dfrac{5}{x^2 + 3x - 10} + \dfrac{3}{x^2 - 3x + 2}$

81. $1 - \dfrac{x}{x - y}$

82. $\dfrac{x}{x - y} - \dfrac{y}{x + y}$

83. $\dfrac{\dfrac{2}{x} + \dfrac{3}{y}}{\dfrac{1}{x} - \dfrac{2}{y}}$

84. $\dfrac{\dfrac{1}{x} - \dfrac{1}{y}}{(x - y)^2}$

85. $\dfrac{\dfrac{1}{x} - \dfrac{1}{x^2}}{\dfrac{1}{x} + \dfrac{1}{x^2}}$

86. $\dfrac{\sqrt{x}}{\sqrt{x} - \sqrt{y}} + \dfrac{\sqrt{y}}{\sqrt{x} + \sqrt{y}}$

In Exercises 87–89, write the statement as an equation.

87. The surface area S of a rectangular box with height h and square base of side s is equal to the sum of twice the square of s and four times the product of s and h.

88. The surface area S of a cylindrical can (including top and bottom) of radius r and height h is the sum of 2π times the radius squared and 2π times the product of the radius and height.

89. The area A of an equilateral triangle is the product of $\sqrt{3}/4$ and the square of the length s of a side.

90. The age of Rob is x, and Rachel is 3 years less than twice as old as Rob. What is the sum of their ages?

91. Jill's salary is \$30,000. Susan's salary can be determined by dividing Jill's salary by 8 and then adding to the result the product of 1982 and 13. Who has the larger salary?

92. What are the possible values of $\dfrac{a-b}{|a-b|}$?

93. Prove that

$$\left(\frac{x-y}{2}\right)^2 = \left(\frac{x+y}{2}\right)^2 - xy$$

94. A positive integer is called a **perfect number** if it is the sum of its positive divisors less than itself. Show that the following are perfect numbers.
    a. 6   b. 28   c. 496

95. Let a and b be real numbers with $b \neq -1$, and let

$$x = \frac{a^2}{1+b^3}, \quad y = bx, \quad \text{and} \quad z = ax$$

Show that $x^3 + y^3 = z^2$.

96. Let u and v be positive numbers with $u > v$, and let

$$a = 2uv, \quad b = u^2 - v^2, \quad \text{and} \quad c = u^2 + v^2$$

The numbers a, b, and c are known as a **Pythagorean triple.**
    a. Show that $c^2 = a^2 + b^2$. (This is the reason for the name of the triple; after all, it follows that there is a right triangle whose sides have lengths a, b, and c.)
    b. Find the Pythagorean triples arising from the choices
       i. $u = 2, v = 1$ ii. $u = 3, v = 1$ iii. $u = 3, v = 2$

EQUATIONS
AND
INEQUALITIES

2

In Chapter 1 we discussed the basic kinds of arithmetic operations on numbers and algebraic expressions. In Chapter 2 we consider equations and inequalities. On the one hand, equations arise by setting two given algebraic expressions equal to one another. On the other hand, inequalities arise by making one given algebraic expression less than another. As you see, equations and inequalities are close relatives of one another; they are also extremely important in mathematics. In fact, it could almost be said that higher mathematics revolves about the study of equations and inequalities.

However, equations and inequalities are of interest not only in mathematics itself but also in applications to many areas outside mathematics. These applications range from the solutions of simple, everyday problems to the formulation of the deepest principles of modern science. In this chapter we will illustrate some of the simpler applications of equations and inequalities.

2.1

LINEAR EQUATIONS

Before we begin our discussion of linear equations, we need some terminology concerning equations in general. First of all, an *equation* is a statement that two expressions are equal. In Chapter 1 we encountered equations such as

$$x^2 + 6x + 9 = (x + 3)^2 \quad \text{and} \quad \frac{x^2 - 4}{x - 2} = x + 2 \tag{1}$$

and in the present chapter we will see equations such as

$$2x + 7 = 5x - 3 \quad \text{and} \quad 3x^2 - 4x + 1 = 0 \tag{2}$$

However, the role the variable x plays in (1) is different from its role in (2). Indeed, the first equation in (1) is valid for all values of x, and the second equation in (1) is valid for all values of x except 2 (for which the expression on the left side is undefined). In contrast, the equations in (2) are valid only for very special values of x. In the course of this chapter we will show that the equation $2x + 7 = 5x - 3$ is valid only for $x = \frac{10}{3}$ and that the equation $3x^2 - 4x + 1 = 0$ is valid only for $x = 1$ or $x = \frac{1}{3}$.

We call an equation an *identity* if it is valid for all values of the variable that make each constituent expression meaningful. Thus the equations in (1) are identities. In contrast, an equation is called *conditional* if it is valid only for special values of the variable. This is true of both equations in (2).

Two conditional equations are said to be *equivalent* if they are valid for precisely the same values of the variable. In particular, if two equations are the same except for the letter serving as the variable, then the equations are equivalent. For instance,

$$x^2 - 6x + 5 = 0 \quad \text{and} \quad t^2 - 6t + 5 = 0$$

are equivalent. Also, if the same constant is added to or subtracted from both sides of an equation, or if both sides are multiplied or divided by the same nonzero constant, the resulting equation is equivalent to the given one. Thus

$$\frac{9}{5}x + 32 = 68 \quad \text{and} \quad \frac{9}{5}x = 36$$

are equivalent because the second is obtained from the first by subtracting 32 from both sides. Similarly,

$$3x = 17 \quad \text{and} \quad x = \frac{17}{3}$$

are equivalent because the second is obtained from the first by dividing both sides by 3.

The equations of main interest in this chapter will be conditional equations, and because conditional equations are valid only for special values of the variable, our goal will be to determine those special values of the variable, called **solutions** (or sometimes **roots**) of the equation. In this terminology, 1 and $\frac{1}{3}$ are solutions of $3x^2 - 4x + 1 = 0$, because if we substitute 1 or $\frac{1}{3}$ for x in the equation, the resulting equation is valid. The process of finding all the solutions of a given equation is called **solving the equation**. Generally speaking, when we set out to solve a given equation whose solutions are not obvious, we will try to obtain an equivalent equation that is more easily solved.

Linear Equations

The simplest of all conditional equations is a **linear equation**, that is, an equation that is equivalent to either

$$ax = b \quad \text{or} \quad ax + b = 0$$

where a and b are fixed real numbers and $a \neq 0$. The following are examples of linear equations.

$$x = 3, \qquad 5x = 2, \qquad \pi x - \frac{7}{2} = 0, \quad \text{and} \quad 2x + 7 = 5x - 3$$

Presently we will show that every linear equation has a solution. Moreover, the solution is unique, that is, there is only one value of x for which the equation is valid.

It may happen that a given equation has the form $x = c$, in which case the unique solution is c. For example, the solution of $x = 3$ is 3. However, for other linear equations of the form $ax = b$ we usually find the solution by the following procedure, consisting of a chain of equivalent equations:

a. $ax = b$ (given equation)

b. $\dfrac{ax}{a} = \dfrac{b}{a}$ (division by $a \neq 0$)

c. $x = \dfrac{b}{a}$ (equation (b) simplified)

Because (a) and (c) are equivalent, we see that the unique solution of the equation $ax = b$ is b/a.

In actually solving equations of the form $ax = b$, we frequently combine steps (b) and (c), passing straight from $ax = b$ to $x = b/a$. Example 1 illustrates this procedure.

Example 1. Solve the equation $-2x = 6$ for x.

Solution. We divide both sides of the equation by -2 and simplify:

$$-2x = 6$$

$$x = \frac{6}{-2} = -3$$

Check: $-2(-3) = 6$

Thus -3 is the solution. $\square$

> **Caution:** It is always possible to make mistakes. For that reason it is a good idea to check the proposed solution by substituting it into the original equation, as we have done in Example 1 and will continue to do in the future.

Sometimes the linear equation to be solved is not given in the form $ax = b$ but can be put into that form by algebraic manipulation and then solved for x. The general procedure is:

> i. Put all terms containing x on one side of the equation.
> ii. Put all other terms on the other side of the equation.
> iii. Simplify the resulting equation to solve for x.

Example 2. Solve the equation $\frac{9}{5}x + 32 = 68$ for x.

Solution. We first subtract 32 from both sides:

$$\frac{9}{5}x + 32 = 68$$

$$\left(\frac{9}{5}x + 32\right) - 32 = 68 - 32$$

$$\frac{9}{5}x = 36$$

Now we have an equation in the form $ax = b$, so we divide by $\frac{9}{5}$ (or, equivalently, multiply by $\frac{5}{9}$) and simplify:

$$\frac{9}{5}x = 36$$

$$\frac{5}{9}\left(\frac{9}{5}x\right) = \frac{5}{9}(36)$$

$$x = 20$$

Check: $\frac{9}{5}(20) + 32 = 36 + 32 = 68$

Thus 20 is the solution. $\square$

The equation solved in Example 2 could arise in the conversion from degrees Fahrenheit to degrees Celsius. If F represents degrees Fahrenheit and C represents degrees Celsius, then F and C are related by the equation

$$\frac{9}{5}C + 32 = F \tag{3}$$

Suppose we wish to determine the number of degrees Celsius corresponding to 68 degrees Fahrenheit. In that case we can substitute 68 for F in (3) to obtain the linear equation

$$\frac{9}{5}C + 32 = 68$$

which is equivalent to the equation solved in Example 2. Its solution is 20, so 20 degrees Celsius corresponds to 68 degrees Fahrenheit.

Example 3. Solve the equation $2x + 7 = 5x - 3$ for x.

Solution. In order to have all terms containing x on the left side and all other terms on the right, we first subtract $5x$ from both sides:

$$2x + 7 = 5x - 3$$
$$(2x + 7) - 5x = (5x - 3) - 5x$$
$$-3x + 7 = -3$$

Then we subtract 7 from both sides, and finally solve the resulting equation as in previous examples:

$$(-3x + 7) - 7 = -3 - 7$$
$$-3x = -10$$
$$x = \frac{-10}{-3} = \frac{10}{3}$$

Check: $2\left(\dfrac{10}{3}\right) + 7 = \dfrac{20}{3} + 7 = \dfrac{41}{3}$ and

$5\left(\dfrac{10}{3}\right) - 3 = \dfrac{50}{3} - 3 = \dfrac{41}{3}$

Therefore the solution is $\frac{10}{3}$. □

It can happen that an equation looks very different from a linear equation but by various manipulations turns out to be equivalent to a linear equation, which can be solved by the process outlined in (i)–(iii).

Example 4. Solve $(x - 3)(x - 4) = (x + 1)(x - 2)$ for x.

Solution. We multiply out both sides and then simplify:

$$(x - 3)(x - 4) = (x + 1)(x - 2)$$
$$x^2 - 7x + 12 = x^2 - x - 2$$
$$-7x + 12 = -x - 2$$
$$-6x = -14$$
$$x = \frac{-14}{-6} = \frac{7}{3}$$

Check: $\left(\frac{7}{3} - 3\right)\left(\frac{7}{3} - 4\right) = \left(-\frac{2}{3}\right)\left(-\frac{5}{3}\right) = \frac{10}{9}$ and

$$\left(\frac{7}{3} + 1\right)\left(\frac{7}{3} - 2\right) = \left(\frac{10}{3}\right)\left(\frac{1}{3}\right) = \frac{10}{9}$$

Thus $\frac{7}{3}$ is the solution. □

Example 5. Solve $\dfrac{x + 2}{x - 1} = 2 - \dfrac{x - 1}{x + 3}$ for x.

Solution. First we multiply both sides by $(x - 1)(x + 3)$ in order to remove the terms $x - 1$ and $x + 3$ from the denominators. Then we manipulate the resulting equation in order to obtain the solution:

$$\frac{x + 2}{x - 1} = 2 - \frac{x - 1}{x + 3}$$

$$\left(\frac{x + 2}{x - 1}\right)(x - 1)(x + 3) = 2(x - 1)(x + 3) - \left(\frac{x - 1}{x + 3}\right)(x - 1)(x + 3)$$

$$(x + 2)(x + 3) = 2(x - 1)(x + 3) - (x - 1)^2$$
$$x^2 + 5x + 6 = 2(x^2 + 2x - 3) - (x^2 - 2x + 1)$$
$$x^2 + 5x + 6 = 2x^2 + 4x - 6 - x^2 + 2x - 1$$
$$x^2 + 5x + 6 = x^2 + 6x - 7$$
$$-x = -13$$
$$x = 13$$

Check: $\dfrac{13 + 2}{13 - 1} = \dfrac{15}{12} = \dfrac{5}{4}$ and

$$2 - \frac{13 - 1}{13 + 3} = 2 - \frac{12}{16} = 2 - \frac{3}{4} = \frac{5}{4}$$

Thus 13 is the solution. □

Equations Containing Absolute Values

Consider the equation $|x - 3| = 2$. If the absolute value were not present, the equation would be $x - 3 = 2$, which we know how to solve. Nevertheless, the equation $|x - 3| = 2$ is not linear, so we need a new method to solve it. Since $|x - 3| = x - 3$ or $|x - 3| = -(x - 3)$, depending on whether $x - 3 \geq 0$ or $x - 3 < 0$, it follows that in order to solve the equation $|x - 3| = 2$ we need to consider two cases: $|x - 3| = x - 3$ and $|x - 3| = -(x - 3)$. Each of these cases yields a linear equation, which we can solve by the method already discussed.

Example 6. Solve the equation $|x - 3| = 2$ for x.

Solution. As we suggested above, we consider the following two cases:

(a) If $|x - 3| = x - 3$, then we obtain

$$x - 3 = 2$$
$$x = 5$$

(b) If $|x - 3| = -(x - 3)$, then we obtain

$$-(x - 3) = 2$$
$$-x + 3 = 2$$
$$-x = -1$$
$$x = 1$$

Check: $|5 - 3| = |2| = 2$
$|1 - 3| = |-2| = 2$

Thus 5 and 1 are the solutions of the given equation. $\square$

Example 7. Solve $|3x - 4| = |2x + 5|$ for x.

Solution. Since $|3x - 4| = 3x - 4$ or $|3x - 4| = -(3x - 4)$, and similarly, since $|2x + 5| = 2x + 5$ or $|2x + 5| = -(2x + 5)$, there appear to be four cases: $3x - 4 = 2x + 5$, $3x - 4 = -(2x + 5)$, $-(3x - 4) = 2x + 5$, and $-(3x - 4) = -(2x + 5)$. But notice that the equations $3x - 4 = 2x + 5$ and $-(3x - 4) = -(2x + 5)$ are equivalent, and the equations $-(3x - 4) = 2x + 5$ and $3x - 4 = -(2x + 5)$ are themselves equivalent. Therefore we only have to consider two cases:

(a) $$3x - 4 = 2x + 5$$

which yields

$$x = 9$$

and

(b) $$3x - 4 = -(2x + 5)$$

which yields

$$3x - 4 = -2x - 5$$

$$5x = -1$$

$$x = -\frac{1}{5}$$

Check: $|3(9) - 4| = 23$ and $|2(9) + 5| = |23| = 23$

$$\left|3\left(-\frac{1}{5}\right) - 4\right| = \left|-\frac{23}{5}\right| = \frac{23}{5} \quad \text{and}$$

$$\left|2\left(-\frac{1}{5}\right) + 5\right| = \left|\frac{23}{5}\right| = \frac{23}{5}$$

Thus 9 and $-\frac{1}{5}$ are the two solutions of the given equation.   □

Equations with Two Variables

Consider the equation

$$2x + 5 = 3y - 7 \tag{4}$$

This equation contains two variables, x and y, and thus is quite different from the other equations we have studied in this section. But suppose we assign a numerical value, say 6, to y. Then (4) becomes

$$2x + 5 = 3(6) - 7$$

which reduces to

$$2x + 5 = 11$$

itself a linear equation in x that can be solved by the methods of this section. In fact, no matter what numerical value we assign to y, we obtain a linear equation in x that can be solved. But rather than solve every such equation, we can treat y in (4) as a constant. We use this idea as we solve the equation in (4) in our final example.

Example 8. Solve the equation

$$2x + 5 = 3y - 7$$

for x.

Solution. We treat y as a constant and solve the equation for x:

$$2x + 5 = 3y - 7$$

$$(2x + 5) - 5 = (3y - 7) - 5$$

$$2x = 3y - 12$$

$$x = \frac{1}{2}(3y - 12) = \frac{3}{2}y - 6 \quad \square$$

EXERCISES 2.1

In Exercises 1–12, solve the equation.

1. $2x = 0$

2. $-4x = 0$

3. $-x = \frac{1}{3}$

4. $x - 2 = 5$

5. $4y + 5 = 9$

6. $\sqrt{2}y + \dfrac{1}{\sqrt{2}} = 3\sqrt{2}$

7. $z + 3 = 2z - 4$

8. $4z - 2 = 3 - 7z$

9. $2(1 + z) = 3z + 5$

10. $\frac{1}{2}(2 - t) = \frac{3}{2} + 6t$

11. $1.3t + 5.2 = 2.6 - 7.8t$

12. $\frac{1}{3}(1 - 2t) + \frac{1}{6}(3t + 6) = 0$

In Exercises 13–30, solve the equation.

13. $x^2 - 5x + 6 = x^2 + 3x + 2$

14. $x^2 - 6 = -(3x - x^2)$

15. $(x - 1)(x + 2) = (x + 3)(-4 + x)$

16. $(y + 2)^2 = y^2 - 4$

17. $(y - 3)^2 = y^2 + 6y + 8$

18. $(y + 2)^3 = (y + 1)^3 + 3y^2 - 1$

19. $(y - 1)^3 + 3(y + 2)^2 = y^3 + 2y + 4$

20. $2 - \dfrac{3}{x} = 4 + \dfrac{2}{3x}$

21. $\dfrac{1}{x} + 3 = \dfrac{2}{x} + 4$

22. $\dfrac{1}{1 - t} = \dfrac{1}{1 + t}$

23. $\dfrac{3}{3 + t} = \dfrac{5}{4t - 1}$

24. $\sqrt{u} + \dfrac{4}{\sqrt{u}} = 2\sqrt{u} - \dfrac{3}{2\sqrt{u}}$

25. $u^{1/3} - u^{-2/3} = 4u^{1/3} + 6u^{-2/3}$

26. $\dfrac{2w + 1}{2w - 5} = \dfrac{w}{w - 6}$

27. $\dfrac{w - 2}{w + 3} = \dfrac{w + 1}{w - 1}$

28. $1 - \dfrac{z}{2z + 1} = \dfrac{\frac{1}{2}z - 1}{z + 3}$

29. $\dfrac{z - 2}{z - 1} + 3 = \dfrac{4z + 1}{z + 2}$

30. $\dfrac{1}{1 - t} + \dfrac{2}{1 - t^2} = \dfrac{3}{1 + t}$

In Exercises 31–40, solve the given equation.

31. $|x - 4| = 3$

32. $|x + 7| = 2$

33. $|x + 1| = \frac{5}{2}$

34. $|-2x + 5| = 7$

35. $|\frac{2}{3}x + 4| = 2$

36. $|2x - 1| = |-x + 5|$

37. $|\frac{1}{2}x + 3| = |\frac{2}{3}x - 3|$

38. $|x| = |x + 1|$

39. $|2x - 3| = x$

40. $|3x - 1| = x$

In Exercises 41–50, solve for x.

41. $x - y = 0$

42. $x + y = 1$

43. $2x = 6y - 1$

44. $3x - 2 = 4y$

45. $4x + 2y = 5$

46. $-\frac{1}{2}x + 1 = 3y + 5$

47. $\frac{1}{4}x - \frac{1}{2}y - 3 = 0$

48. $\frac{4}{7}y = \frac{2}{3}x - \frac{5}{2}$

49. $0.3x - 0.2y = 1.8x + 3.3y - 0.1$
50. $1.7x - 2.4y + 1.2 = -8.3x - 3.7y - 2.3$

In Exercises 51–54, solve for y.

51. $x = 2y + 7$
52. $6x + 3y = -1$
53. $\frac{1}{3}x - \frac{1}{2}y - 2 = 0$
54. $1.4x - 2.3y + 1.8 = 5.3x - 4.9y - 0.3$
55. Show that the equation $|2x + 3| = x$ has no solution.
56. Determine those values of a for which $|2x + a| = x$ has a solution.
57. Find a value of a such that -4 is a solution of the equation $2x + 3 - 4a = x + 7$.
58. Find a value of a such that 3 is a solution of the equation

$$\frac{1}{2 - x} + \frac{a}{x + 2} = \frac{1}{x^2 - 4}$$

59. Find a value of a such that $2x + 4 = a$ and $3 - x = 1$ are equivalent.
60. What relationship must a have to b in order for $\frac{1}{2}$ to be a solution of $ax + b = 0$?
61. If x denotes the length of an object in inches and y the length in centimeters, then x and y satisfy the equation

$$y = 2.54x$$

   a. What is the length in centimeters of a foot-long hot dog?
   b. How long in inches is a baby that is 63.5 centimeters long?
   c. Use a calculator to approximate the number of inches in one meter (100 centimeters).
62. If the temperature is 86 degrees Fahrenheit, what is the temperature in degrees Celsius? (*Hint:* Use (3).)
63. If the temperature is 100 degrees Celsius, what is the temperature in degrees Fahrenheit? (*Hint:* Use (3).)
64. Use (3) to determine a number T for which

$$T \text{ degrees Fahrenheit} = T \text{ degrees Celsius}$$

65. By solving equation (3) for C in terms of F, express the temperature in degrees Celsius in terms of the temperature in degrees Fahrenheit.
66. Light of sufficiently high frequency can dislodge electrons from their associated atoms. This effect is known as the ***photoelectric effect.*** The kinetic energy E of an electron so dislodged is given by

$$E = hv - \omega$$

where h is Planck's constant, v is the frequency of the light, and ω is the binding energy of the electron. Solve the equation for the frequency v.

67. The object distance p, image distance q, and focal length f of a simple lens satisfy the equation

$$\frac{1}{p} + \frac{1}{q} = \frac{1}{f}$$

Solve the equation for p.

2.2

APPLICATIONS OF LINEAR EQUATIONS

One of the reasons algebra is so important is that very often it can be used to help solve problems in such disciplines as physics, engineering, economics, and geometry. Problems like these are frequently formulated verbally. For that reason they are sometimes called *word problems*, but we prefer to call them *applied problems*. Although you may think that some of the problems we will present are rather contrived, such problems are part of mathematical folklore, and serve to illustrate the fact that mathematics can indeed be applied in everyday life.

To use the mathematics we have developed in solving an applied problem, we will translate the problem into mathematical language and then solve the resulting mathematical problem, which in this section will always be a linear equation. Because of the extreme breadth of applied problems, there is no hope of prescribing a single detailed procedure that works for all of them. After discussing and solving various applied problems, we will give a synopsis of our general method of attacking such problems. It should serve as a guideline to help you in solving other applied problems.

An Averaging Problem

Example 1. A student has scores of 67, 90, 76, and 82 on the first four tests in a history course. What grade must the student achieve on the fifth test in order for the average of the scores on all five tests to be 80?

Solution. The average of five tests is the sum of the scores on the five tests divided by 5. If we let

$$x = \text{the score on the fifth test}$$

then the average is

$$\frac{67 + 90 + 76 + 82 + x}{5}$$

Since

$$\frac{67 + 90 + 76 + 82 + x}{5} = \frac{315 + x}{5} = 63 + \frac{1}{5}x$$

it follows that the average will be 80 if

$$63 + \frac{1}{5}x = 80$$

Solving for x, we obtain

$$\frac{1}{5}x = 17$$

$$x = 85$$

Check: $\dfrac{67 + 90 + 76 + 82 + 85}{5} = \dfrac{400}{5} = 80$

Consequently the student must achieve an 85 on the fifth test in order for the average of all five tests to be 80. □

A Simple Interest Problem

Suppose that one invests a sum of money P (called the **principal**) in a savings account that draws simple interest at an annual rate r. This means that after n years the interest I earned by the account is given by the formula

$$I = Prn \qquad (1)$$

Four letters appear in (1). If three of them are assigned specific numerical values, then we can solve for the fourth. Normally the interest rate is expressed as a percentage. Thus if the interest rate is 7%, then $r = 0.07$. As an example of the use of (1), suppose that $2000 is invested in a savings account which draws simple interest at an annual rate of 7%. Then after 1 year the interest I earned is given by

$$I = (2000)(0.07)(1) = 140$$

and after 5 years the interest earned is given by

$$I = (2000)(0.07)(5) = 700$$

Example 2. Suppose that $16,000 is deposited in a bank in the following way: $10,000 into a U.S. Treasury note earning 9% simple interest annually, and $6000 into a passbook account earning 5% simple interest annually. How long will it take to earn $12,000 in interest?

Solution. Let

n = the number of years required to earn $12,000 in interest

In n years the amount of interest in dollars earned from the Treasury note is $(10,000)(0.09)(n)$, and the amount earned from the passbook account is $(6000)(0.05)(n)$. Since the total amount of interest to be earned is $12,000, we have the linear equation

$$(10,000)(0.09)(n) + (6000)(0.05)(n) = 12,000$$

This simplifies to

$$900n + 300n = 12,000$$
$$1200n = 12,000$$
$$n = 10$$

Check: After 10 years the Treasury note will have earned $(10,000)(0.09)(10)$ dollars in interest, and the passbook account will have earned $(6000)(0.05)(10)$ dollars in interest. Therefore their combined earned interest will be given by

$$(10,000)(0.09)(10) + (6000)(0.05)(10) = 9000 + 3000$$
$$= 12,000$$

Thus it will take 10 years to earn $12,000 in interest from the two deposits. □

A Geometric Problem

In our next example we discuss a geometric problem to which algebraic techniques apply.

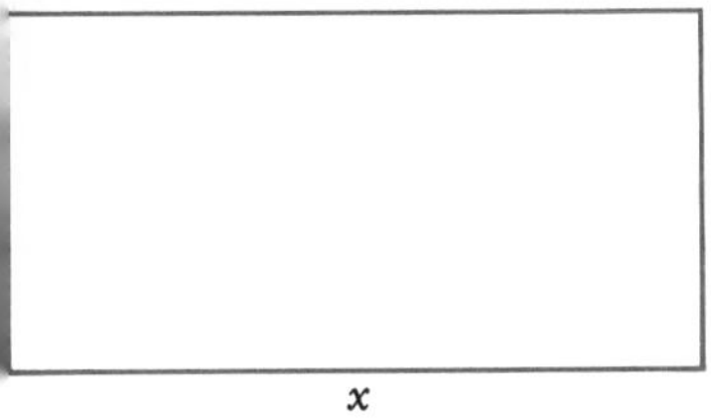

Example 3. A rectangular plot of land has a perimeter of 270 feet, and its length is twice its width (Figure 2.1). Find the dimensions of the plot.

Solution. Let

$$x = \text{the length in feet of the plot}$$
$$y = \text{the width in feet of the plot}$$

x

Length = 2 × Width
Perimeter = 270 feet

FIGURE 2.1

Since the perimeter is 270 feet, we have the equation

$$2x + 2y = 270 \tag{2}$$

What we are told about the length and width implies that

$$x = 2y \tag{3}$$

When we substitute $2y$ for x in (2), we obtain the linear equation

$$2(2y) + 2y = 270$$

which reduces to

$$6y = 270$$
$$y = 45$$

Now it follows from (3) that $x = 2(45) = 90$.

Check: perimeter $= 2x + 2y = 2(90) + 2(45)$

$$= 180 + 90 = 270$$

$$\text{length} = x = 90 = 2y = 2 \cdot \text{width}$$

Therefore the plot is 90 feet long and 45 feet wide. □

A Mixture Problem

Example 4. A radiator contains 6 quarts of fluid consisting of 40% antifreeze and 60% water. How much of the mixture should be drained off and replaced by pure antifreeze in order to obtain a mixture containing 60% antifreeze (and therefore 40% water)?

Solution. Let

$$x = \text{the amount of fluid to be drained off}$$

Then the amount of antifreeze remaining is 40% of $6 - x$ quarts, that is, $(0.4)(6 - x)$ quarts. With x quarts of antifreeze added, the total antifreeze in the radiator will be $x + (0.4)(6 - x)$ quarts, and the goal is for this to be 60% of the total of 6 quarts, that is, $(0.6)(6)$ quarts. Thus we obtain the linear equation

$$x + (0.4)(6 - x) = (0.6)(6)$$

which is solved as follows:

$$x + 2.4 - 0.4x = 3.6$$

$$0.6x = 1.2$$

$$x = \frac{1.2}{0.6} = 2$$

Check: If 2 quarts are removed and replaced by antifreeze, then the amount of antifreeze afterwards will be

$$2 + (0.4)(6 - 2) = 2 + 2.4 - 0.8 = 3.6$$

quarts. Since there are 6 quarts in the radiator and $(3.6)/6 = 0.6$, the percentage of antifreeze in the radiator at the end will be 60%.

Therefore 2 quarts must be drained and replaced by antifreeze to make 60% of the mixture antifreeze. □

Rate Problems

If an object (such as a train or a spaceship) travels at a constant speed or rate r, then the distance d traveled by the object during time t is given by

$$d = rt \tag{4}$$

In order for (4) to apply, the units in which r is measured must be compatible with the units in which d and t are measured. In the problems of this section, distance will be measured in miles, time in hours, and the rate in miles per hour. Notice that if two of the three letters are given specific values, then we can solve for the third.

Example 5. A passenger train averaging 75 miles per hour begins the 200-mile trip from New York to Boston at 2:00 P.M. A freight train traveling at 45 miles per hour sets out from Boston at 4:00 P.M. the same day and travels toward New York on an adjacent set of tracks. At what time will they meet?

Solution. Since we must determine when the trains meet and since they meet sometime after 4 P.M., we let

t = the elapsed time in hours after 4 P.M. until the trains meet

If we let d_1 be the distance in miles the passenger train will have traveled from New York when the trains meet and d_2 the distance in miles the freight train will have traveled from Boston, then $d_1 + d_2$ must be 200, the distance in miles between New York and Boston (Figure 2.2). Thus we have

$$d_1 + d_2 = 200 \qquad (5)$$

Now since the passenger train (which left New York at 2 P.M.) will have traveled $t + 2$ hours at 75 miles per hour and the freight train (which left Boston at 4 P.M.) will have traveled t hours at 45 miles per hour, (4) tells us that

$$d_1 = 75(t + 2) \quad \text{and} \quad d_2 = 45t \qquad (6)$$

Substituting the expressions given in (6) for d_1 and d_2 into (5), we obtain the linear equation

$$75(t + 2) + 45t = 200$$

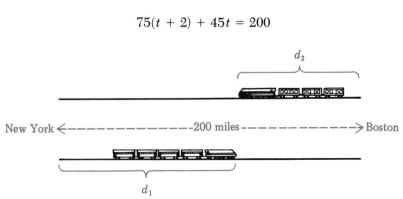

FIGURE 2.2

which is solved as follows:

$$75t + 150 + 45t = 200$$

$$120t = 50$$

$$t = \frac{50}{120} = \frac{5}{12}$$

Check: $\frac{5}{12}$ hour after 4 P.M. the passenger train is

$$\left(2 + \frac{5}{12}\right) 75 = 150 + \frac{125}{4} = 181\frac{1}{4}$$

miles from New York, and the freight train is

$$\frac{5}{12}(45) = \frac{75}{4} = 18\frac{3}{4}$$

miles from Boston. Since New York and Boston are 200 miles apart, when the freight train is $18\frac{3}{4}$ miles from Boston, it is

$$200 - 18\frac{3}{4} = 181\frac{1}{4}$$

miles from New York and thus meets the passenger train at that time.

Thus the trains indeed meet $\frac{5}{12}$ hour after 4 P.M., that is, at 4:25 P.M.   □

Next, suppose that a certain job (such as mowing a lawn) is performed at a constant rate. If r is the rate, then r is the fraction of the job that is performed in one hour (or in some other convenient unit of time). It follows that the amount W of work that is done in t hours is given by

$$W = rt \tag{7}$$

(Notice the similarity between equations (4) and (7).) Again the units in which W, r, and t are measured must be compatible, as they will be in our problems. If we solve equation (7) for r, we find that

$$r = \frac{W}{t} \tag{8}$$

The work W required to perform the entire job is 1, so that by (8),

$$r = \frac{1}{\text{the time required to perform the job}} \tag{9}$$

Example 6. If it takes an adult 2 hours to mow a lawn and a child 3 hours, how long would it take them to mow the lawn at the same time with two lawn mowers?

Solution. Let r_a and r_c be the rates at which the adult and the child can mow the grass, respectively. Then (9) implies that

$$r_a = \frac{1}{2} \quad \text{and} \quad r_c = \frac{1}{3} \tag{10}$$

This means that in 1 hour the adult can finish one half of the lawn and the child one third of the lawn. Let

$$t = \text{the time in hours needed for the adult and the child}$$
$$\text{to mow the whole lawn simultaneously}$$

If W_a denotes the amount of work done by the adult in mowing his portion of the lawn and W_c the corresponding amount of work done by the child, then

$$1 = W_a + W_c \tag{11}$$

By (7) and (10) we have

$$W_a = r_a t = \frac{1}{2} t \quad \text{and} \quad W_c = r_c t = \frac{1}{3} t$$

Substituting for W_a and W_c in (11) yields the linear equation

$$1 = \frac{1}{2} t + \frac{1}{3} t$$

which simplifies to

$$1 = \frac{5}{6} t$$

Thus

$$t = \frac{1}{\frac{5}{6}} = \frac{6}{5}$$

Check: The work done by the adult and the child in $\frac{6}{5}$ hours is given by

$$\frac{1}{2}\left(\frac{6}{5}\right) + \frac{1}{3}\left(\frac{6}{5}\right) = \frac{3}{5} + \frac{2}{5} = 1$$

Therefore it would take $\frac{6}{5}$ hours (that is, 72 minutes) for the adult and the child to mow the lawn simultaneously. □

Solving Applied
Problems

The following guidelines should help you solve other applied
problems, including the exercises that follow.

1. After reading the problem carefully, choose a variable
   for the quantity to be determined. If necessary, choose
   auxiliary variables for other quantities appearing in
   the problem (as in Examples 3, 5, and 6).
2. From the information given in the problem, write
   down any equations that must be satisfied by the variables. This step may require ingenuity. A picture is
   sometimes helpful (as in Examples 3 and 5).
3. Eliminate all the auxiliary variables from the equations. The goal is to obtain an equation that contains
   only the variable for the quantity to be determined
   (see Examples 3, 5, and 6).
4. Solve the equation obtained in step 3.
5. Check the solution to be sure that it really is a solution
   of the given applied problem.

Before going on to the exercises, you might find it helpful to
reread the examples in this section and their solutions, in order to
see how we followed the guidelines listed above. And when you attempt to solve applied problems, don't give up too quickly. Trans-
lating applied problems and the information in them into mathematical equations takes time, and the more you try and practice, the
more successful you will become at solving them.

EXERCISES 2.2

1. A student's grades from three examinations are 82, 64, and 91. In
   order to obtain an average of 80, what must the student's grade on
   the fourth test be?

2. In order to raise the average of the three tests in the preceding
   exercise by 3 points, what must the score on the fourth test be?

3. A student has a 72 average on four examinations. If the score on
   the final examination is to count double, what score will yield an
   average of 80?

4. Jack's mother is three times as old as Jack. In 14 years she will be
   twice as old as Jack is then. How old are they both now?

5. Karen's father is four times as old as Karen. In 6 years his age will
   be 10 years more than double Karen's age at that time. How old
   are they both now?

6. The sum of the ages of two children is 14. Two years ago the age
   of one child exceeded twice the age of the other by one year.
   How old are they now?

7. Find two numbers whose sum is 29 and whose difference is 5.

8. Find three consecutive odd integers whose sum is 147.

9. Receipts from 6500 tickets to a basketball game totaled $14,800. If student tickets cost $2 each and nonstudent tickets cost $3 each, how many of each were sold?

10. A child has 64 coins worth $4.50. If the coins are nickels and dimes, how many of each are there?

11. A newspaper boy collects $112 in dollar bills, quarters, dimes, and nickels. If he has 5 times as many dollar bills as quarters, $\frac{2}{3}$ as many quarters as dimes, and 10 more dimes than nickels, determine the number of each denomination.

12. In a local election, 75% of the eligible voters voted for more primary schools, 15% of those eligible voted against more primary schools, and the remaining 92 eligible voters did not vote on the issue. How many eligible voters were there?

13. A professor has a list of problems to assign to the students in a class. If 3 problems are assigned to each student, there will be 33 problems left over, but 15 more problems would be needed in order to be able to assign 5 problems to each person in the class. How many students are there in the class?

14. A football coach rewards a quarterback with 5¢ for each correctly solved mathematics problem and fines the quarterback 3¢ for each incorrectly solved problem. After working 100 problems, the quarterback has a net gain of $2.20. How many problems were solved correctly?

15. Suppose the cost of a digital clock, including a 4% sales tax, is $26.52. Find the cost of the clock without the tax.

16. According to the provisions of a will, a sum of money is to be divided among four people. One person is to receive one half of the sum, a second person is to receive one third of the sum, and a third is to receive one twelfth of the sum. The fourth person is to receive $1500. How large is the sum of money?

17. In a basketball game the high scorer had 30 points and the rest of the team scored $\frac{14}{16}$ of the team's total points. How many points did the team score?

18. A mutual fund pays an 11% dividend per year. If the dividend is $467.50 after the first year, how much was originally invested?

19. Suppose the amount to be invested in two savings accounts with simple interest is $3000, and assume that the accounts pay 6% and 8%, respectively. If the interest accumulated in one year is to be $196, how much must be deposited in each account?

20. Suppose that $6000 is to be invested in two savings accounts, one at 7% simple interest and the other at 8% simple interest. How much should be put into each account to have both accounts yield the same amount of interest?

21. The interest rate on an investment of $4200 is 1% greater than on a second investment of $2400. If the total annual interest from the two investments is $471, what are the respective interest rates?

22. A bricklayer gets a 6% increase in salary, and this amounts to $1116.
    a. What was the original salary?
    b. What is the new salary?

23. The perimeter of a rectangle is 100 inches, and the width is two thirds of the length. Find the dimensions of the rectangle.

24. The perimeter of a rectangular driveway is 106 feet, and the difference between the length and width is 23 feet. Find the length and width of the driveway.

25. The perimeter of a rectangle is 160 meters. If a new rectangle is formed by doubling the length of one pair of sides and decreasing the length of the remaining pair of sides by 30 meters, the perimeter remains the same. What were the dimensions of the original rectangle?

26. The perimeter of a rectangular rose garden is 56 feet. Determine its dimensions if three such gardens placed side by side would form a square.

27. The perimeter of an isosceles triangle is 33, and one side is three fourths as long as the other two sides. Determine the length of the shortest side.

28. The perimeter of an isosceles trapezoid is 42 inches. The shortest two sides, which are the same length, are 7 inches shorter than the next shortest side and 15 inches shorter than the longest side. Determine the length of the shortest sides.

29. The ratio of the earth's area of land to area of sea is approximately $\frac{7}{18}$, and the total area of the earth is approximately 197,000,000 square miles. Determine the approximate number of square miles of land on the earth.

30. A sample of 20 pounds of sea water has a 10% salt content. How much fresh water must be added to produce a mixture with a 6% salt content?

31. One alloy is 40% silver, and another alloy is 30% silver. How much of each should be used to produce 50 pounds of an alloy that is 36% silver?

32. How much pure alcohol must be added to 4 gallons of a solution that is 60% alcohol to achieve a solution that is 84% alcohol?

33. How much water must be evaporated from 8 gallons of a saline solution with a 20% salt content in order for it to have a 25% salt content?

34. Gasoline for cars has octane ratings (percent by volume of iso-octane in one gallon of gasoline). Normally, the octane ratings have been 89 for regular gasoline and 93 for super, or ethyl, gasoline. Suppose a gas tank contains 6 gallons of regular with an octane rating of 89. To achieve a mixture with an octane rating of 90, how many gallons of super with an octane rating of 93 must be added?

35. Two bicyclists are 4 miles apart and travel toward each other. One bicycles at the rate of 8 miles per hour, and the other bicycles at 12 miles per hour. After how long will they meet?

36. A child hikes along a trail through a forest at 2.5 miles per hour. If a parent sets out after the child half an hour later at a pace of 4 miles per hour, after how long will the parent overtake the child?

37. A jet traveled from Philadelphia to San Francisco at an average velocity of 500 miles per hour, and because of the west-to-east

tailwind it made the return trip at an average velocity of 600 miles per hour. The return trip took 48 minutes less than the trip west. What is the distance between Philadelphia and San Francisco?

38. A farmer sets out to walk to town 13 miles away at a rate of 4 miles per hour. After a while the farmer is picked up and driven the rest of the way at an average speed of 40 miles per hour. If the total trip takes 1 hour, how far does the farmer walk?

39. Two runners in the 26-mile Boston marathon travel at rates of 6 and 8 miles per hour, respectively. Suppose that they begin at the same instant.
    a. How far back will the slower runner be when the faster one completes the course?
    b. How long will it take for the two runners to be 4.5 miles apart?

40. One picker can harvest a strawberry patch in 2 hours, and a second picker can harvest the patch in 2.5 hours. How long would it take for both pickers to harvest the patch together?

41. One gasoline truck can fill a storage tank in 20 minutes. Another can fill the same tank in 30 minutes. How long would it take to fill the tank using both trucks simultaneously?

42. Suppose it takes 4 hours for Pam to shovel the snow from the driveway and 4.5 hours for Sam. How long would it take them to do it together, with one shovel each?

43. Suppose Bram joins Pam and Sam, and Bram alone can shovel the driveway in 6 hours. How long would it take the three to shovel the driveway together, with three shovels?

44. Suppose it would take Mary and Jane 3 hours and 45 minutes to paint a porch together. If Mary could paint it alone in 6 hours, how long would it take Jane to paint it alone?

45. Suppose that Mary, Jane, and Barbara can paint a porch in 3 hours together. If it would take 12 hours for Mary to paint it alone and 9 hours alone for Jane, how long would it take Barbara to paint the porch alone?

46. A gambler began with a certain sum of money. In the first investment $\frac{1}{3}$ of the money was lost. In the second investment $\frac{1}{4}$ of the remaining money was lost. In the third investment $\frac{1}{5}$ of the money remaining after the second investment was lost. If a total of $36,000 was lost, how large was the original sum of money?

47. A silversmith starts with a collection of necklaces and a small amount of savings, and he sells his wares at three crafts shows, where he also purchases other items of silver. At the first show, the silversmith doubles his money and then spends $20. At the second, he triples his new sum of money and then spends $40. At the third show, he quadruples his new sum of money and then spends $150. If the silversmith now has $650, how much did he originally have?

48. Little is known about the personal life of the mathematician Diophantus of Alexandria, who is thought to have lived in the third century A.D. However, in an epitaph to Diophantus, it is stated that he lived one sixth of his life in childhood, one twelfth in his youth, and one seventh more as a bachelor. Five years after his

marriage a son was born who died four years before Diophantus, at half of Diophantus's final age. Assuming that all this is true, determine the age of Diophantus when he died.

49. The **harmonic mean** of two numbers a and b with nonzero sum is

$$\frac{2ab}{a + b}$$

*a. Suppose a car travels from A to B with speed v_1 and returns with speed v_2. Show that the round trip could be made in the same length of time if the car traveled at a constant speed equal to the harmonic mean of v_1 and v_2.

b. If v_1 in part (a) is 40 miles per hour and v_2 is 60 miles per hour, find the constant speed that yields the same time for the round trip.

2.3

QUADRATIC EQUATIONS AND THE QUADRATIC FORMULA

A **quadratic equation** is an equation of the form

$$ax^2 + bx + c = 0, \quad \text{with } a \neq 0 \qquad (1)$$

or any equation that can be put into the form of (1) by shifting all nonzero terms to the left side of the equation. Thus

$$x^2 = 9, \quad x^2 - 4x - 3 = 0, \quad \text{and} \quad 3t^2 = 2t - 1$$

are quadratic equations.

Quadratic equations occur in physical applications. For instance, suppose that a ball is thrown upward at 64 feet per second from a dormitory roof 48 feet above the ground. If the air resistance is negligible, the laws of motion imply that until it hits the ground, the ball's height above the ground x seconds after it is thrown is $-16x^2 + 64x + 48$ feet. Now when the ball hits the ground, its height is 0, so x satisfies

$$-16x^2 + 64x + 48 = 0$$

or equivalently,

$$x^2 - 4x - 3 = 0 \qquad (2)$$

But this is a quadratic equation. What value(s) of x satisfy (2)? You might see if you can answer the question before reading on.

Our goal is to give a method of determining all real solutions of an arbitrary quadratic equation. In contrast to a linear equation, which always has one and only one real solution, a quadratic equation can have two, one, or no distinct real solutions, depending on the constants occurring in the equation.

If we can find real numbers p and q such that $x - p$ and $x - q$ are factors of $ax^2 + bx + c$, that is,

$$ax^2 + bx + c = a(x - p)(x - q) \qquad (3)$$

then p and q are evidently solutions of (1), because

$$ap^2 + bp + c = a\overbrace{(p - p)}^{0}(p - q) = 0$$

and

$$aq^2 + bq + c = a(q - p)\overbrace{(q - q)}^{0} = 0$$

Moreover, p and q are the only solutions of the given equation, because if

$$ax^2 + bx + c = a(x - p)(x - q) = 0$$

then since $a \neq 0$ by hypothesis, the Zero Property (see Section 1.1) implies that either $x - p = 0$ or $x - q = 0$, that is, $x = p$ or $x = q$.

Example 1. Solve the following quadratic equations.

    a. $x^2 + 2x - 3 = 0$ b. $2x^2 - x - 1 = 0$

Solution.

    a. Since

$$x^2 + 2x - 3 = (x - 1)(x + 3)$$

the given equation is equivalent to

$$(x - 1)(x + 3) = 0$$

Therefore the proposed solutions are 1 and -3.

        Check: $1^2 + 2(1) - 3 = 1 + 2 - 3 = 0$
                  $(-3)^2 + 2(-3) - 3 = 9 - 6 - 3 = 0$

Consequently 1 and -3 are the solutions of the given equation.

    b. Since

$$2x^2 - x - 1 = (2x + 1)(x - 1) = 2\left(x + \frac{1}{2}\right)(x - 1)$$

the given equation is equivalent to

$$2\left(x + \frac{1}{2}\right)(x - 1) = 0$$

Therefore the proposed solutions are $-\frac{1}{2}$ and 1.

$$\text{Check: } 2\left(-\frac{1}{2}\right)^2 - \left(-\frac{1}{2}\right) - 1 = \frac{1}{2} + \frac{1}{2} - 1 = 0$$

$$2(1)^2 - 1 - 1 = 2 - 1 - 1 = 0$$

Consequently $-\frac{1}{2}$ and 1 are the solutions of the given equation. $\square$

Solving a quadratic equation by factorization, as we did in Example 1, depends on our ability to factor a quadratic polynomial. Now we will derive a formula for the solutions of a quadratic equation that does not depend on factoring but nevertheless is easily applied.

The Quadratic Formula

In order to find the solutions of the equation

$$x^2 - 4x - 3 = 0 \tag{4}$$

in (2), we begin in what may at first seem a roundabout way by adding 3 to both sides, to obtain

$$x^2 - 4x = 3$$

Then we add a suitable constant to both sides so that the left side is the square $(x + r)^2$ of a linear polynomial. In this case the constant to add is 4, because $x^2 - 4x + 4 = (x - 2)^2$. Adding 4 to both sides, we obtain

$$x^2 - 4x + 4 = 3 + 4$$

which is equivalent to

$$(x - 2)^2 = 7 \tag{5}$$

To solve for x is now simple. By taking square roots of both sides of (5), we find that

$$x - 2 = \sqrt{7} \quad \text{or} \quad x - 2 = -\sqrt{7}$$

which means that

$$x = 2 + \sqrt{7} \quad \text{or} \quad x = 2 - \sqrt{7}$$

Thus $2 + \sqrt{7}$ and $2 - \sqrt{7}$ are the solutions of the equation in (4), which is, as we mentioned, related to the motion of a ball. Since $2 - \sqrt{7} < 0$, the only possible solution giving the length of time until the ball hits the ground is $2 + \sqrt{7}$ seconds, which is approximately 4.6 seconds.

The process leading from the given equation in (4) to the equation in (5) is called *completing the square*, because the terms involving x^2 and x become a part of the complete square of the form $(x + r)^2$.

Example 2. Find all real solutions of the quadratic equation

$$x^2 - 7x + 6 = 0$$

by first completing the square.

Solution. We follow the procedure detailed above:

$$x^2 - 7x = -6$$

$$x^2 - 7x + \frac{49}{4} = -6 + \frac{49}{4} = -\frac{24}{4} + \frac{49}{4} = \frac{25}{4}$$

$$\left(x - \frac{7}{2}\right)^2 = \frac{25}{4} = \left(\frac{5}{2}\right)^2$$

Taking square roots, we find that

$$x - \frac{7}{2} = \frac{5}{2} \quad \text{or} \quad x - \frac{7}{2} = -\frac{5}{2}$$

so that

$$x = \frac{7}{2} + \frac{5}{2} = \frac{12}{2} = 6 \quad \text{or} \quad x = \frac{7}{2} - \frac{5}{2} = \frac{2}{2} = 1$$

Therefore the proposed solutions are 6 and 1.

Check: $6^2 - 7(6) + 6 = 36 - 42 + 6 = 0$
$1^2 - 7(1) + 6 = 1 - 7 + 6 = 0$

Consequently 6 and 1 are the solutions of the given equation. □

Notice that the answer obtained in Example 2 by completing the square is the same as would be obtained by factoring, since

$$x^2 - 7x + 6 = (x - 6)(x - 1)$$

For any quadratic equation of the form

$$ax^2 + bx + c = 0, \quad \text{with } a \neq 0 \tag{6}$$

the procedure for completing the square is quite similar to that used in going from (4) to (5), provided that we first factor out a from the terms involving x^2 and x:

$$ax^2 + bx + c = 0$$

$$ax^2 + bx = -c$$

$$a\left(x^2 + \frac{b}{a}x\right) = -c$$

$$a\left(x^2 + \frac{b}{a}x + \frac{b^2}{4a^2} - \frac{b^2}{4a^2}\right) = -c$$

$$a\left(x^2 + \frac{b}{a}x + \frac{b^2}{4a^2}\right) - \frac{b^2}{4a} = -c$$

$$a\left(x + \frac{b}{2a}\right)^2 = \frac{b^2}{4a} - c = \frac{b^2 - 4ac}{4a}$$

$$\left(x + \frac{b}{2a}\right)^2 = \frac{b^2 - 4ac}{4a^2} \qquad (7)$$

If $b^2 - 4ac \geq 0$, we can find the real solutions of (7) by taking square roots of both sides of (7), which yields

$$x + \frac{b}{2a} = \frac{\sqrt{b^2 - 4ac}}{2a} \quad \text{or} \quad x + \frac{b}{2a} = -\frac{\sqrt{b^2 - 4ac}}{2a}$$

or equivalently

$$x = -\frac{b}{2a} + \frac{\sqrt{b^2 - 4ac}}{2a} \quad \text{or} \quad x = -\frac{b}{2a} - \frac{\sqrt{b^2 - 4ac}}{2a}$$

Thus if $b^2 - 4ac \geq 0$, the real solutions of (7) and hence of (6) are

$$x = \frac{-b + \sqrt{b^2 - 4ac}}{2a} \quad \text{and} \quad x = \frac{-b - \sqrt{b^2 - 4ac}}{2a} \qquad (8)$$

The formulas in (8) are often combined to give one formula:

THE QUADRATIC FORMULA

$$x = \frac{-b \pm \sqrt{b^2 - 4ac}}{2a}$$

Because the square root has been defined only for nonnegative numbers, the quadratic formula yields real solutions of the quadratic equation $ax^2 + bx + c = 0$ only if $b^2 - 4ac$, called the *discriminant* of the quadratic equation, is nonnegative. (In Section 7.1 we will define square roots of negative numbers to be complex numbers and thereby have an interpretation of the quadratic formula for any value of the discriminant.)

The number of real solutions of $ax^2 + bx + c = 0$ given by the quadratic formula depends on the value of the discriminant:

i. If $b^2 - 4ac > 0$, the quadratic formula provides two real solutions.
ii. If $b^2 - 4ac = 0$, the quadratic formula provides one real solution.
iii. If $b^2 - 4ac < 0$, the quadratic formula provides no real solutions.

In practice, we determine the real solutions of $ax^2 + bx + c = 0$ by applying the quadratic formula in the following way. We first substitute the values of a, b, and c into the formula

$$x = \frac{-b \pm \sqrt{b^2 - 4ac}}{2a} \tag{9}$$

regardless of the value of $b^2 - 4ac$, and then we simplify $b^2 - 4ac$, noting whether it is positive, zero, or negative. If $b^2 - 4ac \geq 0$, we further simplify the right side of (9) when possible to find the value of x; if $b^2 - 4ac < 0$, we conclude that no real solutions exist.

The power of the quadratic formula is that it enables us by numerical computation to tell how many real solutions a given quadratic equation has and to compute the values of any real solutions that exist. The following examples illustrate this point.

Example 3. Find all real solutions of the following equations.
  a. $3x^2 - 4x + 1 = 0$ b. $6x^2 + \sqrt{2}x - 2 = 0$

Solution.
  a. By the quadratic formula, any real solutions are given by

$$x = \frac{-(-4) \pm \sqrt{(-4)^2 - 4(3)(1)}}{2(3)} = \frac{4 \pm \sqrt{16 - 12}}{6}$$

$$= \frac{4 \pm \sqrt{4}}{6} = \frac{4 \pm 2}{6}$$

Since

$$\frac{4 + 2}{6} = 1 \quad \text{and} \quad \frac{4 - 2}{6} = \frac{1}{3}$$

there are two proposed solutions of the given equation: 1 and $\frac{1}{3}$.

Check: $3(1)^2 - 4(1) + 1 = 3 - 4 + 1 = 0$
$$3 \left(\frac{1}{3}\right)^2 - 4 \left(\frac{1}{3}\right) + 1 = \frac{1}{3} - \frac{4}{3} + 1 = 0$$

Consequently 1 and $\frac{1}{3}$ are the solutions of the given equation.

b. By the quadratic formula, any real solutions are given by

$$x = \frac{-\sqrt{2} \pm \sqrt{(\sqrt{2})^2 - 4(6)(-2)}}{2(6)} = \frac{-\sqrt{2} \pm \sqrt{2 + 48}}{12}$$

$$= \frac{-\sqrt{2} \pm \sqrt{50}}{12} = \frac{-\sqrt{2} \pm \sqrt{25 \cdot 2}}{12}$$

$$= \frac{-\sqrt{2} \pm \sqrt{25} \sqrt{2}}{12} = \frac{-\sqrt{2} \pm 5\sqrt{2}}{12}$$

Since

$$\frac{-\sqrt{2} + 5\sqrt{2}}{12} = \frac{4\sqrt{2}}{12} = \frac{1}{3}\sqrt{2}$$

and

$$\frac{-\sqrt{2} - 5\sqrt{2}}{12} = \frac{-6\sqrt{2}}{12} = -\frac{1}{2}\sqrt{2}$$

the two proposed solutions of the given equation are $\frac{1}{3}\sqrt{2}$ and $-\frac{1}{2}\sqrt{2}$.

Check: $6\left(\frac{1}{3}\sqrt{2}\right)^2 + \sqrt{2}\left(\frac{1}{3}\sqrt{2}\right) - 2$

$$= 6\left(\frac{2}{9}\right) + \frac{2}{3} - 2 = 0$$

$$6\left(-\frac{1}{2}\sqrt{2}\right)^2 + \sqrt{2}\left(-\frac{1}{2}\sqrt{2}\right) - 2$$

$$= 6\left(\frac{1}{2}\right) - 1 - 2 = 0$$

Consequently $\frac{1}{3}\sqrt{2}$ and $-\frac{1}{2}\sqrt{2}$ are the solutions of the given equation. □

Example 4. Find all real solutions of the following equations.
  a. $16x^2 + 24x + 9 = 0$ b. $3t^2 = 2t - 1$

Solution.
  a. By the quadratic formula, any real solutions are given by

$$x = \frac{-24 \pm \sqrt{(24)^2 - 4(16)(9)}}{2(16)}$$

$$= \frac{-24 \pm \sqrt{576 - 576}}{32} = \frac{-24 \pm 0}{32} = -\frac{3}{4}$$

Thus there is one proposed solution of the given equation: $-\frac{3}{4}$.

Check: $16 \left(-\dfrac{3}{4} \right)^2 + 24 \left(-\dfrac{3}{4} \right) + 9$

$$= 16 \left(\dfrac{9}{16} \right) - 18 + 9 = 0$$

Consequently $-\frac{3}{4}$ is the solution of the equation.

b. If we subtract $2t - 1$ from both sides of the given equation, we obtain the equivalent equation

$$3t^2 - 2t + 1 = 0$$

whose solutions, if any, are precisely the solutions of the given equation. But by the quadratic formula, any real solutions are given by

$$t = \dfrac{-(-2) \pm \sqrt{(-2)^2 - 4(3)(1)}}{2(3)}$$

$$= \dfrac{2 \pm \sqrt{4 - 12}}{6} = \dfrac{2 \pm \sqrt{-8}}{6}$$

Since $\sqrt{-8}$ is not defined, there are no real solutions. $\square$

One consequence of the quadratic formula is the fact that any quadratic equation

$$ax^2 + bx + c = 0$$

for which a and c have opposite signs has two real solutions. The reason is that in this case, $-4ac > 0$, so that the discriminant $b^2 - 4ac > 0$. In contrast, if a and c have the same sign, then the equation can have two, one, or no real solutions. These possibilities occur in Examples 3 and 4.

Occasionally we encounter an equation that is not quadratic but can be transformed into an equivalent quadratic equation by means of an algebraic manipulation.

Example 5. Find all real solutions of $1 + \dfrac{1}{x^2} = \dfrac{3}{x}$.

Solution. We multiply both sides of the equation by x^2 and then solve the resulting quadratic equation:

$$1 + \dfrac{1}{x^2} = \dfrac{3}{x}$$

$$x^2 + 1 = 3x$$

$$x^2 - 3x + 1 = 0$$

By the quadratic formula,

$$x = \dfrac{-(-3) \pm \sqrt{(-3)^2 - 4(1)(1)}}{2(1)} = \dfrac{3 \pm \sqrt{9 - 4}}{2} = \dfrac{3 \pm \sqrt{5}}{2}$$

Therefore the proposed solutions are $(3 + \sqrt{5})/2$ and $(3 - \sqrt{5})/2$. We could check here that these are actually the solutions of the given equation, but the calculations are tedious. $\square$

In conclusion, we emphasize the close connection between nontrivial factors of the quadratic polynomial $ax^2 + bx + c$ and solutions of the quadratic equation $ax^2 + bx + c = 0$. Indeed, if

$$ax^2 + bx + c = a(x - p)(x - q)$$

then p and q are the solutions of the equation $ax^2 + bx + c = 0$. If $b^2 - 4ac \geq 0$, then conversely, the solutions P and Q of the equation $ax^2 + bx + c = 0$ given by

$$P = \frac{-b + \sqrt{b^2 - 4ac}}{2a} \quad \text{and} \quad Q = \frac{-b - \sqrt{b^2 - 4ac}}{2a}$$

provide the factorization

$$ax^2 + bx + c = a(x - P)(x - Q)$$

of the quadratic polynomial $ax^2 + bx + c$.

EXERCISES 2.3

In Exercises 1–24, solve the given equation by rearranging (if necessary) and then factoring.

1. $x^2 = 16$
2. $x^2 - 121 = 0$
3. $4x^2 - 36 = 0$
4. $3x^2 - 25 = 0$
5. $x^2 + 4x + 4 = 0$
6. $x^2 - x - 2 = 0$
7. $x^2 - 50x + 625 = 0$
8. $y^2 - 2y = 3$
9. $9y^2 - 12y + 4 = 0$
10. $y^2 + 3y - 4 = 0$
11. $y^2 - 2\sqrt{2}y = -2$
12. $2y^2 + 18 = -12y$
13. $4x^2 + 2x + \frac{1}{4} = 0$
14. $9x^2 + \frac{1}{9} = 2x$
15. $5x + 4 = -x^2$
16. $6x - 9x^2 = 1$
17. $x^2 + 3x + 5 = 15$
18. $t^2 + t = 0$
19. $2t^2 + 42 = 20t$
20. $3x + 18 = x^2$
21. $5x = \dfrac{3}{x}$
22. $16 + \dfrac{25}{x^2} = \dfrac{40}{x}$
23. $2 + \dfrac{1}{x^2} = \dfrac{3}{x}$
24. $25x + \dfrac{4}{x} = 20$

In Exercises 25–30, solve the given equation by completing the square.

25. $x^2 - 2x - 6 = 0$
26. $y^2 = 4y + 12$

27. $y^2 - 10y = 2$ 28. $3t^2 - 6t = 12$

29. $2x^2 + 4x - 3 = 0$ 30. $4x - x^2 = -3$

In Exercises 31–54, find all real solutions (if any) of the given equation by using the quadratic formula.

31. $x^2 - 3x + 1 = 0$ 32. $x^2 + 3x + 1 = 0$

33. $x^2 + x + 1 = 0$ 34. $x^2 + x - 4 = 0$

35. $x^2 - 2x = 4$ 36. $x^2 = 4\sqrt{3}x - 12$

37. $3 + 2x = 2x^2$ 38. $2x^2 + 3x - 5 = 0$

39. $1 - 6w + 3w^2 = 0$ 40. $6w^2 + 5w = 1$

41. $6x^2 + 5x = -1$ 42. $3x^2 - 4\sqrt{3}x + 4 = 0$

43. $3x^2 - 4\sqrt{3}x - 4 = 0$ 44. $5y^2 + 7y + 5 = 0$

45. $4y^2 + 4\sqrt{2}y = 2$ 46. $4y^2 + 4\sqrt{2}y + 2 = 0$

47. $t^2 - t + 3 = 0$ 48. $t + \dfrac{1}{t} = 6$

49. $\dfrac{5}{t^2} + \dfrac{3}{t} - 1 = 0$ 50. $\dfrac{3}{t^2} - \dfrac{5}{t} = 2$

51. $\dfrac{x - 1}{2x - 3} = \dfrac{x + 2}{x - 2}$ 52. $\dfrac{x}{x^2 - 1} = 4$

53. $(x - 1)(x - 2) = 3(x + 1) + 4$

54. $(x + 3)(x + 7) = 3(x - 2) + 5$

55. Find the values of b for which there is exactly one solution of the equation $x^2 + bx + 9 = 0$.

56. Find all values of b such that $x^2 + bx + 5 = 0$ has exactly one real solution.

57. Find the value of b such that the real solutions of $x^2 + bx - 8 = 0$ are negatives of each other.

58. Show that no matter what the value of b is, there are two distinct real solutions of $x^2 + bx - 9 = 0$.

59. Find the values of b for which the real solutions of $x^2 + bx - 8 = 0$ differ by 6.

60. Show that if x_1 and x_2 are real solutions of the equation $x^2 + bx + c = 0$, then $x_1 + x_2 = -b$ and $x_1 x_2 = c$.

c In Exercises 61–64, use a calculator and the quadratic formula to approximate the solutions of the given equation.

61. $3x^2 - 17x - 23 = 0$ 62. $4x^2 + 100x + 3 = 0$

63. $3.14x^2 - 1.3x - 4.59 = 0$ 64. $x^2 - 3\sqrt{2}x + \sqrt{5} = 0$

65. The area A of a circle of radius r is given by $A = \pi r^2$. Solve the equation for r.

66. The surface area S of a sphere of radius r is given by $S = 4\pi r^2$. Solve the equation for r.

67. The volume V of a cylinder of radius r and height h is given by $V = \pi r^2 h$. Solve the equation for r.

68. The area A of an equilateral triangle of side s is given by $A = \sqrt{3}s^2/4$. Solve the equation for s.

69. The ancient Greeks thought that the rectangles with the most pleasing shapes are those for which the ratio of the width to the length is the same as the ratio of the length to the sum of the length and the width, that is,

$$\frac{w}{l} = \frac{l}{l + w} = \frac{1}{1 + \dfrac{w}{l}}$$

where l is the length and w is the width of the rectangle. If we let $x = w/l$, the above equation becomes

$$x = \frac{1}{1 + x} \tag{10}$$

The ratio x is called the **golden ratio** (or **golden section**) because of the interpretation the ancient Greeks associated with it. Determine the golden ratio by solving (10).

2.4
OTHER TYPES OF EQUATIONS

In the first three sections of this chapter we discussed linear and quadratic equations and found formulas for their solutions. Rather than moving on to third-degree equations, whose solutions are generally much more difficult to obtain, we will devote this section to several general types of equations that can be modified and then solved by methods we have already used.

Quadratic-Type Equations

Each of the equations

$$(3 + x)^2 - 5(3 + x) + 4 = 0,$$

$$x^4 - 7x^2 + 10 = 0,$$

and

$$x - 4\sqrt{x} + 4 = 0$$

has the feature that if we substitute u for a suitable expression in x, the resulting equation in u is a quadratic equation. As a result, such an equation is called an **equation of quadratic type**. It is solved by first solving the associated quadratic equation in u and then solving for x in terms of u.

Example 1. Find all solutions of $(3 + x)^2 - 5(3 + x) + 4 = 0$.

Solution. If we substitute u for $3 + x$, the equation becomes

$$u^2 - 5u + 4 = 0$$

which can be factored to yield

$$(u - 4)(u - 1) = 0$$

Therefore $u = 4$ or $u = 1$. Since $u = 3 + x$, we find that $x = u - 3$, so that if $u = 4$, then $x = 4 - 3 = 1$, and if $u = 1$, then $x = 1 - 3 = -2$. Thus the proposed solutions of the given equation are 1 and -2.

Check: $(3 + 1)^2 - 5(3 + 1) + 4 = 16 - 20 + 4 = 0$

$$(3 + (-2))^2 - 5(3 + (-2)) + 4 = 1 - 5 + 4 = 0$$

Thus 1 and -2 are the solutions. $\square$

Example 2. Find all solutions of $x^4 - 7x^2 + 10 = 0$.

Solution. This time we substitute u for x^2, which in particular implies that $u \geq 0$. The given equation becomes

$$u^2 - 7u + 10 = 0, \quad \text{with } u \geq 0$$

which yields

$$(u - 5)(u - 2) = 0, \quad \text{with } u \geq 0$$

The solutions of the associated equation in u are 5 and 2, both of which are positive. Since $u = x^2$, it follows that $x = \sqrt{u}$ or $x = -\sqrt{u}$, so that the proposed solutions of the given equation are $\sqrt{5}$, $-\sqrt{5}$, $\sqrt{2}$, and $-\sqrt{2}$.

Check: $(\sqrt{5})^4 - 7(\sqrt{5})^2 + 10 = 25 - 35 + 10 = 0$

$(-\sqrt{5})^4 - 7(-\sqrt{5})^2 + 10 = 25 - 35 + 10 = 0$

$(\sqrt{2})^4 - 7(\sqrt{2})^2 + 10 = 4 - 14 + 10 = 0$

$(-\sqrt{2})^4 - 7(-\sqrt{2})^2 + 10 = 4 - 14 + 10 = 0$

Thus $\sqrt{5}$, $-\sqrt{5}$, $\sqrt{2}$, and $-\sqrt{2}$ are the solutions of the given equation. $\square$

If the equation given in Example 2 had been

$$x^4 + 7x^2 + 10 = 0 \tag{1}$$

then we would have substituted u for x^2 as before, and the associated equation in u would have been

$$u^2 + 7u + 10 = 0, \quad \text{with } u \geq 0$$

so that

$$(u + 5)(u + 2) = 0, \quad \text{with } u \geq 0 \tag{2}$$

But both of the solutions -5 and -2 of the equation in (2) are negative, so there are no nonnegative solutions that satisfy (2). Consequently there are no solutions of the equation in (1).

Example 3. Find all solutions of $x - 4\sqrt{x} + 4 = 0$.

Solution. If we substitute u for $\sqrt{x}$ and note that $u \geq 0$, then we find that the given equation is transformed into

$$u^2 - 4u + 4 = 0, \quad \text{with } u \geq 0$$

which yields

$$(u - 2)^2 = 0, \quad \text{with } u \geq 0$$

Therefore $u = 2$. Since $u = \sqrt{x}$, we have $x = u^2$, so that if $u = 2$, then $x = 2^2 = 4$. Thus the proposed solution of the given equation is 4.

Check: $4 - 4\sqrt{4} + 4 = 4 - 8 + 4 = 0$

Thus 4 is the only solution of the given equation. □

Equations Involving Radicals

As illustrated in Example 3, an equation involving radicals occasionally arises. Often the solutions are most easily ascertained if we first eliminate the radicals by substitution or by squaring both sides of the equation and then solve the new equation. But we must be careful to check each solution of the new equation to determine which, if any, yield solutions of the original equation.

Example 4. Find all solutions of $\sqrt{3x + 4} = 5$.

Solution. Let us square both sides of the equation and then solve for x:

$$\sqrt{3x + 4} = 5$$
$$3x + 4 = 25$$
$$3x = 21$$
$$x = \frac{21}{3} = 7$$

Check: $\sqrt{3(7) + 4} = \sqrt{25} = 5$

Therefore 7 is the only solution of the given equation. □

Example 5. Find all solutions of the equation $\sqrt{4 - 3x} = x$.

Solution. Following the same procedure of squaring both sides of the equation, we obtain

$$\sqrt{4 - 3x} = x$$
$$4 - 3x = x^2$$
$$x^2 + 3x - 4 = 0$$
$$(x - 1)(x + 4) = 0$$

Therefore the proposed solutions are 1 and -4.

Check: $\sqrt{4 - 3(1)} = 1$
$\sqrt{4 - 3(-4)} = \sqrt{16} = 4 \neq -4$, so -4 is not a solution.

Consequently the only real solution of the given equation is 1. □

In working the preceding example, we altered the original equation and found two solutions of a new equation, but one of these turned out not to be a real solution of the original equation! What went wrong? The answer is that we squared both sides of the original equation, and as frequently happens, squaring both sides of the equation did *not* lead to an equivalent equation. For a simple example in which squaring leads to a nonequivalent equation, observe that

$$x = 2$$

is *not* equivalent to the equation obtained when both sides are squared:

$$x^2 = 4$$

Indeed, $x = 2$ has one solution, 2, whereas $x^2 = 4$ has two solutions, 2 and -2.

All solutions of a given equation are retained when we square both sides, but other candidates, which actually are not solutions, may be introduced by squaring. A solution of an altered equation that does not satisfy the original equation is called an ***extraneous solution.*** Thus -4 is an extraneous solution of the equation $\sqrt{4 - 3x} = x$ of Example 5. As we have seen, extraneous solutions can be introduced when we square both sides of an equation. The same is true if we raise both sides of an equation to any *even* power (see Exercise 63).

Caution: Because of the possibility of introducing extraneous solutions when we raise both sides of a given equation to an even power, it is doubly important to check proposed solutions in these cases.

In the next example we will cube each side of an equation. Yet no extra candidates for solutions will be introduced. More generally, when both sides of an equation are raised to an *odd* power, no extraneous real solutions are introduced.

Example 6. Find all solutions of $\sqrt[3]{x^2 - 8} = 2$.

Solution. We cube both sides of the equation and then simplify:

$$\sqrt[3]{x^2 - 8} = 2$$
$$x^2 - 8 = 8$$
$$x^2 = 16$$

Therefore the proposed solutions are 4 and -4.

Check: $\sqrt[3]{4^2 - 8} = \sqrt[3]{8} = 2$

$\sqrt[3]{(-4)^2 - 8} = \sqrt[3]{8} = 2$

Consequently 4 and -4 are the solutions of the given equation. □

In the following example we will square both sides of an equation twice, because squaring once does not clear the equation of radicals.

Example 7. Find the solutions of $\sqrt{x + 2} - \sqrt{3x - 5} = 1$.

Solution. First we alter the equation so that one radical appears on each side of the equation. Then we square both sides and simplify:

$$\sqrt{x + 2} - \sqrt{3x - 5} = 1$$
$$\sqrt{x + 2} = 1 + \sqrt{3x - 5}$$
$$x + 2 = 1 + 2\sqrt{3x - 5} + (3x - 5)$$
$$-2x + 6 = 2\sqrt{3x - 5}$$
$$-x + 3 = \sqrt{3x - 5}$$

Squaring again, we obtain

$$x^2 - 6x + 9 = 3x - 5$$
$$x^2 - 9x + 14 = 0$$
$$(x - 2)(x - 7) = 0$$

Thus the proposed solutions are 2 and 7.

Check: $\sqrt{2 + 2} - \sqrt{3(2) - 5} = \sqrt{4} - \sqrt{1} = 2 - 1 = 1$

$\sqrt{7 + 2} - \sqrt{3(7) - 5} = \sqrt{9} - \sqrt{16}$

$= 3 - 4 = -1 \neq 1$

Therefore 2 is a genuine solution of the given equation, whereas 7 is an extraneous solution. We conclude that the given equation has one real solution, 2. □

Equations Involving Integral Exponents

Some equations involving integral exponents can be solved for a variable by taking roots. The next two examples illustrate this feature.

Example 8. Solve the following equations for x.
  a. $(x^2 - 1)^3 = 27$ b. $81x^4 = 16$

Solution.
  a. We take cube roots and then solve for x:

$$x^2 - 1 = \sqrt[3]{27} = 3$$
$$x^2 = 3 + 1 = 4$$
$$x = 2 \text{ or } -2$$

Check: $(2^2 - 1)^3 = 3^3 = 27$
$[(-2)^2 - 1]^3 = 3^3 = 27$

Consequently the solutions are 2 and -2.

  b. Here we divide both sides by 81 and then take fourth roots:

$$x^4 = \frac{16}{81}$$

$$|x| = \sqrt[4]{\frac{16}{81}} = \sqrt[4]{\frac{2^4}{3^4}} = \frac{2}{3}$$

Therefore the proposed solutions are $\frac{2}{3}$ and $-\frac{2}{3}$.

Check: $81(\frac{2}{3})^4 = 81(\frac{16}{81}) = 16$
$81(-\frac{2}{3})^4 = 81(\frac{16}{81}) = 16$

Consequently the solutions are $\frac{2}{3}$ and $-\frac{2}{3}$. □

Example 9. Let T be the time in hours required for a planet to orbit its sun once, and let a be the average distance in miles between the planet and its sun. Kepler's Third Law states that $T^2 = ca^3$ where c is a constant. Solve the equation for a.

Solution. From the equation $T^2 = ca^3$ we have

$$a^3 = \frac{T^2}{c}$$

Therefore by taking cube roots of both sides, we obtain

$$a = \left(\frac{T^2}{c}\right)^{1/3} = \frac{T^{2/3}}{c^{1/3}} □$$

Other Equations

In this final part of the section we will analyze equations that do not fall into any of the earlier categories but whose solutions can be found expeditiously.

Example 10. Find all solutions of $x^3 - 6x^2 + 8x = 0$.

Solution. We can factor out an x from the equation, which yields

$$x(x^2 - 6x + 8) = 0$$

and then factor $x^2 - 6x + 8$ to obtain

$$x(x - 2)(x - 4) = 0$$

Thus by the Zero Property (see Section 1.1) the proposed solutions are 0, 2, and 4.

Check: $0^3 - 6(0)^2 + 8(0) = 0$

$2^3 - 6(2)^2 + 8(2) = 8 - 24 + 16 = 0$

$4^3 - 6(4)^2 + 8(4) = 64 - 96 + 32 = 0$

Therefore 0, 2, and 4 are the solutions of the given equation. □

Example 11. Find all solutions of $(x + 2)(2x^2 - 5x - 7) = 0$.

Solution. By the Zero Property (see Section 1.1) the solutions of this equation consist of all solutions of $x + 2 = 0$ and all those of $2x^2 - 5x - 7 = 0$. Now for $x + 2 = 0$ we have the solution -2. For $2x^2 - 5x - 7 = 0$ we use the quadratic formula (or factor, if you wish):

$$x = \frac{-(-5) \pm \sqrt{(-5)^2 - 4(2)(-7)}}{2(2)} = \frac{5 \pm \sqrt{25 + 56}}{4}$$

$$= \frac{5 \pm \sqrt{81}}{4} = \frac{5 \pm 9}{4}$$

Thus $x = \frac{14}{4} = \frac{7}{2}$ or $x = -\frac{4}{4} = -1$, so the proposed solutions of the given equation are $-2, \frac{7}{2}$, and -1.

Check: $(-2 + 2)[2(-2)^2 - 5(-2) - 7] = (0)(8 + 10 - 7) = 0$

$\left(\frac{7}{2} + 2\right)\left[2\left(\frac{7}{2}\right)^2 - 5\left(\frac{7}{2}\right) - 7\right] = \left(\frac{7}{2} + 2\right)\left(\frac{49}{2} - \frac{35}{2} - 7\right)$

$= \left(\frac{7}{2} + 2\right)(0) = 0$

$(-1 + 2)[2(-1)^2 - 5(-1) - 7] = (-1 + 2)(2 + 5 - 7)$

$= 1(0) = 0$

Therefore $-2, \frac{7}{2}$, and -1 are the solutions of the given equation. □

Example 12. Find all solutions of $x^{3/2} = -2x^{1/2}$.

Solution. We alter the equation so that it becomes

$$x^{3/2} + 2x^{1/2} = 0$$

which is equivalent to

$$x^{1/2}(x + 2) = 0$$

Thus the proposed solutions are 0 and -2.

Check: $0^{3/2} = 0$ and $(-2)(0)^{1/2} = 0$

Since $(-2)^{3/2}$ is undefined, -2 is an extraneous solution.

Consequently 0 is the only real solution of the given equation. □

EXERCISES 2.4

In Exercises 1–22, find all real solutions (if any) of the given equation.

1. $(x + 2)^2 + 11(x + 2) + 18 = 0$
2. $(x^2 - 3)^2 - 5(x^2 - 3) - 14 = 0$
3. $(x - 2)^2 + x - 32 = 0$
4. $(x^2 + x)^2 - 5(x^2 + x) - 6 = 0$

5. $x^4 - 6x^2 + 8 = 0$
6. $x^4 - 8x^2 + 15 = 0$
7. $x^4 - 8x^2 - 9 = 0$
8. $x^4 - 4x^2 - 12 = 0$
9. $x^4 + 5x^2 + 6 = 0$
10. $x^6 + 7x^3 - 8 = 0$
11. $x^6 - 27x^3 - 28 = 0$
12. $x^8 - 4x^4 - 12 = 0$
13. $x - \sqrt{x} - 12 = 0$
14. $x + 7\sqrt{x} + 10 = 0$
15. $x - 5\sqrt{x} - 6 = 0$
16. $x + 6\sqrt{x} - 3 = 0$
17. $x^{1/2} + 3x^{1/4} - 18 = 0$
18. $x^{1/2} - 2x^{1/4} + 1 = 0$
19. $x^{1/2} + 3x^{1/4} - 10 = 0$
20. $x^{2/3} - 2x^{1/3} + 1 = 0$
21. $x^{2/3} - 3x^{1/3} + 2 = 0$
22. $x^{4/3} - 5x^{2/3} + 6 = 0$

In Exercises 23–38, find all real solutions of the given equation.

23. $\sqrt{2x + 4} = 6$
24. $\sqrt{1 - x} = 4$
25. $\sqrt{4 + x^2} = 3$
26. $\sqrt{x^2 + 2x} = 2\sqrt{2}$
27. $\sqrt{x^2 - 5x} = \sqrt{14}$
28. $\sqrt{2x + 3} = x$
29. $\sqrt{6x - 1} = 3x$
30. $\sqrt[3]{x^2 + 2} = 3$
31. $\sqrt[3]{3x^2 - 1} = 2$
32. $\sqrt{3 - x} = \sqrt{5 - x^2}$
33. $\sqrt{4x - 5} = \sqrt{x^2 - 2x}$
34. $\sqrt{2x - 1} = 3 + \sqrt{x - 5}$
35. $\sqrt{5 - x} + 1 = \sqrt{7 + 2x}$
36. $\sqrt{x + 1} + \sqrt{x - 1} = \sqrt{2x + 1}$
37. $\sqrt{3 - 2\sqrt{x}} = \sqrt{x}$
38. $\sqrt{10 + 3\sqrt{x}} = \sqrt{x} + 2$

In Exercises 39–52, find all real solutions of the given equation.

39. $8x^3 = 27$

40. $6x^3 = -16$

41. $27x^3 - 10 = 0$

42. $16x^4 - 0.0081 = 0$

43. $x^n = 2$

44. $n^n = 3^n x^n$

45. $x^{2n-1} + 1 = 0$

46. $(x^2 + 1)^2 = \dfrac{9}{4}$

47. $(x^2 - 1)^2 = 4$

48. $(x^2 - 9)^2 = 4$

49. $(x^3 - 27)^3 = -64$

50. $(16x^2 - 9)^4 = 0$

51. $x^9 + x^4 = 0$

52. $\left(1 + \dfrac{1}{x}\right)^{100} = 2$

In Exercises 53–58, find all real solutions of the given equation.

53. $x^5 + x^3 - 2x = 0$

54. $x^3 - 16x^2 + 48x = 0$

55. $(x^2 - 4)(x^2 - 6x + 8) = 0$

56. $(x^2 - 6)(x^2 + x - 2) = 0$

57. $(x^2 - 9)(x^2 + 16) = 0$

58. $(x^3 + 8)(4x^4 - 2x^2 + \frac{1}{4}) = 0$

In Exercises 59–62, determine all real solutions of the given equation by finding common factors.

59. $x^3 - x^2 + x - 1 = 0$

60. $x^3 - x^2 - x + 1 = 0$

61. $x^4 - 3x^3 - 4x^2 + 12x = 0$

62. $x^4 + 5x^3 - 9x^2 - 45x = 0$

63. a. Determine the extraneous solution of the equation

$$\sqrt[4]{2x^2 - 1} = x$$

that we introduce if we solve it by raising both sides to the fourth power.

b. Let m be a positive integer. Determine the extraneous solution of the equation

$$\sqrt[4m]{2x^{2m} - 1} = x$$

that we introduce if we solve it by raising both sides to the $(4m)$th power.

64. The volume V of a sphere of radius r is given by $V = \frac{4}{3}\pi r^3$. Solve the equation for r.

65. The following equation occurs in the study of electricity.

$$E^2 = \frac{Q^2}{(1 + a^2)^3}$$

Solve the equation for a.

66. The equation

$$y = \frac{100\,kx^n}{1 + kx^n}$$

appears in the study of the saturation of hemoglobin with oxygen. Solve the equation for x.

67. The equation

$$V = \frac{\pi p r^4}{8 \eta l}$$

expresses the rate of volume flow V (volume per unit time) of a fluid through a cylindrical tube in terms of the radius r, the length l, the pressure difference p at the ends of the tube, and the viscosity η of the fluid. Solve the equation for r.

68. About 150 B.C., the ancient Greek astronomer Hipparchus developed a scale for measuring the brightness of stars. The scale gave what is called the *apparent magnitude* of the stars visible to the naked eye, from first magnitude to sixth magnitude. The brightest 20 stars were assigned the first magnitude, the next fainter group the second magnitude, and so on until the very faintest stars visible to the eye were assigned the sixth magnitude. It turns out that, on the average, a first magnitude star is approximately 100 times as bright as a sixth-magnitude star, and moreover, there is a number c such that the average brightness of the stars of any given magnitude is c times the average brightness of stars of the next fainter magnitude.
   a. Determine the number c.
   b. Let d be the ratio of the average brightness of a second-magnitude star to a fifth-magnitude star. Solve for d in terms of c.

C 69. Use a calculator to approximate the numbers c and d in Exercise 68.

2.5

APPLICATIONS OF QUADRATIC EQUATIONS

In Section 2.2 we solved applied problems that led to linear equations when translated into mathematical language. However, not all applied problems can be solved by means of linear equations. In this section we will consider problems that lead to quadratic equations.

The general method of solving applied problems used in Section 2.2 still applies. However, in order to facilitate working the examples, we will not write out our checks of the proposed solutions of the equations. Nevertheless, we will need to see which of any proposed solutions qualify as solutions of the given applied problems.

Vertical Motion

Suppose that an object is thrown or propelled or dropped and as a result moves vertically under the sole influence of gravity. For all practical purposes, this would be the case, for example, if a person jumps from a diving tower or throws a rock straight up.

Let us suppose that time is measured in seconds and that at some convenient moment (such as when the object begins moving) the time is 0. Suppose also that the height is measured in feet, with

h_0 the *initial height* above ground level at the instant $t = 0$, and finally, assume that the velocity is measured in feet per second, with v_0 denoting the velocity at time $t = 0$. Then v_0 is the *initial velocity* of the object. We take v_0 to be positive if the object is moving upward at time $t = 0$ and negative if the object is moving downward at that time. For example, if a ball is hurled upward at 40 feet per second, then $v_0 = 40$, whereas if the ball is thrown downward at 40 feet per second, then $v_0 = -40$. If the ball is dropped (from rest), then $v_0 = 0$.

The laws of motion imply that the height h of such an object above ground level at time $t \geq 0$ is given by

$$h = -16t^2 + v_0 t + h_0 \qquad\qquad (1)$$

It follows that if the constants h_0 and v_0 are known and if h is given, then (1) is a quadratic equation in the variable t. Of course, (1) is valid only until the object hits the ground or some other object or is influenced by other forces. By using (1) we can solve many problems concerning vertical motion.

Example 1. Suppose a diver descends from a platform 64 feet above a deep pool. How long will the diver fall before touching the water?

Solution. We take $t = 0$ at the instant the diver begins to descend, the initial velocity v_0 to be 0, and the initial height h_0 to be 64. Since the height h will be 0 when the diver first touches the water, we must find the time at which $h = 0$. To accomplish this we use (1) with $h = 0$, $v_0 = 0$, and $h_0 = 64$:

$$0 = -16t^2 + (0)t + 64$$

which reduces to

$$16t^2 = 64$$

or

$$t^2 = 4$$

Therefore $t = 2$ or $t = -2$. Consequently 2 and -2 are proposed solutions of the problem. Since $t = 0$ when the diver jumps starts the descent, the proposed solution -2 does not apply. Thus the diver touches the water after 2 seconds. □

Caution: In Example 1 we derived the mathematical equation $t^2 = 4$ from the given physical conditions. After we found the solutions 2 and -2 of the equation, we rejected the solution -2. It was *not* an extraneous solution of the equation but was rejected because it did not satisfy the physical condition that $t \geq 0$. Therefore in solving an applied problem it is critical that we scrutinize solutions arising from the mathematics in order to see which if any are actually solutions of the given applied problem, that is, which solutions actually satisfy the conditions relating to the given applied problem.

Example 2. A rocket loaded with fireworks is to be shot vertically upward from the ground with an initial velocity of 160 feet per second. The fireworks are to be detonated at a height of 384 feet while they are still on the rise. How long after takeoff should detonation occur?

Solution. Let $t = 0$ at takeoff. Our problem is to find the time at which the height h of the rocket is 384 feet and the rocket is rising. We are assuming that the rocket is shot upward from ground level, which means that $h_0 = 0$. By assumption the initial velocity is given by $v_0 = 160$, so that by (1) we need to solve the equation

$$384 = -16t^2 + 160t + 0$$

or

$$16t^2 - 160t + 384 = 0 \qquad (2)$$

To solve (2) we have the following equivalent equations:

$$16t^2 - 160t + 384 = 0$$

$$16(t^2 - 10t + 24) = 0$$

$$t^2 - 10t + 24 = 0$$

$$(t - 4)(t - 6) = 0$$

Therefore $t = 4$ or $t = 6$. Thus the rocket is at a height of 384 feet at 4 seconds and again at 6 seconds. Since the rocket is ascending at the former time and descending at the latter time, detonation should occur 4 seconds after the rocket is launched. □

Geometric Applications

One of the most important results known from antiquity is the Pythagorean Theorem, named after the outstanding Greek mathematician Pythagoras, who lived around 550 B.C. Recall that a right triangle is a triangle two of whose sides are perpendicular to each other. Assume that the lengths of these two sides, called the *legs* of the triangle, are a and b (Figure 2.3) and the length of the third side, called the *hypotenuse* of the triangle, is c. Then the lengths of the three sides are related by the following formula:

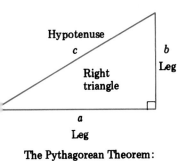

The Pythagorean Theorem:
$$a^2 + b^2 = c^2$$

FIGURE 2.3

PYTHAGOREAN THEOREM

$$a^2 + b^2 = c^2$$

The Pythagorean Theorem is basic to arguments involving right triangles.

Example 3. Suppose that the lengths of the legs of a right triangle differ by one inch, and the hypotenuse is one inch longer than the longer leg. Determine the lengths of all the sides of the triangle.

Solution. Let

$$x = \text{the length in inches of the shorter leg}$$

Then the other leg is $x + 1$ inches long, and the hypotenuse is $x + 2$ inches long. By the Pythagorean Theorem,

$$x^2 + (x + 1)^2 = (x + 2)^2$$

We solve this equation by using the following equivalent equations:

$$x^2 + (x^2 + 2x + 1) = x^2 + 4x + 4$$

$$2x^2 + 2x + 1 = x^2 + 4x + 4$$

$$x^2 - 2x - 3 = 0$$

$$(x - 3)(x + 1) = 0$$

Therefore $x = 3$ or $x = -1$. Since all lengths must be nonnegative, it follows that the shorter leg is 3 inches long. Consequently the other leg is 4 inches long, and the hypotenuse is 5 inches long. $\square$

Miscellaneous
Applications

Example 4. A rectangular painting has an area of 200 square inches. A frame 1 inch wide is added to each side of the painting, making the total area of painting and frame 261 square inches (Figure 2.4). What were the dimensions of the painting before the addition of the frame?

Solution. Let

$$x = \text{the width of the painting before framing}$$
$$y = \text{the height of the painting before framing}$$

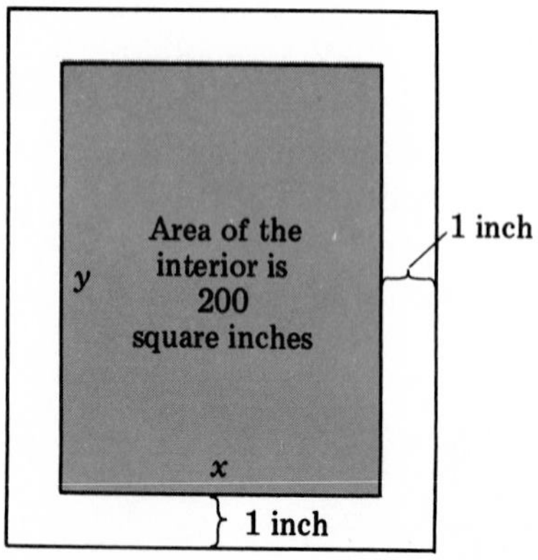

Then

$$xy = 200$$

so that

$$y = \frac{200}{x} \tag{3}$$

FIGURE 2.4

After the frame has been added, the width becomes $x + 2$ and the height becomes $y + 2$. Since the area of the painting plus the frame is by assumption 261 square inches, this means that

$$(x + 2)(y + 2) = 261 \qquad (4)$$

Equation (4) has two variables, x and y. Since they are related by (3), we substitute for y in (4) to obtain

$$(x + 2)\left(\frac{200}{x} + 2\right) = 261$$

which is an equation in the single variable x, and hence an equation we know how to treat. Multiplying out and then multiplying both sides by x, we obtain the following equivalent equations:

$$200 + \frac{400}{x} + 2x + 4 = 261$$

$$2x - 57 + \frac{400}{x} = 0$$

$$2x^2 - 57x + 400 = 0$$

By the quadratic formula,

$$x = \frac{-(-57) \pm \sqrt{(-57)^2 - 4(2)(400)}}{2(2)} = \frac{57 \pm \sqrt{3249 - 3200}}{4}$$

$$= \frac{57 \pm \sqrt{49}}{4} = \frac{57 \pm 7}{4}$$

Since

$$\frac{57 + 7}{4} = \frac{64}{4} = 16 \quad \text{and} \quad \frac{57 - 7}{4} = \frac{50}{4} = \frac{25}{2}$$

if follows that $x = 16$ or $x = \frac{25}{2}$. Now by (3),

$$\text{if } x = 16, \quad \text{then} \quad y = \frac{200}{16} = \frac{25}{2}$$

$$\text{if } x = \frac{25}{2}, \quad \text{then} \quad y = \frac{200}{25/2} = 16$$

Consequently the dimensions of the painting before framing were 16 inches by $\frac{25}{2}$ inches. □

Example 5. Pat drives the 432 miles between Boston and Washington, D.C., in one hour less than Dean and at an average speed of 6 miles per hour more than Dean. How fast does each drive?

Solution. Let

$$v = \text{Dean's speed}$$

so that Pat's speed is $v + 6$. Since $d = rt$ (see (4) of Section 2.2), the time required for Dean to drive 432 miles is $432/v$, and the time for Pat is $432/(v + 6)$. Since Pat drives the distance in one hour less than Dean, this means that

$$\frac{432}{v + 6} = \frac{432}{v} - 1$$

Multiplying each side by $v(v + 6)$, we find that

$$432v = 432(v + 6) - v(v + 6)$$

$$432v = 432v + 2592 - v^2 - 6v$$

$$v^2 + 6v - 2592 = 0$$

By the quadratic formula,

$$v = \frac{-6 \pm \sqrt{6^2 - 4(1)(-2592)}}{2(1)} = \frac{-6 \pm \sqrt{36 + 10{,}368}}{2}$$

$$= \frac{-6 \pm \sqrt{10{,}404}}{2} = \frac{-6 \pm 102}{2}$$

Since

$$\frac{-6 + 102}{2} = \frac{96}{2} = 48 \quad \text{and} \quad \frac{-6 - 102}{2} = -\frac{108}{2} = -54$$

it follows that $v = 48$ or $v = -54$. Because speed must be nonnegative, we conclude that Dean travels at a rate of 48 miles per hour. Pat therefore travels at the rate of $48 + 6 = 54$ miles per hour. □

EXERCISES 2.5

1. A ball is dropped from a balcony 64 feet above the ground. How long will it take for the ball to hit the ground?
2. A skyscraper window cleaner loses a pail 256 feet above the ground. How many seconds later does it pass by a window at the 112-foot level?
3. An acrobat drops from a platform 180 feet above a deep pool. How long will it take for the acrobat to reach the pool?
4. A rock is thrown down from a bridge 96 feet above the water. If the initial velocity is -16 feet per second, how long does it take the rock to hit the water?

5. A paintbrush is thrown straight up with a velocity of 48 feet per second toward a painter 32 feet higher up. How long a wait does the painter have before catching it?

6. A ball is thrown downward from a bridge 192 feet above a river. After 1 second the ball has traveled 80 feet. After how many more seconds will it hit the water?

7. The legs of a right triangle have lengths 5 and 6 inches. What is the length of the hypotenuse?

8. The length of the hypotenuse of a certain right triangle is 1 inch longer than one leg and is 8 inches longer than the other leg. Find the lengths of all three sides of the triangle.

9. The hypotenuse of a right triangle is 10 meters long, and the length of one leg exceeds the length of the other by one meter. How long is each leg?

10. A baseball diamond is a square 90 feet on a side. What is the distance from home plate to second base?

11. A plot of land is in the shape of a right triangle whose hypotenuse is 1300 feet long and one of whose legs is 1200 feet long. How many feet of fencing are required to enclose the plot?

12. A ladder 13 feet long is placed so that its base lies 5 feet from a wall and its top is 12 feet above the ground. If the ladder slips so that the base is 7 feet from the wall, how far does the top of the ladder fall?

13. A car travels 8 miles per hour faster than a truck and travels 160 miles in one hour less than the truck. How fast is the car traveling?

14. Two jets leave St. Louis at the same time, one traveling north at 520 miles per hour and the other traveling west at 450 miles per hour. How far apart are they after 2 hours?

15. Two airplanes begin 1000 miles apart and fly along lines that intersect at right angles. One plane flies an average of 100 miles per hour faster than the other. If the planes meet after 2 hours, how fast do the airplanes fly?

16. A jet leaves Chicago at noon and travels south at 600 miles per hour. One hour later a second jet leaves Chicago and travels east at 520 miles per hour. How far apart are the jets at 2:30 P.M.?

17. A pilot wishes to make a round trip between Los Angeles and San Francisco in 5 hours. The distance between the airports is 420 miles, and it is anticipated that there will be a head wind of 30 miles per hour as the plane flies north and a tail wind of 40 miles per hour when the plane returns. At what constant air speed must the plane be flown to achieve the goal?

18. A baker must make a delivery by truck to a store 12.5 miles from the bakery. By increasing the normal speed of the truck by 5 miles per hour, the baker could reduce the delivery time by 5 minutes. How fast does the baker normally drive?

19. The average speed of a commuter on a 20-mile trip into downtown Chicago is 16 miles per hour slower at rush hour than at midday, and the trip takes 20 minutes longer. What are the two rates?

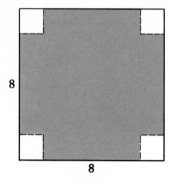

8

8

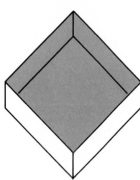

FIGURE 2.5

20. Find two consecutive odd integers whose product is 195.

21. Find two consecutive integers whose product is 1056.

22. One positive number is 3 more than twice a second positive number and the sum of their squares is 194. What are the numbers?

23. Find the two points on the y axis that are a distance of 6 units from the point $(4, 0)$ on the x axis.

24. A rectangular garden has an area of 1750 square feet and is 15 feet longer than it is wide. Find its dimensions.

25. The perimeter of a rectangle is 32 inches, and the area is 63 square inches. What are the dimensions of the rectangle?

26. A rectangular corral adjoins a barn. The corral has an area of 2352 square feet and is enclosed by a fence 140 feet long. If the barn forms one side of the corral, find the possible dimensions of the corral.

27. A rectangular lawn is 80 feet long and 60 feet wide. How wide a strip must be mowed around the lawn for half of the lawn to be cut?

28. A rectangular cloth measures 20 inches by 24 inches. We wish to embroider a strip of equal width on each side of the cloth in such a way that the cloth with embroidery will be rectangular with an area of 672 square inches. How wide a strip must we embroider?

29. A wire 44 inches long is cut into two pieces, each of which is bent into the form of a square. If the sum of the areas of the squares is 65 square inches, how long were the pieces of wire?

30. A square sheet of metal has sides of length 8 inches. A square piece is cut from each of the corners, and the edges are folded up to form a pan (Figure 2.5). If the area of the base of the pan is equal to the sum of the areas of the sides of the pan, what is the length of the sides of the squares that were cut out?

31. The volume of a cylindrical tin can is 48π cubic inches. If the can is 3 inches tall, find the radius of the base. (*Hint:* The volume is given by $V = \pi r^2 h$, where V, r, and h denote the volume, radius of the base, and height, respectively.)

32. A circular pool covers an area of 400π square feet. A path of constant width is to surround the pool. If the enlarged area of pool and path is 676π square feet, how wide must the path be?

33. On a certain rainy day the manager of a store estimates that if the price of umbrellas is set at p dollars, then $10(10 - p)$ umbrellas will be sold. Suppose the store acquires umbrellas at a cost of $3 each.
    a. What are the prices at which the store can sell umbrellas and neither make nor lose money on the sale (that is, at which the profit is 0)? How many umbrellas will be sold at those prices?
    b. What is the minimum price at which the store can sell umbrellas and make a profit of exactly $100? How many umbrellas will be sold at that price?

34. One can calculate the distance from ground level to water level in a well by dropping a stone from the ground and measuring the time t_0 elapsed until the splash is heard at ground level. Now $t_0 = t_1 + t_2$, where t_1 is the time it takes for the stone to hit the

water and t_2 is the time it takes the echo to return to ground level. If the distance is s, then $s = 16t_1^2$, so that

$$t_1 = \sqrt{\frac{s}{16}} = \frac{1}{4}\sqrt{s}$$

With the assumption that sound travels at 1100 feet per second, the distance is also given by $s = 1100t_2$, which means that

$$t_2 = \frac{s}{1100}$$

Consequently

$$t = t_1 + t_2 = \frac{1}{4}\sqrt{s} + \frac{s}{1100}$$

a. Calculate the distance if the time between drop and echo is $6\frac{9}{11}$ seconds.

b. Use a calculator to approximate the distance if the time between drop and echo is 5 seconds.

35. When an 18-foot tall stalk of bamboo is broken, the top portion of the stalk bends over and touches the ground 6 feet from the base of the stalk (Figure 2.6). How far from the ground was the bamboo broken? (This problem appeared in a Chinese algebra book in 1261.)

36. If a cannon is fired at an angle of 30° with respect to the ground, then until the ball hits the ground, the height y of the cannon ball after t seconds is given by $y = \frac{1}{2}v_0t - 16t^2$ where v_0 is the initial speed of the ball. Determine how long the cannon ball is in the air before hitting the ground.

37. A geometric proof of the Pythagorean Theorem is suggested by Figure 2.7. Let the hypotenuse have length c and the legs lengths a and b with $a \le b$. Assuming that a square of side c can be dissected into four copies of the right triangle and a square of side $b - a$, and using the fact that the area of the triangle is $\frac{1}{2}ab$, prove that $c^2 = a^2 + b^2$.

*38. A proof of the Pythagorean Theorem that uses similar triangles is suggested by Figure 2.8. Let the given right triangle have sides of length a, b, and c, with c the length of the hypotenuse. Using the fact that triangle ABC is similar to each of triangles ADC and BCD, prove that $c^2 = a^2 + b^2$. (*Hint:* Find expressions for a^2 and b^2 and then add them, using the fact that $s + t = c$.)

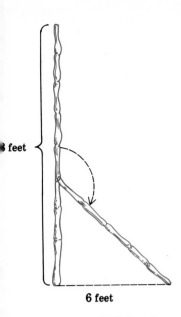

feet

6 feet

FIGURE 2.6

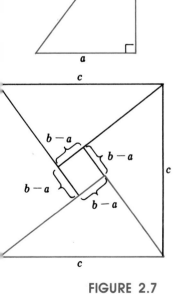

FIGURE 2.7

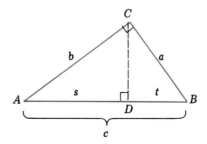

FIGURE 2.8

2.6

INEQUALITIES

So far in this chapter we have studied equations and their solutions. Although equations are fundamental to mathematics, so are inequalities. In the remainder of this chapter we will study inequalities and their solutions. We will first introduce notation to facilitate the description of the solutions.

Intervals

The four basic inequalities involving the real numbers a and b are

$$a < b, \quad a \le b, \quad a > b, \quad \text{and} \quad a \ge b$$

From Section 1.2 we know that

$$a < x < b \quad \text{means} \quad a < x \text{ and } x < b \text{ simultaneously}$$

and

$$a > x \ge b \quad \text{means} \quad a > x \text{ and } x \ge b \text{ simultaneously}$$

Other compound inequalities, which are defined similarly, include combinations of $<$ and $\le$ and combinations of $>$ and $\ge$.

> **Caution:** We *never* use either $<$ or $\le$ together with $>$ or $\ge$ in the same compound inequality. Thus we never write an expression like $2 < 5 \ge 3$. In all compound inequalities the inequality signs must open in the same direction ($<$ and $\le$, for example).

With the basic inequalities and the basic compound inequalities we can describe nine categories of special sets of real numbers called **intervals**. They are listed below. In our list we use the symbols ∞ (read "infinity") and $-\infty$ (read "minus infinity" or "negative infinity"). These two symbols *do not* represent real numbers but merely help us represent certain kinds of intervals.

Type of Interval	*Notation*	*Description*
Open interval	(a, b)	all x such that $a < x < b$
	(a, ∞)	all x such that $a < x$
	$(-\infty, a)$	all x such that $x < a$
	$(-\infty, \infty)$	all real numbers
Closed interval	$[a, b]$	all x such that $a \le x \le b$
	$[a, \infty)$	all x such that $a \le x$
	$(-\infty, a]$	all x such that $x \le a$
Half-open interval	$(a, b]$	all x such that $a < x \le b$
	$[a, b)$	all x such that $a \le x < b$

In the list, the symbols [and] indicate that the corresponding endpoint is included in the set, whereas the symbols (and) indi-

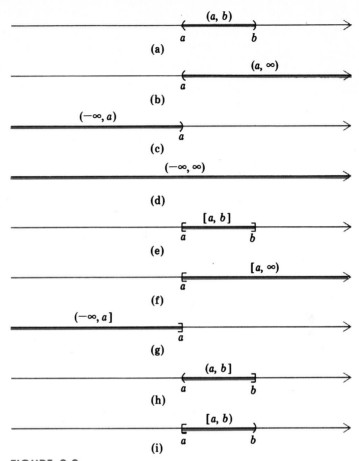

FIGURE 2.9

cate that the corresponding endpoint (if there is one) is not included in the set.

The four intervals (a, b), $[a, b]$, $(a, b]$, and $[a, b)$ are **bounded intervals**, and the remaining ones, each of which involves ∞ or $-\infty$, are **unbounded intervals**. The various kinds of intervals are described graphically on the real line in Figure 2.9.

Example 1. Determine whether each of the following intervals is open, closed, or half-open, and whether it is bounded or unbounded. Then locate the interval on the real line.

a. $(0, 3]$ b. $(-\infty, -2)$ c. $[-4, 1]$
d. $[-1, \infty)$ e. $\left(-\frac{3}{2}, -\frac{1}{3}\right)$

Solution.

a. $(0, 3]$ is half-open and bounded.
b. $(-\infty, -2)$ is open and unbounded.
c. $[-4, 1]$ is closed and bounded.
d. $[-1, \infty)$ is closed and unbounded.
e. $\left(-\frac{3}{2}, -\frac{1}{3}\right)$ is open and bounded.

The five intervals are shown in Figure 2.10.   □

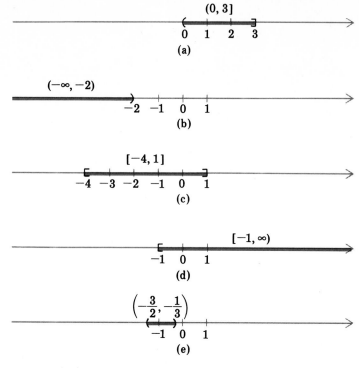

FIGURE 2.10

Basic Laws of Inequalities

In Section 1.2 we presented the law of trichotomy, which says that for any two given real numbers a and b, either $a < b$ or $a = b$ or $a > b$. We are ready now to present four more laws that will form the basis of mathematical computation involving inequalities. Throughout we will assume that a, b, c, and d are real numbers.

> **LAWS OF INEQUALITIES**
> If $a < b$ and $b < c$, then $a < c$.
> If $a < b$, then $a + c < b + c$.
> If $a < b$ and $c > 0$, then $ac < bc$.
> If $a < b$ and $c < 0$, then $ac > bc$.

Interchanging $<$ and $>$, we obtain the following versions.

> If $a > b$ and $b > c$, then $a > c$.
> If $a > b$, then $a + c > b + c$.
> If $a > b$ and $c > 0$, then $ac > bc$.
> If $a > b$ and $c < 0$, then $ac < bc$.

The laws listed above remain valid if $<$ is replaced by $\leq$ and if $>$ is replaced by $\geq$.

Caution: Note carefully that when both sides of an inequality are multiplied by a *negative* number, the sense of the inequality must be reversed (from $<$ to $>$, from $\leq$ to $\geq$, from $>$ to $<$, or from $\geq$ to $\leq$).

Solutions of Linear Inequalities

Consider the inequality

$$5x + 2 < -3$$

with x a variable as usual. If $x = -2$, then x satisfies $5x + 2 < -3$, since

$$5(-2) + 2 = -10 + 2 = -8 < -3$$

so we say that -2 is a solution of the inequality. By contrast, if $x = 1$, then x does not satisfy $5x + 2 < -3$, because

$$5(1) + 2 = 7 > -3$$

so we say that 1 is not a solution of $5x + 2 < -3$. More generally, the **solutions** of a given inequality are the values of the variable that make the inequality valid. We will show below that the solutions of $5x + 2 < -3$ consist of all x such that $x < -1$, that is, they comprise the interval $(-\infty, -1)$. For simplicity, in this chapter we will only consider inequalities whose solutions consist of either an interval or a finite collection of intervals.

When we determine the solutions of a given inequality, we say that we **solve the inequality**. This often involves finding a series of inequalities that are equivalent to the original inequality, by which we mean that they all have the same set of solutions as the original one has.

In this section we will solve **linear inequalities**, that is, inequalities that are equivalent to an inequality having one of the forms

$$ax < b, \quad ax \leq b, \quad ax > b, \quad \text{or} \quad ax \geq b$$

We solve a linear inequality in a way analogous to the way we solved linear equations:

> i. Put all terms containing x on one side of the inequality.
> ii. Put all other terms on the other side of the inequality.
> iii. Simplify the resulting inequality to solve for x.

Example 2. Solve the inequality $x - 6 > 0$.

Solution. Since the term containing x is on the left side, we need only add 6 to each side and then simplify:

$$x - 6 > 0$$

$$(x - 6) + 6 > 0 + 6$$

$$x > 6$$

Since $x > 6$ is equivalent to $x - 6 > 0$, it follows that x is a solution of the given equation if $x > 6$. Therefore the solutions form the interval $(6, \infty)$. □

Example 3. Solve the inequality $5x + 2 < -3$.

Solution. Again, the only term containing x is on the left side, so we need only subtract 2 from each side and then simplify:

$$5x + 2 < -3$$

$$(5x + 2) - 2 < -3 - 2$$

$$5x < -5$$

$$\frac{1}{5}(5x) < \frac{1}{5}(-5)$$

$$x < -1$$

Since $x < -1$ is equivalent to $5x + 2 < -3$, it follows that x is a solution of the given equation if $x < -1$. In other words, the solutions form the interval $(-\infty, -1)$. □

Example 4. Solve the inequality $3x - 2 \geq 8 + 5x$.

Solution. Using (i)–(iii), we obtain the following equivalent inequalities.

$$3x - 2 \geq 8 + 5x$$

$$(3x - 2) - 5x \geq (8 + 5x) - 5x$$

$$-2x - 2 \geq 8$$

$$(-2x - 2) + 2 \geq 8 + 2$$

$$-2x \geq 10$$

$$\left(-\frac{1}{2}\right)(-2x) \leq \left(-\frac{1}{2}\right)(10)$$

$$x \leq -5$$

In other words, the solutions of the original inequality consist of all $x \leq -5$, which means that they form the interval $(-\infty, -5]$. □

Observe that we reversed the sense of the inequality when we multiplied by the negative number $-\frac{1}{2}$ in the solution of Example 4.

Solutions of Composite Inequalities

Recall that the composite inequality

$$-2 < \frac{5 - x}{3} \leq 6x - 1 \tag{1}$$

is a shorthand way of writing the pair of inequalities

$$-2 < \frac{5 - x}{3} \quad \text{and} \quad \frac{5 - x}{3} \leq 6x - 1 \tag{2}$$

Therefore x is a solution of the composite inequality in (1) if and only if it is a solution of both inequalities in (2) simultaneously.

Example 5. Solve the inequality $-2 < \dfrac{5-x}{3} \le 6x - 1$.

Solution. We solve the given composite inequality by working separately on the two inequalities in (2):

$$-2 < \frac{5-x}{3} \qquad\qquad \frac{5-x}{3} \le 6x - 1$$

$$(-2)(3) < \frac{5-x}{3}(3) \qquad \frac{5-x}{3}(3) \le (6x-1)(3)$$

$$-6 < 5 - x \qquad\qquad 5 - x \le 18x - 3$$

$$-6 - 5 < (5-x) - 5 \qquad (5-x) + x \le (18x-3) + x$$

$$-11 < -x \qquad\qquad 5 \le 19x - 3$$

$$(-11)(-1) > (-x)(-1) \qquad 5 + 3 \le (19x-3) + 3$$

$$11 > x \qquad\qquad 8 \le 19x$$

$$\left(\frac{1}{19}\right)(8) \le \left(\frac{1}{19}\right)(19x)$$

$$\frac{8}{19} \le x$$

Consequently the solution consists of all values of x that satisfy $11 > x$ and $\frac{8}{19} \le x$ simultaneously, that is, $\frac{8}{19} \le x < 11$. Thus the solutions form the interval $[\frac{8}{19}, 11)$. □

Sometimes we can solve a composite inequality by performing the same operations on all members of the inequality. Of course, the basic laws of inequalities must be carefully observed. The procedure is illustrated in the next example.

Example 6. Solve the composite inequality $-3 \le \dfrac{2x+3}{-4} < 7$.

Solution. We have the following equivalent inequalities:

$$-3 \le \frac{2x+3}{-4} < 7$$

$$(-3)(-4) \ge \left(\frac{2x+3}{-4}\right)(-4) > (7)(-4)$$

$$12 \ge 2x + 3 > -28$$

$$12 - 3 \ge (2x+3) - 3 > -28 - 3$$

$$9 \ge 2x > -31$$

$$\frac{9}{2} \ge \frac{2x}{2} > -\frac{31}{2}$$

$$\frac{9}{2} \ge x > -\frac{31}{2}$$

Thus the solutions of the given inequality form the interval $(-\frac{31}{2}, \frac{9}{2}]$. ☐

Our final example is an applied problem whose solution involves inequalities.

Example 7. By common agreement a fever is any oral temperature greater than 98.6 degrees Fahrenheit. What temperatures in degrees Celsius correspond to a fever?

Solution. If F and C represent degrees Fahrenheit and Celsius respectively, then F and C are related by the formula

$$F = \frac{9}{5} C + 32$$

Then a fever corresponds to any Fahrenheit temperature F such that $F > 98.6$. The corresponding Celsius temperature must satisfy

$$\frac{9}{5} C + 32 > 98.6 \tag{3}$$

We need to determine the values of C for which (3) is valid, and this we do with the following equivalent inequalities:

$$\frac{9}{5} C + 32 > 98.6$$

$$\left(\frac{9}{5} C + 32\right) - 32 > 98.6 - 32$$

$$\frac{9}{5} C > 66.6$$

$$\frac{5}{9} \left(\frac{9}{5} C\right) > \frac{5}{9} (66.6)$$

$$C > 37$$

Therefore a fever is any temperature greater than 37 degrees Celsius. ☐

In the next section we will continue solving inequalities.

EXERCISES 2.6

In Exercises 1–14, identify the intervals as open, closed, or half-open and as bounded or unbounded.

1. $(-1, 2)$ 2. $(-7, -6]$

3. $(-\infty, 4]$ 4. $[6, 6.1]$

5. $(6, 6.01]$ 6. $(-1, \infty)$

7. $[7, 7]$

8. $(-\pi, \infty)$

9. $(1.9, 2.1)$

10. $[0, \infty)$

11. all x such that $-2 \leq x < 4$

12. all x such that $0 \leq x \leq 0.01$

13. all x such that $x > \dfrac{\sqrt{3}}{4}$

14. all x such that $-4 \leq x < \infty$

In Exercises 15–22, write the inequality in interval form.

15. $-4 < x \leq 3$

16. $5 \leq x \leq 7$

17. $-1.1 < x < -0.9$

18. $-1.01 < x \leq 0.99$

19. $x > -8$

20. $3 \leq x < \infty$

21. $-1 \leq x < 1$

22. $x \leq -2$

In Exercises 23–36, solve for x and then express the solutions as an interval.

23. $2x \leq 6$

24. $-4x < 8$

25. $-12x \geq -3$

26. $0 > 4x - 15$

27. $-7x - 2 \geq 0$

28. $2x + 7 \leq 5x - 3$

29. $4 - 3x > -1 - x$

30. $\frac{1}{2} + 2x \leq \frac{4}{3} - 5x$

31. $12 - 2x < 4(x - 6)$

32. $\frac{1}{3}(2x - 3) > 3(x + \frac{1}{3})$

33. $\dfrac{1 - x}{2} \geq \dfrac{2 + x}{-3}$

34. $\dfrac{5 - 2x}{7} \leq \dfrac{3x + 4}{2}$

35. $x^2 \geq (x + 3)^2$

36. $(x - 1)^2 < (x + 2)^2$

In Exercises 37–46, solve for x and then express the solutions as an interval.

37. $3 > x + 5 > 0$

38. $-2 \leq x - 1 \leq 4$

39. $0 \leq 5(x + 3) < 10$

40. $0 < \frac{1}{2}(2x + 4) < \frac{1}{3}$

41. $6 \geq \dfrac{3 - 3x}{12} \geq 4$

42. $\dfrac{1}{5} > \dfrac{2 - x}{-15} > \dfrac{1}{10}$

43. $-0.01 < x - 2 < 0.01$

44. $-10^{-4} < x - 2 < 10^{-4}$

45. $0 < x - a < d$

46. $-d < x - a < d$

47. Let $a, b, c,$ and d be real numbers with b and d positive.
   a. Prove that

$$\frac{a}{b} < \frac{c}{d} \quad \text{if and only if} \quad ad < bc$$

   (*Hint:* Multiply both sides of the first inequality by bd.)
   b. Use (a) to show that

$$\frac{22}{59} < \frac{3}{8}$$

48. Which temperatures in degrees Celsius correspond to the temperatures larger than 32 and smaller than 212 degrees Fahrenheit?

49. A snack bar manager estimates that if the price of hot dogs is set at x cents, where $20 \leq x \leq 200$, then $1000 - 5x$ hot dogs will be

sold daily. If the manager must sell at least 400 hot dogs each day at a cost of at least 20 cents, what are the possible prices for hot dogs?

50. A farmer is willing to sell s bushels of corn if the price of a bushel is $2 + (s/100{,}000)$ dollars. If the government sets a ceiling of \$4.50 on the price of a bushel of corn, what are the possible numbers of bushels the farmer would sell?

51. A store manager figures that $10(10 - p)$ umbrellas can be sold at p dollars per umbrella. If at least 45 umbrellas are to be sold, what are the possible prices that can be charged for each one?

52. A certain car holds 21 gallons of gasoline and gets 22 miles per gallon. If the car runs out of gasoline after having traveled at least 330 miles during a day, what are the possible amounts of gasoline that were in the tank at the start of the day?

53. Assuming that he charges x dollars per gallon, an ice cream factory manager estimates that he can sell $50{,}000 - 10{,}000x$ gallons of marshmallow ice cream per month. If he wishes to sell at least 30,000 gallons of it per month, what are the possible prices per gallon that he can charge?

54. A student pays \$25 per week for a room. Another room is available for \$22 per week. If it would cost \$15 to move to the second room, for what lengths of stay would it pay for the student to make the transfer to the new room?

2.7

MORE ON INEQUALITIES

This section begins where the preceding section ended. The method of solving the inequalities appearing in this section will rely heavily on the following rules:

$$ac > 0 \quad \text{if} \quad a > 0 \text{ and } c > 0, \quad \text{or if} \quad a < 0 \text{ and } c < 0 \qquad (1)$$

$$ac < 0 \quad \text{if} \quad a < 0 \text{ and } c > 0, \quad \text{or if} \quad a > 0 \text{ and } c < 0 \qquad (2)$$

These rules, which appeared in a slightly different form in Section 1.2, are consequences of the inequality laws of Section 2.6.

We begin by solving *quadratic inequalities*, that is, inequalities that are equivalent to one of the following:

$$ax^2 + bx + c > 0, \qquad ax^2 + bx + c \geq 0,$$
$$ax^2 + bx + c < 0, \quad \text{or} \quad ax^2 + bx + c \leq 0 \qquad (3)$$

Example 1. Solve the inequality $x^2 + x - 12 < 0$.

Solution. Since $x^2 + x - 12 = (x + 4)(x - 3)$, the given inequality is equivalent to

$$(x + 4)(x - 3) < 0$$

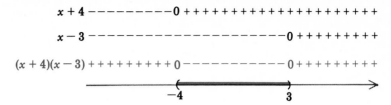

FIGURE 2.11

To solve this inequality we first find the intervals on which the components $x + 4$ and $x - 3$ are positive and those on which they are negative. Then we use this information along with (1) and (2) to determine the intervals on which the product $(x + 4)(x - 3)$ is positive and those on which it is negative. It is convenient to display the results in a diagram such as the one shown in Figure 2.11.

From the diagram we see that $(x + 4)(x - 3) < 0$ for x in $(-4, 3)$. The endpoints of $(-4, 3)$ are not included because $(x + 4)(x - 3) = 0$ for $x = -4$ and for $x = 3$. Thus the solutions of the given inequality form the interval $(-4, 3)$. □

If a quadratic inequality is not given in one of the forms listed in (3), we put it into such a form by transposing all terms to the left side of the inequality.

Example 2. Solve the inequality $x^2 - 7 \geq 6x$.

Solution. We subtract $6x$ from both sides and then factor the left side:

$$x^2 - 7 \geq 6x$$

$$(x^2 - 7) - 6x \geq 6x - 6x$$

$$x^2 - 6x - 7 \geq 0$$

$$(x + 1)(x - 7) \geq 0$$

Now we set up a diagram (Figure 2.12) to determine the intervals on which $(x + 1)(x - 7) \geq 0$. From Figure 2.12 we see that $(x + 1)(x - 7) \geq 0$ for x in $(-\infty, -1]$ or $[7, \infty)$. The endpoints -1 and 7 are included because $(x + 1)(x - 7) = 0$ for $x = -1$ and $x = 7$.

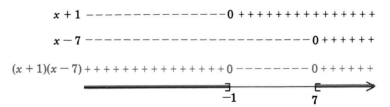

FIGURE 2.12

Thus the solutions of the given inequality form the intervals $(-\infty, -1]$ and $[7, \infty)$. □

The rules in (1) and (2) apply to products. Since any quotient a/c can be written as the product $a(1/c)$, and since $1/c$ has the same sign as c, we have the following rules for quotients:

$$\frac{a}{c} > 0 \quad \text{if} \quad a > 0 \text{ and } c > 0, \quad \text{or if} \quad a < 0 \text{ and } c < 0 \qquad (4)$$

$$\frac{a}{c} < 0 \quad \text{if} \quad a < 0 \text{ and } c > 0, \quad \text{or if} \quad a > 0 \text{ and } c < 0 \qquad (5)$$

Example 3. Solve the inequality $\dfrac{2x - 1}{5x + 3} > 0$.

Solution. Notice that $2x - 1 = 0$ for $x = \frac{1}{2}$ and that $5x + 3 = 0$ for $x = -\frac{3}{5}$. Using this information along with (4), we prepare the diagram shown in Figure 2.13. The endpoint $-\frac{3}{5}$ is not a solution, since $\dfrac{2x - 1}{5x + 3}$ is not defined for $x = -\frac{3}{5}$. The endpoint $\frac{1}{2}$ is not a solution because

$$\frac{2(\frac{1}{2}) - 1}{5(\frac{1}{2}) + 3} = \frac{1 - 1}{\frac{5}{2} + 3} = 0$$

From these observations and the diagram we conclude that the solutions of the given inequality form the intervals $(-\infty, -\frac{3}{5})$ and $(\frac{1}{2}, \infty)$. □

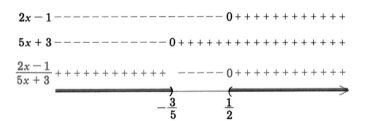

FIGURE 2.13

Example 4. Solve the inequality $\dfrac{2x - 1}{5x + 3} \geq 0$.

Solution. From Example 3 we know that $\dfrac{2x - 1}{5x + 3} > 0$ if and only if x is in $(-\infty, -\frac{3}{5})$ or in $(\frac{1}{2}, \infty)$. Since

$$\frac{2x - 1}{5x + 3} = 0$$

if and only if $2x - 1 = 0$, that is, if and only if $x = \frac{1}{2}$, we conclude that the set of solutions of the given inequality consists of $\frac{1}{2}$ along with the intervals $(-\infty, -\frac{3}{5})$ and $(\frac{1}{2}, \infty)$. Thus the solutions of the given inequality form the intervals $(-\infty, -\frac{3}{5})$ and $[\frac{1}{2}, \infty)$. $\square$

Example 5. Solve the inequality $\dfrac{6}{x - 2} \leq 2$.

Solution. We subtract 2 from both sides so that the right side is 0, and then simplify:

$$\frac{6}{x - 2} \leq 2$$

$$\frac{6}{x - 2} - 2 \leq 0$$

$$\frac{6 - 2(x - 2)}{x - 2} \leq 0$$

$$\frac{10 - 2x}{x - 2} \leq 0$$

$$\frac{2(5 - x)}{x - 2} \leq 0$$

Now we prepare the diagram shown in Figure 2.14. From the diagram we see that

$$\frac{2(5 - x)}{x - 2} \leq 0$$

for x in $(-\infty, 2)$ and $[5, \infty)$. Thus the solutions of the given inequality form the intervals $(-\infty, 2)$ and $[5, \infty)$. $\square$

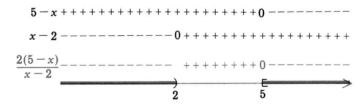

FIGURE 2.14

Another way of solving inequalities such as the one in Example 5 involves eliminating the denominator. Since $x - 2$ can be either positive or negative, we multiply by $(x - 2)^2$, which is always non-negative, and then rearrange:

$$\frac{6}{x - 2} \leq 2$$

$$\frac{6}{x - 2}(x - 2)^2 \leq 2(x - 2)^2$$

$$6(x - 2) \le 2(x^2 - 4x + 4)$$
$$6x - 12 \le 2x^2 - 8x + 8$$
$$0 \le 2x^2 - 14x + 20$$
$$0 \le x^2 - 7x + 10$$
$$0 \le (x - 5)(x - 2)$$

From a diagram analogous to Figure 2.14 we obtain the same solution as before (noting that 2 is not a solution of the original inequality).

So far all of our examples have contained two factors. However, the same general method works for any number of factors. In the next example, there are three factors.

Example 6. Solve the inequality $(x + 3)(x + 1)(x - 2) < 0$.

Solution. We prepare the diagram shown in Figure 2.15. From it we see that the solutions of the given inequality form the intervals $(-\infty, -3)$ and $(-1, 2)$. □

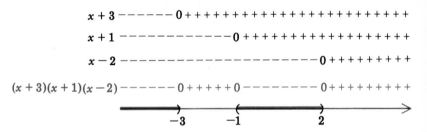

FIGURE 2.15

Applications of Inequalities

In Sections 2.2 and 2.5 we solved applied problems by means of equations. However, some applied problems can be solved by means of inequalities, as the next two examples illustrate.

Example 7. A rock is thrown vertically upward from a height of 6 feet above ground with an initial velocity of 96 feet per second. During what time interval will the rock be more than 134 feet above ground?

Solution. By (1) in Section 2.5, with $v_0 = 96$ and $h_0 = 6$, the height h of the rock at time t is given by

$$h = -16t^2 + 96t + 6$$

We wish to find the values of t for which $h > 134$, which means that we must solve the inequality

$$-16t^2 + 96t + 6 > 134$$

Proceeding as in earlier solutions of inequalities we find that

$$-16t^2 + 96t - 128 > 0$$

$$-16(t^2 - 6t + 8) > 0$$

$$t^2 - 6t + 8 < 0$$

$$(t - 2)(t - 4) < 0$$

Using this inequality and the facts that $t - 2 = 0$ if $t = 2$ and $t - 4 = 0$ if $t = 4$, we prepare the diagram shown in Figure 2.16. From the diagram we see that the solutions of $(t - 2)(t - 4) < 0$ form the interval $(2, 4)$. Thus the rock will be more than 134 feet high during the time interval between 2 seconds and 4 seconds. □

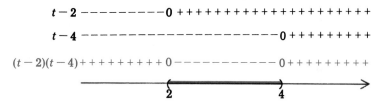

FIGURE 2.16

Example 8. A store manager figures that x chess sets can be sold each month if the price per set is $10 - 0.1x$ dollars. Suppose that chess sets can be purchased from the wholesaler at \$3 per set. If the manager wishes to make a profit of at least \$120 per month from the sale of chess sets, how many sets can be sold per month, and what are the possible prices?

Solution. If x chess sets are sold per month, then by hypothesis the resulting monthly revenue is $x(10 - 0.1x)$ and the monthly cost to the store is $3x$. Therefore the monthly profit, which is the difference between the monthly revenue and the monthly cost, is given by

$$x(10 - 0.1x) - 3x$$

Since the manager wishes the monthly profit to be at least \$120, this means that x must satisfy

$$x(10 - 0.1x) - 3x \geq 120$$

which is equivalent to the following inequalities:

$$10x - 0.1x^2 - 3x \geq 120$$

$$-0.1x^2 + 7x - 120 \geq 0$$

$$x^2 - 70x + 1200 \leq 0$$

$$(x - 30)(x - 40) \leq 0$$

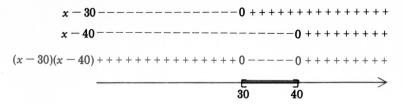

FIGURE 2.17

Using this inequality and the facts that $x - 30 = 0$ for $x = 30$ and $x - 40 = 0$ for $x = 40$, we prepare the diagram shown in Figure 2.17. From the diagram we see that the solutions of the inequality $(x - 30)(x - 40) \leq 0$ form the interval $[30, 40]$. We conclude that between 30 and 40 chess sets must be sold monthly in order to have a monthly profit of at least \$120.

Now that we know that the number x of chess sets must satisfy $30 \leq x \leq 40$, we will determine the possible values of the price $10 - 0.1x$. That is, we wish to determine numbers a and b such that $a \leq 10 - 0.1x \leq b$. Notice that if $30 \leq x \leq 40$, then $3 \leq 0.1x \leq 4$, so that $-4 \leq -0.1x \leq -3$, and consequently $6 \leq 10 - 0.1x \leq 7$. Therefore the manager must price each chess set at between \$6 and \$7. □

EXERCISES 2.7

In Exercises 1–44, solve the given inequality and express the solutions in terms of intervals.

1. $(x - 1)(x - 2) \geq 0$ 2. $(x + 5)(x - 1) < 0$

3. $(x + 1)(x + 3) \leq 0$ 4. $(x - \sqrt{2})(x + \sqrt{3}) \leq 0$

5. $(x - 3)^2 > 0$ 6. $(x + 2)^2 \geq 0$

7. $x^2 < 4$ 8. $x^2 \geq 9$

9. $(x - 2)^2 \leq 1$ 10. $(x^2 - 3)^2 \leq 9$

11. $(x^2 - 5)^2 < 16$ 12. $(x^2 - 3)^2 < 16$

13. $x^2 - 5x + 6 < 0$ 14. $x^2 + 3x - 10 \geq 0$

15. $x^2 - 2x - 15 > 0$ 16. $x^2 + 5x \leq 14$

17. $x^2 \leq -9x$ 18. $x^2 > \sqrt{2}\, x$

19. $-2x^2 + 7x > -4$ 20. $2x^2 + x < 1$

21. $\dfrac{x - 4}{x + 5} > 0$ 22. $\dfrac{2 - x}{3 + x} < 0$

23. $\dfrac{2x - 3}{3x + 6} \leq 0$ 24. $\dfrac{2x - 1}{x} \geq 5$

25. $\dfrac{x}{2x - 1} \geq 5$ 26. $\dfrac{-x + 5}{2x + 1} < -2$

27. $\dfrac{-5x + 3}{15x} > 1$ 28. $\dfrac{1}{x - 1} > \dfrac{1}{x + 1}$

29. $\dfrac{x}{x-1} > \dfrac{x}{x+1}$ 30. $\dfrac{2x}{x-3} \le \dfrac{2x}{x-6}$

31. $x^3 \ge 0$ 32. $x^3 < 8$

33. $(2-x)^3 < 0$ 34. $(x-1)(x-2)(x-3) > 0$

35. $x(x+3)(x+5) \le 0$ 36. $(x-7)(x^2+4) > 0$

37. $(x^2-1)(x^2-9) \le 0$

38. $-5(x-1)(x+\frac{3}{2})(x+2)(x-3) > 0$

39. $(x^2-x)(x^2-x+2) < 0$ 40. $\dfrac{(2-x)(3+x)^2}{x-\frac{1}{2}} \ge 0$

41. $\dfrac{(3x+4)(\frac{1}{6}-x)}{7x+2} < 0$ 42. $\dfrac{x+1}{x^2-x} < 0$

43. $\dfrac{x+1}{x^2-x} \ge -1$ 44. $\dfrac{2x+3}{6x^2+1} \le 3$

In Exercises 45–52, solve the given inequality and express the solutions in terms of intervals. You may need to use the quadratic formula to factor the left side.

45. $x^2 + x < 1$ 46. $x^2 - 3x + 1 \ge 0$

47. $2x^2 + 4x + 1 < 0$ 48. $3x^2 - 2x - 1 \le 0$

49. $\dfrac{1}{2}x^2 + \sqrt{2}\,x - 7 > 0$ 50. $\dfrac{2-x}{1+x^2} > -1$

*51. $\dfrac{3x-1}{2x^2-1} < 1$ *52. $\dfrac{3+4x}{2-x^2} \le 1$

53. Find the values of a such that 2 is a solution of the inequality

$$\frac{x-a}{x+a} \le 3$$

54. Find the values of a such that -3 is a solution of the inequality

$$\frac{-3x-2a}{x-2a} < -4$$

55. Show that if $a < b$, then $a < (a+b)/2 < b$. (The number $(a+b)/2$ is called the **arithmetic mean** of a and b.)

56. Let $0 < a < b$.
    a. Show that $(\sqrt{b} - \sqrt{a})^2 > 0$.
    b. Using part (a), prove that $a < \sqrt{ab} < \dfrac{a+b}{2}$. (The number $\sqrt{ab}$ is called the **geometric mean** of a and b.)

57. a. Show that $x^2 > x$ for $x > 1$.
    b. Show that $x^2 < x$ for $0 < x < 1$.

58. a. If $x^2 \ge 36$, is it necessarily true that $x \ge 6$? Explain.
    b. If $x^3 \ge 64$, is it necessarily true that $x \ge 4$? Explain.

59. For what values of x is $1/x < x$?

60. A ball is thrown vertically upward from a height of 100 feet with an initial velocity of 80 feet per second. During what time interval is the height of the ball at least 4 feet?

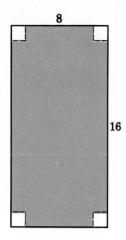

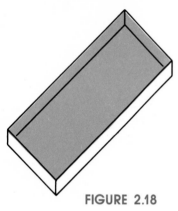

FIGURE 2.18

61. A ball is dropped from a window 100 feet above ground. Between what times will the ball be between 84 and 36 feet above ground?

62. If the ball in Exercise 61 were thrown upward at 48 feet per second, when would the ball be between 36 and 100 feet above ground?

63. In Example 8, if the manager would be satisfied to make a profit of at least $100, what would be the range in the number of chess sets that could be sold?

64. Suppose a bicycle shop can sell x bicycle replacement seats per month if the price is set at $12 - 0.2x$ dollars per seat. If the purchase price from the wholesaler is $4 per seat and the shop manager desired to make a profit of at least $75 per month, what are the possible numbers of seats that can be sold monthly, and at what prices?

*65. Assume, as in Exercise 64, that the price per seat is $12 - 0.2x$ and the purchase price from the wholesaler is $4 per seat. Show that the manager cannot make a profit of more than $80.

66. A baker estimates that he can bake up to 1800 loaves of rye bread during a week and that he can sell x loaves weekly if he charges $100 - \frac{1}{20}x$ cents per loaf. If it costs 25 cents per loaf to produce rye bread, what are the possible numbers of loaves that will net a profit for the baker?

67. A rectangular sheet of metal is 16 inches long and 8 inches wide. A pan is to be made from the sheet by cutting out four square pieces, one from each corner of the sheet (Figure 2.18). If the area of the base is to be at least 48 square inches, what are the possible heights of the sides created by folding up the edges?

2.8
INEQUALITIES INVOLVING ABSOLUTE VALUES

In addition to the kinds of inequalities we have already encountered, inequalities involving absolute values appear from time to time in advanced mathematics, primarily because of the relation of absolute value to distance between real numbers. In this section we will discuss and solve inequalities involving absolute values.

Recall from Section 1.2 that

$$|x| = \begin{cases} x & \text{if } x \geq 0 \\ -x & \text{if } x < 0 \end{cases} = \text{the distance between } x \text{ and } 0 \quad (1)$$

It follows from (1) that if c is any positive number, then

$$|x| < c \text{ means that } \begin{cases} x < c & \text{if } x \geq 0 \\ -x < c & \text{if } x < 0 \text{ (so } x > -c \text{ if } x < 0) \end{cases} \quad (2)$$

Combining the two parts of the right side of (2), we conclude that

$$|x| < c \quad \text{if and only if} \quad -c < x < c \quad (3)$$

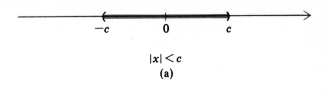

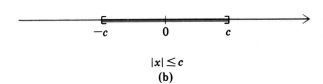

FIGURE 2.19

The inequality $|x| < c$ simply says that the distance between x and 0 is less than the positive number c (Figure 2.19a). For example, from (3) it follows that $|x| < 3$ means that $-3 < x < 3$, or equivalently, that the distance between x and 0 is less than 3. By similar reasoning we find that for any nonnegative number c,

$$|x| \leq c \quad \text{if and only if} \quad -c \leq x \leq c \tag{4}$$

The inequality $|x| \leq c$ means that the distance between x and 0 is less than or equal to c (Figure 2.19b).

Example 1. Solve the following inequalities.
  a. $|x| < \frac{1}{3}$ b. $|x| \leq 2$

Solution.
  a. By (3), the inequality $|x| < \frac{1}{3}$ is equivalent to $-\frac{1}{3} < x < \frac{1}{3}$, so the solutions form the open interval $(-\frac{1}{3}, \frac{1}{3})$.
  b. By (4), the inequality $|x| \leq 2$ is equivalent to $-2 \leq x \leq 2$, so the solutions form the closed interval $[-2, 2]$. □

Now we turn to the inequalities $|x| > c$ and $|x| \geq c$. Since $|x| > c$ means that $|x| \leq c$ is false, (4) implies that for any nonnegative number c,

$$|x| > c \quad \text{if and only if} \quad x > c \text{ or } x < -c \tag{5}$$

Similarly, (3) implies that for any nonnegative number c,

$$|x| \geq c \quad \text{if and only if} \quad x \geq c \text{ or } x \leq -c \tag{6}$$

The inequality $|x| > c$ means that the distance between x and 0 is greater than c (Figure 2.20a), and the inequality $|x| \geq c$ means that

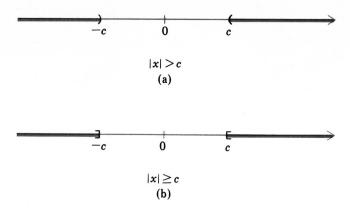

FIGURE 2.20

the distance between x and 0 is greater than or equal to c (Figure 2.20b).

Example 2. Solve the following inequalities.
    a. $|x| > 4$ b. $|x| \geq \frac{1}{2}$

Solution.
    a. By (5), the inequality $|x| > 4$ is equivalent to the statement that $x > 4$ or $x < -4$, so the solutions form the two open intervals $(-\infty, -4)$ and $(4, \infty)$.
    b. By (6), the inequality $|x| \geq \frac{1}{2}$ is equivalent to the statement that $x \geq \frac{1}{2}$ or $x \leq -\frac{1}{2}$, so the solutions form the two closed intervals $(-\infty, -\frac{1}{2}]$ and $[\frac{1}{2}, \infty)$.   □

Recall that $|x - a|$ is the distance between the numbers x and a. Thus geometrically the inequality $|x - a| < c$ means that the distance between x and a is less than c. There are analogous geometric interpretations of the inequalities

$$|x - a| \leq c, \qquad |x - a| > c, \quad \text{and} \quad |x - a| \geq c$$

Example 3. Solve the following inequalities.
    a. $|x - 7| < 2$ b. $|x - 7| \leq 2$

Solution by the algebraic method.
    a. By (3), the inequality $|x - 7| < 2$ is equivalent to

$$-2 < x - 7 < 2$$

Adding 7 to all three expressions, we obtain

$$5 < x < 9$$

Thus the solutions form the open interval $(5, 9)$.

b. By (4), the inequality $|x - 7| \leq 2$ is equivalent to

$$-2 \leq x - 7 \leq 2$$

Adding 7 to each expression yields

$$5 \leq x \leq 9$$

Thus the solutions form the closed interval $[5, 9]$. □

Solution by the geometric method.

a. Notice that x satisfies the inequality $|x - 7| < 2$ if and only if the distance between x and 7 is less than 2 units. On the real line we locate 7 and mark off 2 units to either side (Figure 2.21a). The numbers 5 and 9 are not solutions, since they both lie exactly 2 units from 7, but all numbers between 5 and 9 are solutions. Thus the solutions form the open interval $(5, 9)$.

b. Notice that x satisfies the inequality $|x - 7| \leq 2$ if and only if the distance between x and 7 is less than or equal to 2. Thus 5 and 9 are solutions of the inequality, as are all numbers in between. Consequently the solutions comprise the closed interval $[5, 9]$ (Figure 2.21b). □

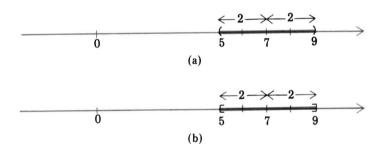

(a)

(b)

FIGURE 2.21

Other inequalities involving absolute values can be solved by similar methods.

Example 4. Solve the inequality $|3x - 5| < \frac{1}{2}$.

Solution. By (3), the inequality $|3x - 5| < \frac{1}{2}$ is equivalent to

$$-\frac{1}{2} < 3x - 5 < \frac{1}{2}$$

We can either treat this compound inequality as the pair of inequalities

$$-\frac{1}{2} < 3x - 5 \quad \text{and} \quad 3x - 5 < \frac{1}{2}$$

and solve the pair for x separately, or we can perform our alterations on the compound inequality itself. We will do the latter:

$$-\frac{1}{2} < 3x - 5 < \frac{1}{2}$$

$$-\frac{1}{2} + 5 < (3x - 5) + 5 < \frac{1}{2} + 5 \quad \bullet$$

$$\frac{9}{2} < 3x < \frac{11}{2}$$

$$\frac{9}{2}\left(\frac{1}{3}\right) < 3x\left(\frac{1}{3}\right) < \frac{11}{2}\left(\frac{1}{3}\right)$$

$$\frac{3}{2} < x < \frac{11}{6}$$

Therefore the solutions comprise the open interval $(\frac{3}{2}, \frac{11}{6})$.   □

EXERCISES 2.8

In Exercises 1–6, solve the given inequality and locate the solutions on the real line.

1. $|x| < 4$
2. $|x| < 1.5$
3. $|x| \le \frac{1}{5}$
4. $|x| \le 3.2$
5. $|x| \ge \frac{9}{2}$
6. $|-x| > 3$

In Exercises 7–38, solve the given inequality.

7. $|x| > 0.01$
8. $|x| \ge 100$
9. $|x - 5| < 3$
10. $|x - 10| < 2$
11. $|x + 3| \le 3$
12. $|x + 1| \le 4$
13. $|7 - x| > 1$
14. $|\frac{1}{2} - x| > \frac{1}{4}$
15. $|x - 2| \ge \frac{1}{3}$
16. $|4 + x| \ge 6$
17. $|\frac{5}{2} - x| < \frac{3}{4}$
18. $|\frac{5}{2} + x| \ge \frac{3}{4}$
19. $|2x - 1| < 3$
20. $|3x - 2| < \frac{1}{2}$
21. $|3x - 2| > 0$
22. $|3x - 2| \ge 0$
23. $|3x - 2| > -1$
24. $|\frac{1}{2}x + 5| \le 1$
25. $|\frac{2}{3}x - \frac{1}{6}| \le \frac{1}{4}$
26. $|5x + 3| > 1$
27. $\left|\dfrac{3x - 4}{2}\right| > 7$
28. $\left|\dfrac{5 + 4x}{3}\right| \le 2$
29. $|4x - 7| \ge 3$
30. $|-2x + 5| \ge 3$
31. $|-\frac{1}{3}x - 2| < 4$
32. $|x^2 - 5| < 4$
33. $|x^2 - 9| < 27$
34. $|x^3 - 13| < 14$
35. $1 \le |x - 5| < 2$
36. $0 < |x + 3| \le 4$
37. $1 \le |6x + 4| \le 3$
38. $\frac{1}{2} < |\frac{3}{2} - 5x| < \frac{3}{4}$

39. Find all values of b for which the equation $x^2 + bx + 5 = 0$ has
   a. no real solution
   b. two real solutions.

40. Find all values of b for which the equation $x^2 - 2bx + 6 = 0$ has
    a. no real solution
    b. two real solutions.

KEY TERMS

equation
  identity
  conditional equation
  equivalent equations
  linear equation
  quadratic equation
solution (root) of an equation
extraneous solution
completing the square

interval
  open interval
  closed interval
  half-open interval
  bounded interval
  unbounded interval
solution of an inequality
linear inequality
quadratic inequality

KEY FORMULAS

$$x = \frac{-b \pm \sqrt{b^2 - 4ac}}{2a} \quad \text{quadratic formula}$$

$a^2 + b^2 = c^2$ Pythagorean Theorem

KEY LAWS

If $a < b$ and $c > 0$, then $ac < bc$.

If $a < b$ and $c < 0$, then $ac > bc$.

$ac > 0$ if $a > 0$ and $c > 0$, or if $a < 0$ and $c < 0$.

$ac < 0$ if $a < 0$ and $c > 0$, or if $a > 0$ and $c < 0$.

REVIEW EXERCISES

In Exercises 1–34, find all real solutions (if any) of the given equation.

1. $\frac{1}{2}x - 7 = 6$

2. $-3t + 2 = -4$

3. $\frac{1}{3}(x - 3) + 4(-x + 2) = -4$

4. $\frac{3}{x + 3} = \frac{-9}{x - 1}$

5. $\frac{3x + 8}{8x - 3} = -2$

6. $\frac{4x - 5}{3 - 7x} = 2$

7. $(3y + 5)(-2y + 1) = 0$

8. $2\sqrt{y} = \frac{1}{\sqrt{y}}$

9. $|3x - 5| = 2$

10. $|\frac{1}{2}x + \frac{1}{3}| = \frac{2}{3}$

11. $5x^2 - 15 = 0$

12. $s^2 - s - 380 = 0$

13. $3y^2 + y + 7 = 0$

14. $4s^2 = 13s + 3$

15. $2s^2 - s - 2 = 0$

16. $4x^2 - 20x + 25 = 0$

17. $\dfrac{5}{x - 2} = x + 2$

18. $3x - \dfrac{2}{x} = 6$

19. $2x(x + 4) = (x - 1)(x + 3)$

20. $2x(x + 1) = (x - 1)(x + 3)$

21. $x^4 - x^2 = 2$

22. $x^{5/2} + 2x^{3/2} + \sqrt{x} = 0$

23. $(2x - 1)^2 - 2x + 1 = 0$

24. $\sqrt{2x - 1} = 1 - \sqrt{4x - 1}$

25. $\sqrt{2 - 3x} = 1 + \sqrt{1 - 2x}$

26. $t^3 - 2t^2 = t$

27. $t^3 + 6t^2 = 9t$

28. $(3x - 5)(2x^2 + 13x - 7) = 0$

29. $2x^4 - x^3 - 2x^2 + x = 0$

30. $\dfrac{4x^2 - 3}{-17x + 7} = 0$

31. $(6 - x^2)^2 - 4 = 0$

32. $(\tfrac{1}{8} - x^3)^3 = -729$

33. $(x^2 - 16)^4 = 256$

34. $2^{n-1} = x^n$

In Exercises 35–52, solve the given inequality and express the solutions in terms of intervals.

35. $4x - 7 \le 3$

36. $-2x + 3 > 6$

37. $-2x + 3 > -6$

38. $-7 \le -3x + 2 < 0$

39. $y^2 \le (y + 1)^2$

40. $(2y + 3)(y - \sqrt{3}) < 0$

41. $x^2 - 9x + 20 > 0$

42. $x^2 - 4x + 2 \le 0$

43. $3t^2 + 7t < -2$

44. $t(2t - 1) \ge (t + 1)^2 + 3$

45. $\dfrac{2x + 1}{-5x + 2} \ge 0$

46. $\dfrac{5x - 2}{2x + 3} < -1$

47. $\dfrac{1 + 8x}{x - 3} \ge 2$

48. $(2y + 1)(y + 2)(y - 5) > 0$

49. $|x| \ge \sqrt{2}$

50. $|x - 3| < 0.5$

51. $|-2x + 5| \le 13$

52. $4 < |-3x - 8| \le 6$

53. Find the value of a such that 0 is a solution of the equation

$$\frac{1}{2x - 1} + \frac{3a}{x + 1} = 4a$$

54. Find the values of a such that 1 is a solution of the equation $a^2x - 2ax = 3$.

55. For what values of a is -1 a solution of the following inequality?

$$\frac{2x - a}{3x + a} < -2$$

56. If the sales tax is 5% and amounts to $3.18 on a pair of sunglasses, what is the total cost of the glasses (item cost plus tax)?

57. The hottest and coldest outdoor temperatures ever recorded on the surface of the earth were 136.4 and -126.9 degrees Fahrenheit, respectively. What are the corresponding temperatures in degrees Celsius? Round your answer off to the nearest tenth of a degree. (*Hint:* Use (3) of Section 2.1.)

58. Two boats traveling at the same speed depart from the same port at the same time. One travels north for a while and then heads straight for a second port. The other travels 10 miles south and then 20 miles west and arrives at the second port at exactly the same time as the first boat. How far north did the first boat travel before heading for the second port?

59. A Coast Guard boat is 5 miles north of Bermuda, and a boat suspected of carrying smugglers is 25 miles west of Bermuda and moving directly toward Bermuda. If both boats travel at the same speed, how far will the Coast Guard boat have to travel in order to intercept the other boat? (*Hint:* See Figure 2.22.)

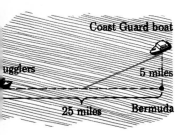

FIGURE 2.22

60. A dog chases a cat that has a lead of 20 feet. If the dog jumps 5 feet each time the cat jumps 3 feet, how many jumps does the dog make before catching the cat?

61. If 3 people all mowing at the same rate can mow 4 acres in 8 hours, how long will it take 4 people all mowing at the same rate to mow 3 acres?

62. Bill and John can do a job in 20 minutes. If Bill worked at twice his normal rate, they could do the job in 15 minutes. How long would it take John to do the job alone?

63. A ball is thrown vertically upward from a height of 12 feet with an initial velocity of 48 feet per second. During what time interval is the ball at least 44 feet high?

c 64. Two square pictures have a combined area of 1000 square inches. The length of a side of one picture is 10 inches less than two thirds the length of a side of the other picture. Use a calculator to find the area of the larger picture.

65. A group of students plan to buy a refrigerator costing $180 and to divide the cost equally. If they can convince two more students to join them (with all paying equal amounts), the cost to each of the original students will decrease by $15. How many students are in the original group?

66. A plane has a cruising speed of 420 miles per hour, and the wind velocity averages 60 miles per hour. Assuming that the flight is with the wind in one direction and against it in the other, how far can the plane travel in a round trip of 7 hours?

67. A printer charges for producing any desired number of copies of a book in the following way. The printing charge depends only on the number of copies printed, and the charge for typesetting is fixed, independent of the number of books to be printed. If the printer is willing to print 20,000 copies of a certain book for $26,000 and to print 30,000 copies of the same book for $34,000, find the cost of typesetting the book.

68. A circular pond has a radius of 8 feet, and there is a reed in the middle of the pond that protrudes 4 feet above the water. When the reed is pulled to the side (without bending), its tip just reaches the edge of the pond. How tall is the reed?

*69. A person delivers telephone books to three apartment buildings. The first building requires one more than half of all the telephone books. The second building requires one more than half of all the remaining telephone books. The third building requires one more than half of all the telephone books still remaining. If there are five telephone books left over, how many were there to begin with?

70. A tank contained 4 gallons of a mixture of antifreeze and water. After 1 gallon of the mixture was drained off and replaced by 1 gallon of pure antifreeze, the new mixture consisted of 60% antifreeze and 40% water. What percentage of the original mixture was antifreeze?

71. The volume V of a cone of radius r and height h is given by $V = \frac{1}{3}\pi r^2 h$. Solve the equation for r.

72. Let $0 < a < b$. Then the harmonic mean h of a and b is given by the equation

$$\frac{1}{h} = \frac{1}{2}\left(\frac{1}{a} + \frac{1}{b}\right)$$

Solve the equation for b.

73. Recall from Section 2.2 that if P dollars are deposited into an account earning simple interest at an annual rate r, then the amount of interest after n years is Prn, and therefore the amount A_n in the account after n years is given by

$$A_n = P + Prn$$

Solve the equation for r.

74. If, in Exercise 73, the interest were compounded annually, then the formula for A_n would be

$$A_n = P(1 + r)^n$$

Solve the equation for r.

75. The heat Q that results when x moles of sulfuric acid are mixed with y moles of water is given by

$$Q = \frac{17{,}860xy}{1.798x + y}$$

Solve the equation for x.

76. Under certain conditions the percentage efficiency E of an internal combustion engine is given by

$$E = 100\left(1 - \frac{v}{V}\right)^{0.4}$$

where V and v are, respectively, the maximum and minimum volumes of air in each cylinder.
a. Solve the equation for v.
b. Solve the equation for V.

FUNCTIONS AND THEIR GRAPHS

3

It is common to associate various pairs of quantities together. For example, one can associate the volume of a spherical balloon with the corresponding radius. Thus by knowing the radius, one can compute the volume of the balloon. Similarly, the weather service at an airport such as O'Hare International Airport charts the temperature during the day. To determine what the temperature was at the airport yesterday noon, we would only need to locate yesterday noon on the chart and then read off the temperature at that instant. The dependence of one quantity on another, such as volume on radius or temperature on the time of day, is described mathematically by the concept of function, which is the main topic of this chapter.

One of the most important ways of describing a function involves a mathematical picture called a graph. So before we define and study functions, we will discuss notions related to graphs.

3.1

CARTESIAN COORDINATES FOR THE PLANE

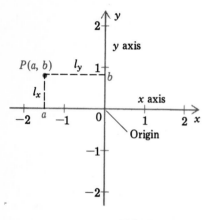

FIGURE 3.1

In Section 1.2 we associated the points on a line with real numbers. Now we will associate the points in a plane with pairs of real numbers. This association will aid us in describing functions.

We begin by drawing two perpendicular lines in the plane, a horizontal one called the ***x axis*** and a vertical one called the ***y axis*** (Figure 3.1). These are the ***coordinate axes***, and they cross at the ***origin***, denoted by 0. We now set up a coordinate system for the x axis so that points on it to the right of the origin correspond to positive numbers and points on it to the left of the origin correspond to negative numbers. Similarly we set up a coordinate system for the y axis so that points on it above the origin correspond to positive numbers and points on it below the origin correspond to negative numbers. The positive directions on the coordinate axes are indicated by arrows (Figure 3.1).

To associate points in the plane with pairs of real numbers, we consider an arbitrary point P in the plane. Through P we draw the line l_x perpendicular to the x axis and the line l_y perpendicular to the y axis. Then l_x crosses the x axis at a point that corresponds to a real number a, called the ***x coordinate*** (or ***abscissa***) of P. Similarly, l_y crosses the y axis at a point corresponding to a real number b, called the ***y coordinate*** (or ***ordinate***) of P (Figure 3.1). Thus P determines the ***ordered pair*** of numbers a and b, written (a, b) and called an ordered pair because the x coordinate, a, precedes the y coordinate, b, in the pair.

Conversely, if (a, b) is a given ordered pair of real numbers, then the vertical line l_x that crosses the x axis at a and the horizontal line l_y that crosses the y axis at b meet at a single point P, and the x and y coordinates of P are a and b, respectively. Thus we have a correspondence between points in the plane and ordered pairs of numbers. If (a, b) corresponds to P, then we refer to a and b as the ***coordinates*** of P; we will occasionally write $P(a, b)$ for the point P whose coordinates are (a, b) and refer to (a, b) as the coordinates of P.

Such an association of the points in a plane with ordered pairs of real numbers is called a ***Cartesian*** (or ***rectangular***) ***coordinate system***, and a plane endowed with a Cartesian coordinate system is called a ***Cartesian plane***. The name "Cartesian" honors René Descartes (1596–1650), the French mathematician who became famous for applying algebraic techniques to the discipline of geometry. From now on the only planes we will consider will have Cartesian coordinate systems, and we will simply call them planes. Moreover, we will frequently refer to points and ordered pairs of real numbers interchangeably. For example, we might refer to the ordered pair (3, 2) as a point.

The spacing between successive integers on an axis of the plane is called the ***scale*** of the axis. We can adjust the scales of the

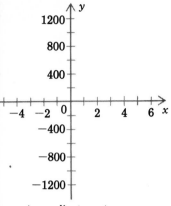

A coordinate system
with different scales on the axes

FIGURE 3.2

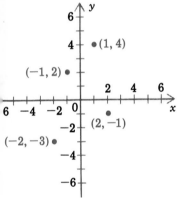

FIGURE 3.3

two coordinate axes to suit our needs; as Figure 3.2 illustrates, the scales of the two axes may be very different from one another.

When we draw a point in the plane corresponding to a given ordered pair of numbers, we say that we *plot*, or *sketch*, the point. In our first few examples, we will plot single points and also sets of points in the plane.

Example 1. Determine the x and y coordinates of the following points, and then plot the points.

    a. $(1, 4)$ b. $(-1, 2)$ c. $(2, -1)$ d. $(-2, -3)$

Solution.

    a. The x coordinate is 1; the y coordinate is 4.
    b. The x coordinate is -1; the y coordinate is 2.
    c. The x coordinate is 2; the y coordinate is -1.
    d. The x coordinate is -2; the y coordinate is -3.
The points are plotted in Figure 3.3. □

Turning to sets of points in the plane, we first observe that there are infinitely many points (x, y) such that $x = 0$ (that is, whose x coordinate is 0), and they constitute the y axis. Likewise, the points (x, y) such that $y = 0$ (that is, whose y coordinate is 0) constitute the x axis.

Example 2. Sketch the set of points (x, y) in the plane satisfying

    a. $x = 2$ b. $y = -1$

Solution.

    a. If $x = 2$, then the x coordinate of (x, y) is 2, so the point lies on the vertical line 2 units to the right of the y axis (Figure 3.4a).
    b. If $y = -1$, then the y coordinate of (x, y) is -1, so the point lies on the horizontal line 1 unit below the x axis (Figure 3.4b). □

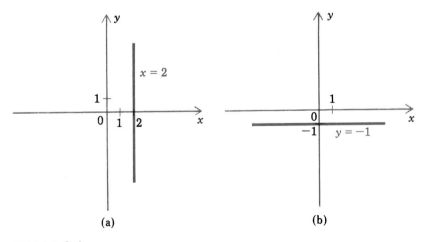

(a)   (b)

FIGURE 3.4

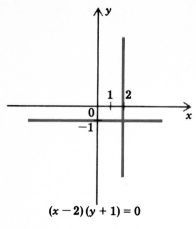

$(x - 2)(y + 1) = 0$

FIGURE 3.5

In general, if a is any real number, then the set of all points (x, y) satisfying the equation $x = a$ is a vertical line, and the set of all points (x, y) satisfying the equation $y = a$ is a horizontal line.

Example 3. Sketch the set of points (x, y) in the plane satisfying $(x - 2)(y + 1) = 0$.

Solution. Observe that $(x - 2)(y + 1) = 0$ if and only if either $x - 2 = 0$ or $y + 1 = 0$, that is, if and only if $x = 2$ or $y = -1$. But the points (x, y) that satisfy $x = 2$ and those that satisfy $y = -1$ are sketched in Figure 3.4a and b. Thus (x, y) satisfies the given equation if and only if it lies on either of the two lines shown in Figure 3.5. □

We can describe more extensive regions of the plane by means of inequalities involving coordinates of the points.

Example 4. Sketch the set of points (x, y) in the plane satisfying
a. $x > 2$ b. $y > -1$

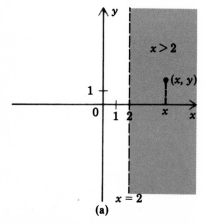

(a)

Solution.
a. If (x, y) satisfies $x > 2$, then the x coordinate of (x, y) is greater than 2, so that (x, y) lies to the right of the vertical line $x = 2$. Thus the points (x, y) satisfying $x > 2$ are the points to the right of, but not on, the line $x = 2$ (Figure 3.6a).
b. If $y > -1$, then the y coordinate of (x, y) is greater than -1, so that (x, y) lies above the horizontal line $y = -1$. Thus the points (x, y) satisfying $y > -1$ are the points above, but not on, the line $y = -1$ (Figure 3.6b). □

Had we desired the points (x, y) satisfying $x \geq 2$ in part (a) of Example 4, we would have found them to be the points either to the right of or on the line $x = 2$.

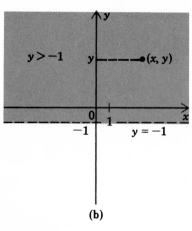

(b)

FIGURE 3.6

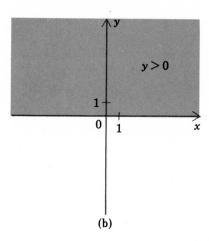

(a)   (b)

FIGURE 3.7

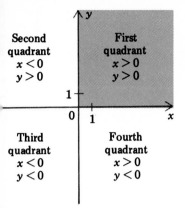

FIGURE 3.8

The points (x, y) satisfying $x > 0$ are the points to the right of the y axis (Figure 3.7a), and the points satisfying $y > 0$ are the points above the x axis (Figure 3.7b). Thus the points in the plane satisfying *both* $x > 0$ and $y > 0$ are the points in the upper right-hand fourth of the plane, determined by the coordinate axes and called the *first quadrant*, or *quadrant I*. The other three quadrants are also identified by the positivity or negativity of the coordinates; all four quadrants are identified in Figure 3.8.

Example 5. Sketch the set of points (x, y) in the plane satisfying $(x - 2)(y + 1) > 0$.

Solution. Notice that $(x - 2)(y + 1) > 0$ if and only if either $x - 2 > 0$ and $y + 1 > 0$, or $x - 2 < 0$ and $y + 1 < 0$. This is equivalent to $x > 2$ and $y > -1$, or $x < 2$ and $y < -1$. From Figures 3.6a and b we deduce that the points (x, y) satisfying $x > 2$ and $y > -1$ are those shaded in Figure 3.9a. In a similar way we find that the points (x, y) satisfying $x < 2$ and $y < -1$ are those shaded in Figure 3.9b. Combining the information obtained so far, we find that the points satisfying the given inequality are those shaded in Figure 3.9c. □

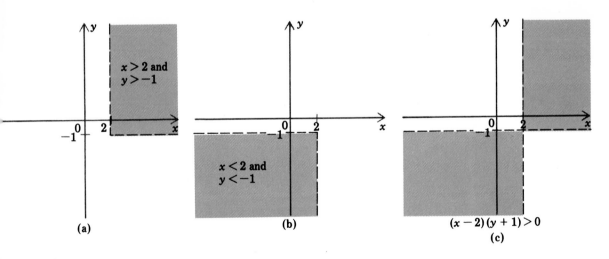

(a)

(b)

$(x - 2)(y + 1) > 0$
(c)

FIGURE 3.9

The Distance Between Two Points

If two points are on the same horizontal or vertical line in the plane, then the distance between them is defined to be the distance between their x coordinates or their y coordinates, respectively. Thus the distance between $P(x_1, y_1)$ and $Q(x_2, y_1)$ is $|x_2 - x_1|$, and the distance between $P(x_1, y_1)$ and $R(x_1, y_2)$ is $|y_2 - y_1|$ (Figure 3.10).

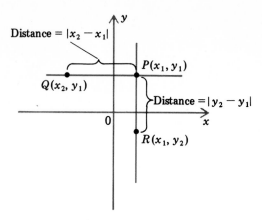

Distance = $|x_2 - x_1|$

$Q(x_2, y_1)$

$P(x_1, y_1)$

Distance $= |y_2 - y_1|$

$R(x_1, y_2)$

FIGURE 3.10

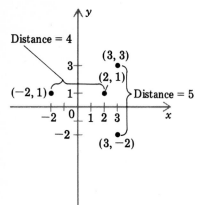

Distance = 4

$(3, 3)$

$(2, 1)$

$(-2, 1)$

Distance = 5

$(3, -2)$

FIGURE 3.11

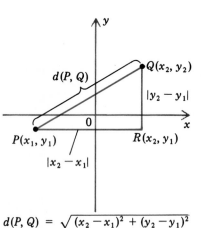

$Q(x_2, y_2)$

$d(P, Q)$

$|y_2 - y_1|$

$P(x_1, y_1)$

$R(x_2, y_1)$

$|x_2 - x_1|$

$d(P, Q) = \sqrt{(x_2 - x_1)^2 + (y_2 - y_1)^2}$

FIGURE 3.12

Example 6. Find the distances between the following pairs of points, illustrated in Figure 3.11.

  a. $(-2, 1)$ and $(2, 1)$ b. $(3, 3)$ and $(3, -2)$

Solution.

  a. The distance between $(-2, 1)$ and $(2, 1)$ is $|2 - (-2)| = |4| = 4$.

  b. The distance between $(3, 3)$ and $(3, -2)$ is $|-2 - 3| = |-5| = 5$. □

We define the *distance* $d(P, Q)$ between any two points $P(x_1, y_1)$ and $Q(x_2, y_2)$ in the plane by the following formula:

DISTANCE FORMULA

$$d(P, Q) = \sqrt{(x_2 - x_1)^2 + (y_2 - y_1)^2}$$

This definition is based on the Pythagorean Theorem, which relates the lengths of the sides of a right triangle (see Figure 3.12).

Example 7. Find the distance between $(2, 6)$ and $(4, -1)$.

Solution. By the distance formula, with $P = (2, 6)$ and $Q = (4, -1)$, we have

$$d(P, Q) = \sqrt{(4 - 2)^2 + (-1 - 6)^2}$$
$$= \sqrt{2^2 + (-7)^2}$$
$$= \sqrt{53} □$$

The Midpoint of a Line Segment

Let $P(x_1, y_1)$ and $Q(x_2, y_2)$ be two distinct points in the plane, so that P and Q determine a line segment (Figure 3.13). Then the coordinates (x, y) of the midpoint M of that line segment can be determined from x_1, y_1, x_2, and y_2. We will not derive the formula for the coordinates but will simply state it:

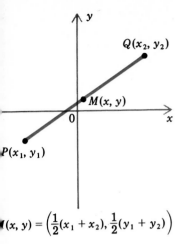

$P(x_1, y_1)$

$M(x, y)$

$Q(x_2, y_2)$

MIDPOINT FORMULA

$$M(x, y) = \left(\frac{1}{2}(x_1 + x_2), \frac{1}{2}(y_1 + y_2) \right)$$

Thus the x coordinate of the midpoint is the average of the x coordinates of the two points P and Q, and the y coordinate of the midpoint is the average of the y coordinates of the two points.

$(x, y) = \left(\frac{1}{2}(x_1 + x_2), \frac{1}{2}(y_1 + y_2) \right)$

FIGURE 3.13

Example 8. Find the coordinates of the midpoint of the line segment joining $(3, 4)$ and $(5, -2)$ (Figure 3.14).

Solution. By the midpoint formula the coordinates are given by

$$\left(\frac{1}{2}(3 + 5), \frac{1}{2}[4 + (-2)] \right)$$

which reduces to $(4, 1)$. $\square$

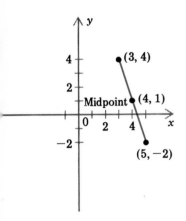

FIGURE 3.14

EXERCISES 3.1

1. Set up a coordinate system, and plot the points $(-2, 0)$, $(1, 2)$, $(-3, 1)$, $(-1.5, -2.5)$, $(0, -\sqrt{2})$, $(1, -2)$.

2. Set up a coordinate system, and plot the points $(100, 200)$, $(-150, 100)$, $(-200, 200)$, $(300, -150)$, $(400, 0)$, $(0, -300)$.

In Exercises 3–22, set up a coordinate system and sketch the set of points (x, y) in the plane satisfying the given equation or inequality.

3. $y = 5$

4. $x = -2$

5. $x > 1$

6. $y < -1$

7. $xy = 0$

8. $(x - 3)(y + 2) = 0$

9. $(\frac{1}{2}x + 1)(2 - y) = 0$

10. $x^2 = 1$

11. $y^2 = 4$

12. $x^2 + x = 0$

13. $y^3 - y = 0$

14. $y^2 - 8y + 16 = 0$

15. $x^2 - 3x - 4 = 0$

16. $x^2 + 5x - 14 = 0$

17. $y^2 = 4y + 12$

18. $x^2 \geq 1$

19. $x^2 < 1$

20. $xy > 0$

21. $xy < 0$

22. $x^2 + y^2 = 0$

In Exercises 23–34, find the distance between the two given points.

23. $(-1, 2)$ and $(-1, 4)$

24. $(10, 12)$ and $(5, 12)$

25. $(5, 12)$ and $(0, 0)$

26. $(3, 3)$ and $(5, 1)$

27. $(4, -3)$ and $(-2, -5)$

28. $(-1, -3)$ and $(-7, 4)$

29. $(4, -3)$ and $(-1, 9)$

30. $(\frac{1}{4}, \frac{1}{4})$ and $(-\frac{1}{4}, -\frac{1}{4})$

31. $(3, \sqrt{2})$ and $(-\sqrt{2}, 4)$

32. $(0.1, -0.1)$ and $(0.4, 0.3)$

33. $(0, 0)$ and $(3a, 4a)$, where $a \geq 0$

34. $(0, 0)$ and $(3a, 4a)$, where $a < 0$

In Exercises 35–42, find the coordinates of the midpoint of the line segment joining the two given points.

35. $(4, 2)$ and $(8, 10)$

36. $(-1, 3)$ and $(5, -1)$

37. $(0, 4)$ and $(0, 10)$

38. $(-2, 4)$ and $(3, 10)$

39. $(3.2, 1.4)$ and $(-1.8, 4.3)$

40. $(-2.2, -0.8)$ and $(3.6, 2.5)$

41. $(-a, a)$ and (a, a)

42. $(1, 1)$ and $(\frac{2}{3}a, \frac{1}{4}a)$

43. Determine the points on the x axis that are 2 units from the point $(4, 1)$.

44. Determine the points on the y axis that are 3 units from the point $(2, 0)$.

45. Let $P(a, b)$ and $Q(c, d)$ be distinct points. Find a formula for the coordinates of the point two thirds of the distance from P to Q.

46. A *rhombus* is a polygon whose four sides are equal in length. Show that the points $(0, 0)$, $(3, 0)$, $(2, \sqrt{5})$, and $(5, \sqrt{5})$ are the vertices of a rhombus.

47. A triangle is *isosceles* if two of its sides have equal length. Show that the points $(0, 4)$, $(3, -1)$, and $(3 + 3\sqrt{2}, 3)$ are the vertices of an isosceles triangle.

Exercises 48–50 illustrate how algebra can be used to prove geometric facts.

48. Show that the line segments joining the midpoints of the sides of an equilateral triangle form an equilateral triangle. (*Hint:* Let the three vertices of the given triangle be $(0, 0)$, $(a, 0)$, and $(a/2, \sqrt{3}\,a/2)$.)

49. Show that the diagonals of a square intersect at their midpoints. (*Hint:* Set up a coordinate system as in Figure 3.15. Then compute the coordinates of the midpoint M_1 of the diagonal joining P_1 and P_3, and the midpoint M_2 of the diagonal joining P_2 and P_4. Finally, show that $M_1 = M_2$.)

50. The following instructions were found at the intersection of Walnut and Broad Streets in Philadelphia: "To find the treasure chest, go west 2 kilometers, then south 1 kilometer, then east 3 kilometers, then south 2 kilometers, and, finally, go half the distance in the direction of the starting point at the intersection of Walnut and Broad." What is the (straight-line) distance between the intersection and the treasure chest?

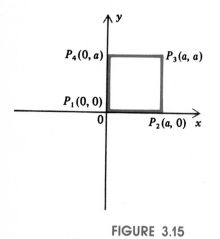

FIGURE 3.15

3.2

FUNCTIONS

Frequently temperature in degrees Fahrenheit is associated with the corresponding temperature in degrees Celsius, and the area of a circle is associated with the radius of the circle. In mathematics such associations are called functions.

DEFINITION 3.1

A *function* consists of a domain and a rule. The *domain* is a collection of real numbers. The *rule* assigns to each number in the domain one and only one number.

The total collection of numbers that a function assigns to the numbers in its domain is called the *range* of the function.

Functions are normally denoted by the letters f or g, and occasionally by other letters such as h, F, G, or H (in the same part of the alphabet). The value assigned by a function f to a member x of its domain is written $f(x)$ and is read "f of x" or "the value of f at x." If f assigns $\sqrt{2}$ to the number -1, we would write $f(-1) = \sqrt{2}$, whereas if f assigns $\frac{2}{3}$ to 0.4, then we would write $f(0.4) = \frac{2}{3}$. A function f is like a machine that applies the rule of f to each number x in the domain of f and thereby produces the number $f(x)$ in the range of f (Figure 3.16).

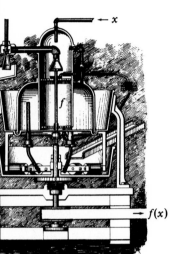

FIGURE 3.16

Caution: A function must make an assignment to *each* number in its domain. For example, if the domain of a function f is $(-\infty, \infty)$, then f must make an assignment to each real number. Moreover, a function assigns *only one* number to any given number in its domain. Thus a function cannot assign both -2 and 3 to a single number in its domain. Nor can a function assign $\pm\sqrt{x}$ to each nonnegative number x, since each positive number x would then be assigned two numbers, $\sqrt{x}$ and $-\sqrt{x}$. Simply remember: A function assigns *one and only one* number to each number in its domain.

Examples of Functions

Our first example of a function f is obtained by assigning the number 4 to each real number. Thus we write

$$f(x) = 4 \quad \text{for all real } x \tag{1}$$

For the particular values $-\pi$, 0, and $\sqrt{3}$ of x we have

$$f(-\pi) = 4, \quad f(0) = 4, \quad \text{and} \quad f(\sqrt{3}) = 4$$

The range of f consists of the single number 4, which is assigned to each number.

More generally, if a function f assigns the same number c to each real number x, then we write

$$f(x) = c \quad \text{for all real } x$$

and call f a *constant function.* Its range consists of the single number c. The function described in (1) is thus a constant function, with $c = 4$. Notice that although Definition 3.1 stipulates that no single member of the domain of a function can be assigned more than one number in the range, it does permit more than one number in the domain of a function to be assigned the same number in the range, as a constant function amply illustrates.

Now let f be the function that assigns $x^2 + 3x - 1$ to each real number x. Then we write

$$f(x) = x^2 + 3x - 1 \quad \text{for all } x \tag{2}$$

Example 1. For the function given in (2), determine the value of
  a. $f(-4)$ 　　　　　　　　b. $f(x + h)$

Solution.
  a. We let $x = -4$ in (2):

$$f(-4) = (-4)^2 + 3(-4) - 1 = 16 - 12 - 1 = 3$$

  b. We substitute $x + h$ for x in (2):

$$f(x + h) = (x + h)^2 + 3(x + h) - 1$$
$$= x^2 + 2xh + h^2 + 3x + 3h - 1 \quad \square$$

Next let f be the function that assigns $\sqrt{x}$ to each nonnegative real number x. Then f is given by

$$f(x) = \sqrt{x} \quad \text{for all } x \geq 0 \tag{3}$$

This is the **square root function**. Its domain consists of all nonnegative numbers, as does its range.

Example 2. For the function f given in (3), determine the value of
  a. $f(2)$ 　　　　b. $f(\tfrac{1}{9})$ 　　　　c. $f(x^2)$

Solution.
  a. Letting $x = 2$ in (3), we have $f(2) = \sqrt{2}$.
  b. Letting $x = \tfrac{1}{9}$ in (3), we have

$$f\left(\frac{1}{9}\right) = \sqrt{\frac{1}{9}} = \frac{1}{3}$$

  c. Substituting x^2 for x in (3), we have

$$f(x^2) = \sqrt{x^2} = |x| \quad \square$$

We remark that although $\sqrt{x}$ is defined for nonnegative values of x, by contrast $\sqrt{x^2}$ is defined for all numbers x. Thus from part (c) of Example 2 we find that

$$\sqrt{(-3)^2} = |-3| = 3$$

As another example of a function, let g be the function that assigns $(x - 1)/(x + 3)$ to each real number x except -3. Then

$$g(x) = \frac{x-1}{x+3} \quad \text{for } x \neq -3 \tag{4}$$

Example 3. For the function g given in (4), determine the value of

    a. $g(0)$ b. $g\left(\frac{1}{x}\right)$

Solution.

    a. Here we let $x = 0$ in (4) and obtain

$$g(0) = \frac{0-1}{0+3} = -\frac{1}{3}$$

    b. We substitute $\frac{1}{x}$ for x in (4):

$$g\left(\frac{1}{x}\right) = \frac{\frac{1}{x} - 1}{\frac{1}{x} + 3} = \frac{1-x}{1+3x} \quad \square$$

    Notice that -3 is not in the domain of the function g defined by (4), so $g(-3)$ is undefined. If we attempted to compute $g(-3)$ by means of the formula in (4), we would obtain

$$g(-3) \overset{?}{=} \frac{-3-1}{-3+3} \overset{?}{=} \frac{-4}{0}$$

but $-4/0$ is undefined.

    We say that two functions are **equal** if they have the same domains and their rules associate the same number with any given number in their common domain. Thus the functions g and h given by

$$g(x) = \frac{x-1}{x+3} \quad \text{for } x \neq -3 \qquad \text{and} \qquad h(t) = \frac{t-1}{t+3} \quad \text{for } t \neq -3$$

are equal, since their domains both consist of all real numbers different from -3 and their rules associate the same number with any given number in their common domain. As these examples illustrate, the specific letter that is used for the variable in describing a function is irrelevant.

    Two functions are *not equal* if either their domains or their rules are distinct. Consequently if

$$g(x) = \frac{x-1}{x+3} \quad \text{for } x \neq -3$$

and

$$G(x) = \frac{x-1}{x+3} \quad \text{for } x \geq 0$$

(5)

then g and G are *not* equal, because the domain of g contains all negative numbers except -3, whereas the domain of G contains no negative numbers.

The rule of a function can be given in two or more parts. For example, consider the function f whose rule is given by

$$f(x) = \begin{cases} x^5 - 3 & \text{for } x < 2 \\ \sqrt{x} & \text{for } x \geq 4 \end{cases} \tag{6}$$

The domain of f consists of the intervals $(-\infty, 2)$ and $[4, \infty)$. The rule by which a number is assigned to a number x in the domain of f depends on whether x is in $(-\infty, 2)$ or in $[4, \infty)$.

Example 4. For the function f given by (6), find the value of

    a. $f(-1)$ b. $f(16)$

Solution.

    a. Since $-1 < 2$, we let $x = -1$ in the top expression on the right of (6):

$$f(-1) = (-1)^5 - 3 = -4$$

    b. Since $16 > 4$, we let $x = 16$ in the bottom expression on the right of (6):

$$f(16) = \sqrt{16} = 4 \quad \square$$

An Alternative Way of Describing a Function

An alternative way of describing a function is obtained by writing y in place of $f(x)$. Thus the function f which associates the area of a circle with its radius x and is defined by

$$f(x) = \pi x^2 \quad \text{for } x \geq 0$$

can be given by

$$y = \pi x^2 \quad \text{for } x \geq 0 \tag{7}$$

In (7) each nonnegative value of x determines a particular value of y, and in this way y "depends" on x, so y is called a ***dependent variable***. In contrast, x is an ***independent variable***.

Functions that describe physical relationships are frequently presented in variable notation; moreover, the letters used for variables usually relate to the physical aspects of the quantities. Since (7) represents the area corresponding to any given radius, the function is frequently given in variable notation by

$$A = \pi r^2 \quad \text{for } r \geq 0 \tag{8}$$

where A stands for area and r for radius. In a like manner, we could describe the function that assigns the temperature C in degrees Celsius to any given temperature F in degrees Fahrenheit by

$$C = \frac{5}{9}(F - 32) \quad \text{for } F \geq -459.67 \tag{9}$$

Notice that the domain of this function consists of all numbers greater than or equal to -459.67, which represents "absolute zero," theoretically the lowest Fahrenheit temperature possible. The variables in (8) and (9) were chosen to remind us of the physical quantities they represent.

Most functions we will encounter express $f(x)$ (or y) in terms of x. When giving such a formula, we normally specify the numbers in the domain directly after the formula, as in (1)–(9). However, when the domain is to consist of all numbers for which the formula is meaningful, we normally omit mention of the domain. Thus we may write

$$f(x) = x^2 + 3x - 1 \quad \text{and} \quad g(x) = \frac{x - 1}{x + 3} \tag{10}$$

respectively, for the functions presented in (2) and (4), since each of their domains consists of all numbers for which the expression on the right side of the equation makes sense. Nevertheless we cannot omit mention of the domain of G in

$$G(x) = \frac{x - 1}{x + 3} \quad \text{for } x \geq 0$$

(which appeared in (5)) because $(x - 1)/(x + 3)$ makes sense for negative as well as nonnegative numbers. If the domain is understood and the rule of the function is simple, we sometimes designate the function by an expression only. Thus we could refer to the functions in (10) as $x^2 - 3x + 1$ and $\frac{x - 1}{x + 3}$, respectively.

When we do not specify the domain outright, we still need to be aware of the domain. Sometimes it is important to determine precisely those real numbers that are members of the domain.

Example 5. Let

$$f(x) = \frac{x + 2}{x^2 - 9}$$

Determine the domain of f.

Solution. A number x is in the domain of f if and only if it does not make the denominator 0, that is, if and only if $x^2 - 9 \neq 0$. Since

$x^2 - 9 = 0$ means that $x^2 = 9$, and thus $x = 3$ or $x = -3$, it follows that the domain consists of all real numbers except 3 and -3. □

Example 6. Find the domain of the function given by

$$f(x) = \frac{1}{\sqrt{x^2 - 16}}$$

Solution. Since the square root is defined only for nonnegative numbers, $\sqrt{x^2 - 16}$ is defined only for x such that $x^2 - 16 \geq 0$. But

$$\frac{1}{\sqrt{x^2 - 16}}$$

is not defined when the denominator is 0, that is, when $x^2 - 16 = 0$. Consequently the domain of f consists of all x such that $x^2 - 16 > 0$, or equivalently, $x^2 > 16$. This set consists of all numbers in the intervals $(-\infty, -4)$ and $(4, \infty)$. □

Example 6 illustrates two facts that help in determining domains of functions defined by formulas:

1. Division by zero is undefined.
2. Square roots of negative numbers are undefined.

An Alternative Definition of Function

An alternative definition of function that is in common use employs the notion of ordered pair introduced in Section 3.1. To set the stage for the alternative definition, let f be a function. For any number x in the domain of f, we combine the numbers x and $f(x)$ to form the ordered pair $(x, f(x))$. Notice that because f is a function, there is precisely one pair whose first entry is x, namely, $(x, f(x))$. The collection of all such pairs $(x, f(x))$ for x in the domain of f identifies the function f.

Conversely, any set of ordered pairs (x, y) of numbers with the property that the first entries in any two distinct pairs are different defines a unique function f: The domain of f is the collection of first entries in the pairs, and for each such pair (x, y), $f(x)$ is the number y.

These comments lead to the following alternative definition of function.

DEFINITION 3.2

A *function* is a nonempty collection of ordered pairs of real numbers no two of which have the same first entry.

In this framework the collection of ordered pairs (x, x^2) defines the function given by the formula $f(x) = x^2$, and the collection of ordered pairs $(x, \sqrt{x})$ for $x \geq 0$ defines the square root function. These sets are sometimes written as

$$\{(x, x^2)|x \text{ is a real number}\}$$

and

$$\{(x, \sqrt{x})|x \geq 0\}$$

respectively.

Although we have two definitions of function, we prefer to rely only on the first, the one appearing as Definition 3.1.

EXERCISES 3.2

In Exercises 1–12, find the indicated values of the given function.

1. $f(x) = -14; f(-2), f(\sqrt{3}), f(-x)$
2. $f(x) = x^2 - 3; f(5), f(-4), f(x - 2)$
3. $f(t) = t^2 - 3t + 4; f(0), f(-1), f(t^2)$
4. $f(t) = \dfrac{t + 4}{t - 1}; f(2), f\left(\dfrac{1}{3}\right), f\left(\dfrac{1}{t}\right)$
5. $f(u) = \dfrac{4u^2 - 2u}{u^2 - 0.09}; f(10^{-1}), f(a)$
6. $g(x) = \dfrac{\dfrac{1}{x} - 1}{\dfrac{1}{x} + 2}; g(1), g(-2), g\left(\dfrac{1}{x}\right)$
7. $g(x) = \sqrt{x + 5}; g(-1), g(2), g(x^2 - 5)$
8. $h(z) = \sqrt[3]{z - 4}; h(12), h(-4), h(z + a)$
9. $h(z) = 2z^{1/3} - z^{3/2}; h(64), h(z^6)$
10. $h(z) = \sqrt{\dfrac{2z^2 - 1}{5z^2 - 1}}; h(-1), h(1), h(\sqrt{z})$
11. $f(x) = \begin{cases} -x & \text{for } x < 1 \\ 0 & \text{for } x > 3 \end{cases}; f(-5), f(\pi)$
12. $f(x) = \begin{cases} \dfrac{x - 1}{x + 2} & \text{for } x \neq -2 \\ 1 & \text{for } x = -2 \end{cases}; f(2), f(-2)$

In Exercises 13–20, find the values of y at the given values of x.

13. $y = -2x + 3; x = -4, x = 0$
14. $y = 3x^2 - 2x + 7; x = -2, x = -\frac{1}{3}$
15. $y = \dfrac{3x - 4}{2x^2 + 3x - 5}; x = 0, x = \frac{1}{2}$
16. $y = \dfrac{x^2 - 1}{x^2 + 1}; x = 1, x = -3$
17. $y = \dfrac{1}{x} - \dfrac{1}{x^2}; x = -1, x = \frac{1}{3}$
18. $y = \sqrt{\dfrac{x}{x - 5}}; x = 9, x = 6$

19. $y = \dfrac{\sqrt[3]{x}}{\sqrt[3]{x} - 1}$; $x = -8$, $x = 54$

20. $y = 3|x|\sqrt{1 - x}$; $x = 1, -2$

In Exercises 21–42, find the domain of the given function.

21. $f(x) = 4x^6 + \sqrt{3}x^3 - \dfrac{1}{3}x^2 + 1$

22. $f(x) = \dfrac{1}{x - 7}$

23. $g(x) = \dfrac{3}{4x} - \dfrac{4}{2x - 5}$

24. $g(x) = \dfrac{x + 1}{x^2 - 4}$

25. $g(x) = \dfrac{x + 2}{x^2 + 4}$

26. $g(x) = \dfrac{x - 3}{x^2 - 9}$

27. $h(z) = \dfrac{1}{(z - 1)(z + 6)}$

28. $h(z) = \dfrac{1}{z^2 + 2z + 1}$

29. $h(z) = \dfrac{1}{z^4 + z^2 + 3}$

30. $f(x) = \dfrac{1}{\sqrt{x}}$

31. $g(x) = \dfrac{1}{\sqrt{-x}}$

32. $h(x) = \dfrac{1}{\sqrt{x} - 5}$

33. $y = \sqrt{x^2 - 4}$

34. $y = \dfrac{1}{\sqrt{x^2 - 4}}$

35. $y = \sqrt{\dfrac{x}{x - 1}}$

36. $y = \sqrt{x(x - 3)}$

37. $y = \sqrt[3]{x - 1}$

38. $y = \sqrt{\sqrt{x} - 1}$

39. $y = \sqrt{x - \dfrac{1}{\sqrt{x}}}$

40. $y = \sqrt{\dfrac{x^2 - 1}{x^2 + 1}}$

41. $f(x) = \dfrac{|x|}{x}$

42. $f(x) = \begin{cases} 2x - 3 & \text{for } x \le 0 \\ x + \dfrac{5}{x} & \text{for } x > 1 \end{cases}$

43. Determine which of the following functions have a range that contains -2.
    a. $f(x) = \sqrt{x + 1}$
    b. $f(x) = \sqrt{1 - x}$
    c. $f(x) = \dfrac{x + 1}{x^2 + 3x + 2}$
    d. $f(x) = \dfrac{x + 2}{x^2 - 4}$
    e. $f(x) = \dfrac{x - 2}{x^2 - 4}$
    *f. $f(x) = \dfrac{x - 2}{x^2 + 4}$

44. Determine which of the following functions have a range that contains all numbers greater than -1.
    a. $f(x) = x^2 - 8$
    b. $f(x) = x^2 + 4$
    c. $f(x) = \sqrt{x + 1}$
    d. $f(x) = \sqrt[3]{x + 1}$

45. Determine which of the following collections of pairs of real numbers define a function according to Definition 3.2.
    a. all (x, y) with $x^2 + y^2 = 1$ and $y \ge 0$
    b. all (x, y) with $x^2 + y^2 = 1$ and $x \ge 0$
    c. all (x, y) with $x = y^3$
    d. all (x, y) with $x = \sqrt{y}$

46. Let $f(x) = (x - 2)^2 + 3(x - 2) - 1$ and $g(x) = x^2 - 3x - 3$. Determine whether or not $f = g$.

47. Let

$$f(x) = \frac{|x|}{x} \quad \text{and} \quad g(x) = \begin{cases} -1 & \text{for } x < 0 \\ 1 & \text{for } x > 0 \end{cases}$$

Determine whether or not $f = g$.

48. Let

$$f(x) = \frac{1}{x - \sqrt{x^2 - 1}} \quad \text{and} \quad g(x) = x + \sqrt{x^2 - 1}$$

Show that $f = g$.

49. Let $f(x) = \dfrac{2x - 3}{x - 2}$. Show that $f(f(x)) = x$.

50. Let $f(x) = \dfrac{1}{x + 1}$. Show that $f\left(\dfrac{1}{x}\right) = xf(x)$.

51. Let $f(x) = \dfrac{1 + x}{1 - x}$. Show that $f\left(\dfrac{1}{x}\right) = -f(x)$.

[C] 52. Let $f(x) = 2.3x^3 + \pi x - 1.7$. Use a calculator to approximate
   a. $f(3.5)$ b. $f(-0.2)$ c. $f(13.9)$

[C] 53. Let $g(t) = 6.54t^2 - \dfrac{0.55}{t}$. Use a calculator to approximate

   a. $g(1.21)$ b. $g(6.01)$
   c. $g(4.97)$ d. $g(0.03)$

[C] 54. Let $h(z) = \sqrt{z^2 + 3}$. Use a calculator to approximate

   a. $h(2.3)$ b. $h\left(\dfrac{4}{3}\right)$ c. $h(\sqrt{2} + 1)$

55. Define a function that expresses the circumference of a circle as a function of
   a. the radius
   b. the diameter

56. Define a function that expresses the area of a square as a function of
   a. the length of a side of the square
   b. the length of a diagonal

57. One inch is 2.54 centimeters. Define a function that converts
   a. from inches to centimeters
   b. from centimeters to inches

3.3

FUNCTIONS AS MODELS

Experimental data collected by scientists usually consist of long lists of numbers. For example, if a ball is dropped into a deep well and the distance traveled by the ball is ascertained each tenth of a second, then the following data would result:

Time elapsed (in seconds):	.1	.2	.3	.4	.5	.6
Depth (in feet):	.16	.64	1.44	2.56	4.00	5.76

Aside from the fact that measurements are at best only approximations to reality, it is unwieldy to handle long lists of numbers. Moreover, it is hard to extrapolate precisely most values not listed in a table. For instance, how far will the ball have traveled in 1.43 seconds or in $\sqrt{2}$ seconds?

For these reasons scientists usually construct "models" for the phenomena they investigate. This means that they formulate physical laws, formulas, or functions that express the dependence of certain quantities on other quantities. One such formula gives the distance an object dropped from rest will fall in t seconds if influenced only by gravity:

$$f(t) = 16t^2 \quad \text{for } t \geq 0 \tag{1}$$

Here t is measured in seconds and $f(t)$ in feet. With such a formula we can confidently compute the values mentioned in the preceding paragraph.

Example 1. Using the formula in (1), determine the distance traveled by the ball during

    a. 1.43 seconds   b. $\sqrt{2}$ seconds

Solution.

    a. By (1) with $t = 1.43$, the distance traveled in 1.43 seconds is given by

$$f(1.43) = 16(1.43)^2 = 32.7184 \text{ (feet)}$$

    b. By (1) with $t = \sqrt{2}$, the distance traveled in $\sqrt{2}$ seconds is given by

$$f(\sqrt{2}) = 16(\sqrt{2})^2 = 16(2) = 32 \text{ (feet)} \quad \square$$

A second model arises from Sir Isaac Newton's famous Second Law of Motion, which states that if an object has a mass of m kilograms, then the force F (in newtons) required to accelerate it at a rate of a (meters per second squared) is the product of the mass and the acceleration. This is expressed by the formula

$$F = ma \tag{2}$$

Under the assumption that the mass is constant, the force F is a function of the acceleration.

Example 2. Use (2) to determine the force required to accelerate a moped with a mass of 100 kilograms at a rate of 3 meters per second squared.

Solution. By (2), with $m = 100$ and $a = 3$, we have

$$F = (100)(3) = 300 \text{ (newtons)} \quad \square$$

Variation and Proportionality

Certain types of models have occurred so frequently in the sciences that they have prompted some special terminology. For example, if c is a fixed nonzero number and if x and y are related by the formula

$$y = cx \qquad (3)$$

for all x in a given collection of real numbers, then y is said to be *proportional* to (or to be **directly proportional** to or to *vary directly* as) x. In this case c is called the **constant of proportionality**; sometimes the letter k is used in place of c. Thus length measured in centimeters is proportional to length measured in inches, because if x denotes the length of an object in inches and y the length in centimeters, then x and y satisfy the equation

$$y = 2.54x \quad \text{for } x \geq 0$$

Likewise, Hooke's Law states that within certain bounds the force y needed in order to keep a spring stretched x units beyond its natural length is given by the formula

$$y = kx$$

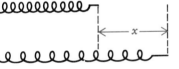

FIGURE 3.17

where k is the so-called *spring constant* of the spring. (See Figure 3.17.) In this case the force is proportional to the distance the spring is stretched.

Example 3. Ohm's Law states that the current i flowing in an electrical circuit is proportional to the voltage V applied to the circuit. Express Ohm's Law by means of a formula.

Solution. Using (3) with V and i replacing x and y, respectively, we may write

$$i = cV$$

where c is a constant. $\quad \square$

In passing, we mention that the voltage V in Example 3 can be either positive or negative (corresponding to the two directions the current can flow through the circuit). The constant c is the **conductivity** of the wire, and its reciprocal is the **resistance** of the wire.

Example 4. One of Albert Einstein's most important contributions to physics was his observation that mass and energy are equivalent.

According to Einstein's theory, when mass is converted to energy, the amount E of energy released is proportional to the amount m of mass converted. Express this result by means of a formula.

Solution. Using (3) and realizing that mass must be nonnegative, we have

$$E = km \quad \text{for } m \geq 0 \qquad (4)$$

where we have used k for the constant of proportionality. Actually, if appropriate units are used, then $k = c^2$, where c is the speed of light in a vacuum. Thus (4) becomes

$$E = mc^2 \quad \text{for } m \geq 0$$

perhaps the most famous physical formula of all time.   □

If c is a nonzero constant and x and y satisfy

$$y = \frac{c}{x} \qquad (5)$$

for all x in a given set of real numbers, then y is said to be ***inversely proportional*** to x. Notice that if $c > 0$, then an increase in x corresponds to a decrease in y. Again, c is the **constant of proportionality**.

Example 5. Boyle's Law states that if the temperature of a gas is constant, then the pressure p of the gas is inversely proportional to the volume V of the gas. Express Boyle's Law by means of a formula.

Solution. Using formula (5) and realizing that the volume of gas must be positive, we may express Boyle's Law as

$$p = \frac{c}{V} \quad \text{for } V > 0$$

where, as usual, c is a constant.   □

Joint Proportionality

Frequently a physical quantity depends on several other quantities, rather than just one. The terms "proportional" and "inversely proportional" may also be employed in such cases. For example, if c is a nonzero constant and if x, y, and z are related by the formula

$$z = cxy$$

then we say that z is ***jointly proportional*** (or more simply, ***proportional***) to x and y. If

$$z = \frac{cx}{y}$$

then we say that z is **proportional** to x and **inversely proportional** to y.

Example 6. The volume V of a cylinder is jointly proportional to the square of the radius r of the base and to the height h of the cylinder. Express this fact by means of a formula involving r and h.

Solution. We have

$$V = cr^2 h$$

In fact, the constant c of proportionality is π, which yields the formula

$$V = \pi r^2 h$$

for the volume of the cylinder. □

The terminology of joint proportionality or inverse proportionality extends to any number of variables. In the next example we have a quantity that is jointly proportional to two other quantities and inversely proportional to a third.

Example 7. Newton's Law of Gravitation states that the force of attraction F between two objects is proportional to each of their masses m_1 and m_2 and is inversely proportional to the square of the distance r between them. Express this law by means of a formula.

Solution. Using G as the constant of proportionality, we have

$$F = \frac{Gm_1 m_2}{r^2}$$

The constant G is known as the **universal gravitational constant**. □

EXERCISES 3.3

1. The perimeter p of a square is proportional to the length s of a side. Express this fact by means of a formula. (The constant is 4.)

2. The circumference C of a circle is proportional to the diameter d. Express this fact by means of a formula. (The constant is π.)

3. The surface area S of a cube is proportional to the square of the side length s. Express this fact by means of a formula. (The constant is 6.)

4. The area A of an equilateral triangle is proportional to the square of the length s of a side.
   a. Express this fact by means of a formula. (The constant is $\sqrt{3}/4$.)
   b. If the area of an equilateral triangle is $9\sqrt{3}$ square inches, determine the length of a side.

5. The volume V of a cube is proportional to the cube of the length s of a side.
   a. Express this fact by means of a formula. (The constant is 1.)
   b. By how much is the volume increased when the length of a side is increased from 3 inches to 4 inches?

6. The volume V of a spherical ball is proportional to the cube of the radius r.
   a. Express this fact by means of a formula. (The constant is $\frac{4}{3}\pi$.)
   b. Would three spherical lead balls of radius 2 inches produce enough lead for a single spherical lead ball of radius 3 inches? Explain your answer.
   c. Would the three spherical balls of part (b) produce enough lead for one cube of side length 4.7 inches? Explain your answer.

7. The length l_f in feet of an object is proportional to the length l_m in meters.
   a. Express this fact by means of a formula. (The constant is $\frac{1250}{381}$.)
   b. Express the fact that the length l_y in yards of an object is proportional to the length l_m in meters by means of a formula. Using (a), determine the constant of proportionality.

8. Distance D_k in kilometers is proportional to distance D_m in miles.
   a. Express this fact by means of a formula. (The constant is 1.609344.)
   b. Express by means of a formula the fact that the distance in miles is proportional to distance in kilometers. Using a calculator, approximate the value of the constant of proportionality.
   c. Use (a), with the numerical value for the constant, to determine the approximate number of kilometers in 93×10^6 miles (which is often taken as the average distance between the earth and sun).

9. The water pressure p at a point is proportional to the depth h of the point below the surface of the water. Express this fact by means of a formula.

10. Planck's quantum theory of light states (in part) that a light ray may be considered as a particle called a photon. The energy E of the photon is proportional to the frequency ν of the light ray. Express this result by means of a formula. (The constant of proportionality, known as **Planck's constant**, is denoted by h.)

11. For each gram of hydrogen that is converted to helium in a star, approximately 0.0277 gram is converted to energy. Use the result of Example 4 to compute the amount of energy released for each gram of hydrogen converted to helium. (*Hint:* The speed of light is approximately 3×10^{10} centimeters per second. Your answer will be in gram centimeters squared per second squared, commonly called "ergs.")

12. According to Hubble's Law, the velocity v with which a galaxy is receding from any other galaxy is proportional to the distance d between the galaxies. (Hubble's Law is sometimes viewed as evidence for the "big bang" theory, which holds that the universe originated with a gigantic explosion.)
    a. Express Hubble's Law by means of a formula.
    b. If a galaxy whose distance from the earth is 10 million light years (that is, 9.46×10^{19} kilometers) is receding from the earth with a velocity of 170 kilometers per second, find the approximate value of the constant of proportionality.
    c. How fast would a galaxy recede if it were located 15 million light years from earth?

13. Joule's Law states that the rate P at which heat is produced in a wire is proportional to the square of the current i flowing in the wire. Express Joule's Law by means of a formula.

14. The kinetic energy K of an object of given mass m is proportional to the square of the speed v of the object.
    a. Express this fact by means of a formula. (The constant of proportionality in this case is equal to one half the mass of the object, if appropriate units are used.)
    b. Suppose two bundles of newspapers, one weighing 10 kilograms and the other 30 kilograms, hit a wall with velocities v_1 and v_2 meters per second, respectively. If both bundles possess the same kinetic energy, what is the relationship between their velocities?

15. In the study of motion it is sometimes assumed that the air resistance R on an object is proportional to the square of the velocity v of the object. Express this assumption by means of a formula.

16. To a very good approximation, the time (or period) T it takes a pendulum to make one complete swing is proportional to the square root of the length l of the pendulum.
    a. Express this fact by means of a formula.
    b. To double the period, how must the length of the pendulum be changed?

17. The wavelength λ of a light ray is inversely proportional to its frequency ν (the number of waves per second). Express this fact by means of a formula. (The constant of proportionality in this case is c, the velocity of light.)

18. Kepler's Third Law of Planetary Motion states that the square of the period T of a planet (the time required for the planet to make one revolution about the sun) is proportional to the cube of the average distance a from the planet to the sun.
    a. Express Kepler's Third Law by means of a formula.
    C b. If the period is 365 days and the average distance is 93×10^6 miles, use a calculator to determine c.
    C c. Use a calculator to determine the constant c in (b) if the period is given in seconds and the average distance in kilometers (see Exercise 8).

19. Coulomb's Law states that the electric force F exerted by one charged particle on another is proportional to the charges q_1 and q_2 on the two particles and is inversely proportional to the square

of the distance r between the two particles. Express Coulomb's Law by means of a formula.

20. The *escape velocity* on a planet is the velocity that a spaceship must possess when leaving the surface of the planet in order to escape the gravitational field of the planet without firing any of the ship's rockets. The *escape energy* of a spaceship on a planet is the amount of energy required to impart escape velocity to the spaceship.

   a. The escape energy E of a spaceship is proportional to the square of the escape velocity v_e and to the mass m of the spaceship. Express this fact by means of a formula.

   b. The escape velocity on a star is defined as for a planet. The escape velocity v_e on a star that is contracting is inversely proportional to the square root of the radius r of the star. Express this fact by means of a formula. (If the radius of the star becomes so small that the escape velocity exceeds the velocity of light, not even light can escape from the star, and the star becomes a "black hole.")

3.4

THE GRAPH OF A FUNCTION

When a function is described by means of a formula, the formula can tell us much about the function. But pictures are also associated with functions and can themselves give much information about the function. A pictorial representation of a function is called a graph.

DEFINITION 3.3

Let f be a function. Then the set of all points $(x, f(x))$ such that x is in the domain of f is called the *graph* of f (Figure 3.18), and we say that such a point $(x, f(x))$ is *on the graph of f*.

Since no two distinct points on the graph of a function can have the same x coordinate (see Definition 3.2), and since a vertical line is characterized by all its points having the same x coordinate, it follows that no vertical line can intersect the graph of a function more than once (Figure 3.19). You might look quickly at the graphs that follow and see that this is true.

Example 1. Let $f(x) = 2$. Sketch the graph of f.

Solution. Notice that if x is any real number, then $(x, f(x))$ is the point $(x, 2)$, which means that the y coordinate of each point on the graph of f is 2. Thus the graph is the horizontal line drawn in Figure 3.20. $\square$

If f is any constant function, say $f(x) = c$, then as in Example 1, the graph of f is a horizontal line that crosses the y axis at the point whose y coordinate is c.

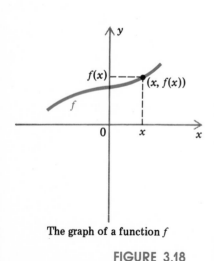

The graph of a function f

FIGURE 3.18

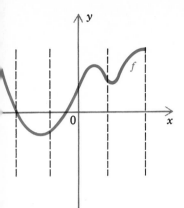

Vertical lines intersect the graph of function f at most once

FIGURE 3.19

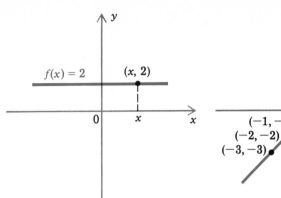

FIGURE 3.20

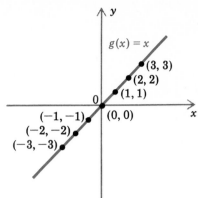

FIGURE 3.21

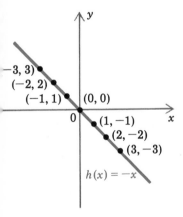

FIGURE 3.22

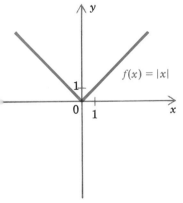

The graph of the absolute value function

FIGURE 3.23

Example 2. Let $g(x) = x$. Sketch the graph of g.

Solution. To get an idea of what the graph looks like, let us plot a few of its points with the help of the following table:

x	-3	-2	-1	0	1	2	3
$g(x)$	-3	-2	-1	0	1	2	3

Then we smoothly connect the points plotted. The graph appears to be a straight line bisecting the first and third quadrants (Figure 3.21). □

Example 3. Let $h(x) = -x$. Sketch the graph of h.

Solution. Again we draw a smooth line through a few points plotted with the help of the following table:

x	-3	-2	-1	0	1	2	3
$h(x)$	3	2	1	0	-1	-2	-3

The graph appears to be a diagonal line bisecting the second and fourth quadrants (Figure 3.22). □

Example 4. Sketch the graph of the *absolute value function*:

$$f(x) = |x|$$

Solution. Since $f(x) = x$ for $x \geq 0$, it follows that the part of the graph of f to the right of the y axis coincides with the corresponding part of the graph of g in Example 2. Since $f(x) = -x$ for $x < 0$, it follows that the part of the graph of f to the left of the y axis coincides with the corresponding part of the graph of h in Example 3. Combining this information, we obtain the graph of f (Figure 3.23). □

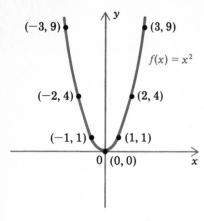

$(-3, 9)$ $(3, 9)$

$f(x) = x^2$

$(-2, 4)$ $(2, 4)$

$(-1, 1)$ $(1, 1)$

$0 \mid (0, 0)$

FIGURE 3.24

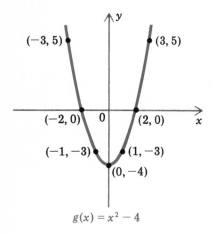

$(-3, 5)$ $(3, 5)$

$(-2, 0)$ 0 $(2, 0)$

$(-1, -3)$ $(1, -3)$

$(0, -4)$

$g(x) = x^2 - 4$

FIGURE 3.25

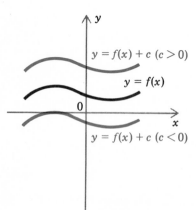

$y = f(x) + c \ (c > 0)$

$y = f(x)$

0

$y = f(x) + c \ (c < 0)$

FIGURE 3.26

Example 5. Let $f(x) = x^2$. Sketch the graph of f.

Solution. As before, we first make a table of values for f:

x	-3	-2	-1	0	1	2	3
$f(x)$	9	4	1	0	1	4	9

Not only do all the points in the graph of f appear to be on or above the x axis but also, as x grows, the value $f(x)$ seems to grow ever faster. We obtain the graph sketched in Figure 3.24. □

Example 6. Let $g(x) = x^2 - 4$. Sketch the graph of g.

Solution. First we fill in the table below:

x	-3	-2	-1	0	1	2	3
$g(x)$	5	0	-3	-4	-3	0	5

Notice that for each value of x, $g(x)$ is exactly 4 units less than the corresponding value for $f(x)$ obtained in the solution of Example 5. Therefore we conclude that the graph of g has the same shape as the graph in Example 5 but is shifted down four units. Figure 3.25 reflects this observation. □

We can go a step beyond Example 6. Suppose we know the graph of a function f, and let

$$h(x) = f(x) + c$$

Then the graph of h has the same shape as the graph of f but is raised by c units if $c > 0$ and is lowered $|c|$ units if $c < 0$ (Figure 3.26).

Example 7. Let $f(x) = -x^2$. Sketch the graph of f.

Solution. The table we make is related to the one in the solution of Example 5:

x	-3	-2	-1	0	1	2	3
$f(x)$	-9	-4	-1	0	-1	-4	-9

As before, we plot the corresponding points and join them with a smooth curve (Figure 3.27). □

Example 8. Sketch the graph of the function given by $y = \sqrt{x}$.

Solution. We prepare the following table for values of x that are nonnegative and hence in the domain:

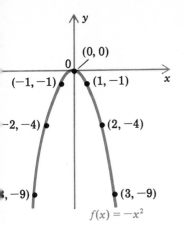

$f(x) = -x^2$

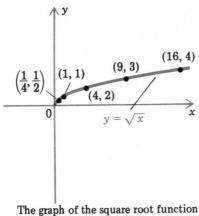

The graph of the square root function

FIGURE 3.28

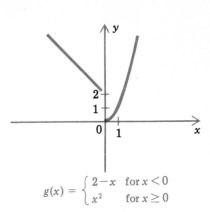

$$g(x) = \begin{cases} 2-x & \text{for } x < 0 \\ x^2 & \text{for } x \geq 0 \end{cases}$$

FIGURE 3.29

x	0	$\frac{1}{4}$	1	4	9	16
y	0	$\frac{1}{2}$	1	2	3	4

Connecting the corresponding points with a smooth curve, we obtain the graph of the square root function (Figure 3.28). □

Example 9. Let

$$g(x) = \begin{cases} 2 - x & \text{for } x < 0 \\ x^2 & \text{for } x \geq 0 \end{cases}$$

Sketch the graph of g.

Solution. Since the formula for g is in two parts, one for x in $(-\infty, 0)$ and the other for x in $[0, \infty)$, we will determine the corresponding parts of the graph one at a time. First, $g(x) = 2 - x$ for $x < 0$, so the value of $g(x)$ for any such x is two units greater than the value of $h(x)$ in Example 3 (where $h(x) = -x$). Therefore the part of the graph of g to the left of the y axis has the same shape as the corresponding part of the graph of h but is raised 2 units. Second, $g(x) = x^2$ for $x \geq 0$, so that the part of the graph of g to the right of the y axis coincides with the corresponding part of the graph of f in Example 5 (where $f(x) = x^2$). Combining this information, we obtain the graph of g (Figure 3.29). □

Symmetry

Symmetry is a quality of art and plays a central role in architecture. One of the most famous examples of symmetry in architecture is the Taj Mahal. The notion of symmetry also relates to graphs of functions in mathematics.

The Taj Mahal, Agra, India

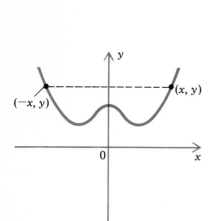

Symmetry with respect to the *y* axis

FIGURE 3.30

We say that the graph of a function is **symmetric with respect to the y axis** if the portion of the graph to the right of the *y* axis is mirrored to the left of the *y* axis. In mathematical terms this means that if (x, y) is on the graph, so is $(-x, y)$ (Figure 3.30). The graphs of the functions in Examples 4–7 have this property, as you can check. In each case the reason is that

$$f(-x) = f(x) \quad \text{for all } x \text{ in the domain of } f \tag{1}$$

A function that satisfies (1) is called an **even function**. Thus $|x|$, x^2, $x^2 - 4$, and $-x^2$ are all even functions.

Suppose that n is an even integer, so that $n = 2m$ for some integer m. Then

$$(-x)^n = (-x)^{2m} = [(-x)^2]^m = (x^2)^m = x^{2m} = x^n$$

This means that any function that can be described by a formula containing only even powers of x is an even function, and its graph is symmetric with respect to the *y* axis.

Our next function has a different pattern of behavior.

Example 10. Let $f(x) = x^3$. Sketch the graph of f.

Solution. We prepare the following table:

x	-3	-2	-1	0	1	2	3
$g(x)$	-27	-8	-1	0	1	8	27

Plotting the corresponding points and connecting them smoothly, we obtain the graph in Figure 3.31. ☐

If

$$f(-x) = -f(x) \quad \text{for all } x \text{ in the domain of } f \tag{2}$$

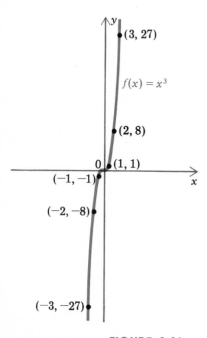

FIGURE 3.31

then f is called an **odd function**. The functions x and x^3 both are odd functions, as is any function of the form x^n, where n is odd. The graph of an odd function is said to be **symmetric with respect to the origin**, by which we mean that corresponding pairs of points on opposite sides of the origin are either both on or both not on the graph. Mathematically this means that if (x, y) is on the graph, then so is $(-x, -y)$ (Figure 3.32).

Intercepts

Our custom so far has been to plot several points before trying to sketch the graph of a function. This raises the question of which points we should plot. Since we always begin a figure with the coordinate axes, it is reasonable to plot the points of the graph that lie on either of the coordinate axes.

A **y intercept** of the graph of a function f is the y coordinate of any point where the graph of f meets the y axis (Figure 3.33). Since $x = 0$ for any point (x, y) on the y axis, it follows that the graph of a function has a y intercept if and only if 0 is in the domain of f, in which case $f(0)$ is the one and only y intercept. Analogously, an **x intercept** of the graph of a function is the x coordinate of any point where the graph meets the x axis (Figure 3.33). Since $y = 0$ for any point (x, y) on the x axis, it follows that the x intercepts of the graph of f are the values of x (if any) for which $f(x) = 0$. In brief,

> i. $f(0)$ is the y intercept (if 0 is in the domain of f).
> ii. x is an x intercept if $f(x) = 0$.

Example 11. Let $g(x) = x^2 - 4$. Find the y and x intercepts of the graph of g.

Solution. Since $g(0) = -4$, we know the y intercept is -4. To find the x intercepts we must find those values of x for which $g(x) = 0$, that is, for which $x^2 - 4 = 0$. But these are the values of x for which $x^2 = 4$, which yields $x = 2$ and $x = -2$. Consequently there are two x intercepts, 2 and -2; Figure 3.34 supports this claim. □

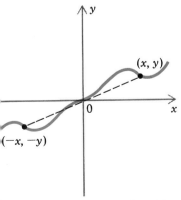

Symmetry with respect to the origin

FIGURE 3.32

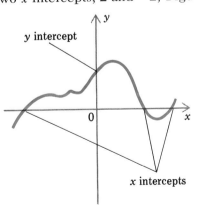

FIGURE 3.33

FIGURE 3.34

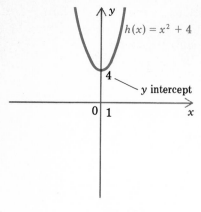

FIGURE 3.35

If we consider the function given by $h(x) = x^2 + 4$, we reach very different conclusions about intercepts. On the one hand, $h(0) = 4$, so the y intercept of h is 4. On the other hand, $h(x) = 0$ if and only if $x^2 + 4 = 0$. But since there are no real solutions of $x^2 + 4 = 0$, the graph of h has no x intercepts (see Figure 3.35). Thus we see that there may or may not be x intercepts for the graph of a given function. Similarly, there may or may not be a y intercept, depending on whether or not 0 is in the domain.

Example 12. Let $f(x) = x^2 + 2x - 3$. Find the x and y intercepts of the graph of f.

Solution. Since $f(0) = 0^2 + 0 - 3 = -3$, the y intercept is -3. To find the x intercepts we solve the equation $f(x) = 0$ for x. This leads to the equation

$$x^2 + 2x - 3 = 0$$

which, after factorization, becomes

$$(x + 3)(x - 1) = 0$$

Thus we see that $f(x) = 0$ if $x = -3$ or $x = 1$. This means that the x intercepts are -3 and 1. The graph of f, which we will learn how to sketch in Section 4.1, is shown in Figure 3.36. □

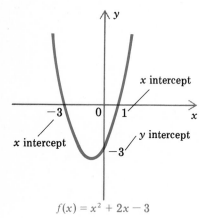

FIGURE 3.36

Example 13. Let $f(x) = 1/x$. Show that the graph of f has no x or y intercepts.

Solution. Since division by 0 is undefined, 0 is not in the domain of f, so there is no y intercept. Next, notice that $1/x$ is never 0, so that there are no values of x for which $f(x) = 0$. This means that there are no x intercepts. The graph of f, which we will be able to sketch later (see Section 4.4), is shown in Figure 3.37 and supports the claim that no intercepts exist. □

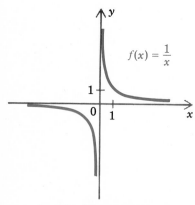

A graph with no intercepts

FIGURE 3.37

EXERCISES 3.4

In Exercises 1–18, sketch the graph of the given function and indicate any intercepts.

1. $f(x) = x + 3$
2. $f(x) = -x - \frac{1}{3}$
3. $f(x) = 2(x - 1)$
4. $f(x) = \frac{1}{3}x - 3$
5. $g(x) = 5 - |x|$
6. $g(x) = |x + 2|$

7. $g(x) = |2 - x|$

8. $h(x) = x^2 + 1$

9. $h(x) = 3 - x^2$

10. $y = \dfrac{1}{x} - 3$

11. $y = \dfrac{x + 1}{x} = 1 + \dfrac{1}{x}$

12. $g(x) = -\dfrac{1}{x}$

13. $g(x) = \sqrt{x} - 1$

14. $g(x) = \sqrt{x} + 3$

15. $h(x) = \begin{cases} -x & \text{for } x \le 0 \\ -2x & \text{for } x > 0 \end{cases}$

16. $h(x) = \begin{cases} x^2 & \text{for } x \le 0 \\ x & \text{for } x > 1 \end{cases}$

17. $h(x) = \begin{cases} \dfrac{1}{x} & \text{for } x < -1 \\ 3x & \text{for } x \ge -1 \end{cases}$

18. $h(x) = \begin{cases} |x| & \text{for } x < 1 \\ x^2 & \text{for } x \ge 1 \end{cases}$

In Exercises 19–30, determine all intercepts of the graph of the given function.

19. $f(x) = (x + 1)^2$

20. $f(x) = (x + 1)^2 - 4$

21. $f(x) = x^2 - 4x$

22. $f(x) = x^2 - 6x + 8$

23. $g(x) = x^2 - 6x + 9$

24. $g(x) = x^2 - 6x + 10$

25. $g(x) = \dfrac{1}{x - 1}$

26. $h(x) = \dfrac{1}{x^2}$

27. $h(x) = \dfrac{x + 1}{x + 2}$

28. $f(x) = \sqrt{x + 5}$

29. $f(x) = \sqrt{2 - x}$

30. $f(x) = ||x| - 2|$

In Exercises 31–38, sketch the graph and indicate whether the graph is symmetric with respect to the y axis, the origin, or neither.

31. $f(x) = -5$

32. $f(x) = 2x$

33. $f(x) = 2x - 1$

34. $g(x) = -|x|$

35. $g(x) = |x| - 4$

36. $g(x) = \dfrac{|x|}{x}$

37. $h(x) = x^2 - \dfrac{1}{2}$

38. $h(x) = \dfrac{1}{x} + 2$

In Exercises 39–48, determine which functions are even, which are odd, and which are neither.

39. $f(x) = x^2 + x$

40. $f(x) = \dfrac{x^2}{x^4 + 2}$

41. $f(x) = x^3 + 4x^5$

42. $f(x) = x - \dfrac{1}{x}$

43. $g(x) = x^2 - \dfrac{1}{x}$

44. $g(x) = \sqrt{x^2 - 1}$

45. $g(x) = \dfrac{x + 3}{x - 5}$

46. $h(x) = \dfrac{x^2 + 3}{x - 5}$

47. $h(x) = \dfrac{x^2 + 1}{x^2 - 1}$

48. $h(x) = \sqrt{x^2 + 3}$

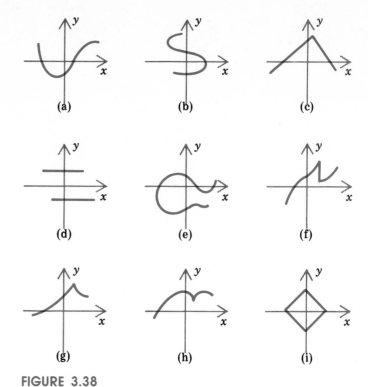

FIGURE 3.38

49. Determine which of the graphs in Figure 3.38 are graphs of functions.

50. Are there any nonzero functions that are both even and odd? Give reasons for your answer.

51. Let $f(x) = (x - 1)^2$. By plotting several points, sketch the graph of f. What is the relationship between the graph of f and the graph of $g(x) = x^2$?

52. Let $f(x) = |x + 1|^2$. By plotting several points, sketch the graph of f. What is the relationship between the graph of f and the graph of $g(x) = |x|$?

53. For any real number x, let $[x]$ denote the largest integer less than or equal to x. Thus $[2] = 2$, $[-\frac{7}{5}] = -2$, and $[\pi] = 3$. Let

$$f(x) = [x]$$

Then f is called the **greatest integer function**, or the **staircase function**. Sketch the graph of f.

3.5

LINES
IN THE PLANE

In this section we analyze straight lines in the plane and show that any nonvertical line can be associated with a function. For convenience we will separate the collection of straight lines in the plane into two categories: vertical and nonvertical lines.

Consider the vertical line drawn in Figure 3.39. Notice that the x coordinate of each point on the line is -1. Indeed, a point (x, y) is on the line if and only if $x = -1$. As a result, we say that the equation $x = -1$ is an equation of the given vertical line.

More generally, if c is any fixed number, then we call the equation $x = c$ an equation of the vertical line whose x intercept is c.

Example 1. Find equations for the vertical lines l_1, l_2, and l_3 appearing in Figure 3.40.

Solution. We have the following:

$$l_1: x = -4, \qquad l_2: x = \frac{5}{3}, \quad \text{and} \quad l_3: x = \pi \quad \square$$

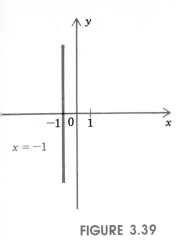

FIGURE 3.39

$x = -1$

This is really all there is to say about vertical lines. Henceforth we will restrict our discussion to nonvertical lines.

Slope of a Nonvertical Line

Let l be any nonvertical line in the plane, and let $P_1(x_1, y_1)$, $P_2(x_2, y_2)$, $P_3(x_3, y_3)$, and $P_4(x_4, y_4)$ be any points on l, with P_1 distinct from P_2, and P_3 distinct from P_4 (Figure 3.41). Since l is not vertical, it follows that $x_1 \neq x_2$ and $x_3 \neq x_4$. From Figure 3.41 we see that triangles $P_1 Q P_2$ and $P_3 R P_4$ are similar, so that

$$\frac{y_2 - y_1}{x_2 - x_1} = \frac{y_4 - y_3}{x_4 - x_3}$$

Consequently the ratio

$$\frac{y_2 - y_1}{x_2 - x_1}$$

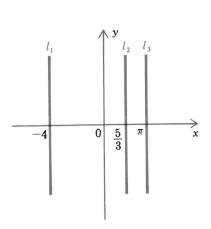

FIGURE 3.40

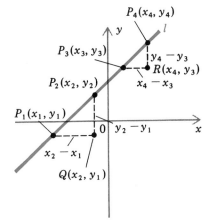

FIGURE 3.41

arising from the coordinates of two distinct points $P_1(x_1, y_1)$ and $P_2(x_2, y_2)$ on l, depends only on the line and *not* on the particular points. This ratio is called the *slope* of l and is usually denoted by m. Thus if $P_1(x_1, y_1)$ and $P_2(x_2, y_2)$ are any two distinct points on l, then the slope of l is given by the following equation:

SLOPE OF A LINE

$$m = \frac{y_2 - y_1}{x_2 - x_1} \qquad (1)$$

Example 2. Find the slope of the line l that contains the points $(-1, 2)$ and $(3, 4)$ (Figure 3.42).

Solution. If we let $(-1, 2)$ be $P_1(x_1, y_1)$ and $(3, 4)$ be $P_2(x_2, y_2)$, then by (1) the slope m of l is given by

$$m = \frac{4 - 2}{3 - (-1)} = \frac{2}{4} = \frac{1}{2} \quad \square$$

Caution: We have not defined the slope of a vertical line, nor will we do so. If one tries to apply (1) to a vertical line $x = c$, one obtains the meaningless expression

$$\frac{y_2 - y_1}{c - c}$$

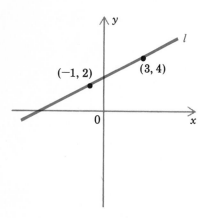

FIGURE 3.42

Next we will use the notion of slope to lead us to two types of equations for nonvertical lines—the point–slope equation and the slope–intercept equation.

Point–Slope Equation of a Line

Suppose that we are given the slope m and a point $P_1(x_1, y_1)$ on a nonvertical line l. We would like to find an equation that is satisfied by any point $P(x, y)$ on l. If we apply (1) with $P_2(x_2, y_2)$ replaced by $P(x, y)$, we find that

$$m = \frac{y - y_1}{x - x_1}$$

which is equivalent to the following equation:

POINT–SLOPE EQUATION OF A LINE

$$y - y_1 = m(x - x_1) \qquad (2)$$

A point–slope equation involves the slope m and a single point $P_1(x_1, y_1)$ of the line. A point $P(x, y)$ is on the line if and only if x and y satisfy (2).

Example 3. Find a point–slope equation of the line l that has slope 0 and passes through the point $(-2, -1)$. Then sketch the line.

Solution. Using (2) with $m = 0$, $x_1 = -2$, and $y_1 = -1$, we obtain the point–slope equation

$$y - (-1) = 0(x - (-2))$$

To sketch l we first simplify the equation:

$$y + 1 = 0$$
$$y = -1$$

Consequently l is horizontal and crosses the y axis at the point $(0, -1)$ (Figure 3.43). ☐

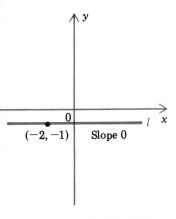

FIGURE 3.43

In Example 3 we found that the given line, which has slope 0, is horizontal. By the same method we deduce that any line having slope 0 and passing through, say (x_1, y_1), has an equation of the form

$$y - y_1 = 0(x - x_1)$$

which simplifies to

$$y - y_1 = 0$$

and then to

$$y = y_1 \tag{3}$$

It follows from (3) that a line with slope 0 is horizontal.

Example 4. Find a point–slope equation of the line that passes through the point $(3, 2)$ and has the given slope. Then sketch the line.

a. $\frac{1}{2}$ b. 3 c. -1

Solution. In each case we use (2):

a. We have $m = \frac{1}{2}$, $x_1 = 3$, and $y_1 = 2$, so that a point–slope equation of the line is

$$y - 2 = \frac{1}{2}(x - 3)$$

which simplifies to

$$y - 2 = \frac{1}{2}x - \frac{3}{2}$$

or

$$y = \frac{1}{2}x + \frac{1}{2}$$

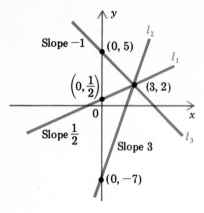

FIGURE 3.44

In order to sketch the graph, we find a second point on the line by setting $x = 0$ and noting that $y = \frac{1}{2}$. Therefore $(0, \frac{1}{2})$ is a second point, and the line, l_1, can be drawn (Figure 3.44).

b. This time $m = 3$, $x_1 = 3$, and $y_1 = 2$, so that

$$y - 2 = 3(x - 3)$$

which simplifies to

$$y - 2 = 3x - 9$$

or

$$y = 3x - 7$$

For the graph we again find a second point on the line by letting $x = 0$ and noting that $y = -7$. Therefore $(0, -7)$ is a second point, and the line, l_2, can be drawn (Figure 3.44).

c. Now we have $m = -1$, $x_1 = 3$, and $y_1 = 2$, so that

$$y - 2 = -1(x - 3)$$

which simplifies to

$$y - 2 = -x + 3$$

or

$$y = -x + 5$$

As before, we let $x = 0$ and find that $y = 5$, so that $(0, 5)$ is a second point on the line. The line, l_3, is drawn in Figure 3.44.   □

Let us tie together the results of Examples 3 and 4 as illustrated in Figures 3.43 and 3.44. First, if $m = 0$, then the line is horizontal. Next, if $m > 0$, then the line moves upward from left to right. Finally, if $m < 0$, then the line moves downward from left to right. Moreover, the larger the absolute value of the slope, the steeper the line is.

If we know the slope m and a point (x_1, y_1) on a given line l, then by (2) the point on l whose x coordinate is $x_1 + 1$ satisfies

$$y - y_1 = m(x_1 + 1 - x_1) = m$$

so that

$$y = y_1 + m$$

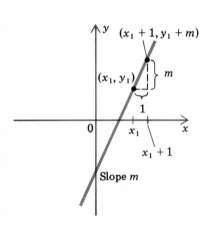

FIGURE 3.45

Therefore an increase of 1 unit in the value of x (from x_1 to $x_1 + 1$) causes a change of m units in the value of y (from y_1 to $y_1 + m$), and the point $(x_1 + 1, y_1 + m)$ also lies on l (Figure 3.45). We will use these ideas to help sketch the line discussed in the next example.

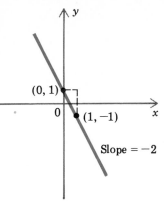

(0, 1)

0

(1, −1)

Slope = −2

y

x

FIGURE 3.46

Example 5. Sketch the line that contains the point (0, 1) and has slope −2.

Solution. We can sketch the line once we have two points on it. We are given one point on the line, (0, 1). To find a second point we use the fact that the line has slope −2. Thus if we start at (0, 1) and *increase* the value of *x* by 1, then the value of *y* must *decrease* by 2. This yields the point (0 + 1, 1 − 2), or (1, − 1), on the line. Now we can sketch the line (Figure 3.46). □

In order to find an equation of the line *l* that contains two given distinct points, we first use (1) to determine the slope of *l*, and then use a point–slope equation.

Example 6. Find an equation of the line *l* that contains the points (− 1, 2) and (3, 4). Then show that (7, 6) is on *l*.

Solution. By the solution of Example 2 we know that *l* has slope $\frac{1}{2}$. Thus if we apply the point–slope equation of the line with $m = \frac{1}{2}$ and $P_1(x_1, y_1) = (-1, 2)$, we obtain

$$y - 2 = \frac{1}{2}(x - (-1))$$

which is equivalent to

$$y - 2 = \frac{1}{2}(x + 1)$$

To show that (7, 6) is on *l* we let $x = 7$ and $y = 6$ and see that the equation is valid:

$$6 - 2 = \frac{1}{2}(7 + 1) \quad \square$$

Slope–Intercept Equation of a Line

A nonvertical line *l* must cross the *y* axis at some point. The *y* coordinate of that point is called the **y intercept** of *l* and is usually denoted by *b*. Thus (0, *b*) is a point on *l*. Now suppose that we are given the slope *m* and the *y* intercept *b* of a nonvertical line *l*. If we let $P_1(x_1, y_1) = (0, b)$ and apply the formula for the point–slope equation of a line, we find that $P(x, y)$ is on the line *l* provided that

$$y - b = m(x - 0)$$

which simplifies to give us the following equation:

SLOPE–INTERCEPT EQUATION OF A LINE

$$y = mx + b \qquad (4)$$

The slope–intercept equation of a line involves the slope m and the y intercept b of the line. As before, (x, y) is on the line if and only if (4) is satisfied.

Example 7. Find the slope–intercept equation of the line that has slope -2 and y intercept $\frac{1}{2}$. Then sketch the line.

Solution. By (4) with $m = -2$ and $b = \frac{1}{2}$, the slope–intercept equation is

$$y = -2x + \frac{1}{2} \tag{5}$$

Since the y intercept is $\frac{1}{2}$, we know that $(0, \frac{1}{2})$ is one point on the line. Taking $x = 1$ in equation (5), we find that

$$y = -2(1) + \frac{1}{2} = -\frac{3}{2}$$

Thus $(1, -\frac{3}{2})$ is a second point on the line, which is sketched in Figure 3.47. $\square$

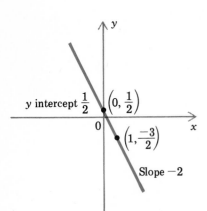

FIGURE 3.47

We have presented two types of equations for a line: point–slope and slope–intercept. Other types of equations will appear in Exercises 53, 56, and 58. However, all equations for a given line are equivalent, and information that may be explicit in one equation of the line may be obtained from any other. For example, the y intercept b of a line is explicitly written in the slope–intercept equation $y = mx + b$, but it does not appear explicitly in a point–slope equation $y - y_1 = m(x - x_1)$ unless $x_1 = 0$ and $y_1 = b$. Nevertheless we can determine the y intercept from *any* equation of the line by setting $x = 0$ and solving for y.

Example 8. Find the y intercept of the line with equation $y - 5 = -4(x + 2)$.

Solution. Setting $x = 0$ and solving for y, we obtain

$$y - 5 = -4(0 + 2) = -8$$
$$y = -8 + 5 = -3$$

Thus the y intercept is -3. $\square$

Parallel and Perpendicular Lines

Among the geometric properties of lines that are easily studied by means of slopes of lines, the two most important are the properties of parallelism and perpendicularity:

> i. Two nonvertical lines are parallel if and only if their slopes are equal.
> ii. Two nonvertical lines are perpendicular if and only if the product of their slopes is -1.

We will not prove the results just stated, which depend only on the definition of slope and basic geometric facts.

Example 9. Show that the lines

$$l_1: y = -2x + 6 \quad \text{and} \quad l_2: y + 4 = -2(x - 4)$$

are parallel.

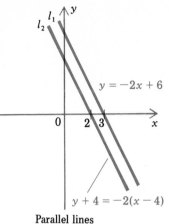

$y = -2x + 6$

$y + 4 = -2(x - 4)$

Parallel lines

FIGURE 3.48

Solution. The equation of l_1 is the slope–intercept equation of l_1; from it we see that the slope of l_1 is -2. The equation of l_2 is a point–slope equation of l_2; from it we see that the slope of l_2 is also -2. Since the two slopes are equal, the lines are parallel by (i) (Figure 3.48). ☐

Example 10. Show that the lines

$$l_1: y = \frac{1}{3}(x - 5) \quad \text{and} \quad l_2: y = -3x + 4$$

are perpendicular.

Solution. The equation of l_1 is a point–slope equation; from it we see that l_1 has slope $\frac{1}{3}$. The equation of l_2 is the slope–intercept equation of l_2; from it we see that l_2 has slope -3. Since

$$\frac{1}{3}(-3) = -1$$

it follows from (ii) that l_1 and l_2 are perpendicular (Figure 3.49). ☐

$y = -3x + 4$

$y = \frac{1}{3}(x - 5)$

Perpendicular lines

FIGURE 3.49

Example 11. Find an equation of the line l that contains the point $(\frac{1}{2}, \frac{2}{3})$ and is parallel to the line with equation $y = -4x + 7$.

Solution. The slope of the line with equation $y = -4x + 7$ is -4. It follows from (i) that l also has slope -4. Since we are given a point on l and we now know the slope of l, we use the point–slope equation in (2) with $m = -4$, $x_1 = \frac{1}{2}$, and $y_1 = \frac{2}{3}$ to obtain the equation

$$y - \frac{2}{3} = -4\left(x - \frac{1}{2}\right)$$

for l. ☐

Linear Functions

The slope–intercept equation of a nonvertical line is

$$y = mx + b \tag{6}$$

Such an equation yields the function f described by

$$f(x) = mx + b \tag{7}$$

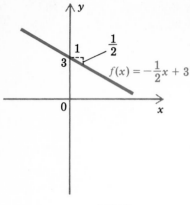

FIGURE 3.50

Accordingly, any function of the form given in (6) or (7) is called a *linear function*. Thus the graph of a linear function is a nonvertical line with appropriate slope and y intercept. (Since a vertical line cannot be the graph of a function, a straight line is the graph of a function if and only if it is nonvertical.)

Example 12. Let $f(x) = -\frac{1}{2}x + 3$. Sketch the graph of f.

Solution. We know already that the graph of f is a line. Since the equation is in slope–intercept form, we know the slope is $-\frac{1}{2}$ and the y intercept is 3. We place $(0, 3)$ on the graph, and note that an increase of 1 unit in the value of x yields a decrease of $\frac{1}{2}$ unit in the y value. From this information we can sketch the graph (Figure 3.50). $\square$

EXERCISES 3.5

In Exercises 1–6, find the slope of the line that contains the points P_1 and P_2.

1. $P_1 = (4, 3), P_2 = (6, 9)$ 2. $P_1 = (0, 4), P_2 = (-3, 10)$
3. $P_1 = (-3, -3), P_2 = (-5, 2)$ 4. $P_1 = (-5, -4), P_2 = (-9, -7)$
5. $P_1 = (\frac{1}{4}, \frac{1}{6}), P_2 = (\frac{3}{4}, \frac{1}{2})$
6. $P_1 = (1.57, 2.86), P_2 = (1.84, 2.56)$

In Exercises 7–12, find a point–slope equation of the line that has slope m and passes through the point P. Then sketch the line.

7. $m = 0, P = (3, -4)$ 8. $m = 2, P = (-2, -1)$
9. $m = \frac{2}{3}, P = (3, 0)$ 10. $m = -4, P = (-\frac{1}{2}, 2)$
11. $m = -1, P = (0, \frac{3}{2})$ 12. $m = \sqrt{2}, P = (\sqrt{2}, 3)$

In Exercises 13–18, find an equation of the line that contains the points P_1 and P_2.

13. $P_1 = (0, 1), P_2 = (2, 5)$ 14. $P_1 = (0, -1), P_2 = (3, 8)$
15. $P_1 = (-1, -2), P_2 = (-2, 2)$ 16. $P_1 = (2, 3), P_2 = (0, 0)$
17. $P_1 = (2, -1), P_2 = (2, 6)$ 18. $P_1 = (4, -3), P_2 = (2, 4)$

In Exercises 19–24, find the slope–intercept equation of the line that has slope m and y intercept b. Then sketch the line.

19. $m = 0, b = -2$ 20. $m = 3, b = 4$
21. $m = -\frac{4}{5}, b = -1$ 22. $m = 1, b = \pi$
23. $m = -\sqrt{3}, b = \frac{1}{2}$ 24. $m = 0.25, b = 0.4$

In Exercises 25–32, find the slope m and the y intercept b (if any) of the line with the given equation. Then sketch the line.

25. $y = 3$ 26. $x = -4$

27. $y = 3x - \frac{1}{2}$

28. $3y = x - 4$

29. $2x + 3y = 6$

30. $\frac{1}{2}y - 2x = 2$

31. $3.6x - 1.2y = 0$

32. $y - 7 = 3(x - 2)$

In Exercises 33–44, determine whether the two lines having the given equations are parallel, perpendicular, or neither.

33. $x = 0; y = 0$

34. $y = 3; x = -4$

35. $y = x; y = -x$

36. $y = 2x; y = -\frac{1}{2}x$

37. $y = 2x; y = -2x$

38. $2x - 3y = 4; y = -\frac{3}{2}x$

39. $\frac{1}{2}y = x; y = 2(x - 4)$

40. $x = -y; y = 4x + 2$

41. $3x - 5y = 4; 2x - 6y = 3$

42. $4x + 6y = 5; 6x + 9y = 1$

43. $y = 0.01(x - 1); y + 100x = 50$

44. $y - 5 = \frac{1}{2}(x + 3); x - 2y = 6$

In Exercises 45–52, sketch the graph of f.

45. $f(x) = -2$

46. $f(x) = \frac{1}{3}x$

47. $f(x) = -4x$

48. $f(x) = -x + 3$

49. $f(x) = 3(x - 2)$

50. $f(x) = \frac{1}{2}x - 3$

51. $f(x) = -\frac{2}{3}x + 4$

52. $f(x) = -0.5x + 1.5$

53. Let $P_1(x_1, y_1)$ and $P_2(x_2, y_2)$ be two distinct points not on the same vertical line, and let the line l pass through P_1 and P_2. Then $P(x, y)$ is on l provided that

$$(x_2 - x_1)(y - y_1) = (y_2 - y_1)(x - x_1) \qquad (8)$$

This equation is called a *two-point equation* of l.
a. Show that P_1 and P_2 satisfy (8).
b. Use (8) to obtain a point–slope equation of l.

54. Find a two-point equation of the line that passes through each of the following pairs of points. Then sketch the line.
a. $(4, 2)$ and $(3, -3)$
b. $(-1, -2)$ and $(-5, 0)$
c. $(\frac{1}{2}, \frac{3}{2})$ and $(-\frac{5}{2}, \frac{3}{2})$

55. Let l be a nonhorizontal line. The x coordinate of the point at which l crosses the x axis is the *x intercept* of l, and is usually denoted by a. Thus $(a, 0)$ is on l. The x intercept is determined by setting $y = 0$ in an equation of l and then solving for x. Determine the x intercept of the lines having the following equations.
a. $y = 6x - 2$ b. $y - 2 = \frac{1}{2}(x + 1)$ c. $x = -\frac{3}{4}$

56. Let l be a line that is neither horizontal nor vertical, and let its x and y intercepts be a and b, respectively. Then (x, y) is on l provided that

$$\frac{x}{a} + \frac{y}{b} = 1$$

This equation is called the *two-intercept equation* of l. Show that the slope of l is $-b/a$ by writing the equation in the equivalent slope-intercept form.

57. Find the two-intercept equation of the line that has x intercept a and y intercept b. Then sketch the line.
    a. $a = -1, b = 2$
    b. $a = 3, b = \frac{1}{2}$
    c. $a = 0.7, b = 0.3$
    d. $a = -2, b = -3$

58. *Any* line l can be described by an equation of the form

$$Ax + By = 1$$

where either $A \neq 0$ or $B \neq 0$. This equation is called a **general linear equation**.
    a. Determine the x intercept a (if any) and the y intercept b (if any) of l.
    b. Determine the slope (if any) of l.

59. Determine which of the following lines contain the point $(-2, 4)$.
    a. $y = 2x + 8$
    b. $y = 2x - 8$
    c. $y + 2 = -2(x - 1)$
    d. $y - 2 = 2(x + 1)$
    e. $3x + 4y = 22$
    f. $3x + 4y = 10$

60. Determine which of the following lines contain the point $(-3, -1)$.
    a. $y = 2x - 5$
    b. $y = 2x + 5$
    c. $y + 1 = \frac{1}{2}(x + 3)$
    d. $y - 1 = \frac{1}{2}(x - 3)$
    e. $-2x + y = -5$
    f. $2x - y = -5$

61. Suppose that a line l is to pass through $(2, 3)$. What slope must it have in order to pass through $(5, -1)$ as well?

62. Show that a line whose equation is $Ax + By = 0$ passes through the origin. If $B \neq 0$, what is the slope of the line?

63. Find a formula for the linear function whose graph passes through the point $(-1, -4)$ and has slope -2.

64. Find a formula for the linear function whose graph passes through the point $(3, 0)$ and has y intercept $-\frac{1}{3}$.

65. Find a formula for the linear function whose graph has x and y intercepts -1 and 4, respectively.

66. Line l_1 passes through the points $(-2, 1)$ and $(4, -1)$, and line l_2 passes through the points $(5, -3)$ and $(-7, -6)$. Which line has the greater slope?

67. Find an equation of the line that passes through the point $(-2, 5)$ and is parallel to the line with equation $4x + 3y = 1$.

68. Find an equation of the line that passes through the point $(3, -3)$ and is perpendicular to the line with equation $y = 6x - 2$.

69. Find the point of intersection of the lines having equations $2x + y = 0$ and $x - y = -3$. (*Hint:* First solve for y in terms of x in the first equation, and then substitute for y in the second equation.)

70. Find the point of intersection of the lines having equations $2x - 4y = 3$ and $x - y = 1$.

71. Show that the lines having the following equations determine a rectangle: $y = 2x + 1$, $y = 2x + 5$, $y = -\frac{1}{2}x + 1$, $y = -\frac{1}{2}x - 3$.

72. Show that the figure determined by the four points $(2, 1)$, $(3, 3)$, $(2, 5)$, and $(1, 3)$ is a parallelogram by proving that the opposite sides are parallel.

73. Show that the figure determined by the four points $(3, 2)$, $(4, 3)$, $(3, 4)$, and $(2, 3)$ is a square.

74. Show that the figure determined by the three points $(1, 1)$, $(5, 8)$, and $(-1, 5)$ is a right triangle.

3.6
COMPOSITION OF FUNCTIONS

When we compute the value $h(x)$ of a function h we frequently perform the computation in more than one step. For example, if

$$h(x) = \sqrt{x + 5} \qquad (1)$$

then we obtain $h(x)$ for $x \geq -5$ by first computing $x + 5$ and then taking the square root of $x + 5$. Thus

$$h(4) = \sqrt{4 + 5} = \sqrt{9} = 3$$

If we let

$$f(x) = x + 5 \quad \text{and} \quad g(x) = \sqrt{x}$$

then for $x \geq -5$ we have

$$\sqrt{x + 5} = \sqrt{f(x)} = g(f(x))$$

so that

$$h(x) = g(f(x))$$

Similarly, if

$$h(x) = \frac{1}{x - 4} \qquad (2)$$

then to compute $h(x)$ for any x such that $x - 4 \neq 0$, that is, for $x \neq 4$, we first compute $x - 4$ and then take its reciprocal. Therefore, if

$$f(x) = x - 4 \quad \text{and} \quad g(x) = \frac{1}{x}$$

then for $x \neq 4$ we have

$$\frac{1}{x - 4} = \frac{1}{f(x)} = g(f(x))$$

so once again

$$h(x) = g(f(x))$$

In general, if f and g are two functions, then we define the *composite function* $g \circ f$ to be the function whose rule is given by

$$(g \circ f)(x) = g(f(x)) \tag{3}$$

and whose domain consists of all the numbers x in the domain of f such that $f(x)$ is in the domain of g. (This is just the set of all numbers x for which the right side of (3) is meaningful.) Thus each of the functions h defined in (1) and (2) may be expressed as the composite of two functions f and g.

Example 1. Let $h(x) = \dfrac{1}{(x + 3)^2}$. Express h as the composite $g \circ f$ of two functions f and g.

Solution. If we let

$$f(x) = x + 3 \quad \text{and} \quad g(x) = \frac{1}{x^2}$$

then for $x \neq -3$ we have

$$h(x) = \frac{1}{(x + 3)^2} = \frac{1}{(f(x))^2} = g(f(x))$$

Thus $h = g \circ f$. □

In the solution of Example 1, if we had let

$$f(x) = (x + 3)^2 \quad \text{and} \quad g(x) = \frac{1}{x}$$

then it would still have been true that $h = g \circ f$. In general, there are many ways of writing a given function as the composite of two other functions.

Example 2. Let $f(x) = \sqrt{x}$ and $g(x) = x^2 - 1$. Find the domain and rule of each of the following composite functions.
    a. $g \circ f$ b. $f \circ g$

Solution.
    a. The domain of $g \circ f$ consists of all numbers x in the domain of f such that $f(x)$ is in the domain of g. But since the domain of g consists of all numbers, it follows that $f(x)$ is in the domain of g for any number x in the domain of f. Therefore the domain of $g \circ f$ is the set of all nonnegative numbers. The rule of $g \circ f$ is given by

$$(g \circ f)(x) = g(f(x)) = g(\sqrt{x}) = (\sqrt{x})^2 - 1 = x - 1 \tag{4}$$

b. The domain of $f \circ g$ consists of all numbers x in the domain of g such that $g(x)$ is in the domain of f, which consists of all nonnegative numbers. Thus the domain of $f \circ g$ consists of all numbers x such that $x^2 - 1 \geq 0$, which is equivalent to $x^2 \geq 1$. But $x^2 \geq 1$ for x in $(-\infty, -1]$ or $[1, \infty)$. We conclude that the domain of $f \circ g$ consists of the intervals $(-\infty, -1]$ and $[1, \infty)$. The rule of $f \circ g$ is given by

$$(f \circ g)(x) = f(g(x)) = f(x^2 - 1) = \sqrt{x^2 - 1} \quad \square$$

Caution: Notice that in Example 2 the functions $g \circ f$ and $f \circ g$ are not the same. This is usually the case. Notice also that the right side of (4) is meaningful for all real numbers x, whereas the domain of $g \circ f$ consists only of the nonnegative numbers. Therefore we must be careful when specifying the domain of a composite function.

EXERCISES 3.6

In Exercises 1–6, express h as the composite $g \circ f$ of two functions f and g (neither of which is equal to h).

1. $h(x) = 2(x - 1)$
2. $h(x) = -7\sqrt{x}$
3. $h(x) = \dfrac{3}{x^4}$

4. $h(x) = \dfrac{-5}{x + 4}$
5. $h(t) = \dfrac{1}{2\sqrt{t}}$
6. $h(t) = (t + 1)^{1/3}$

In Exercises 7–20, find the domain and rule of $g \circ f$.

7. $f(x) = x - 1$; $g(x) = 2x^2 + x + 1$
8. $f(x) = 2x - 1$; $g(x) = x^2$
9. $f(x) = \dfrac{1}{2x}$; $g(x) = \sqrt{x}$
10. $f(x) = \sqrt{x}$; $g(x) = \dfrac{1}{2x}$
11. $f(x) = x - 1$; $g(x) = \sqrt{x} + 1$
12. $f(x) = \sqrt{x}$; $g(x) = \dfrac{1}{2x - 4}$
13. $f(x) = \dfrac{1}{2x}$; $g(x) = \dfrac{1}{x^2 - 1}$
14. $f(x) = \dfrac{x^2 - 1}{x^2 + 1}$; $g(x) = \dfrac{1}{x}$
15. $f(x) = \dfrac{1}{x}$; $g(x) = \dfrac{1}{x}$
16. $f(x) = \dfrac{x - 1}{x + 1}$; $g(x) = \dfrac{x + 1}{x - 1}$

17. $f(x) = \dfrac{x-1}{x+1}$; $g(x) = \dfrac{x+3}{x-2}$

18. $f(x) = \sqrt{x^2+1}$; $g(x) = \sqrt{x^2-1}$

19. $f(x) = \sqrt{x^2+1}$; $g(x) = \sqrt{x^2-4}$

20. $f(x) = \dfrac{2x}{x-1}$; $g(x) = \sqrt{2x-4}$

21. Let $f(x) = x$. Show that $g \circ f = g$ and $f \circ g = g$ for any function g.

22. Let $f(x) = \dfrac{x}{x-1}$. Show that $(f \circ f)(x) = x$ for all $x \neq 1$.

23. Let $f(x) = \dfrac{1}{1-x}$. Show that $(f \circ f \circ f)(x) = x$ for all x except 0 and 1. (By definition, $(f \circ f \circ f)(x) = [f \circ (f \circ f)](x)$.)

24. Suppose the domain of f is $[0, 1)$ and $g(x) = f(x + 10)$. Find the domain of g.

25. Suppose the domain of f is $[0, 1)$ and $g(x) = f(10 - x)$. Find the domain of g.

26. Suppose the domain of f is $(-\pi/2, \pi/2)$ and $g(x) = f(x - \pi/6)$. Find the domain of g.

27. Let $f(x) = mx + b$, where m and b are constants. Let $g(x) = f(x + 1) - f(x)$. Show that g is a constant function, and determine the constant.

* 28. Let $f(x) = ax^2 + bx + c$, where a, b, and c are constants. Show that if

$$g(x) = f(x + 1) - f(x) \quad \text{and} \quad G(x) = g(x + 1) - g(x)$$

then G is a constant function that is independent of b and c.

29. The volume $V(r)$ of a spherical balloon of radius r is given by

$$V(r) = \frac{4}{3}\pi r^3$$

If the radius of the balloon is increasing with time t according to the formula

$$r(t) = \frac{3}{2}t^2 \quad \text{for } t \geq 0$$

find a formula for the volume of the balloon at any time $t \geq 0$.

3.7
INVERSES OF FUNCTIONS

We can convert from degrees Celsius to degrees Fahrenheit by means of the formula

$$F = \frac{9}{5}C + 32 \tag{1}$$

By solving this equation for C we obtain a formula for converting from degrees Fahrenheit to degrees Celsius:

$$C = \frac{5}{9}(F - 32) \tag{2}$$

In order to explore further the relationship between (1) and (2), we view (1) and (2) as defining the functions f and g given by

$$f(x) = \frac{9}{5}x + 32 \quad \text{and} \quad g(x) = \frac{5}{9}(x - 32) \tag{3}$$

Notice that if $y = f(x)$, then

$$y = \frac{9}{5}x + 32$$

so that

$$g(y) = g\left(\frac{9}{5}x + 32\right) = \frac{5}{9}\left[\left(\frac{9}{5}x + 32\right) - 32\right] = \frac{5}{9}\left(\frac{9}{5}x\right) = x$$

Thus if $y = f(x)$, then $x = g(y)$. A similar calculation shows that the converse is true: if $x = g(y)$, then $y = f(x)$. We express this relationship between f and g by saying that g is an inverse of f.

DEFINITION 3.4

Let f be a function. Then f **has an inverse** if there is a function g such that
 i. the domain of g is the range of f
ii. for all x in the domain of f and all y in the range of f,

$$f(x) = y \text{ if and only if } g(y) = x \tag{4}$$

Under these conditions g is an **inverse** of f.

Notice that in Definition 3.4, g is a *function*. Its domain is specified by (i) and its rule by (ii). From this it follows that g is unique. To emphasize its connection with f, we write f^{-1} for g and call f^{-1} *the inverse of f*. With g replaced by f^{-1}, (4) becomes

$$f(x) = y \quad \text{if and only if} \quad f^{-1}(y) = x \tag{5}$$

for all x in the domain of f and all y in the range of f. We also find that if f has an inverse, then f^{-1} has f as inverse, so f and f^{-1} are inverses of each other.

Let us observe that in (3) the function g is the inverse of f.

Caution: Not every function has an inverse. At the end of this section we will show that, among other functions, x^2 does not have an inverse.

From (5) we obtain two important formulas relating a function f and its inverse f^{-1}:

$$f^{-1}(f(x)) = x \text{ for all } x \text{ in the domain of } f \qquad (6)$$

$$f(f^{-1}(y)) = y \text{ for all } y \text{ in the range of } f \qquad (7)$$

Example 1. Let $f(x) = 5x^3$, and assume that we already know that f has an inverse and that

$$f^{-1}(y) = \sqrt[3]{\frac{y}{5}}$$

Verify (6) and (7) from the formulas given for f and f^{-1}.

Solution. For (6) we have

$$f^{-1}(f(x)) = f^{-1}(5x^3) = \sqrt[3]{\frac{5x^3}{5}} = \sqrt[3]{x^3} = x$$

and for (7) we have

$$f(f^{-1}(y)) = f\left(\sqrt[3]{\frac{y}{5}}\right) = 5\left(\sqrt[3]{\frac{y}{5}}\right)^3 = 5\left(\frac{y}{5}\right) = y \quad \square$$

Because we usually write formulas for functions in terms of x (rather than y), we will also usually write the formula for f^{-1} in terms of x. Thus in Example 1 the formula for the inverse f^{-1} would normally be written as

$$f^{-1}(x) = \sqrt[3]{\frac{x}{5}}$$

Finding Formulas for Inverses

In Example 1 we assumed the formula for f^{-1} on faith. Now we will describe a procedure that frequently allows us to determine a formula for the inverse of a function f, provided that f is given by a simple formula. The individual steps in the procedure are:

Step 1. Write $y = f(x)$.
Step 2. Solve the equation in Step 1 for x in terms of y.
Step 3. In the formula for x arising from Step 2, replace x by $f^{-1}(y)$.
Step 4. Replace each y in the result of Step 3 by x.

Let us see how the method works for the temperature conversion function f appearing in (3).

Example 2. Let $f(x) = \frac{9}{5}x + 32$. Find a formula for f^{-1}.

Solution. Following the steps listed above, we obtain

Step 1: $y = \dfrac{9}{5}x + 32$

Step 2: $\begin{cases} 5y = 9x + 160 \\[4pt] 9x = 5y - 160 = 5(y - 32) \\[4pt] x = \dfrac{5}{9}(y - 32) \end{cases}$

Step 3: $f^{-1}(y) = \dfrac{5}{9}(y - 32)$

Step 4: $f^{-1}(x) = \dfrac{5}{9}(x - 32)$ □

Notice that f^{-1} is the function g appearing in (3), so g is shown to be the inverse of f in a second way.

Example 3. Let $f(x) = 2 - x^3$. Find a formula for f^{-1}.

Solution. Again we use the steps given above:

Step 1: $y = 2 - x^3$

Step 2: $\begin{cases} x^3 = 2 - y \\[4pt] x = \sqrt[3]{2 - y} \end{cases}$

Step 3: $f^{-1}(y) = \sqrt[3]{2 - y}$

Step 4: $f^{-1}(x) = \sqrt[3]{2 - x}$ □

Because mistakes are easy to make, it is advisable to check either that $f(f^{-1}(x)) = x$ or that $f^{-1}(f(x)) = x$ after obtaining a formula for f^{-1}. Using this advice for the function in Example 3, we have

$$f(f^{-1}(x)) = 2 - (f^{-1}(x))^3 = 2 - (\sqrt[3]{2 - x})^3 = 2 - (2 - x) = x$$

Graphs of Inverses

To find a method of graphing the inverse of a function f, we begin by observing that if $(1, 3)$ is on the graph of f, then $f(1) = 3$; therefore by the definition of f^{-1}, it follows that $f^{-1}(3) = 1$, which in turn means that $(3, 1)$ is on the graph of f^{-1}. More generally, for any real numbers a and b, if (a, b) is on the graph of f, then (b, a) is on the graph of f^{-1}, and vice versa. In particular, $(x, f(x))$ is on the graph of f for any x in the domain of f, so that by the preceding comments $(f(x), x)$ is on the graph of f^{-1}. Notice that $(x, f(x))$ and $(f(x), x)$ are opposite one another with respect to the line $y = x$ (Figure 3.51). For this reason we say that $(x, f(x))$ and $(f(x), x)$ are **symmetric with**

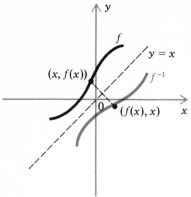

The graphs of f and f^{-1} are symmetric with respect to the line $y = x$.

FIGURE 3.51

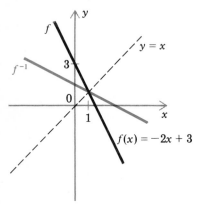

FIGURE 3.52

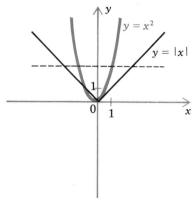

The functions x^2 and $|x|$ do not have inverses

FIGURE 3.53

respect to the line y = x, and when we pass from one of these points to the other, we say that we *reflect* the point through the line $y = x$. Therefore we can obtain the graph of f^{-1} by reflecting all the points on the graph of f through the line $y = x$, or as we usually say, by reflecting the graph of f through the line $y = x$ (Figure 3.51). The graphs of f and f^{-1} are said to be symmetric with respect to the line $y = x$.

Example 4. Let $f(x) = -2x + 3$. First sketch the graph of f, and then obtain the graph of f^{-1} by reflecting the graph of f through the line $y = x$.

Solution. The graph of f is the line with slope -2 and y intercept 3. It is sketched in Figure 3.52. Next we draw the line $y = x$, and finally we sketch the graph of f^{-1} by reflecting through the line $y = x$ (Figure 3.52). □

There is a geometric method for telling which functions f have inverses: f has an inverse if and only if no horizontal line intersects the graph of f more than once. As you can easily see from Figure 3.53, the functions x^2 and $|x|$ do not have inverses. In algebraic terms, a function has an inverse if and only if the following condition holds:

For any x and z in the domain of f

$$\text{if } x \neq z, \quad \text{then} \quad f(x) \neq f(z)$$

In this case f is said to be *one-to-one*.

EXERCISES 3.7

In all exercises in this section, assume the existence of the inverse. In Exercises 1–22, find a formula for f^{-1}.

1. $f(x) = 2x$

2. $f(x) = 2x + 1$

3. $f(x) = -\dfrac{1}{2}(x - 3)$

4. $f(x) = \pi x - \sqrt{2}$

5. $f(x) = x^3$

6. $f(x) = x^5$

7. $f(x) = 3x^3 - 5$

8. $f(x) = \pi x^2$ for $x \geq 0$

9. $f(x) = -x^5 + 1$

10. $f(x) = \dfrac{1}{x}$

11. $f(x) = \sqrt{x}$

12. $f(x) = \sqrt{x} + 5$

13. $f(x) = \sqrt{x - 3}$

14. $f(x) = 7 + \sqrt{2x - 1}$

15. $f(x) = \dfrac{1}{\sqrt{x}}$

16. $f(x) = \sqrt[3]{x + 5}$

17. $f(x) = \dfrac{1}{\sqrt[3]{4 - 2x}}$

18. $f(x) = \dfrac{-4}{x^3}$

19. $f(x) = \dfrac{x - 1}{x + 1}$

20. $f(x) = \dfrac{-x + 3}{2x - 5}$

21. $f(x) = \dfrac{2x^3 - 1}{x^3 + 3}$

22. $f(x) = \dfrac{x^5}{-x^5 + 3}$

In Exercises 23–32, sketch the graphs of f and f^{-1} in the same coordinate system.

23. $f(x) = x + 1$

24. $f(x) = 3x$

25. $f(x) = 3(x + 1)$

26. $f(x) = 3x + 1$

27. $f(x) = -2x + 4$

28. $f(x) = \sqrt{x}$

29. $f(x) = \dfrac{1}{x}$

30. $f(x) = -\dfrac{1}{x}$

31. $f(x) = \sqrt{x} + 1$

32. $f(x) = \sqrt{x} - 1$

33. Show that each of the following functions is equal to its own inverse:

a. $f(x) = x$

b. $f(x) = \dfrac{1}{x}$

c. $f(x) = -x$

d. $f(x) = \sqrt{1 - x^2}$ for $0 \le x \le 1$

34. Does a constant function have an inverse? Explain your answer.

3.8

GRAPHS OF EQUATIONS

Thus far in Chapter 3 we have concentrated on functions and their graphs. The importance of functions and their graphs cannot be overemphasized; yet equations involving x and y that do not necessarily represent functions play a role in mathematics, and so do their graphs; it is therefore well to study some of the more common ones. The **graph of an equation** in x and y is the collection of all ordered pairs (x, y) of real numbers that satisfy the equation.

We begin with circles. A **circle** of radius r centered at the point (a, b) in the plane is by definition the collection of all points (x, y) whose distance from (a, b) is r. From the distance formula (in Section 3.1), the distance between (a, b) and (x, y) is

$$\sqrt{(x - a)^2 + (y - b)^2}$$

Therefore (x, y) is on the circle if and only if

$$\sqrt{(x - a)^2 + (y - b)^2} = r$$

or, when both sides are squared,

$$(x - a)^2 + (y - b)^2 = r^2 \qquad (1)$$

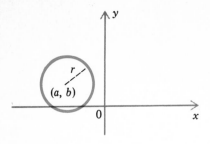

FIGURE 3.54

In summary, the graph of (1) is the circle of radius r centered at (a, b) (Figure 3.54). If the center of the circle is $(0, 0)$, then (1) becomes

$$x^2 + y^2 = r^2 \qquad (2)$$

Example 1. Find an equation of the circle of radius 1 centered at $(0, 0)$. Then sketch the circle.

Solution. Using (2) with $r = 1$, we obtain the equation

$$x^2 + y^2 = 1$$

The circle is sketched in Figure 3.55. □

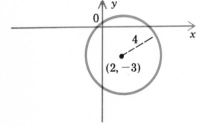

The unit circle

FIGURE 3.55

The circle $x^2 + y^2 = 1$ is usually called the **unit circle**.

Example 2. Find an equation of the circle of radius 4 centered at the point $(2, -3)$. Then sketch the circle.

Solution. By (1) an equation is given by

$$(x - 2)^2 + (y - (-3))^2 = 4^2$$

which simplifies to

$$(x - 2)^2 + (y + 3)^2 = 16$$

The circle is sketched in Figure 3.56. □

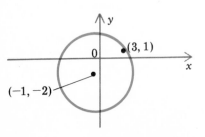

FIGURE 3.56

Example 3. Find an equation of the circle whose center is $(-1, -2)$ and which passes through the point $(3, 1)$.

Solution. The radius r of the circle is the distance between the center $(-1, -2)$ and the point $(3, 1)$ on the circle. By the distance formula,

$$r = \sqrt{(3 - (-1))^2 + (1 - (-2))^2} = \sqrt{16 + 9} = \sqrt{25} = 5$$

Thus by (1) the equation we seek is

$$(x - (-1))^2 + (y - (-2))^2 = 5^2$$

which simplifies to

$$(x + 1)^2 + (y + 2)^2 = 25$$

FIGURE 3.57

The circle is sketched in Figure 3.57. □

We begin our discussion of symmetry by sketching the graph of a simple equation.

Example 4. Sketch the graph of $y^2 = x$.

Solution. First we make a table:

x	0	1	4	9
y	0	1 and -1	2 and -2	3 and -3

Then we plot the corresponding points (x, y) on the graph and connect the points with a smooth curve (Figure 3.58). □

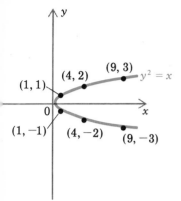

FIGURE 3.58

The graph of $y^2 = x$ in Figure 3.58 has the feature that it is symmetric with respect to the x axis, that is, the part above the x axis is mirrored below the x axis, and vice versa. Indeed, if the point (x, y) is on the graph, then the point $(x, -y)$, which lies opposite to (x, y) across the x axis, also lies on the graph (Figure 3.58). More generally, we say that the graph of an equation is *symmetric with respect to the x axis* if $(x, -y)$ is on the graph whenever (x, y) is (Figure 3.59).

The definition of symmetry with respect to the y axis given in Section 3.4 for graphs of functions applies equally well to graphs of equations. The graph of an equation is *symmetric with respect to the y axis* if $(-x, y)$ is on the graph whenever (x, y) is (Figure 3.60). For example, the circle $x^2 + (y - 1)^2 = 2$ is symmetric with respect to the y axis (Figure 3.61).

Similarly, the definition of symmetry with respect to the origin given in Section 3.4 for graphs of functions transfers to graphs of equations: The graph of an equation is *symmetric with respect to*

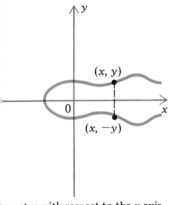

mmetry with respect to the x axis

FIGURE 3.59

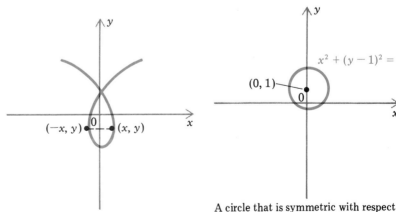

Symmetry with respect to the y axis

FIGURE 3.60

A circle that is symmetric with respect to the y axis.

FIGURE 3.61

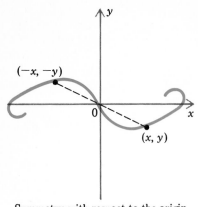

Symmetry with respect to the origin

FIGURE 3.62

the origin if $(-x, -y)$ is on the graph whenever (x, y) is on the graph (Figure 3.62).

Example 5. Sketch the graph of $x = y^3$.

Solution. With the following table

x	-8	-1	0	1	8
y	-2	-1	0	1	2

we are able to plot the corresponding points and connect them with a smooth curve (Figure 3.63). We notice that the graph is symmetric with respect to the origin, since if $x = y^3$, then $-x = -y^3 = (-y)^3$. □

The graph of an equation need not be symmetric with respect to either of the axes or the origin (Figure 3.64). In contrast, the graph of an equation can be symmetric with respect to *both* axes as well as to the origin (see Figure 3.55). Finally, we mention that symmetry can facilitate sketching of graphs, because a portion of the graph then determines the remainder of the graph.

Intercepts

As we sketch the graph of an equation, the points where the graph of an equation meets the coordinate axes are of interest. A *y intercept* of the graph of an equation is the y coordinate of a point at which the graph meets the y axis (Figure 3.65). Since any point on the y axis has x coordinate 0, the y intercepts (if any) can be found by setting $x = 0$ in the equation and then solving for y. Similarly, an *x intercept* of the graph of an equation is the x coordinate of a point at which the graph meets the x axis (Figure 3.65), and the x intercepts (if any) can be found by setting $y = 0$ in the equation and then solving for x.

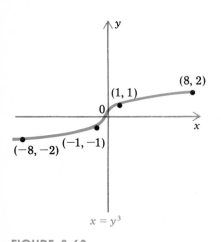

FIGURE 3.63

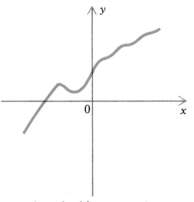

A graph with no symmetry

FIGURE 3.64

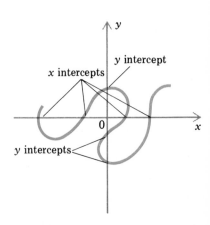

FIGURE 3.65

Example 6. Find the intercepts of the graph of the equation $4x^2 + y^2 = 4$.

Solution. To find the y intercepts we set $x = 0$ and solve for y:

$$4(0)^2 + y^2 = 4$$

$$y^2 = 4$$

$$y = 2 \quad \text{or} \quad y = -2$$

To find the x intercepts we set $y = 0$ and solve for x:

$$4x^2 + (0)^2 = 4$$

$$x^2 = 1$$

$$x = 1 \quad \text{or} \quad x = -1$$

Thus the y intercepts are 2 and -2, and the x intercepts are 1 and -1. □

Example 7. Find the intercepts of the graph of the equation $x^2 - y^2 = 1$.

Solution. To find the y intercepts (if any) we set $x = 0$ and obtain

$$0^2 - y^2 = 1$$

$$y^2 = -1$$

Since this equation has no real solutions, there are no y intercepts. (Thus the graph of $x^2 - y^2 = 1$ does not meet the y axis.) To find the x intercepts we set $y = 0$ and solve for x. We find that

$$x^2 - (0)^2 = 1$$

$$x^2 = 1$$

$$x = 1 \quad \text{or} \quad x = 1$$

Thus the x intercepts are 1 and -1. □

Example 8. Show that the graph of $y^2 - 5 = x + \dfrac{10}{x}$ has no intercepts.

Solution. There are no y intercepts because $10/x$ is not defined for $x = 0$, and hence the given equation is meaningless for $x = 0$. To search for x intercepts, we set $y = 0$ and attempt to solve the equation for x:

$$(0)^2 - 5 = x + \frac{10}{x}$$

$$-5x = x^2 + 10$$

$$x^2 + 5x + 10 = 0 \tag{3}$$

If we attempt to use the quadratic formula, we obtain

$$x = \frac{-5 \pm \sqrt{5^2 - 4(1)(10)}}{2(1)} = \frac{-5 \pm \sqrt{-15}}{2}$$

Since $\sqrt{-15}$ is undefined, (3) has no real solutions, and consequently the given equation has no x intercepts. □

EXERCISES 3.8

In Exercises 1–8, find an equation of the given circle and then sketch the circle.

1. The circle with radius 3 and center $(0, 0)$.
2. The circle with radius 5 and center $(3, 6)$.
3. The circle with radius 2 and center $(-1, 4)$.
4. The circle with radius $\frac{1}{2}$ and center $(-1, -2)$.
5. The circle with center $(0, 0)$ that passes through $(4, -1)$.
6. The circle with center $(5, 12)$ that passes through $(0, 0)$.
7. The circle with center $(-2, 3)$ that passes through $(1, -1)$.
8. The circle with center $(-3, -1)$ that passes through $(0, 4)$.

In Exercises 9–20, sketch the graph of the given equation and label all intercepts. Determine whether the graph possesses symmetry with respect to either axis or the origin.

9. $y^2 - 6y + 9 = 0$

10. $6x^2 + x - 2 = 0$

11. $x = |y|$

12. $|y| = |x|$

13. $y^2 = x - 1$

14. $x^2 = y^4$

15. $x = -y^3$

16. $x^3 = y^2$

17. $x^2 + y^2 = 4$

18. $(x - 1)^2 + (y + 2)^2 = 4$

19. $x = \sqrt{y}$

20. $x = \begin{cases} \sqrt{-y} & \text{for } y < 0 \\ \sqrt{y} & \text{for } y \geq 0 \end{cases}$

In Exercises 21–32, determine the intercepts (if any) of the graph of the given equation.

21. $x^2 + 4y^2 = 1$

22. $2y^2 - 3x^2 = 1$

23. $x - y^2 = 3$

24. $\dfrac{1}{x} + \dfrac{1}{y} = 1$

25. $y^2 = \sqrt{x^2 - 1}$

26. $y^2 - 4 = x + \dfrac{4}{x}$

27. $y^2 = |x - 5|$

28. $y + \dfrac{6}{y} = x^2 - 1$

29. $y - \dfrac{6}{y} = x^2 + 1$

30. $\dfrac{2}{x} - \dfrac{4}{x^2} = \pi xy - 5y^2 - 1$

31. $|x - 2| = |y + 1|$

32. $|x - 2| + y^2 = 4$

3.9

RELATIONS

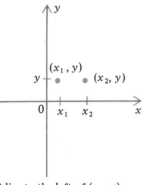

(x_1, y) lies to the left of (x_2, y)

(a)

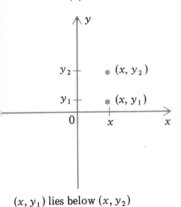

(x, y_1) lies below (x, y_2)

(b)

FIGURE 3.66

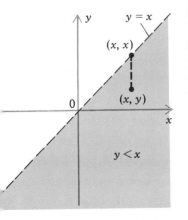

FIGURE 3.67

The alternative definition of function given in Definition 3.2 implies that a function f can be regarded as the collection of ordered pairs $(x, f(x))$ of numbers for x in the domain of f. It is possible to go a step further and give a name to any arbitrary collection of ordered pairs of numbers. A *relation* is *any* collection of ordered pairs of real numbers. The following are examples of relations:

The pairs $(0, 2)$, $(3, 4)$, $(3, \frac{1}{2})$, $(0, 4)$
All (x, y) such that $x = y$
All (x, y) such that $x \neq y^2 - 3$
All (x, y) such that $y \geq x^2$
All (x, y) such that $x^2 + y^2 \leq 25$
All (x, y) such that x/y is irrational
All ordered pairs

In particular, any function is a relation, and furthermore, the collection of all pairs of numbers satisfying any equation represents a relation.

The *graph of a relation* is the collection of points in the plane corresponding to the ordered pairs comprising the relation. We have already sketched the graphs of many relations—some that are graphs of functions and equations, and a few that do not correspond to functions or equations (see especially Examples 4 and 5 of Section 3.1).

Since inequalities give rise to some of the most important relations, we present several of them and their graphs now. Notice that if (x_1, y) and (x_2, y) are two points in the plane such that $x_1 < x_2$, then (x_1, y) lies to the left of (x_2, y) (Figure 3.66a). Similarly, if (x, y_1) and (x, y_2) are two points such that $y_1 < y_2$, then (x, y_1) lies below (x, y_2) (Figure 3.66b). These observations will help us draw graphs of relations.

Example 1. Sketch the graph of all (x, y) with $y < x$.

Solution. Notice that the inequality $y < x$ holds if and only if the point (x, y) lies below the point (x, x). Now the set of points (x, x) comprises the line $y = x$. Therefore the collection of points (x, y) with $y < x$ is the set of points lying below the line $y = x$, and is shaded in Figure 3.67. ☐

Example 2. Sketch the graph of all (x, y) with $x \geq y + 2$.

Solution. Here we notice that $x \geq y + 2$ if and only if the point (x, y) lies to the right of or coincides with the point $(y + 2, y)$, which is on the line $x = y + 2$. This line has slope 1 and y intercept -2. Consequently the collection of all points (x, y) with $x \geq y + 2$ is the

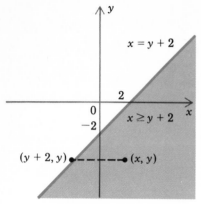

$x = y + 2$

$x \geq y + 2$

$(y + 2, y)$ •‒ ‒ ‒ ‒• (x, y)

FIGURE 3.68

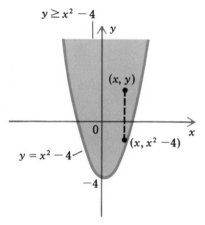

$y \geq x^2 - 4$

(x, y)

$(x, x^2 - 4)$

$y = x^2 - 4$

FIGURE 3.69

set of points lying to the right of or on the line $x = y + 2$. The set is shaded in Figure 3.68. ☐

Example 3. Sketch the graph of all (x, y) with $y \geq x^2 - 4$.

Solution. Observe that the inequality $y \geq x^2 - 4$ holds if and only if the point (x, y) lies above or coincides with the point $(x, x^2 - 4)$. But the set of points $(x, x^2 - 4)$ forms the graph of $y = x^2 - 4$, which is the curve drawn in Figure 3.69 (see Figure 3.25). Thus the collection of points (x, y) with $y \geq x^2 - 4$ is the set of points lying above or on the curve $y = x^2 - 4$ and is shaded in Figure 3.69. ☐

Example 4. Sketch the graph of all (x, y) with $y < x$ and $y \geq x^2 - 4$.

Solution. The graph of all (x, y) with $y < x$ is sketched in Figure 3.67, and the graph of all (x, y) with $y \geq x^2 - 4$ is sketched in Figure 3.69. It follows that the collection of all points (x, y) with $y < x$ *and* $y \geq x^2 - 4$ consists of the points common to the two shaded regions in Figure 3.67 and 3.69. It is shaded in Figure 3.70. ☐

Example 5. Sketch the graph of all (x, y) with $x^2 + y^2 \leq 25$.

Solution. Recall that $x^2 + y^2$ can be interpreted as the square of the distance between the point (x, y) and the origin. Therefore the point (x, y) satisfies $x^2 + y^2 \leq 25$ if and only if the square of the distance between (x, y) and the origin is less than or equal to 25, or equivalently, the distance between (x, y) and the origin is less than or equal to 5. Consequently the collection of points (x, y) satisfying $x^2 + y^2 \leq 25$ is the set of points that lie inside or on the circle of radius 5 centered at the origin (Figure 3.71). ☐

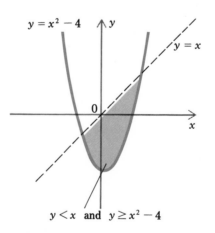

$y = x^2 - 4$

$y = x$

$y < x$ and $y \geq x^2 - 4$

FIGURE 3.70

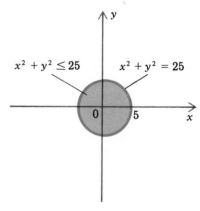

$x^2 + y^2 \leq 25$

$x^2 + y^2 = 25$

FIGURE 3.71

EXERCISES 3.9

In Exercises 1–18, sketch the graph of the relation.

1. $(4, 1), (2, \sqrt{2}), (-3, -5), (1, 1), (-3, \frac{1}{2})$
2. All (x, y) with $x = -2$ or $y = 3$
3. All (x, y) with $x \geq y - 2$
4. All (x, y) with $5x + 2y \leq 3$
5. All (x, y) with $3x - y < 1$
6. All (x, y) with $x < 4y^2$
7. All (x, y) with $x^2 > -y$
8. All (x, y) with $x^2 + y^2 > \frac{1}{4}$
9. All (x, y) with $x^2 + y^2 < 9$
10. All (x, y) with $x^2 + y^2 \geq 16$
11. All (x, y) with $(x - 2)^2 + (y - 3)^2 < 1$
12. All (x, y) with $x > 2$ and $y < 1$
13. All (x, y) with $|x| \leq 3$ and $|y| \leq 1$
14. All (x, y) with $|x| > |y|$
15. All (x, y) with $x > |y|$
16. All (x, y) with $0 \leq x < 2$ and $-1 \leq y \leq 3$
17. All (x, y) with $y \geq x^2$ and $y \leq 2x$
18. All (x, y) with $x^2 + y^2 \leq 4$ and $y \geq 2 - x^2$

KEY TERMS

coordinate axes
  x axis
  y axis
coordinates
  x coordinate (abscissa)
  y coordinate (ordinate)
origin
ordered pair
Cartesian coordinate system
quadrant
distance between points
function
  domain
  rule
  range
  even
  odd
  composite
  inverse

variable
  dependent variable
  independent variable
proportion
  directly proportional
  inversely proportional
  jointly proportional
  constant of proportionality
circle
  unit circle
relation
graph
  of a function
  of an equation
  of a relation
slope
point–slope equation
slope–intercept equation
linear equation

symmetry	intercept
with respect to the x axis	x intercept
with respect to the y axis	y intercept
with respect to the origin	

KEY FORMULAS

$$d(P, Q) = \sqrt{(x_2 - x_1)^2 + (y_2 - y_1)^2}$$
$$M(x, y) = (\tfrac{1}{2}(x_1 + x_2), \tfrac{1}{2}(y_1 + y_2))$$
$$m = \frac{y_2 - y_1}{x_2 - x_1}$$
$$f^{-1}(f(x)) = x \quad \text{and} \quad f(f^{-1}(y)) = y$$

REVIEW EXERCISES

1. Determine the distance between each of the following pairs of points.
   a. $(-1, -3)$ and $(-4, 3)$
   b. $(\sqrt{2}, \tfrac{1}{2}\sqrt{2})$ and $(3\sqrt{2}, -2\sqrt{2})$

2. For each of the following functions, determine which of the numbers $-2, 2, -4,$ and 4 are in the domain.
   a. $f(x) = \dfrac{x + 2}{x - 2}$ b. $f(x) = \dfrac{x + 2}{x^2 - 4}$

In Exercises 3–10, determine the domain of the function.

3. $f(x) = \dfrac{1}{x + 3}$ 4. $f(x) = \dfrac{x - 2}{x + 5} - \dfrac{x + 1}{x - 2}$

5. $g(x) = \dfrac{1}{x^2 + 4x - 7}$ 6. $g(x) = \sqrt{x + \pi}$

7. $h(x) = \sqrt{6x - 4}$ 8. $h(x) = |x - 3|$

9. $h(x) = \sqrt{|x| - 5}$

10. $f(x) = \begin{cases} x^2 & \text{for } x < -2 \\ 3 & \text{for } -2 < x \le 1 \\ \sqrt{x - 4} & \text{for } 5 < x \end{cases}$

11. Show that if $a \ne 0$, then the domain and range of $\sqrt{x + a}$ are distinct.

12. Let $f(x) = \dfrac{3x - 4}{2x + 5}$. Find
    a. $f(0)$ b. $f(-1)$ c. $f(\tfrac{4}{3})$ d. $f\left(-\dfrac{2}{x}\right)$

13. Let $f(x) = x^2 + 3ax - 5$, and suppose that $f(-2) = 1$. Determine a.

14. Let $f(t) = 1 - t^2$. Find $\dfrac{f(b) - f(a)}{b - a}$.

15. Let $f(x) = \dfrac{16}{x^2 - \sqrt{x^4 - 16}}$.
    a. Find the domain of f.
    b. Let $g(x) = x^2 + \sqrt{x^4 - 16}$. Show that $f = g$.

16. Let

$$f(x) = \sqrt{x+1} - \sqrt{x-2} \quad \text{and} \quad g(x) = \frac{3}{\sqrt{x+1} + \sqrt{x-2}}$$

Determine whether or not $f = g$.

In Exercises 17–26, sketch the graph of the function, noting any intercepts and symmetry.

17. $f(x) = 2x - 5$

18. $f(x) = -\frac{1}{2}x - \frac{3}{2}$

19. $f(x) = x^2 - 6$

20. $f(x) = -2x^2$

21. $f(x) = ||x| - 2|$

22. $f(x) = \frac{3}{x}$

23. $g(x) = \sqrt{-x}$

24. $g(x) = \sqrt{x+1}$

25. $g(x) = \begin{cases} x^2 + 4 & \text{for } x \le -1 \\ -4 - x^2 & \text{for } x \ge 1 \end{cases}$

26. $g(x) = \begin{cases} -x & \text{for } x < -1 \\ 2 & \text{for } x = -1 \\ 3x & \text{for } x \ge 0 \end{cases}$

In Exercises 27–34, sketch the graph of the equation, noting any intercepts and symmetry.

27. $(x + \sqrt{2})(y - \sqrt{3}) = 0$

28. $x^2 = y^2$

29. $|x - 4| = |y + 1|$

30. $x^2 + y^2 = 16$

31. $(x - 1)^2 = 4 - (y + 3)^2$

32. $7 + y^2 = x$

33. $y^2 = x + 3$

34. $x^2 + 4x - 12 = 0$

In Exercises 35–40, sketch the graph of the relation.

35. $xy > 1$

36. $-2 \le x \le 1$ and $-\frac{1}{2} \le y \le 3$

37. $1 \le x^2 < 2$

38. $|x| + |y| \le 2$

39. $x^2 + y^2 > 9$

40. $(x + 1)^2 + (y - 2)^2 \le 6$

In Exercises 41–46, find $f \circ g$ and $g \circ f$ for the given functions f and g.

41. $f(x) = \frac{25}{x^2}$ and $g(x) = 5x$

42. $f(x) = \frac{1}{x+2}$ and $g(x) = x - 2$

43. $f(x) = x^3$ and $g(x) = \sqrt{2x + 3}$

44. $f(x) = \frac{x+4}{x-3}$ and $g(x) = x^2 + 3$

45. $f(x) = |x|$ and $g(x) = \sqrt{x^2 - 4}$

46. $f(x) = \frac{2 - 3x}{4 + x} = g(x)$

47. Let $f(x) = \frac{x+1}{x-1}$. Show that $(f \circ f)(x) = x$ for all $x \ne 1$, and thus that f is its own inverse.

48. Let $f(x) = \frac{1}{x}$ and $g(x) = \frac{x^2 - 1}{x^2 + 1}$. Show that $(g \circ f)(x) = -g(x)$.

49. Let $f(x) = \sqrt{x^2 - 3}$ and $g(x) = \sqrt{x^2 + 9}$. Determine whether or not $f \circ g = g \circ f$. Explain your answer.

C 50. Let $f(x) = \sqrt{2 + x}$.
   a. Use a calculator to approximate the following numbers.
      i. $f(\sqrt{2})$ ii. $(f \circ f)(\sqrt{2})$ iii. $(f \circ f \circ f)(\sqrt{2})$

b. Can you guess the value that

$$\underbrace{(f \circ f \circ \cdots \circ f)}_{n \text{ of these}} (\sqrt{2})$$

approaches as n increases without bound?

In Exercises 51–58, either find a formula for the inverse of the given function or explain why there is none.

51. $f(x) = \dfrac{1}{x + 3}$

52. $f(x) = \dfrac{5}{x^{1/3} + 4}$

53. $f(x) = \dfrac{4x - 1}{3x + 2}$

54. $f(x) = \dfrac{1}{x^2 + 1}$

55. $f(x) = \dfrac{1}{x^2 - 1}$

56. $f(x) = \sqrt{x + 3}$

57. $f(x) = \sqrt{x - 2}$

58. $f(x) = \sqrt{x^2 + 4}$

59. Determine the distance between the points $(-\sqrt{a}, -\sqrt{b})$ and $(\sqrt{a}, \sqrt{b})$.

60. Find the points on the line $y = 2x + 1$ that are 1 unit from the point $(0, 3)$.

61. Find the points on the line $x - 2y = 3$ that are 5 units from the point $(7, 2)$.

62. Find an equation of all points (x, y) that are 4 units from the point $(-2, -3)$.

63. Find an equation of the circle centered at $(\frac{1}{2}, -\frac{1}{4})$ and passing through $(0, -\frac{1}{4} + \frac{3}{4}\sqrt{7})$.

64. A line l has x intercept -2 and y intercept -4.
    a. Write the slope–intercept equation of l.
    b. Write a point–slope equation of l.

65. Find an equation of the line that is parallel to the line $x + 2y = 3$ and passes through the point $(-1, -3)$.

66. Find an equation of the line that is perpendicular to the line $3x - 2y = 6$ and passes through the point $(4, 6)$.

67. Find an equation of the line consisting of all points (x, y) that are equidistant from the points $(3, 1)$ and $(-1, 4)$.

68. Determine whether or not the points $(-1, -1)$, $(1, 3)$, and $(57, 115)$ lie on a straight line.

69. A converse of the Pythagorean Theorem states that if the lengths a, b, and c of the sides of a triangle satisfy the equation $c^2 = a^2 + b^2$, then the triangle is a right triangle. Use this result to determine whether or not the three points $(2, 1)$, $(1, -2)$, and $(-2, 3)$ form the vertices of a right triangle.

70. Let M be the midpoint of the line segment joining the points $(-2, 1)$ and $(-1, 3)$. Find the distance between M and the origin.

71. Prove that the midpoint of the hypotenuse of a right triangle is equidistant from all three vertices. (*Hint:* Let the two legs lie on the coordinate axes.)

72. Prove that the line joining the midpoints of two sides of a triangle is

parallel to the third side. (*Hint:* Set up the coordinate system so that the vertices are $P_1(0, 0)$, $P_2(a, 0)$, and $P_3(b, c)$.)

73. Prove that the points $(-2, 2)$, $(-1, -1)$, $(1, 3)$, and $(2, 0)$ are the vertices of a square.

74. A cylindrical can has a height of 5 inches, and a top and bottom. In terms of the radius, the surface area S of the can is given by

$$S = 2\pi(r^2 + 5r) \quad \text{for } r \geq 0$$

Write a formula for the radius in terms of the surface area. (*Hint:* $r^2 + 5r = (r + \frac{5}{2})^2 - \frac{25}{4}$.)

75. The illumination L from a lamp is directly proportional to the intensity I of the light bulb and is inversely proportional to the square of the distance D from the lamp.
    a. Express this fact by means of a formula.
    b. If one moves from 12 feet to 4 feet away from a light source, how is the illumination of the person affected?

76. The safe load L of a horizontal beam supported at both ends is jointly proportional to the width w and the square of the depth d and is inversely proportional to the length l of the beam.
    a. Express this fact by means of a formula.
    b. If the safe load on such a beam 2 inches wide, 8 inches deep, and 10 feet long is 1000 pounds, find the safe load on a beam 4 inches wide, 10 inches deep, and 12 feet long.

POLYNOMIAL AND RATIONAL FUNCTIONS

4

Any polynomial expression defines a function whose domain consists of all real numbers. For example, the polynomial expression $2x^4 - x^3 + 3x^2 + \sqrt{2}$ defines the function f whose rule is given by

$$f(x) = 2x^4 - x^3 + 3x^2 + \sqrt{2}$$

In general, if n is a nonnegative integer and $a_n, a_{n-1}, a_{n-2}, \ldots, a_1$, and a_0 are constants with $a_n \neq 0$, then a function f is defined by the formula

$$f(x) = a_n x^n + a_{n-1} x^{n-1} + a_{n-2} x^{n-2} + \cdots + a_1 x + a_0$$

Such a function is called a ***polynomial function***. This chapter will be mainly devoted to the graphs of polynomial functions and quotients of polynomial functions.

4.1

QUADRATIC FUNCTIONS

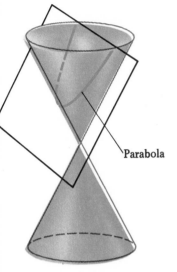

Because they have the lowest degree, constant functions and linear functions are the simplest polynomial functions. Nonzero constant functions have degree 0, and linear functions have degree 1. The next simplest kind of polynomial functions are those of degree 2, called *quadratic functions*. A quadratic function is usually given in the form

$$f(x) = ax^2 + bx + c$$

where a, b, and c are constants with $a \neq 0$. The graph of a quadratic function is called a *parabola*, and this type of curve has been famous since antiquity as one of those that can result when a cone is sliced by a plane (Figure 4.1). The emphasis in this section will be on sketching parabolas.

Parabola

Let f and g be the quadratic functions defined by

$$f(x) = x^2 \quad \text{and} \quad g(x) = -x^2$$

The graphs of these functions were sketched in Section 3.4 and are reproduced in Figure 4.2a, b. Notice that the graph of x^2 is cupped upward, whereas the graph of $-x^2$ is cupped downward. Both graphs are symmetric with respect to the y axis, which is referred to as the *axis* of each parabola. The point $(0, 0)$ is the lowest point on the graph of f and the highest point on the graph of g; it is called the *vertex* of each parabola.

FIGURE 4.1

If

$$f(x) = ax^2 \quad \text{with } a \neq 0$$

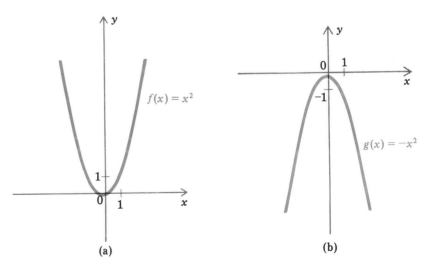

(a)

(b)

FIGURE 4.2

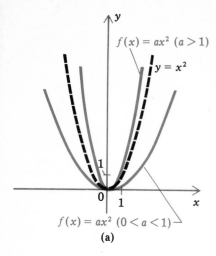

$f(x) = ax^2 \ (a > 1)$

$y = x^2$

$f(x) = ax^2 \ (0 < a < 1)$

(a)

$f(x) = ax^2$
$(-1 < a < 0)$

$y = -x^2$

$f(x) = ax^2$
$(a < -1)$

(b)

FIGURE 4.3

then the graph of f has the same general shape as that of x^2 or $-x^2$, depending on whether $a > 0$ or $a < 0$. However, the graph is flatter if $|a| < 1$ and is more pointed if $|a| > 1$ (Figure 4.3a, b).

Example 1.

a. Let $f(x) = 2x^2$. Sketch the graph of f.
b. Let $g(x) = -\frac{1}{3}x^2$. Sketch the graph of g.

Solution.

a. The y coordinate of each point on the graph of f is twice the y coordinate of the corresponding point on the graph of x^2 (see Figure 4.2a). This leads us to the sketch in Figure 4.4a, with a few points plotted for assistance.

b. The y coordinate of each point on the graph of g is $\frac{1}{3}$ the y coordinate of the corresponding point on the graph of $-x^2$ (see Figure 4.2b). This leads us to the sketch in Figure 4.4b, again with a few points plotted for assistance. □

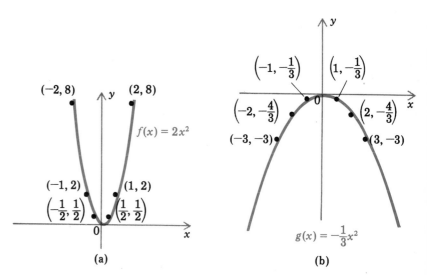

(a)

(b)

FIGURE 4.4

Vertical Shifts in Graphs

Next we let g be a given function, and let

$$f(x) = g(x) + k$$

with k a fixed nonzero number. Since $f(x)$ is obtained from $g(x)$ by adding k, it follows that the graph of f is k units above the graph of g if $k > 0$, and is $-k$ units below the graph of g if $k < 0$ (Figure 4.5). In either case, the graph of f is obtained by shifting the graph of g vertically. If g happens to be a quadratic function, then its graph is a parabola. When we shift the graph of g, the axis of the parabola remains the y axis, but the vertex shifts from the origin to $(0, k)$. It is

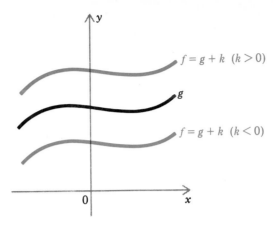

FIGURE 4.5

then immediate that the y intercept of the graph of f is k. In contrast, the existence of x intercepts for the graph of f depends on the sign of k. As the next two examples illustrate, there are two x intercepts if $k < 0$ and no x intercepts if $k > 0$.

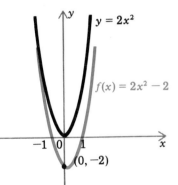

FIGURE 4.6

Example 2. Let $f(x) = 2x^2 - 2$. Sketch the graph of f, and determine the vertex and the x intercepts of the parabola.

Solution. We have $k = -2$, so the graph of f is obtained by shifting the graph of $2x^2$ down 2 units (Figure 4.6). Thus the vertex is $(0, -2)$. For the x intercepts we solve the equation $f(x) = 0$ for x:

$$2x^2 - 2 = 0$$
$$2x^2 = 2$$
$$x^2 = 1$$

Therefore

$$x = 1 \quad \text{or} \quad x = -1$$

Consequently the x intercepts are 1 and -1.   □

Example 3. Let $y = 2x^2 + \frac{1}{2}$. Sketch the graph of the equation, and determine the vertex and any x intercepts of the parabola.

FIGURE 4.7

Solution. In this case $k = \frac{1}{2}$, so the graph of the equation is obtained by shifting the graph of $2x^2$ up $\frac{1}{2}$ unit (Figure 4.7). Thus the vertex is $(0, \frac{1}{2})$. Concerning the x intercepts, we observe that

$$y = 2x^2 + \frac{1}{2} > 0 \quad \text{for all } x$$

so there are no x intercepts.   □

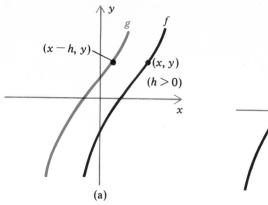 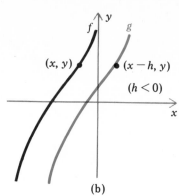

(a)   (b)

FIGURE 4.8

**Horizontal Shifts
in Graphs**

Now we study horizontal shifts of a graph. Let g be a given function, and let

$$f(x) = g(x - h)$$

where h is a fixed nonzero number. Then $y = f(x)$ if and only if $y = g(x - h)$, so (x, y) is on the graph of f if and only if $(x - h, y)$ is on the graph of g. But observe that the point (x, y) lies h units to the right of the point $(x - h, y)$ if $h > 0$ and $-h$ units to the left if $h < 0$. We conclude that if $h > 0$, then the graph of f is h units to the right of the graph of g, and if $h < 0$, then the graph of f is $-h$ units to the left of the graph of g (Figure 4.8).

Example 4. Let $f(x) = (x - 2)^2$. Sketch the graph of f.

Solution. Here $h = 2 > 0$, so by the comments above, the graph of f is 2 units to the right of the graph of x^2 (Figure 4.9a). $\square$

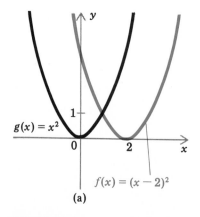 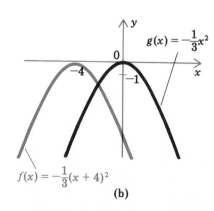

(a)   (b)

FIGURE 4.9

Example 5. Let $f(x) = -\frac{1}{3}(x + 4)^2$. Sketch the graph of f.

Solution. Since $x + 4 = x - (-4)$, we have $h = -4 < 0$, so by the comments above, the graph of f is 4 units to the left of the graph of $-\frac{1}{3}x^2$ (see Figure 4.4b). The graph of f is sketched in Figure 4.9b. □

The Graph of a General Quadratic Function

Suppose that we have an arbitrary quadratic function, given by

$$f(x) = ax^2 + bx + c \quad \text{with } a \neq 0 \tag{1}$$

In order to sketch the graph of f, we complete the square in $ax^2 + bx + c$, as we did in Section 2.3, and obtain

$$f(x) = a\left(x + \frac{b}{2a}\right)^2 + c - \frac{b^2}{4a} \tag{2}$$

At first glance it seems no easier to draw the graph of f from (2) than from (1). However, let us simplify the right side of (2) by letting

$$h = -\frac{b}{2a} \quad \text{and} \quad k = c - \frac{b^2}{4a} \tag{3}$$

Then (2) becomes

$$f(x) = a(x - h)^2 + k$$

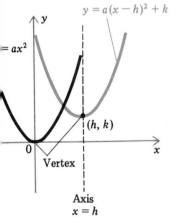

$y = a(x - h)^2 + k$, $y = ax^2$, (h, k), Vertex, Axis $x = h$

FIGURE 4.10

The graph of $a(x - h)^2 + k$ is obtained from the graph of ax^2 by a vertical shift of $|k|$ units (upward if $k > 0$ and downward if $k < 0$) and a horizontal shift of $|h|$ units (to the right if $h > 0$ and to the left if $h < 0$) (Figure 4.10). For example, if

$$f(x) = -\frac{1}{3}(x + 4)^2 + 2$$

then the graph of f can be obtained by shifting the graph of $-\frac{1}{3}x^2$ upward 2 units and to the left 4 units.

When we shift from the graph of ax^2 to the graph of $a(x - h)^2 + k$, the vertex $(0, 0)$ of ax^2 is shifted to the point (h, k) and the axis $x = 0$ is shifted to the line $x = h$ (Figure 4.10). Thus (h, k) is called the **vertex** of the graph of $a(x - h)^2 + k$, and the line $x = h$ is called the **axis** of the graph. The vertex is the lowest point on the graph if $a > 0$ and is the highest point if $a < 0$.

Example 6. Let $f(x) = 2(x + \frac{1}{2})^2 + 1$. Sketch the graph of f, and locate the vertex of the parabola.

Solution. The graph of f can be obtained by shifting the graph of $2x^2$ upward 1 unit and to the left $\frac{1}{2}$ unit (Figure 4.11). The vertex is $(-\frac{1}{2}, 1)$. □

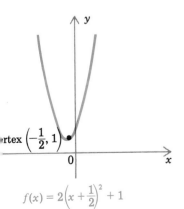

vertex $\left(-\frac{1}{2}, 1\right)$, $f(x) = 2\left(x + \frac{1}{2}\right)^2 + 1$

FIGURE 4.11

Let us return to the formula

$$f(x) = ax^2 + bx + c$$

and use the results of the preceding discussion to outline a procedure for sketching the graph of f. As is usually the case, it helps to plot a few points, so our outline includes finding the intercepts.

  i. Find the y intercept by calculating $f(0)$. Plot the corresponding point.
 ii. Find the x intercepts (if any) by solving $f(x) = 0$ for x. Plot the corresponding points (if any).
iii. Complete the square of the right side to express $f(x)$ in the form

$$f(x) = a(x - h)^2 + k \qquad (4)$$

 iv. From the new equation notice that the graph of f is a parabola whose vertex is (h, k) and whose axis is $x = h$. Plot the vertex, (h, k).
  v. Obtain the graph of f by shifting the graph of ax^2 vertically $|k|$ units (upward if $k > 0$ and downward if $k < 0$) and horizontally $|h|$ units (to the right if $h > 0$ and to the left if $h < 0$). The graph will contain the points already plotted and will be cupped upward if $a > 0$ and downward if $a < 0$.

From (3) it follows that the vertex and axis of the parabola are given (in terms of the original coefficients a, b, and c) by

$$\text{Vertex: } \left(-\frac{b}{2a}, c - \frac{b^2}{4a} \right)$$

$$\text{Axis: } \quad x = -\frac{b}{2a}$$

But the vertex and axis are much simpler to find after we have completed the square. As a result, when we set out to sketch the graph of a quadratic function, we usually first complete the square, and then sketch the graph by reading off the information we need from the expression for $f(x)$.

Example 7. Let $f(x) = x^2 + 2x - 3$. Sketch the graph of f, and locate the vertex of the parabola.

Solution. Since $f(0) = -3$, the y intercept is -3. To find the x intercepts we solve the equation $f(x) = 0$ for x:

$$x^2 + 2x - 3 = 0$$

$$(x + 3)(x - 1) = 0$$

Consequently

$$x = -3 \quad \text{or} \quad x = 1$$

Therefore the x intercepts are -3 and 1. Now we complete the square:

$$
\begin{aligned}
f(x) &= (x^2 + 2x) - 3 \\
&= (x^2 + 2x + 1 - 1) - 3 \\
&= (x^2 + 2x + 1) - 1 - 3 \\
&= (x^2 + 2x + 1) - 4 \\
&= (x + 1)^2 - 4
\end{aligned}
$$

In the notation of (4) we have $a = 1$, $h = -1$, and $k = -4$. Thus we obtain the graph of f by shifting the graph of x^2 downward 4 units and to the left 1 unit (Figure 4.12). ☐

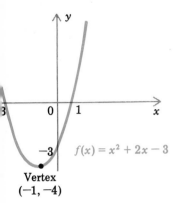

$f(x) = x^2 + 2x - 3$

Vertex
$(-1, -4)$

FIGURE 4.12

Example 8. Sketch the graph of

$$y = -3x^2 + 12x - 8 \tag{5}$$

and locate the vertex of the parabola.

Solution. Since $y = -8$ for $x = 0$, the y intercept is -8. The x intercepts are obtained by setting $y = 0$ and solving for x by the quadratic formula:

$$-3x^2 + 12x - 8 = 0$$

$$
\begin{aligned}
x &= \frac{-12 \pm \sqrt{(12)^2 - 4(-3)(-8)}}{2(-3)} \\
&= \frac{-12 \pm \sqrt{48}}{-6} = \frac{-12 \pm 4\sqrt{3}}{-6} = 2 \pm \frac{2}{3}\sqrt{3}
\end{aligned}
$$

Therefore the x intercepts are $2 + \frac{2}{3}\sqrt{3}$ and $2 - \frac{2}{3}\sqrt{3}$. Completing the square in (5), we find that

$$
\begin{aligned}
y &= -3x^2 + 12x - 8 \\
&= -3(x^2 - 4x) - 8 \\
&= -3(x^2 - 4x + 4 - 4) - 8 \\
&= -3(x^2 - 4x + 4) + 12 - 8 \\
&= -3(x - 2)^2 + 4
\end{aligned}
$$

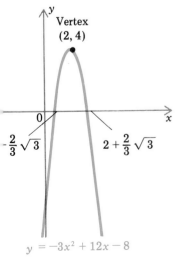

Vertex
$(2, 4)$

$-\frac{2}{3}\sqrt{3}$

$2 + \frac{2}{3}\sqrt{3}$

$y = -3x^2 + 12x - 8$

FIGURE 4.13

Using the notation of (4), we have $a = -3$, $h = 2$, and $k = 4$. Thus we obtain the graph of the given equation by shifting the graph of $-3x^2$ upward 4 units and to the right 2 units (Figure 4.13). ☐

In Section 4.2 we will continue our study of quadratic functions. However, the main point of interest will be the largest, or smallest, value assumed by such a function.

EXERCISES 4.1

In Exercises 1–20, sketch the graph of the given function. In each case, determine the vertex and all intercepts of the parabola.

1. $f(x) = \frac{3}{4}x^2$

2. $f(x) = -3x^2$

3. $f(x) = x^2 + 1$ •

4. $f(x) = x^2 - \frac{9}{4}$

5. $f(x) = -2x^2 + 8$

6. $f(x) = 1 - 4x^2$

7. $f(x) = -\frac{3}{2}(x - 1)^2$

8. $f(x) = \frac{1}{2}(x + 3)^2$

9. $f(x) = \frac{3}{2}(x + \frac{1}{3})^2 - \frac{1}{3}$

10. $f(x) = 2(x - \frac{1}{2})^2 + \frac{3}{2}$

11. $g(x) = x^2 + 2x$

12. $g(x) = x^2 - 2x$

13. $g(x) = x^2 - 2x + 2$

14. $f(x) = x^2 + 2x + 2$

15. $f(x) = -x^2 + 6x - 5$

16. $f(x) = 2x^2 + 4x + 5$

17. $f(x) = 1 - 2x - 3x^2$

18. $y = \frac{1}{2}x^2 + x + 1$

19. $y = (x + 1)(2x - 1)$

20. $y = (3x + 1)(x - 1)$

In Exercises 21–24, find an equation of the sketched parabola. (*Hint:* Use (4).)

21. The parabola in Figure 4.14a.

22. The parabola in Figure 4.14b.

23. The parabola in Figure 4.14c.

24. The parabola in Figure 4.14d.

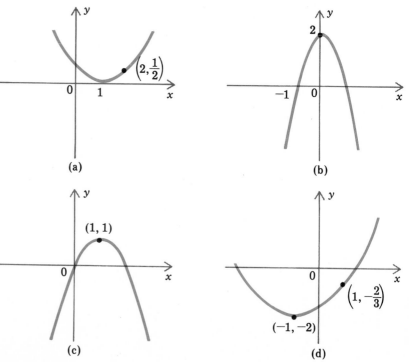

FIGURE 4.14

25. Let $f(x) = -2x^2 + 3$. Find an equation of the function g whose graph is obtained from the graph of f by
    a. shifting upward 1 unit
    b. shifting downward 3 units
    c. shifting to the right $\frac{1}{2}$ unit
    d. shifting to the left 4 units

26. Let $g(x) = -x^2 + 2x - 5$. Find an equation of the function f whose graph is obtained from the graph of g by
    a. shifting upward $\frac{3}{2}$ units and to the left 4 units
    b. shifting downward 2 units and to the right $\frac{1}{2}$ unit

27. For what value of a does the graph of the equation $y = ax^2 + 6x + 3$ pass through the point $(-1, 0)$?

28. For what value of b does the graph of the equation $y = -\frac{1}{2}x^2 + bx + \frac{1}{3}$ pass through the point $(-2, -1)$?

29. For what value of c does the graph of the equation $y = -x^2 + x + c$ pass through the point $(2, 6)$?

30. In each part find the value of c for which the graph of f touches but does not cross the x axis.
    a. $f(x) = \frac{1}{2}x^2 - 2x + c$ b. $f(x) = x^2 - 3x + c$
    c. $f(x) = 3x^2 + 10x + c$

31. Let $f(x) = ax^2 + c$. Prove that the graph of f has
    a. two x intercepts if a and c have opposite signs
    b. no x intercept if a and c have the same sign

32. Let $f(x) = 2x^2 - 8x + 3$. Determine the points on the graph of f that lie on the horizontal line 18 units above the vertex.

33. A parabolic arch is 16 feet tall at its highest point and spans 36 feet at its base. Find its height 9 feet from the plane on which its axis is located.

4.2

MAXIMUM AND MINIMUM VALUES OF QUADRATIC FUNCTIONS

From Section 4.1 we know that the graph of any quadratic function f is a parabola. If

$$f(x) = ax^2 + bx + c \quad \text{with } a \neq 0 \tag{1}$$

then we also know that the vertex of the parabola is

$$\left(-\frac{b}{2a}, \ c - \frac{b^2}{4a}\right) \tag{2}$$

The vertex is the lowest point on the graph if $a > 0$ and is the highest point if $a < 0$. It follows that

$$\text{if } a > 0, \quad \text{then} \quad f\left(-\frac{b}{2a}\right) \leq f(x) \quad \text{for all } x$$

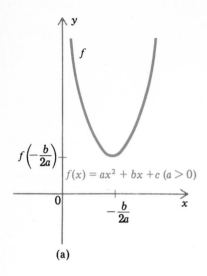

(a)

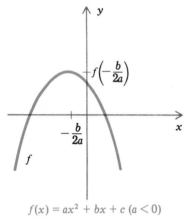

$f(x) = ax^2 + bx + c \ (a < 0)$

(b)

FIGURE 4.15

so $f(-b/2a)$ is called the **minimum value** of f (Figure 4.15a). Analogously,

$$\text{if } a < 0, \quad \text{then} \quad f\left(-\frac{b}{2a}\right) \geq f(x) \quad \text{for all } x$$

and $f(-b/2a)$ is the **maximum value** of f (Figure 4.15b). Thus every quadratic function has either a minimum or a maximum value, depending on the sign of the leading coefficient (a in (1)).

To compute the minimum or maximum value of such a function f, we can calculate $f(-b/2a)$, either by substituting $-b/2a$ for x in (1) or by calculating $c - b^2/4a$ (see (2)).

Example 1. Let $f(x) = 2x^2 - 3x + 1$. Verify that f assumes a minimum value, and compute the minimum value.

Solution. In this example $f(x)$ has the form of (1) with $a = 2$, $b = -3$, and $c = 1$. Since $a > 0$, it follows that $f(-b/2a)$ is the minimum value. Now

$$-\frac{b}{2a} = -\frac{(-3)}{2(2)} = \frac{3}{4}$$

Consequently to find the minimum value of $f(x)$, we calculate $f(\tfrac{3}{4})$:

$$f\left(\frac{3}{4}\right) = 2\left(\frac{3}{4}\right)^2 - 3\left(\frac{3}{4}\right) + 1 = \frac{9}{8} - \frac{9}{4} + 1 = -\frac{1}{8}$$

Therefore the minimum value of f is $-\tfrac{1}{8}$. □

Example 2. Let $f(x) = -x^2 + \sqrt{2}x + \tfrac{1}{2}$. Verify that f assumes a maximum value, and compute the maximum value.

Solution. Here $f(x)$ has the form of (1) with $a = -1$, $b = \sqrt{2}$, and $c = \tfrac{1}{2}$. Since $a < 0$, $f(-b/2a)$ is the maximum value of f. Next,

$$-\frac{b}{2a} = -\frac{\sqrt{2}}{2(-1)} = \frac{\sqrt{2}}{2}$$

Therefore to find the maximum value of f we calculate $f(\sqrt{2}/2)$:

$$f\left(\frac{\sqrt{2}}{2}\right) = -\left(\frac{\sqrt{2}}{2}\right)^2 + \sqrt{2}\left(\frac{\sqrt{2}}{2}\right) + \frac{1}{2} = -\frac{1}{2} + 1 + \frac{1}{2} = 1$$

Consequently the maximum value of f is 1. □

Applications

In applications it is frequently desirable to maximize or minimize a variable quantity, such as area, profit, or cost. If the quantity to be maximized or minimized can be expressed as a quadratic function of another quantity, the discussion in this section provides us with a method for finding the maximum or minimum value.

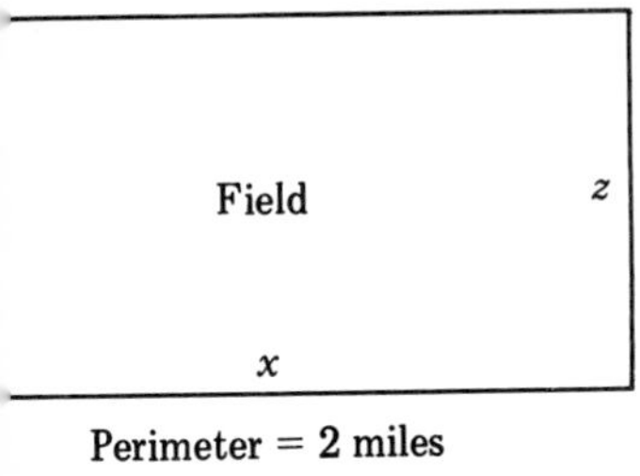

Example 3. A rancher has 2 miles of fencing and wishes to fence in a rectangular grazing field with area as large as possible. What should the dimensions of the rectangle be?

Field z

x

Perimeter = 2 miles

FIGURE 4.16

Solution. Since the rancher has 2 miles of fencing, the perimeter of the rectangle will be 2 miles. Let x and z be the dimensions in miles of any rectangle with a perimeter of 2 miles (Figure 4.16). Then

$$2x + 2z = 2$$

so that

$$x + z = 1$$

or equivalently,

$$z = 1 - x$$

This implies that the area xz of the rectangle can be rewritten solely in terms of x:

$$xz = x(1 - x) = x - x^2 = -x^2 + x$$

If we let

$$f(x) = -x^2 + x \quad \text{for} \quad 0 \le x \le 1$$

then $f(x)$ represents the area of a field one of whose sides has length x. Thus we wish to maximize $f(x)$. Now $f(x)$ has the form of (1) with $a = -1, b = 1$, and $c = 0$. Since $a < 0$, we know that there is a maximum value and that the maximum value occurs for

$$x = -\frac{b}{2a} = -\frac{1}{2(-1)} = \frac{1}{2}$$

But if $x = \frac{1}{2}$, then

$$z = 1 - \frac{1}{2} = \frac{1}{2}$$

Thus the maximum area results if the rectangle is a square with sides of length $\frac{1}{2}$ mile. □

Example 4. On a rainy day the manager of a store estimates that she can sell 20 umbrellas at $9 each, and that for each nickel decrease in price, one more umbrella can be sold (which corresponds to 20 more umbrellas sold for each dollar decrease in price). Assuming that this is true and that she acquires umbrellas at a cost of $3 each, determine the price at which she should sell umbrellas in order to maximize her profit. How many umbrellas will be sold at that price?

Solution. We will express all monetary quantities in dollars. Let x be the price of each umbrella and z the number of umbrellas sold. It follows that the amount that the original price of $9 is decreased is $9 - x$, and thus according to our assumptions, the sales over and above 20 umbrellas is $20(9 - x)$. Therefore the number of umbrellas sold is given by

$$z = 20 + 20(9 - x)$$

or equivalently,

$$z = 200 - 20x$$

Now the revenue R resulting from the sale of the z umbrellas at the price x is given by

$$R = xz = x(200 - 20x) = 200x - 20x^2$$

Since each umbrella costs $3, the cost C of the z umbrellas sold is given by

$$C = 3z = 3(200 - 20x) = 600 - 60x$$

As a result, the profit P is given by

$$P = R - C = (200x - 20x^2) - (600 - 60x)$$
$$= -20x^2 + 260x - 600$$

Thus if we let

$$f(x) = -20x^2 + 260x - 600$$

then we wish to find the maximum value of f. Since $f(x)$ has the form of (1) with $a = -20$, $b = 260$, and $c = -600$, it follows that the maximum value of f occurs for

$$x = -\frac{b}{2a} = -\frac{260}{2(-20)} = 6.5$$

Consequently each umbrella should be priced at $6.50 in order to maximize profit. If $x = 6.5$, then

$$z = 200 - 20(6.5) = 70$$

so 70 umbrellas will be sold at the optimal price of $6.50. □

Example 5. Find the point on the line $y = 3x$ that is closest to the point $(5, 0)$.

Solution. The distance between a point (x, y) on the line $y = 3x$ and the point $(5, 0)$ is given by

$$\sqrt{(x - 5)^2 + (3x - 0)^2}$$

so we let

$$
\begin{aligned}
f(x) &= \sqrt{(x - 5)^2 + (3x - 0)^2} \\
&= \sqrt{(x^2 - 10x + 25) + 9x^2} \\
&= \sqrt{10x^2 - 10x + 25}
\end{aligned}
$$

We desire to find the minimum value of $f(x)$. However, f is not a quadratic function. But if we let

$$g(x) = 10x^2 - 10x + 25$$

then since $g(x) = (f(x))^2$, the minimum value of f occurs for the same number x as the minimum value of g. Now g is a quadratic function, and $g(x)$ has the form of (1) with $a = 10$, $b = -10$, and $c = 25$. Since $a > 0$, there is a minimum value, which occurs for

$$x = -\frac{b}{2a} = -\frac{-10}{2(10)} = \frac{1}{2}$$

Therefore $y = 3(\frac{1}{2}) = \frac{3}{2}$, so the closest point is $(\frac{1}{2}, \frac{3}{2})$. $\square$

Notice that in the preceding example, we did not need to find the minimum value of f, but only the number x at which the minimum value occurs.

Problems concerning the motion of an object moving vertically under the sole influence of gravity involve quadratic functions. Recall from (1) in Section 2.5 that the height at time t of an object moving solely under the influence of gravity is given by

$$h(t) = -16t^2 + v_0 t + h_0 \tag{3}$$

where v_0 is the initial velocity and h_0 is the initial height. Since h is a quadratic function, if we know the initial velocity and initial height of an object, we can determine the maximum height.

Example 6. A tape measure is thrown vertically upward from an initial height of 6 feet with an initial velocity of 32 feet per second toward a third floor balcony 21 feet above ground. Does the tape measure reach the balcony?

Solution. Taking $v_0 = 32$ and $h_0 = 6$ in (3), we have

$$h(t) = -16t^2 + 32t + 6$$

which has the form of (1) with $a = -16$, $b = 32$, and $c = 6$, and with t and h replacing x and f, respectively. Since $a < 0$, we know that h assumes a maximum value for

$$t = -\frac{b}{2a} = -\frac{32}{2(-16)} = 1$$

Thus the tape measure attains its maximum height after 1 second. The fact that

$$h(1) = -16(1)^2 + 32(1) + 6$$
$$= -16 + 32 + 6 = 22$$

implies that the maximum height attained by the tape measure is 22 feet. Therefore the tape measure does reach the balcony. □

In solving maximum and minimum problems, the following approach is frequently effective:

i. After reading the problem carefully, choose a letter for the quantity to be maximized or minimized. Also choose auxiliary variables for the other quantities appearing in the problem.
ii. Express the quantity to be maximized or minimized in terms of the auxiliary variables.
iii. Choose one auxiliary variable x to serve as master variable, and use the information given in the problem to express all other auxiliary variables in terms of x.
iv. Use the results of steps (ii) and (iii) to express the quantity to be maximized or minimized in terms of x alone.
v. Use the theory of this section to find the desired maximum or minimum value.

EXERCISES 4.2

In Exercises 1–10, determine whether the given quadratic function assumes a maximum value or a minimum value. Then find the maximum or the minimum value.

1. $f(x) = 3x^2 + 5$
2. $f(x) = x^2 + x - \frac{1}{2}$
3. $f(x) = 20x^2 - 40x$
4. $f(x) = -x^2 + 2x + 3$
5. $g(x) = x - \frac{1}{4}x^2$
6. $g(x) = -5x^2 - x + 1$
7. $g(t) = 16t^2 + 64t + 36$
8. $g(t) = -\frac{1}{4}t^2 + t - 13$
9. $y = t^2 + 3t$
10. $y = \frac{1}{2}x^2 - 3x + \frac{3}{2}$

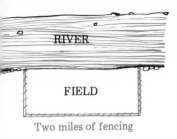

RIVER

FIELD

Two miles of fencing

FIGURE 4.17

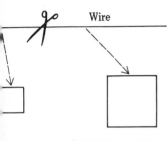

Wire

FIGURE 4.18

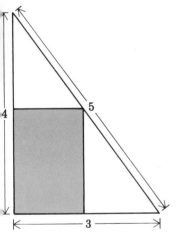

4

5

3

FIGURE 4.19

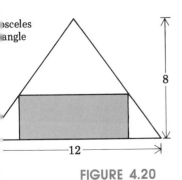

osceles
angle

8

12

FIGURE 4.20

11. A rancher has 2 miles of fencing and wishes to fence in a rectangular grazing field with an area as large as possible by erecting a fence on only 3 sides, a river forming the other side (Figure 4.17). What should the dimensions of the rectangle be?

12. A farmer plans to fence in two adjacent rectangular grazing fields of the same dimensions, one for cows and the other for horses, with a common fence separating the two fields. If the farmer has 900 meters of fencing, how large can the area of each field be?

13. A piece of wire 1 meter long is to be cut into two pieces, each of which is then to be bent into the shape of a square (Figure 4.18). How should the wire be cut if the sum of the areas of the two squares is to be minimized?

14. Find the dimensions of the largest rectangle that can be inscribed in the triangle sketched in Figure 4.19.

15. A rectangle is to have a perimeter of 12 inches. Determine the lengths of the sides that will yield the largest area.

16. The base of a given isosceles triangle is 12, and the height is 8. Find the dimensions of the rectangle with maximum area that can be inscribed in the triangle if the base of the rectangle is to lie along the base of the triangle (Figure 4.20).

17. Find the dimensions of the rectangle whose perimeter is 16 and whose diagonal is as small as possible.

18. Find two numbers whose sum is 24 and whose product is as large as possible.

19. Find two numbers whose sum is -16 and whose product is as large as possible.

20. Find two numbers whose difference is 6 and whose product is as small as possible.

21. Find two numbers whose product is as large as possible, under the condition that the sum of one of the numbers and 3 times the other is 48.

22. Find two numbers whose sum is 36 and the sum of whose squares is as small as possible. Then determine the sum of the two squares.

23. Find the point on the line $2x + y = 2$ that is closest to the point $(0, -3)$.

24. Find the point on the line $y = 10 - 3x$ that is closest to the origin.

25. Find the points on the parabola $y = x^2$ that are closest to the point $(0, 1)$. (*Hint:* The square of the distance is a quadratic function of x^2. First find the value of x^2 for which the distance is minimal.)

26. The Hot-Cake Company can sell 100 pounds of pancake mix a day at $1.80 a pound, and it believes that sales will decrease $\frac{1}{5}$ pound for each penny increase in price per pound. For what price would the revenue be maximized? What will the corresponding revenue be?

27. The Fuzzy-Wuzzy Fabric Shop can sell 36 yards of denim per day if it charges $2 per yard, and it figures that each increase of

$.25 per yard will reduce sales of denim by 3 yards. Determine the price per yard that will maximize the daily revenue from the sale of denim.

28. A construction company estimates that it can build and sell x houses in a certain development if it sets the price of each home at $100{,}000 - 1000x$ dollars. If each house costs \$60,000 to construct, how many houses must the company sell to maximize its profit? How much must the company charge for each house in order to maximize its profit?

29. Suppose a baseball is thrown vertically upward from a height of 64 feet with an initial velocity of 48 feet per second. Find the maximum height attained by the baseball and the length of time it takes the ball to attain that height.

4.3

GRAPHS OF HIGHER-DEGREE POLYNOMIAL FUNCTIONS

In Section 4.1 we discussed the graph of a quadratic function. The graph is a parabola, and with a minimum of calculation we can determine its vertex, intercepts, and general shape. Now we turn to higher-degree polynomial functions. The higher the degree, the less accurate our graph is apt to be. Nevertheless, there are techniques that allow us to sketch the general shape of many such functions.

First we will analyze the graphs of polynomials of the form ax^n, and then we will study the graphs of other higher-degree polynomial functions.

Example 1. Let $f(x) = x^3$. Sketch the graph of f.

Solution. We begin by observing that

$$f(-x) = (-x)^3 = (-1)^3 x^3 = -x^3 = -f(x)$$

which means that the graph of f is symmetric with respect to the origin. Next, we plot a few points on the graph, using the following table:

x	0	$\frac{1}{2}$	1	$\frac{3}{2}$	2
$f(x)$	0	$\frac{1}{8}$	1	$\frac{27}{8}$	8

Now we connect the points with a smooth curve and then use the symmetry mentioned above to finish the graph (Figure 4.21).   □

Example 2. Let $f(x) = x^4$. Sketch the graph of f.

Solution. Since

$$f(-x) = (-x)^4 = (-1)^4 x^4 = x^4 = f(x)$$

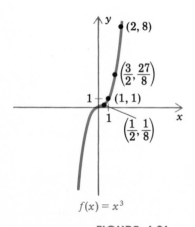

$f(x) = x^3$

FIGURE 4.21

the graph of f is symmetric with respect to the y axis. We plot a few points on the graph, using the following table:

x	0	$\frac{1}{2}$	1	$\frac{3}{2}$	2
$f(x)$	0	$\frac{1}{16}$	1	$\frac{81}{16}$	16

Next we draw a smooth curve through the points we have plotted, and finally we use the symmetry of the graph to obtain the sketch in Figure 4.22. □

The shapes of the graphs of x^2 and x^4 appear generally similar (Figure 4.23), and in fact the shapes of the graphs of x^2 and x^n are generally similar for any *positive even* integer n. More particularly, they are symmetric with respect to the y axis (since $(-x)^n = x^n$ if n is even), have their minimum value of 0 at 0, and lie above the x axis (because $x^n \geq 0$ if n is even).

A similar pattern emerges with respect to the graph of x^3 and the graph of x^n for any odd integer $n \geq 3$. Here the graphs are symmetric with respect to the origin (since $(-x)^n = -x^n$ if n is odd) and have neither maximum nor minimum values.

If a is a nonzero constant, the graphs of ax^n and x^n are related to each other in the same way as the graphs of ax^2 and x^2 are. The following example illustrates this fact.

Example 3. Sketch the graphs of the following functions.

a. $f(x) = 2x^3$ b. $g(x) = -\dfrac{1}{3} x^4$

Solution.

a. Since we already have the graph of x^3 in Figure 4.21, to obtain the graph of f we need only multiply the y coordinate of each point on the graph of x^3 by 2, which has the effect of "pulling" the graph of x^3 away from the x axis by a factor of 2 (Figure 4.24).

b. The graph of x^4 is shown in Figure 4.22. To obtain the graph of g, just multiply the y coordinate of each point on the graph of x^4 by $-\frac{1}{3}$, which has the effect of reflecting the graph of x^4 through the x axis and "pulling" it toward the x axis by a factor of 3 (Figure 4.25). □

Sometimes the graph of a polynomial can be obtained by a vertical and/or horizontal shift of the graph of a polynomial of the form ax^n.

Example 4. Let $f(x) = (x - 2)^3 + 1$. Sketch the graph of f.

Solution. The graph of f can be obtained by shifting the graph of x^3 upward 1 unit and to the right 2 units (Figure 4.26). □

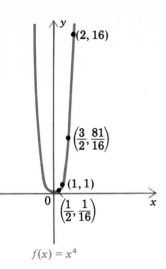

$f(x) = x^4$

FIGURE 4.22

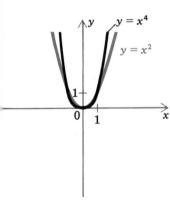

$y = x^4$

$y = x^2$

FIGURE 4.23

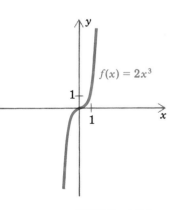

$f(x) = 2x^3$

FIGURE 4.24

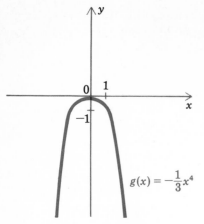

$$g(x) = -\frac{1}{3}x^4$$

FIGURE 4.25

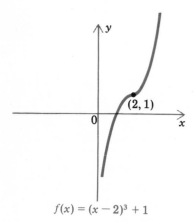

(2, 1)

$$f(x) = (x - 2)^3 + 1$$

FIGURE 4.26

Now we turn to graphs of other polynomial functions of degree greater than 2. We could plot as many points on the graph of such a function as we like by calculating the values of the function for various numbers in the domain, but this would be tedious and, moreover, would not indicate any possible relationship between the graphs of different polynomial functions. In the discussion that follows we will concentrate instead on such features as how a graph looks far away from the y axis and where it is above and where below the x axis.

Let

$$f(x) = a_n x^n + a_{n-1} x^{n-1} + a_{n-2} x^{n-2} + \cdots + a_1 x + a_0$$

with $a_n \neq 0$ (1)

Far away from the y axis the graph of f has the general shape of the graph of the first term, $a_n x^n$. The reason is that when x is large, $a_n x^n$ dominates $a_{n-1} x^{n-1} + a_{n-2} x^{n-2} + \cdots + a_1 x + a_0$. Thus, far away from the y axis the graphs of $x^3 - x$ and x^3 have the same general shape, and likewise, far away from the y axis the graphs of $2x^4 - 8x^3 + 8x^2$ and $2x^4$ have the same general shape.

To sketch the portions of the graph of the function in (1) nearer the y axis, we are helped enormously if we can factor $f(x)$ as

$$f(x) = a_n(x - c_1)(x - c_2) \cdots (x - c_n)$$

(2)

As a result, we will assume henceforth that we can find such a factorization of each polynomial function under consideration. The numbers $c_1, c_2, \ldots, c_n$ appearing in (2) are the x intercepts of f, because $f(x) = 0$ if and only if

$$a_n(x - c_1)(x - c_2) \cdots (x - c_n) = 0$$

which happens if and only if one of the factors, say $x - c_k$, is 0, which means that $x = c_k$. In other words, $f(x) = 0$ if and only if x is one of the numbers $c_1, c_2, \ldots, c_n$.

To illustrate, we let

$$f(x) = x^3 - x$$

Factoring, we find that

$$x^3 - x = x(x^2 - 1) = x(x - 1)(x + 1)$$

If we write x as $x - 0$, then

$$f(x) = (x - 0)(x - 1)(x + 1)$$

which is in the form of (2), with $n = 3$, $a_3 = 1$, $c_1 = 0$, $c_2 = 1$, and $c_3 = -1$. It follows that the x intercepts of f are 0, 1, and -1, so the graph of f intersects the x axis at $(0, 0)$, $(1, 0)$, and $(-1, 0)$.

Knowing the x intercepts allows us to determine the intervals on which $f(x)$ is positive and those on which $f(x)$ is negative. We see how this is done in the following example.

Example 5. Let $f(x) = x^3 - x$. Sketch the graph of f.

Solution. First of all, the y intercept is 0, since $f(0) = 0$. Next,

$$f(x) = x(x - 1)(x + 1)$$

so the x intercepts are 0, 1, and -1 (as we said above). Now the x intercepts divide the x axis into the subintervals $(-\infty, -1)$, $(-1, 0)$, $(0, 1)$, and $(1, \infty)$. To determine whether $f(x)$ is positive or negative on each of these subintervals, we study the factors x, $x - 1$, and $x + 1$ individually on each subinterval. Figure 4.27 gives the results.

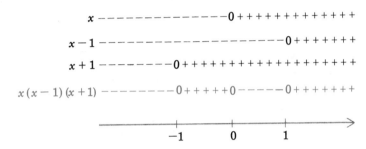

FIGURE 4.27

In constructing Figure 4.27 we used the fact that $x - c < 0$ if $x < c$, and $x - c > 0$ if $x > c$. We also used the fact that $f(x) > 0$ if an even number of the factors of $f(x)$ are negative and the rest are positive, and $f(x) < 0$ if an odd number of the factors are negative and the rest are positive. At this point we know the x intercepts and the intervals on which $f(x)$ is positive or negative. Moreover, because

$$f(-x) = (-x)^3 - (-x) = -x^3 + x = -(x^3 - x) = -f(x)$$

the graph is symmetric with respect to the origin. Finally, far away from the y axis the graph has the general shape of the graph of x^3 (see Figure 4.21). Consequently we can draw a smooth curve for the graph of f (Figure 4.28). □

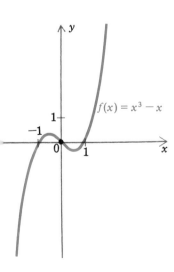

FIGURE 4.28

The fact that the "hilltop" to the left of the y axis in Figure 4.28 occurs for $x = -\sqrt{3}/3$, and the "valley bottom" to the right of the y axis occurs for $x = \sqrt{3}/3$, can be proved by methods of calculus. Even without this specific information we were able to give a reasonably accurate sketch of the graph of f.

Notice that the graph of f in Example 5 *crosses* the x axis at the points corresponding to the x intercepts. Example 6 shows that the graph of a function may touch, but not cross, the x axis at points corresponding to an x intercept.

Example 6. Let $g(x) = x^3 + 2x^2 + x$. Sketch the graph of g.

Solution. Since $g(0) = 0$, the y intercept is 0. Next we factor $g(x)$:

$$g(x) = x^3 + 2x^2 + x = x(x^2 + 2x + 1) = x(x + 1)^2$$

Therefore the x intercepts are 0 and -1, and we prepare the chart shown in Figure 4.29, which gives the signs of x and $(x + 1)^2$, as well as the resulting signs of $g(x)$, on the intervals $(-\infty, -1)$, $(-1, 0)$, and $(0, \infty)$:

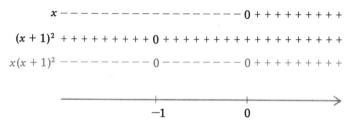

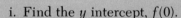

	$(-\infty, -1)$	-1	$(-1, 0)$	0	$(0, \infty)$
x	− − − − − − − −		− − − − − − − −	0	+ + + + + + + + +
$(x+1)^2$	+ + + + + + + +	0	+ + + + + + + +		+ + + + + + + + +
$x(x+1)^2$	− − − − − − − −	0	− − − − − − − −	0	+ + + + + + + + +

−1 0

FIGURE 4.29

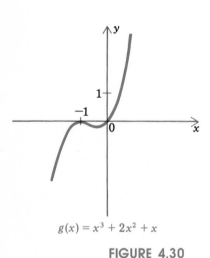

$g(x) = x^3 + 2x^2 + x$

FIGURE 4.30

Using the information in Figure 4.29, along with the fact that far away from the y axis the graph of g has the same general shape as the graph of x^3, we sketch the graph of g in Figure 4.30. □

Through these examples we have developed a procedure that facilitates drawing a rough sketch of the graph of any polynomial function f that we can factor:

i. Find the y intercept, $f(0)$.
ii. Factor $f(x)$ to obtain a formula in the form of (2).
iii. Obtain the x intercepts from the factorization in (ii), and note the intervals into which the x intercepts divide the real line.
iv. Prepare a chart that tells on which of the intervals in (iii) the value $f(x)$ is positive and on which negative.
v. Determine the general shape of the graph far away from the y axis (the same as the shape of the graph of $a_n x^n$).
vi. Using the information obtained in (i)–(v), sketch the graph of f.

Notice in Figure 4.30 that the graph of g *touches*, but does not *cross*, the x axis at -1. The reason is that in the factorization of g, $x + 1$ appears *twice*, that is, $(x + 1)^2$ appears in the factorization. Because $(x + 1)^2 \geq 0$ for all x, $g(x)$ does not change sign at -1. In

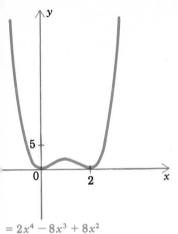

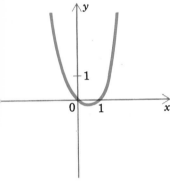

$= 2x^4 - 8x^3 + 8x^2$

FIGURE 4.31

general, if c is an x intercept of a polynomial function f and if $x - c$ appears an odd number of times in the factorization of f, then $f(x)$ changes sign at c and the graph of f crosses the x axis at c, whereas if $x - c$ appears an even number of times, then $f(x)$ does not change sign at c and the graph of f merely touches the x axis at c. We make use of this observation in the solution of the following example.

Example 7. Let $f(x) = 2x^4 - 8x^3 + 8x^2$. Sketch the graph of f.

Solution. The y intercept is 0, since $f(0) = 0$. For the x intercepts we factor $f(x)$:

$$f(x) = 2x^4 - 8x^3 + 8x^2 = 2x^2(x^2 - 4x + 4) = 2x^2(x - 2)^2$$

Therefore the x intercepts are 0 and 2. Since x appears twice, and similarly $x - 2$ appears twice, the graph touches but does not cross the x axis at $x = 0$ and at $x = 2$. In fact, $f(x) \geq 0$ for all x, so there is no need for a chart in this case. Finally, far from the y axis the graph of f has the general shape of the graph of $2x^4$. We are now ready to sketch the graph (Figure 4.31).   □

Caution: The methods of this section do not give precise, detailed information about the local shape of a graph. For example, if

$$f(x) = x^4 - x^3$$

then

$$f(x) = x^3(x - 1)$$

so by using the procedure outlined above we would draw the graph appearing in Figure 4.32a. A more precise sketch of the graph appears in Figure 4.32b. Such refinements in the graph can be obtained by methods of calculus, or by calculating a large number of values of f.

A plausible graph of $x^4 - x^3$

(a)

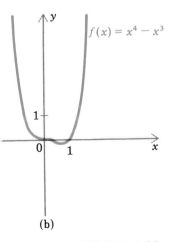

$f(x) = x^4 - x^3$

(b)

FIGURE 4.32

EXERCISES 4.3

In Exercises 1–24, sketch the graph of the given function, and indicate all intercepts.

1. $f(x) = x^3 + 1$
2. $f(x) = 1 - x^4$
3. $f(x) = \frac{1}{4} + 2x^3$
4. $f(x) = 4 - \frac{1}{4}x^4$
5. $f(x) = -2(x + 1)^3$
6. $g(x) = \frac{1}{8}(x - 2)^4 - 2$
7. $g(x) = x^5 + 1$
8. $g(x) = -\frac{1}{81}(x - 3)^6$
9. $g(x) = (x + 1)(x - 1)(x - 2)$
10. $f(x) = x(x + 1)^3$
11. $f(x) = -x^2(x + 1)(x + 2)$
12. $f(x) = 2x^2(x - 4)$
13. $f(x) = x^2(x - 1)^2$
14. $f(x) = x^3(x - 1)^2$

15. $g(x) = (x^2 - 1)(x + 1)^2$

16. $g(x) = (x^2 - 3x + 2)(x^2 - 4)$

17. $g(x) = x^3 - x^2 - 2x$

18. $g(x) = x^3 - 4x^2 + 4x$

19. $f(x) = x^4 - x^2$

20. $f(x) = -x^5 - x^4 + 6x^3$

21. $f(x) = x^5 - x^4$

22. $y = x^5 - x^3$

23. $y = x^4 - 5x^2 + 4$

24. $y = -x^4 + 2x^3 + x^2 - 2x$

25. Find the number c such that the graph of the equation $y = x^4 - x^2 + 3x + c$ passes through $(1, 5)$.

26. Find the number c such that the graph of the equation $y = 3x^3 - x^2 + 2x + c$ has y intercept 7.

27. Let $f(x) = ax^3 - 2x^2 + x - 6$. If -1 is an x intercept of the graph of f, find the value of a.

28. Let $f(x) = x^3 + bx^2 + 3x + 9$. If 3 is an x intercept of the graph of f, find the value of b.

4.4

GRAPHS OF RATIONAL FUNCTIONS

Now that we have studied graphs of polynomial functions at some length, we turn to graphs of quotients of polynomials. The quotient of two polynomials is called a *rational function.* In other words, if P and Q are polynomials, then the function f defined by

$$f(x) = \frac{P(x)}{Q(x)}$$

is a rational function. The domain of f consists of all real numbers x such that $Q(x) \neq 0$ (since we cannot divide by 0). Ironically, those numbers that are not in the domain of a rational function f frequently play an essential role in sketching the graph of f. Let us illustrate this with the graph of $1/x$.

Example 1. Let $f(x) = 1/x$. Sketch the graph of f.

Solution. The domain of f consists of all real numbers x except 0. There are no x or y intercepts, as we showed in Example 13 of Section 3.4. Moreover,

$$f(-x) = \frac{1}{-x} = -\frac{1}{x} = -f(x)$$

and thus the graph of f is symmetric with respect to the origin. Next we construct tables of values of $f(x)$ for small and for large positive values of x:

x	1	$\frac{1}{10}$	$\frac{1}{100}$	$\frac{1}{1000}$
$f(x)$	1	10	100	1000

x	1	10	100	1000
$f(x)$	1	$\frac{1}{10}$	$\frac{1}{100}$	$\frac{1}{1000}$

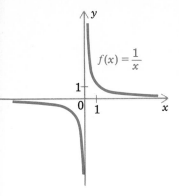

f(x) = $\frac{1}{x}$

FIGURE 4.33

The first table suggests that as x shrinks toward 0, $f(x)$ grows without bound; accordingly, the part of the graph of f corresponding to small positive values of x is far from the x axis. From the second table we conclude that as x grows without bound, $f(x)$ shrinks toward 0; thus the part of the graph of f corresponding to large positive values of x is close to the x axis. Combining this information and then using the symmetry with respect to the origin, we obtain the graph of f shown in Figure 4.33. □

Because $1/x$ becomes arbitrarily large if x is positive and approaches 0 (that is, the point $(x, 1/x)$ on the graph gets arbitrarily far from the x axis as x approaches 0 from the right), the vertical line $x = 0$ is called a vertical asymptote of the graph. In general, we say that a vertical line $x = a$ is a ***vertical asymptote*** of the graph of a function g if either (or both) of the following conditions hold:

1. As x approaches a from the left or from the right (or both), $g(x)$ gets arbitrarily large and is positive (Figure 4.34a, b).
2. As x approaches a from the left or from the right (or both), $g(x)$ becomes arbitrarily large in absolute value and is negative (Figure 4.34c, d).

The graph in Figure 4.34e satisfies both condition 1 and condition 2.

It can be proved that if P and Q are polynomial functions such that $P(a) \neq 0$ and $Q(a) = 0$, then the line $x = a$ is a vertical asymptote of the graph of the rational function P/Q. Thus both the rational functions

$$\frac{x}{(x-2)^3} \quad \text{and} \quad \frac{x^3 + 2x - 1}{x^2 - 6x + 8}$$

have the same vertical asymptote $x = 2$, and the second function also has the vertical asymptote $x = 4$.

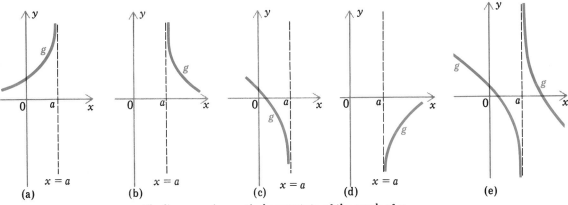

(a) (b) (c) (d) (e)

The line $x = a$ is a vertical asymptote of the graph of g.

FIGURE 4.34

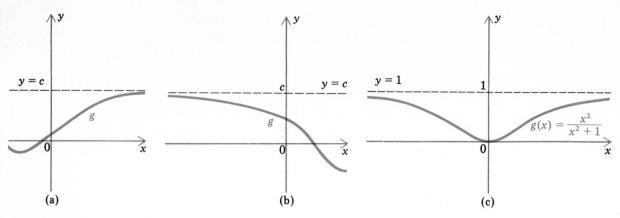

FIGURE 4.35

Recall that $1/x$ approaches 0 if x is positive and becomes arbitrarily large (so that far away from the y axis the graph approaches the horizontal line $y = 0$). We say that the horizontal line $y = 0$ is a horizontal asymptote of the graph. In general, we say that a horizontal line $y = c$ is a **_horizontal asymptote_** of the graph of a function g if either (or both) of the following conditions hold:

   3. If x is positive and becomes arbitrarily large, $g(x)$ gets close to c (Figure 4.35a).
   4. If x is negative and becomes arbitrarily large in absolute value, $g(x)$ gets close to c (Figure 4.35b).

If $g(x) = x^2/(x^2 + 1)$, then g satisfies both condition 3 and condition 4 (Figure 4.35(c)).

Asymptotes play a major role in graphing most rational functions that are not polynomials, as we will presently see.

Example 2. Let $f(x) = 1/x^2$. Sketch the graph of f.

Solution. The domain consists of all $x \neq 0$, and $1/x^2$ is never 0. Consequently there are no x or y intercepts. Next, observe that

$$f(-x) = \frac{1}{(-x)^2} = \frac{1}{x^2} = f(x)$$

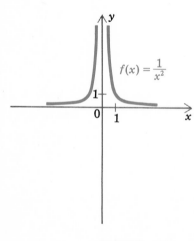

FIGURE 4.36

so the graph of f is symmetric with respect to the y axis. Notice that if x is positive and gets close to 0, so does x^2, and consequently $1/x^2$ becomes very large. Therefore the vertical line $x = 0$ is a vertical asymptote of the graph. Similarly, if x is positive and becomes arbitrarily large, so does x^2, and thus $1/x^2$ approaches 0. Consequently the horizontal line $y = 0$ is a horizontal asymptote of the graph. Combining this information and using the symmetry of the graph with respect to the y axis, we draw the graph of f in Figure 4.36. $\square$

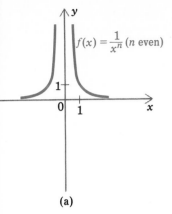

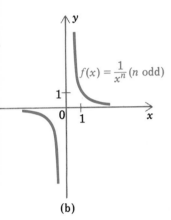

FIGURE 4.37

When n is any positive integer, the graph of $1/x^n$ may be determined in the same way that the graphs of $1/x$ and $1/x^2$ are. It is only necessary to notice whether n is even or odd. On the one hand, if n is even, then the graph of $1/x^n$ is symmetric with respect to the y axis and resembles the graph in Figure 4.37a. On the other hand, if n is odd, then the graph of $1/x^n$ is symmetric with respect to the origin and resembles the graph in Figure 4.37b.

Finally, if a is a nonzero constant, then the graph of $1/(x - a)^n$ has the same appearance as the graph of $1/x^n$ does, except that the vertical asymptote is $x = a$ (instead of $x = 0$), and the graph is shifted horizontally so as to be centered about the line $x = a$.

Example 3. Sketch the graphs of the following functions, and identify their vertical asymptotes.

$$\text{a. } f(x) = \frac{1}{(x - 2)^3} \qquad \text{b. } g(x) = \frac{1}{(x + 4)^4}$$

Solution.

a. The graph of f has the vertical asymptote $x = 2$, and the graph is as sketched in Figure 4.38a, with y intercept $-\frac{1}{8}$, since

$$f(0) = \frac{1}{(-2)^3} = -\frac{1}{8}$$

b. Since $x + 4 = x - (-4)$, the vertical asymptote is $x = -4$. The graph is as sketched in Figure 4.38b. The y intercept is $\frac{1}{256}$, since

$$g(0) = \frac{1}{4^4} = \frac{1}{256} \quad \square$$

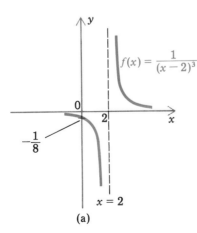

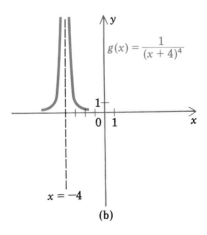

FIGURE 4.38

Graphs of Factored Rational Functions

If we can factor both numerator and denominator of a rational function, we can sometimes obtain a reasonably accurate sketch of the graph with the techniques already at hand. These include finding the domain and any intercepts, determining where the function is positive and where negative, and locating any asymptotes.

Example 4. Let $f(x) = \dfrac{1}{(x-2)(x+3)}$. Sketch the graph of f.

Solution. The domain consists of all numbers except 2 and -3. The y intercept is $-\frac{1}{6}$, since

$$f(0) = \frac{1}{(-2)(3)} = -\frac{1}{6}$$

Since the numerator of $f(x)$ cannot be 0, so that $f(x) \neq 0$ for all x in the domain, there are no x intercepts. The vertical asymptotes are determined from the values at which the denominator is 0, which are $x = 2$ and $x = -3$. Next, we prepare the chart in Figure 4.39 to determine the sign of $f(x)$ on the three intervals $(-\infty, -3)$, $(-3, 2)$, and $(2, \infty)$, into which -3 and 2 divide the x axis.

$$x - 2 \quad - - - - - - - - - - - - - - - - - 0 + + + + + + + + +$$

$$x + 3 \quad - - - - - - - - 0 + + + + + + + + + + + + + + + + + + +$$

$$\frac{1}{(x-2)(x+3)} \quad + + + + + + + + + \quad - - - - - - - - \quad + + + + + + + + +$$

$$\xrightarrow{\qquad\qquad\qquad\qquad\qquad\qquad\qquad\qquad}$$
$$\qquad\qquad\qquad -3 \qquad\qquad\qquad 2$$

FIGURE 4.39

As our final preparation for sketching the graph, we observe that $f(x)$ approaches 0 if x is positive and becomes large, or if x is negative and becomes large in absolute value. This yields the horizontal asymptote $y = 0$. Utilizing this information, we sketch the graph of f in Figure 4.40. □

That the "hilltop" in Figure 4.40 occurs at $x = -\frac{1}{2}$ can conveniently be found by methods of calculus. Nevertheless, the general appearance of the graph can be determined by the methods we are using.

Now we will see when an arbitrary rational function P/Q has horizontal asymptotes. Let

$$P(x) = a_n x^n + a_{n-1} x^{n-1} + \cdots + a_1 x + a_0 \quad \text{with } a_n \neq 0$$

and

$$Q(x) = b_m x^m + b_{m-1} x^{m-1} + \cdots + b_1 x + b_0 \quad \text{with } b_m \neq 0$$

so that

$$\frac{P(x)}{Q(x)} = \frac{a_n x^n + a_{n-1} x^{n-1} + \cdots + a_1 x + a_0}{b_m x^m + b_{m-1} x^{m-1} + \cdots + b_1 x + b_0}$$

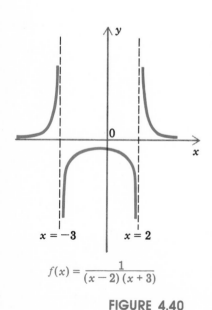

$$f(x) = \frac{1}{(x-2)(x+3)}$$

FIGURE 4.40

Then

  i. If $n < m$, then the line $y = 0$ is a horizontal asymptote.
  ii. If $n = m$, then the line $y = a_n/b_m$ is a horizontal asymptote.
  iii. If $n > m$, then there is no horizontal asymptote. Instead, far away from the y axis the graph of P/Q has the general appearance of the graph of

$$\frac{a_n}{b_m}\, x^{n-m}$$

whose graph was described in Section 4.3.

Example 5. Let $f(x) = \dfrac{(1 - x^2)(x + 2)}{x^2 + 2x}$. Sketch the graph of f.

Solution. The denominator $x^2 + 2x$ can be rewritten as $x(x + 2)$, which implies that the domain of f consists of all numbers except 0 and -2. Thus there is no y intercept. Now we factor the numerator and denominator:

$$f(x) = \frac{(1 - x^2)(x + 2)}{x^2 + 2x} = \frac{(1 - x)(1 + x)(x + 2)}{x(x + 2)}$$

Canceling the factor $x + 2$ in the numerator and denominator yields

$$f(x) = \frac{(1 - x)(1 + x)}{x} \quad \text{for } x \neq 0,\, -2 \tag{1}$$

From (1) we see that -1 and 1 are x intercepts. Since the denominator is 0 for $x = 0$ but the numerator is not 0 for $x = 0$, the line $x = 0$ is a vertical asymptote. Using (1), we assemble the chart in Figure 4.41 that shows the sign of $f(x)$ on each of the intervals $(-\infty, -1)$, $(-1, 0)$, $(0, 1)$, and $(1, \infty)$ determined by the vertical asymptote and the x intercepts.

By (1),

$$f(x) = \frac{(1 - x)(1 + x)}{x} = \frac{1 - x^2}{x} \tag{2}$$

$1 - x$ $+ 0 - - - - - -$

$1 + x$ $- - - - - - 0 +$

x $- - - - - - - - - - - 0 + + + + + + + + + + + + +$

$\dfrac{(1 - x)(1 + x)}{x}$ $+ + + + + + 0 - - - - - - \quad + + + + + + + 0 - - - - - -$

$$\xrightarrow{\qquad \overset{\displaystyle |}{-1} \qquad\quad \overset{\displaystyle |}{0} \qquad\quad \overset{\displaystyle |}{1} \qquad}$$

FIGURE 4.41

so that $f(x)$ satisfies possibility (iii). Therefore far from the y axis the graph of f resembles the graph of $-x^{2-1} = -x$, and consequently there is no horizontal asymptote. Finally, from (2) we find that

$$f(-x) = \frac{1 - (-x)^2}{-x} = -\frac{1 - x^2}{x} = -f(x)$$

so that the graph of f is symmetric with respect to the origin. Now we are ready to draw the graph of f (Figure 4.42). ☐

The line $y = -x$ in Figure 4.42 is called an ***oblique asymptote*** of the graph of f, because far away from the y axis the graph of f approaches the line $y = -x$, which is neither horizontal nor vertical. The graph of any rational function for which the degree of the numerator is one more than the degree of the denominator has an oblique asymptote.

In our final example of this section we will sketch the graph of a rational function f that has the factor $x^2 + 1$ that itself cannot be factored. The polynomial $x^2 + 1$ is positive for every value of x, and hence does not create any intercepts or vertical asymptotes in the graph and, in particular, does not affect the sign of $f(x)$.

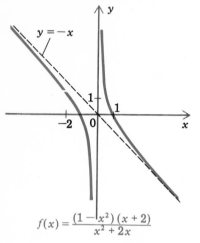

$$f(x) = \frac{(1 - x^2)(x + 2)}{x^2 + 2x}$$

FIGURE 4.42

Example 6. Let $f(x) = \dfrac{x^2 + 1}{4x^2 - 1}$. Sketch the graph of f.

Solution. The y intercept is -1, because $f(0) = -1$. There are no x intercepts, because $x^2 + 1 > 0$ for all x in the domain. Since

$$f(x) = \frac{x^2 + 1}{4x^2 - 1} = \frac{x^2 + 1}{4(x^2 - \frac{1}{4})} = \frac{x^2 + 1}{4(x + \frac{1}{2})(x - \frac{1}{2})}$$

it follows that the lines $x = -\frac{1}{2}$ and $x = \frac{1}{2}$ are vertical asymptotes (and that $-\frac{1}{2}$ and $\frac{1}{2}$ are not in the domain of f). The sign of $f(x)$ on each of the intervals $(-\infty, -\frac{1}{2})$, $(-\frac{1}{2}, \frac{1}{2})$, and $(\frac{1}{2}, \infty)$ determined by the vertical asymptotes is given in the chart in Figure 4.43. We also find that f satisfies possibility (ii) above, so that since $\frac{1}{4}x^{2-2} = \frac{1}{4}$, it follows that $y = \frac{1}{4}$ is a horizontal asymptote. Finally,

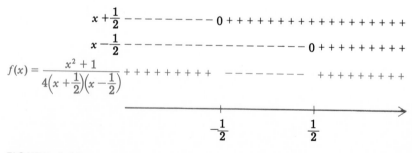

FIGURE 4.43

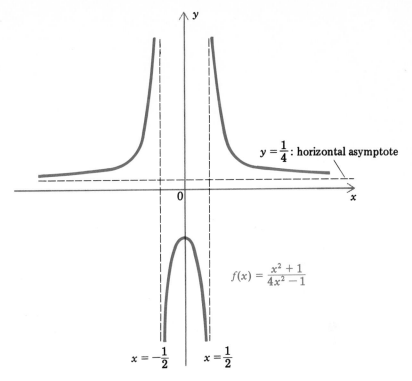

FIGURE 4.44

$f(-x) = f(x)$, as you can easily verify, so the graph of f is symmetric with respect to the y axis. With all the assembled information we can draw the graph of f (Figure 4.44). □

The steps listed below, which we have followed in the preceding examples, can help in graphing a rational function $f = P/Q$, where P and Q are polynomial functions.

1. If it has not been done already, factor the denominator $Q(x)$ into linear polynomials of the form $x - a$, quadratic polynomials that cannot be factored (such as $x^2 + 1$), and a constant if different from 1. Then determine the domain of f, which consists of all x for which $Q(x) \neq 0$. (Notice that a number a is not in the domain of f if and only if $x - a$ is a factor of $Q(x)$.)
2. Factor the numerator $P(x)$ in the same way as $Q(x)$. Then cancel any common factors in the numerator and the denominator.
3. Determine the intercepts (if any). The y intercept is $f(0)$ if 0 is in the domain of f. The x intercepts are the numbers a such that $x - a$ is a factor of the numerator after all cancellation has been completed in step 2.
4. Determine the vertical asymptotes (if any). The line $x = a$ is

a vertical asymptote if a factor of $x - a$ remains in the denominator after all cancellation has been completed in step 2. Such a number a is not in the domain of f.
5. Determine the sign of $f(x)$ on each of the intervals into which the x intercepts and the vertical asymptotes divide the x axis.
6. Determine the general appearance of the graph far away from the y axis as follows:
Let

$$f(x) = \frac{a_n x^n + a_{n-1} x^{n-1} + \cdots + a_1 x + a_0}{b_m x^m + b_{m-1} x^{m-1} + \cdots + b_1 x + b_0}$$

a. If $n < m$, then the line $y = 0$ is a horizontal asymptote.
b. If $n = m$, then the line $y = a_n/b_n$ is a horizontal asymptote.
c. If $n > m$, then far away from the y axis the graph of f resembles the graph of

$$y = \frac{a_n}{b_m} x^{n-m}$$

In case $n - m = 1$, the latter graph is a straight line.
7. Use the information in steps 1–6 to sketch the graph of f.

EXERCISES 4.4

In Exercises 1–24, sketch the graph of the given function. Indicate all intercepts and asymptotes, as well as the behavior for large values of the variable.

1. $f(x) = -\dfrac{1}{x}$

2. $f(x) = \dfrac{1}{x - 1}$

3. $f(x) = \dfrac{4}{2 - x}$

4. $f(x) = 2 + \dfrac{1}{x^2}$

5. $f(x) = -\dfrac{1}{(x + 1)^2}$

6. $f(x) = \dfrac{1}{x^3} - 1$

7. $g(x) = -\dfrac{1}{4(x + \frac{1}{2})^4}$

8. $g(x) = \dfrac{3}{x^2 - 1}$

9. $g(x) = \dfrac{x}{x - 2}$

10. $g(x) = \dfrac{x - 1}{x + 1}$

11. $g(x) = \dfrac{-x}{x^2 - 4}$

12. $f(x) = -\dfrac{3}{x^2 + x}$

13. $f(x) = \dfrac{2}{(x - 1)(x + 2)}$

14. $f(x) = \dfrac{x(x + 2)}{(x - 1)(x^2 - 4)}$

15. $f(x) = \dfrac{x}{x^2 + 4x + 3}$

16. $f(x) = \dfrac{2x - 1}{x^2 - x - 2}$

17. $f(x) = \dfrac{x^2}{2x^2 - 3x - 2}$

18. $y = \dfrac{4x^2 + 4x + 1}{x}$

19. $y = \dfrac{4x^2 + 4x + 1}{x^2}$

20. $f(t) = \dfrac{t^2 + 1}{t + 1}$

21. $g(t) = \dfrac{-3t - 1}{t^2 + 4}$

22. $g(t) = \dfrac{t^4 - 1}{t^2 - 1}$

23. $f(x) = \dfrac{3x^2 + 6x + 3}{x^3 + 4x^2 + x + 4}$

24. $f(x) = \dfrac{x + \frac{1}{2}}{4x^3 + 4x^2 + x}$

4.5

THE PARABOLA

Conic sections have been of interest to mathematicians since the third century B.C., when the Greek geometer Apollonius wrote eight treatises about conic sections and became known as the "Great Geometer." A *conic section* is formed when a double right circular cone is sliced by a plane. The intersection can be a *parabola*, an *ellipse* (a special case of which is a circle), or a *hyperbola*, and in very special cases it can be a point, a line, or two intersecting lines (called *degenerate* conic sections). Figure 4.45 depicts the various kinds of conic sections.

A conic section is a curve in a plane that cuts a double cone. If we think of the plane as the Cartesian plane, then the conic section can be described as the graph of an equation in x and y. In this and the next two sections we will associate the three main kinds of

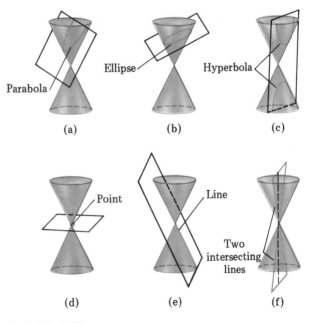

FIGURE 4.45

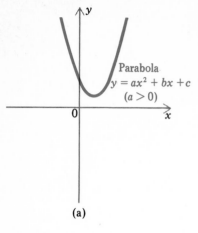

(a)

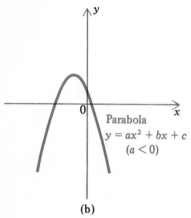

(b)

FIGURE 4.46

conic sections—parabola, ellipse, and hyperbola—with special types of equations, and we will sketch the graphs of such equations.

The graph of an equation of the form $y = ax^2 + bx + c$, with $a \neq 0$, is a parabola, as defined in Section 4.1, and has a graph as in Figure 4.46a, b. If the graph is turned on its side, then it remains a parabola and has an equation of the form

$$x = ay^2 + by + c \quad \text{with } a \neq 0 \qquad (1)$$

We sketched the graph of one such equation, $x = y^2$, in Figure 3.58. In general, to sketch the graph of (1) we complete the square on the right side of (1) to obtain an equation of the form

$$x = a(y - k)^2 + h, \quad \text{or equivalently,} \quad x - h = a(y - k)^2 \qquad (2)$$

Then the graph opens to the right of the point (h, k) if $a > 0$ and to the left of (h, k) if $a < 0$ (Figure 4.47a, b). The point (h, k) is the *vertex* of the parabola. Moreover, the graph is symmetric with respect to the line $y = k$, called the *axis* of the parabola.

Example 1. Sketch the parabola having equation

$$x = 2y^2 - 4y + 5$$

and note the intercepts, vertex, and axis of the parabola.

Solution. The x intercept is 5 because if $y = 0$, then $x = 5$. To determine any y intercepts, we set $x = 0$ in the given equation and solve for y:

$$2y^2 - 4y + 5 = 0 \qquad (3)$$

$$y = \frac{-(-4) \pm \sqrt{(-4)^2 - 4(2)(5)}}{2(2)}$$

$$= \frac{4 \pm \sqrt{16 - 40}}{4} = \frac{4 \pm \sqrt{-24}}{4}$$

Since $\sqrt{-24}$ is not defined, there is no real number y satisfying equation (3), and hence there are no y intercepts. To help locate the vertex of the parabola, we complete the square:

$$x = 2y^2 - 4y + 5$$
$$= 2(y^2 - 2y) + 5$$
$$= 2(y^2 - 2y + 1 - 1) + 5$$
$$= 2(y^2 - 2y + 1) - 2 + 5$$
$$= 2(y - 1)^2 + 3$$

In the notation of (2) we have $a = 2$, $h = 3$, and $k = 1$. Therefore the vertex is the point $(3, 1)$, the axis is the line $y = 1$, and the

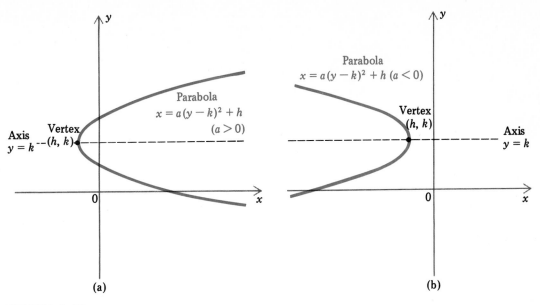

Parabola
$x = a(y - k)^2 + h$
$(a > 0)$

Axis
$y = k$ --- Vertex --(h, k)

(a)

Parabola
$x = a(y - k)^2 + h \ (a < 0)$

Vertex
(h, k)

Axis
$y = k$

(b)

FIGURE 4.47

parabola opens to the right because $a = 2 > 0$. The parabola is sketched in Figure 4.48. □

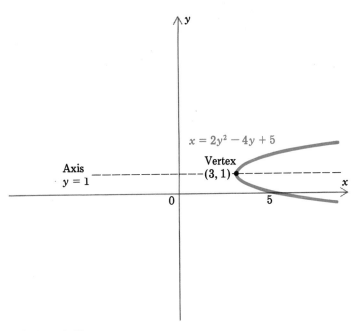

$x = 2y^2 - 4y + 5$

Vertex
(3, 1)

Axis
$y = 1$

5

FIGURE 4.48

Example 2. Sketch the parabola having equation

$$x = -y^2 - 4y - 3$$

and note the intercepts, vertex, and axis of the parabola.

Solution. The x intercept is -3 because if $y = 0$, then $x = -3$. To find any y intercepts, we set $x = 0$ in the given equation and solve for y:

$$-y^2 - 4y - 3 = 0$$
$$y^2 + 4y + 3 = 0$$
$$(y + 3)(y + 1) = 0$$
$$y = -3 \quad \text{or} \quad y = -1$$

As a result, the y intercepts are -3 and -1. Next we complete the square on the right side of the given equation:

$$\begin{aligned} x &= -y^2 - 4y - 3 \\ &= -(y^2 + 4y) - 3 \\ &= -(y^2 + 4y + 4 - 4) - 3 \\ &= -(y^2 + 4y + 4) + 4 - 3 \\ &= -(y + 2)^2 + 1 \end{aligned}$$

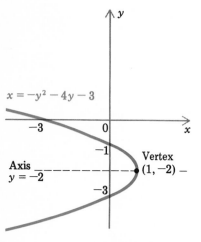

$x = -y^2 - 4y - 3$

Axis
$y = -2$

This time when we use the notation of (2), we find that $a = -1$, $h = 1$, and $k = -2$. Therefore the vertex is $(1, -2)$, and the axis of the parabola is the line $y = -2$. Since $a = -1 < 0$, the parabola opens to the left. The parabola is sketched in Figure 4.49. □

The graph of any equation of the form

$$Ax^2 + By^2 + Cx + Dy = E$$

with $A = 0$ and $B \neq 0$, or $A \neq 0$ and $B = 0$ (4)

FIGURE 4.49

is a parabola. By using the technique of completing the square, one can rewrite (4) as either

$$x = a(y - k)^2 + h \tag{5}$$

or

$$y = a(x - h)^2 + k \tag{6}$$

In each case the vertex of the parabola is (h, k). For (5) the axis is the horizontal line $y = k$, and for (6) the axis is the vertical line $x = h$. In either case the parabola is symmetric with respect to its axis.

Example 3. Sketch the parabola having equation

$$2x^2 - 12x - 3y = -6$$

noting the vertex and axis of the parabola.

Solution. The given equation has the form of (4) with $A = 2$, $B = 0$, $C = -12$, $D = -3$, and $E = -6$. Since $B = 0$, we can put the given equation into the form of (6) by completing the square:

$$2x^2 - 12x = 3y - 6$$

$$2(x^2 - 6x) = 3y - 6$$

$$2(x^2 - 6x + 9 - 9) = 3y - 6$$

$$2(x^2 - 6x + 9) - 18 = 3y - 6$$

$$2(x - 3)^2 = 3y + 12$$

$$2(x - 3)^2 = 3(y + 4)$$

$$\frac{2}{3}(x - 3)^2 = y + 4$$

$$y = \frac{2}{3}(x - 3)^2 - 4$$

Thus the vertex is $(3, -4)$, and the axis is the vertical line $x = 3$. The graph is sketched in Figure 4.50. $\square$

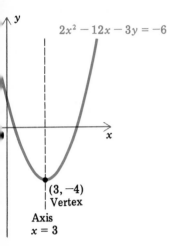

$2x^2 - 12x - 3y = -6$

$(3, -4)$
Vertex

Axis
$x = 3$

FIGURE 4.50

There is an alternative way of defining the parabola that is geometric in nature. Let l be a fixed line in the plane and F a fixed point not on l. The set of all points P in the plane equidistant from l and F is a parabola (Figure 4.51). The line l is called the **directrix** and the point F the **focus** of the parabola.

There are many theoretical and practical applications of parabolas. We list several of them here.

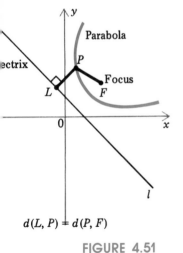

Parabola

Focus

$d(L, P) = d(P, F)$

FIGURE 4.51

1. When an object moves under the sole influence of the earth's gravity and eventually strikes the earth, its path is normally parabolic (or linear if the motion is vertical). Thus the path of a golf ball, baseball, or football in flight would be parabolic if there were no air resistance.

2. The shapes of certain mirrors can be obtained by revolving a parabola about its axis. Such mirrors, called **parabolic mirrors**, are found in automobile headlights and reflecting telescopes. In a headlight all light rays emitted from a light source at the focus are reflected from the mirror along lines parallel to the axis of the parabola. In a reflecting telescope, such as the one at Mt. Palomar and various radio telescopes, all incoming rays from a star, which are essentially parallel to the axis, are reflected by the mirror and then pass through the focus, where the eyepiece of the telescope should be located (see Figure 4.52a).

3. A suspended cable tends to hang in the shape of a parabola if the weight of the cable is small compared to the weight it supports and if the weight is uniformly distributed along the cable. Bridges that are supported by cables in this fashion

(a)

(b)

FIGURE 4.52

are called *suspension bridges* (see Figure 4.52b). All the larger bridges in the world, and most of the more famous ones, are suspension bridges; examples are the Golden Gate Bridge, the Brooklyn Bridge, and the Delaware Memorial Bridge.

EXERCISES 4.5

In Exercises 1–14, sketch the parabola, noting the vertex and the axis of the parabola.

1. $x = 3y^2$

2. $x + \dfrac{1}{2} y^2 = 0$

3. $x + 5 = -\dfrac{1}{3} (y - 1)^2$

4. $x - 1 = 4(y + 2)^2$

5. $y^2 = x - y$

6. $x = y^2 + 2y - 4$

7. $x = 2y^2 - 2y + 1$

8. $5x - 19 = -y(y + 2)$

9. $y + 4x^2 - 24x + 34 = 0$

10. $y + x^2 - 8x + 22 = 0$

11. $2x - y^2 + 6y - 7 = 0$

12. $3x + 2y^2 + 4y + 8 = 0$

13. $x^2 + 4x = 3y + 5$

14. $(x - y)^2 + (x + y)^2 = 2x^2 + 4y - 2x$

In Exercises 15–20, find an equation of the parabola having the given properties.

15. Vertex $(0, 0)$; axis: y axis; $(2, 5)$ on the parabola

16. Vertex $(0, 0)$; axis: x axis; $(2, 5)$ on the parabola

17. Vertex $(-2, 5)$; axis: $x = -2$; $(0, 7)$ on the parabola

18. Vertex $(1, -4)$; axis: $y = -4$; $(-1, -1)$ on the parabola

19. Vertex $(3, 4)$; $(2, 5)$ and $(4, 5)$ on the parabola

20. Vertex $(-1, 1)$; $(3, 0)$ and $(0, \frac{1}{2})$ on the parabola

In Exercises 21–24, sketch the graph of the given equation. In each case the graph is a degenerate conic section.

21. $x^2 - 9y^2 + 2x + 18y = 8$

22. $4x^2 + 3y^2 - 24x - 18y + 63 = 0$
23. $36x^2 + 36x + 11 = -18y^2 + 12y$
24. $3x^2 + 6\sqrt{3}x = 2y^2 - 16y + 23$
25. For what value of c will the parabola with equation $cx = (y + 2)^2 - 6$ pass through the point $(-2, 4)$?
* 26. For what value of c will the axis of the parabola with equation $2x - 5 = (1/c)y^2 + 4y + 9$ be the line $y = 3$?
27. Determine all points on the parabola with equation $3x - 4y^2 + 2y = 7$ that have x coordinate 3.
28. Determine all points on the parabola with equation $x^2 - 2x - y = -\frac{1}{2}$ that have y coordinate 3.
29. In a suspension bridge the horizontal distance between the supports is the **span** of the bridge, and the vertical distance between the points where the cable is attached to the supports and the lowest point of the cable is the **sag** (Figure 4.53). The span of the George Washington Bridge is 3500 feet, and its sag is 316 feet. Find an equation of the parabola that represents the cable.

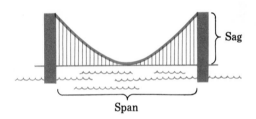

Suspension bridge

FIGURE 4.53

4.6

THE ELLIPSE

The graph of an equation of the form

$$Ax^2 + By^2 = E \quad \text{with } A, B, \text{ and } E \text{ of the same sign} \quad (1)$$

is an ellipse. Dividing each side of the equation in (1) by E allows us to rewrite the equation in the standard form

$$\frac{x^2}{a^2} + \frac{y^2}{b^2} = 1 \quad \text{or} \quad \frac{x^2}{b^2} + \frac{y^2}{a^2} = 1 \quad \text{where } a \geq b > 0 \quad (2)$$

Ellipses given in standard form are symmetric with respect to both axes as well as the origin, which is called the **center** of the ellipse (Figure 4.54a, b). Setting $y = 0$ in the first equation of (2), and $x = 0$ in the second equation, yields $-a$ and a for two intercepts of the ellipse. The corresponding points of the ellipse are called **vertices** of the ellipse, and the line segment joining the vertices is the

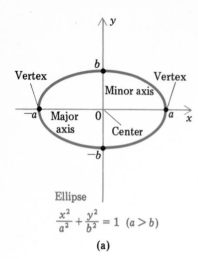

Ellipse

$$\frac{x^2}{a^2} + \frac{y^2}{b^2} = 1 \quad (a > b)$$

(a)

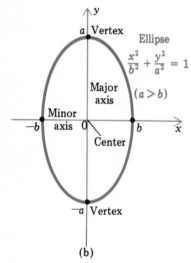

(b)

FIGURE 4.54

major axis of the ellipse. The length of the major axis is $2a$ (Figure 4.54a, b). The remaining two intercepts of the ellipse are obtained by setting $x = 0$ in the first equation of (2) and $y = 0$ in the second, which results in the intercepts $-b$ and b. Although the corresponding points have no special name, the line segment joining those points is called the *minor axis* of the ellipse. The minor axis has length $2b$ (Figure 4.54a, b). Since $a \geq b$ by assumption in (2),

the length of the major axis $\geq$ the length of the minor axis

Example 1. Sketch the ellipse having equation

$$\frac{x^2}{9} + \frac{y^2}{4} = 1$$

and note the major and minor axes.

Solution. Since the coefficient in the denominator of x^2 is larger than that of y^2, we compare the given equation with the first equation of (2). This yields $a^2 = 9$, so that $a = \sqrt{9} = 3$, and it also yields $b^2 = 4$, so that $b = \sqrt{4} = 2$. It follows that the x intercepts are -3 and 3 and the y intercepts are -2 and 2. The ellipse is sketched in Figure 4.55. □

Example 2. Sketch the ellipse having equation

$$16x^2 + 2y^2 = 32$$

and note the major and minor axes.

Solution. In order to use (2), we first divide each side by 32 to obtain

$$\frac{x^2}{2} + \frac{y^2}{16} = 1$$

Since the coefficient in the denominator of y^2 is larger than that of x^2, we compare our new equation with the second equation of (2). We find that $a = \sqrt{16} = 4$ and $b = \sqrt{2}$. Therefore the x intercepts are $-\sqrt{2}$ and $\sqrt{2}$, and the y intercepts are -4 and 4. The ellipse is sketched in Figure 4.56. □

Example 3. Find an equation in standard form for the ellipse whose vertices are $(-2, 0)$ and $(2, 0)$ and which passes through the point $(1, -\frac{3}{2})$ (Figure 4.57).

Solution. Since the vertices lie on the x axis, the equation we seek has the form

$$\frac{x^2}{a^2} + \frac{y^2}{b^2} = 1$$

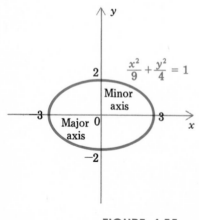

FIGURE 4.55

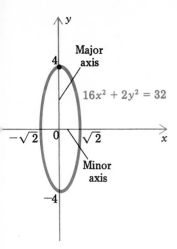

Major
axis

$16x^2 + 2y^2 = 32$

$-\sqrt{2}$ 0 $\sqrt{2}$

Minor
axis

−4

FIGURE 4.56

The value of a is 2, because the vertices are $(-2, 0)$ and $(2, 0)$. We will be done once we find the numerical value of b^2. To that end we notice that the point $(1, -\frac{3}{2})$ is on the ellipse, so that

$$\frac{1^2}{4} + \frac{(-3/2)^2}{b^2} = 1$$

Solving for b^2, we find that

$$\frac{(-3/2)^2}{b^2} = 1 - \frac{1}{4} = \frac{3}{4}$$

and consequently

$$b^2 = \left(-\frac{3}{2}\right)^2 \cdot \frac{4}{3} = \frac{9}{4} \cdot \frac{4}{3} = 3$$

Therefore the equation we seek is

$$\frac{x^2}{4} + \frac{y^2}{3} = 1 \quad \square$$

The ellipses in Figures 4.55 and 4.57 are wider than they are tall, whereas the ellipse in Figure 4.56 is taller than it is wide. One can tell immediately from an equation of an ellipse in standard form whether it is wider or taller: if the number under x^2 is larger than the number under y^2, the ellipse is wider; if the number under y^2 is larger than the number under x^2, the ellipse is taller. In the event that the numbers under x^2 and y^2 are the same, the ellipse is a circle, because in that case the equation becomes

$$\frac{x^2}{a^2} + \frac{y^2}{a^2} = 1$$

or equivalently,

$$x^2 + y^2 = a^2$$

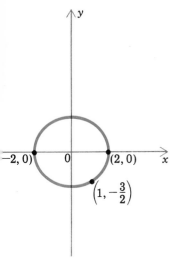

−2, 0)   0   (2, 0)

$\left(1, -\frac{3}{2}\right)$

FIGURE 4.57

Thus a circle is indeed a special case of an ellipse.

Except in very special cases that will not concern us here, the graph of an equation of the form

$$Ax^2 + By^2 + Cx + Dy = E \quad \text{with } A \text{ and } B \text{ of the same sign} \quad (3)$$

is an ellipse whose center can be any point in the plane, depending on the values of the constants. By completing squares, one can rearrange (3) so that it has the form

$$\frac{(x - h)^2}{a^2} + \frac{(y - k)^2}{b^2} = 1 \quad \text{or} \quad \frac{(x - h)^2}{b^2} + \frac{(y - k)^2}{a^2} = 1$$

$$\text{where } a \geq b > 0 \quad (4)$$

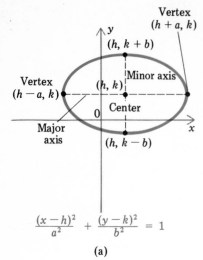

$$\frac{(x-h)^2}{a^2} + \frac{(y-k)^2}{b^2} = 1$$

(a)

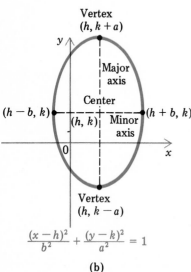

$$\frac{(x-h)^2}{b^2} + \frac{(y-k)^2}{a^2} = 1$$

(b)

FIGURE 4.58

The *center* of the ellipse is the point (h, k), and the ellipse is symmetric with respect to the two lines $x = h$ and $y = k$. For an ellipse defined by the first equation in (4), the *vertices* are $(h - a, k)$ and $(h + a, k)$ (Figure 4.58a), and the line segment joining the vertices is the *major axis* of the ellipse, whereas the line segment joining the points $(h, k - b)$ and $(h, k + b)$ is the *minor axis* of the ellipse. Analogously, for an ellipse defined by the second equation in (4), the vertices are $(h, k - a)$ and $(h, k + a)$ (Figure 4.58b), with the line joining them the *major axis*, and the line joining the points $(h - b, k)$ and $(h + b, k)$ the *minor axis* of the ellipse. Again the length of the major axis is $2a$, and the length of the minor axis is $2b$.

Example 4. Sketch the ellipse having equation

$$9x^2 + 54x + y^2 - 4y = -76$$

and note the center and vertices, as well as the major and minor axes.

Solution. We complete squares and then rearrange so that the right side is 1:

$$(9x^2 + 54x) + (y^2 - 4y) = -76$$

$$9(x^2 + 6x) + (y^2 - 4y) = -76$$

$$9(x^2 + 6x + 9 - 9) + (y^2 - 4y + 4 - 4) = -76$$

$$9(x^2 + 6x + 9) - 81 + (y^2 - 4y + 4) - 4 = -76$$

$$9(x + 3)^2 + (y - 2)^2 = -76 + 81 + 4 = 9$$

$$\frac{(x + 3)^2}{1} + \frac{(y - 2)^2}{9} = 1$$

This equation has the form of the second equation in (4), with $h = -3, k = 2, a = 3$, and $b = 1$. Consequently the center of the ellipse is $(-3, 2)$. The vertices are $(-3, 2 - 3)$ and $(-3, 2 + 3)$, that is, $(-3, -1)$ and $(-3, 5)$, and these points serve as the endpoints of the major axis. The minor axis is bounded by the points $(-3 - 1, 2)$ and $(-3 + 1, 2)$, that is, $(-4, 2)$ and $(-2, 2)$. Now we are ready to sketch the graph, with all pertinent information on it (Figure 4.59). □

Example 5. Find an equation of the ellipse whose major axis extends from $(-6, -6)$ to $(4, -6)$ and whose minor axis extends from $(-1, -8)$ to $(-1, -4)$ (Figure 4.60).

Solution. The center of the ellipse is the midpoint of the major (or minor) axis, which is $(-1, -6)$. Since the major axis is horizontal, the equation we seek has the form

$$\frac{(x - (-1))^2}{a^2} + \frac{(y - (-6))^2}{b^2} = 1$$

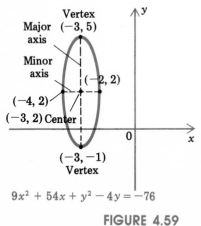

$$9x^2 + 54x + y^2 - 4y = -76$$

FIGURE 4.59

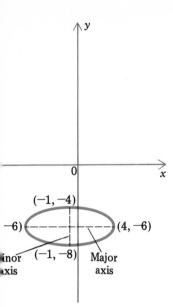

(−1, −4)

(−6) (4, −6)

minor (−1, −8) Major
axis   axis

FIGURE 4.60

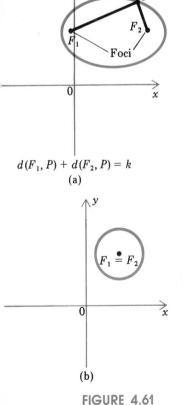

$d(F_1, P) + d(F_2, P) = k$
(a)

$F_1 \overset{\bullet}{=} F_2$

(b)

FIGURE 4.61

or equivalently,

$$\frac{(x + 1)^2}{a^2} + \frac{(y + 6)^2}{b^2} = 1$$

The distance between $(-6, -6)$ and $(4, -6)$ is $4 - (-6) = 10$, and this is the length of the major axis, which is also $2a$. Therefore $2a = 10$, so that $a = 5$. Similarly, the distance between $(-1, -8)$ and $(-1, -4)$ is $-4 - (-8) = 4$, and this is the length of the minor axis, which is also $2b$. Therefore $2b = 4$, so that $b = 2$. Consequently the equation we seek is

$$\frac{(x + 1)^2}{25} + \frac{(y + 6)^2}{4} = 1 \quad \square$$

For a geometric description of an ellipse, we let F_1 and F_2 be points in the plane and k a number greater than the distance between F_1 and F_2. The set of all points P in the plane such that

$$d(F_1, P) + d(F_2, P) = k$$

is an ellipse (Figure 4.61a). The points F_1 and F_2 are called the *foci* of the ellipse. In the event that F_1 and F_2 are the same point, the ellipse is a circle (Figure 4.61b).

We end our discussion of ellipses with a few applications:

1. According to Kepler's laws of planetary motion, planets move in elliptical orbits. Some comets, such as Halley's comet, also move in elliptical orbits about the sun. Halley's comet returns to the earth's vicinity every 76 years and should be visible from earth in May 1986.
2. An electron in an atom normally moves in an approximately elliptical orbit.
3. The dome of a whispering gallery has the shape of an ellipse that has been revolved about its major axis. Sound emanating from one focus bounces from anywhere on the dome and is then directed to the second focus. It is said that John C. Calhoun used this phenomenon in the Statuary Hall of the Capitol in Washington, D.C., in order to eavesdrop on his adversaries.
4. Ellipses are aesthetically pleasing and are used in the design of buildings and the construction of formal gardens.

EXERCISES 4.6

In Exercises 1–14, sketch the ellipse, noting the center, the major and minor axes, and the vertices.

1. $\dfrac{x^2}{16} + \dfrac{y^2}{4} = 1$

2. $x^2 + \dfrac{y^2}{4} = 1$

3. $\dfrac{x^2}{3} + \dfrac{y^2}{3} = 1$

4. $4x^2 + y^2 = 1$

5. $9x^2 + 4y^2 = 36$

6. $2x^2 + y^2 = 2$

7. $\dfrac{(x - 4)^2}{9} + \dfrac{(y + 3)^2}{16} = 1$

8. $\dfrac{(x + 4)^2}{2} + \dfrac{(y - 1)^2}{25} = 1$

9. $(x + 2)^2 + 4(y + 1)^2 = 16$

10. $4x^2 + y^2 + 6y + 5 = 0$

11. $4x^2 - 8x + y^2 + 4y = 8$

12. $9x^2 + 18x + 4y^2 + 16y = 11$

13. $25x^2 + 16y^2 + 150x - 96y = 31$

14. $4x^2 + 9y^2 - 16x + 54y + 96 = 0$

In Exercises 15–24, find an equation of the ellipse having the given properties.

15. Center: $(0, 0)$; x intercept: 2; y intercept: -3

16. Center: $(0, 0)$; x intercept: -4; y intercept: 1

17. Center: $(0, 0)$; vertex: $(0, 4)$; $(\sqrt{3}, 2)$ on the ellipse

18. Center: $(-1, 2)$; vertex: $(3, 2)$; $(0, 4)$ on the ellipse

19. Center: $(3, 2)$; $(3, 7)$ and $(-4, 2)$ on the ellipse

20. Center: $(-1, -3)$; $(-1, 5)$ and $(3, -3)$ on the ellipse

21. Vertices: $(-2, -3)$ and $(-2, -7)$; length of minor axis: 2

22. Vertices: $(6, -1)$ and $(-10, -1)$; length of minor axis: $\frac{1}{4}$

23. Center: $(-\frac{3}{2}, \frac{1}{2})$; length of major axis: 5; length of minor axis: 1; minor axis horizontal

24. Major axis from $(1, 1)$ to $(1, 10)$; minor axis from $(0, \frac{11}{2})$ to $(2, \frac{11}{2})$

25. Find equations for the two distinct ellipses each of which has center $(0, 4)$, major axis of length 3, and minor axis of length 2.

26. Find equations for the two distinct ellipses each of which has center $(-3, -2)$, major axis of length $\frac{3}{2}$, and minor axis of length $\frac{1}{2}$.

27. Show that if the major and minor axes of an ellipse have the same length, the ellipse is a circle.

28. For what value of c does the ellipse with equation $9(x - 1)^2 + 4(y + 2)^2 = 36c$ pass through the origin?

29. For what values of h does the ellipse with equation $\dfrac{(x - h)^2}{9} + \dfrac{(y + 5)^2}{4} = 1$ pass through the point $(4, -4)$?

30. Let a and b be positive numbers. The area of the ellipse having equation

$$\frac{x^2}{a^2} + \frac{y^2}{b^2} = 1$$

is πab.

a. Show that the above result implies that the area of the circle with radius r and center at the origin is πr^2.

b. Find the area of the ellipse having equation

$$\frac{x^2}{2} + \frac{y^2}{8} = 1$$

31. a. Let a and b be positive numbers, and let u and v be numbers such that $u^2 + v^2 \neq 0$. Show that the point

$$\left(\frac{au}{\sqrt{u^2 + v^2}}, \frac{bv}{\sqrt{u^2 + v^2}} \right)$$

lies on the ellipse having equation

$$\frac{x^2}{a^2} + \frac{y^2}{b^2} = 1$$

b. Taking $u = 4$ and $v = -3$ in part (a), find the corresponding point on the ellipse having equation

$$\frac{x^2}{36} + \frac{y^2}{16} = 1$$

32. The *eccentricity* e of an ellipse with equation in the form of (4) is given by

$$e = \sqrt{1 - \frac{b^2}{a^2}}$$

  a. Show that $0 \le e < 1$.
  b. Show that $e = 0$ if and only if the ellipse is a circle.
  c. Find a formula for b in terms of a and e.

33. A planetary orbit is elliptical and is frequently identified by the length $2a$ of the major axis and by the eccentricity e of the orbit. For the earth's orbit around the sun, the numbers $2a$ and e are approximately 299,000,000 (kilometers) and 0.17, respectively.

  c a. Use part (c) of Exercise 32 and a calculator to determine an approximate value of b.
  b. Using the given information, find an equation in the form

$$\frac{x^2}{a^2} + \frac{y^2}{b^2} = 1$$

  that approximately represents the earth's orbit around the sun.

4.7

THE HYPERBOLA

The last of the major kinds of conic sections is the hyperbola. The graph of an equation of the form

$$Ax^2 + By^2 = E \quad \text{with } A \text{ and } B \text{ of different signs}$$

is a hyperbola, and by arranging the constants suitably we can write it in one of the two standard forms

$$\frac{x^2}{a^2} - \frac{y^2}{b^2} = 1 \quad \text{or} \quad \frac{y^2}{a^2} - \frac{x^2}{b^2} = 1 \quad \text{with } a > 0 \text{ and } b > 0$$

The corresponding hyperbolas are sketched in Figure 4.62a, b. They are symmetric with respect to both axes and the origin, which is called the *center* of the hyperbolas.

Notice that the hyperbola given by

$$\frac{x^2}{a^2} - \frac{y^2}{b^2} = 1 \qquad (1)$$

has no y intercepts but has x intercepts a and $-a$. The points $(a, 0)$ and $(-a, 0)$ are called the **vertices** of the hyperbola, and the line segment determined by the vertices is called the **transverse axis** of the hyperbola. The hyperbola also has two oblique asymptotes, as we now show. Let us first consider the part of the graph in the first quadrant, corresponding to $x > 0$ and $y > 0$. Solving the equation in (1) for y (with $y > 0$), we obtain

$$\frac{y^2}{b^2} = \frac{x^2}{a^2} - 1$$

$$y^2 = b^2 \left(\frac{x^2}{a^2} - 1 \right)$$

$$y = b\sqrt{\frac{x^2}{a^2} - 1} = b\sqrt{\frac{x^2}{a^2}\left(1 - \frac{a^2}{x^2}\right)} = \frac{bx}{a}\sqrt{1 - \frac{a^2}{x^2}}$$

If x is positive and becomes arbitrarily large, a^2/x^2 gets close to 0, so $1 - (a^2/x^2)$ and hence $\sqrt{1 - (a^2/x^2)}$ get close to 1. Thus for large positive values of x, y is close to bx/a. This means that the line

$$y = \frac{b}{a}x$$

is an oblique asymptote of the part of the graph in the first quadrant.

Now let us consider the portions of the graph in the other quadrants. By the symmetry of the hyperbola with respect to the origin, the same line is an oblique asymptote of the portion of the hyperbola in the third quadrant (Figure 4.62a). It now follows from the symmetry of the graph with respect to the axes that the line

$$y = -\frac{b}{a}x$$

is an oblique asymptote of the part of the graph in the second and fourth quadrants, as is also indicated in Figure 4.62a.

The hyperbola given by

$$\frac{y^2}{a^2} - \frac{x^2}{b^2} = 1 \qquad (2)$$

has no x intercepts but has the two y intercepts a and $-a$. In this case the points $(0, a)$ and $(0, -a)$ are called the **vertices** of the hyperbola and the line segment determined by the vertices is called the **transverse axis.** In addition, the hyperbola has the two oblique asymptotes

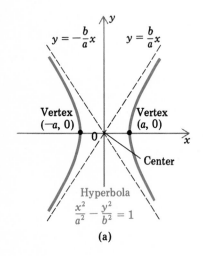

$y = -\frac{b}{a}x$ $y = \frac{b}{a}x$

Vertex $(-a, 0)$ Vertex $(a, 0)$

Center

Hyperbola
$\frac{x^2}{a^2} - \frac{y^2}{b^2} = 1$

(a)

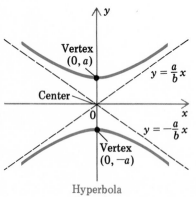

Vertex $(0, a)$

$y = \frac{a}{b}x$

Center

$y = -\frac{a}{b}x$

Vertex $(0, -a)$

Hyperbola
$\frac{y^2}{a^2} - \frac{x^2}{b^2} = 1$

(b)

FIGURE 4.62

$$y = \frac{a}{b} x \quad \text{and} \quad y = -\frac{a}{b} x$$

as indicated in Figure 4.62b.

One way to associate equations (1) and (2) with the corresponding graphs is this: If x comes first (as in (1)), then the graph crosses the x axis; if y comes first (as in (2)), then the graph crosses the y axis. It is also useful to note that, whereas the vertices help in graphing points closest to the center of the hyperbola, the asymptotes help in graphing points on the hyperbola far from the center.

Example 1. Sketch the hyperbola having equation

$$\frac{x^2}{4} - y^2 = 1$$

and note the vertices and asymptotes.

Solution. Since the given equation is equivalent to

$$\frac{x^2}{4} - \frac{y^2}{1} = 1$$

it has the form of equation (1) with $a = 2$ and $b = 1$. Therefore the vertices are $(-2, 0)$ and $(2, 0)$, and the asymptotes are

$$y = \frac{1}{2} x \quad \text{and} \quad y = -\frac{1}{2} x$$

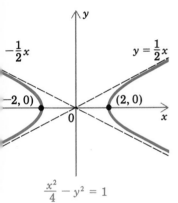

$$\frac{x^2}{4} - y^2 = 1$$

FIGURE 4.63

The hyperbola is sketched in Figure 4.63. □

Example 2. Sketch the hyperbola having equation

$$16y^2 - 9x^2 = 144$$

and note the vertices and asymptotes.

Solution. Dividing both sides of the equation by 144, we obtain

$$\frac{y^2}{9} - \frac{x^2}{16} = 1$$

This equation has the form of (2) with $a = 3$ and $b = 4$. The vertices are $(0, -3)$ and $(0, 3)$, and the asymptotes are

$$y = \frac{3}{4} x \quad \text{and} \quad y = -\frac{3}{4} x$$

$16y^2 - 9x^2 = 144$

FIGURE 4.64

The hyperbola is sketched in Figure 4.64. □

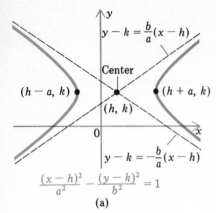

$$\frac{(x-h)^2}{a^2} - \frac{(y-k)^2}{b^2} = 1$$

(a)

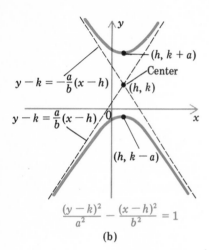

$$\frac{(y-k)^2}{a^2} - \frac{(x-h)^2}{b^2} = 1$$

(b)

FIGURE 4.65

Except in very special cases, the graph of any equation of the form

$$Ax^2 + By^2 + Cx + Dy = E \quad \text{with } A \text{ and } B \text{ of different signs} \quad (3)$$

is a hyperbola whose center can be any point in the plane, depending on the values of the constants. After completing squares and rearranging the constants, we can rewrite (3) either as

$$\frac{(x-h)^2}{a^2} - \frac{(y-k)^2}{b^2} = 1 \qquad (4)$$

whose graph has vertices $(h - a, k)$ and $(h + a, k)$ and asymptotes

$$y - k = \frac{b}{a}(x - h) \quad \text{and} \quad y - k = -\frac{b}{a}(x - h)$$

(Figure 4.65a), or as

$$\frac{(y-k)^2}{a^2} - \frac{(x-h)^2}{b^2} = 1 \qquad (5)$$

whose graph has vertices $(h, k - a)$ and $(h, k + a)$ and asymptotes

$$y - k = \frac{a}{b}(x - h) \quad \text{and} \quad y - k = -\frac{a}{b}(x - h)$$

(Figure 4.65b). In either case, the point (h, k) is the *center* of the hyperbola, and the hyperbola is symmetric with respect to the lines $x = h$ and $y = k$.

Example 3. Sketch the hyperbola having equation

$$9x^2 - 25y^2 - 18x - 50y = -209$$

and note the center, vertices, and the asymptotes.

Solution. In order to put the equation into the form of (4) or (5), we complete squares:

$$(9x^2 - 18x) - (25y^2 + 50y) = -209$$

$$9(x^2 - 2x) - 25(y^2 + 2y) = -209$$

$$9(x^2 - 2x + 1 - 1) - 25(y^2 + 2y + 1 - 1) = -209$$

$$9(x^2 - 2x + 1) - 9 - 25(y^2 + 2y + 1) + 25 = -209$$

$$9(x - 1)^2 - 25(y + 1)^2 = -209 + 9 - 25 = -225$$

$$\frac{(y + 1)^2}{9} - \frac{(x - 1)^2}{25} = 1 \qquad (6)$$

Equation (6) has the form of (5) with $h = 1$, $k = -1$, $a = 3$, and $b = 5$. As a result, the center is $(1, -1)$ and the vertices are $(1, -1 - 3) = (1, -4)$ and $(1, -1 + 3) = (1, 2)$. The asymptotes are

$$y + 1 = \frac{3}{5}(x - 1) \quad \text{and} \quad y + 1 = -\frac{3}{5}(x - 1)$$

The hyperbola is sketched in Figure 4.66. □

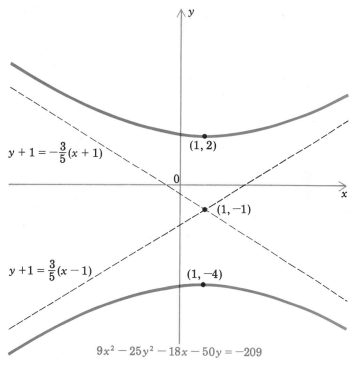

FIGURE 4.66

Like the parabola and ellipse, the hyperbola has a geometric description. Let F_1 and F_2 be two distinct points in the plane, and let k be a positive number less than the distance between F_1 and F_2. Then the set of all points P in the plane such that

$$|d(F_1, P) - d(F_2, P)| = k$$

is a hyperbola (Figure 4.67). The points F_1 and F_2 are called the *foci* of the hyperbola.

We mention a few applications of hyperbolas.

1. Comets that do not move in elliptical orbits around the sun almost always move in hyperbolic orbits. (In theory they can also move in parabolic orbits.) Such comets could be seen at most once from earth.

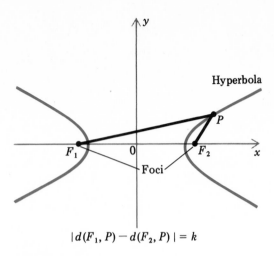

$$|\,d(F_1, P) - d(F_2, P)\,| = k$$

FIGURE 4.67

2. Boyle's Law, relating the pressure p and the volume V of a perfect gas at constant temperature, states that $pV = c$, where c is a constant. The graph of p as a function of V is a hyperbola.
3. Hyperbolas can be used to locate the source of a sound heard at three different locations.

Summary

As we mentioned at the outset of Section 4.5, the conic sections are generated by cutting a double cone by various planes (see Figure 4.45). Complementing this relationship between the different conic sections is the fact that our equations for the conic sections can be described in a unified way. More precisely, the graph of any equation of the form

$$Ax^2 + By^2 + Cx + Dy = E$$

is a conic section:

1. If $A \neq 0$ and $B = 0$, or if $A = 0$ and $B \neq 0$, then the graph is a parabola.
2. If A and B have the same sign, then the graph is an ellipse (or is degenerate).
3. If A and B have different signs, then the graph is a hyperbola (or is degenerate).

Our short introduction to conic sections gives you some of the basic information about them and their associated equations. Most books on calculus have a more extensive treatment of conic sections.

EXERCISES 4.7

In Exercises 1–14, sketch the hyperbola, noting the center, vertices, and asymptotes.

1. $x^2 - y^2 = 1$

2. $y^2 - x^2 = 4$

3. $\dfrac{x^2}{9} - \dfrac{y^2}{4} = 1$

4. $\dfrac{y^2}{25} - \dfrac{x^2}{16} = 1$

5. $25y^2 - 4x^2 = 100$

6. $4x^2 - y^2 = 4$

7. $\dfrac{(x + 2)^2}{16} - \dfrac{(y + 1)^2}{25} = 1$

8. $\dfrac{(y - 1)^2}{1/4} - x^2 = 1$

9. $4y^2 - x^2 + 8y = 0$

10. $16x^2 - y^2 - 32x - 6y = 57$

11. $9x^2 - 16y^2 - 54x + 225 = 0$

12. $x^2 - 4y^2 - 4x - 8y = 4$

13. $9x^2 - 4y^2 - 18x - 24y = 63$

14. $9x^2 - 4y^2 - 18x - 24y = 26$

In Exercises 15–20, find an equation of the hyperbola with the given properties.

15. Center: $(0, 0)$; x intercept: 2; asymptote: $y = -3x$

16. Center: $(0, 0)$; y intercept: $-\frac{1}{2}$; asymptote: $y = \frac{3}{2}x$

17. Center: $(0, 0)$; vertex: $(0, 4)$; $(-4, -3\sqrt{3})$ on the hyperbola

18. Center: $(-1, 2)$; vertex: $(-1, -3)$; $(-3, -4)$ on the hyperbola

19. Center: $(2, -3)$; $(0, -3)$ and $(6, -3 + 2\sqrt{3})$ on the hyperbola

20. Center: $(7, -6)$; $(7, -1)$ on the hyperbola; asymptote: $y + 6 = \frac{5}{7}(x - 7)$

21. For what values of c does the hyperbola having equation $x^2 + 2x - c^2y^2 + y = 2$ pass through the point $(1, 2)$?

22. For what values of h does the hyperbola having equation

$$\frac{(x - h)^2}{16} - \frac{(y + 1)^2}{36} = 1$$

pass through the point $(\sqrt{10}, 1)$?

23. a. Let a and b be positive numbers, and let u and v be numbers such that $u^2 - v^2 > 0$. Show that the point

$$\left(\frac{au}{\sqrt{u^2 - v^2}}, \frac{bv}{\sqrt{u^2 - v^2}} \right)$$

lies on the hyperbola having equation

$$\frac{x^2}{a^2} - \frac{y^2}{b^2} = 1$$

b. Taking $u = 13$ and $v = -5$ in part (a), find the corresponding point on the hyperbola having equation

$$\frac{x^2}{4} - \frac{y^2}{9} = 1$$

24. A hyperbola having an equation of the form

$$\frac{x^2}{a^2} - \frac{y^2}{b^2} = 1 \quad \text{or} \quad \frac{y^2}{b^2} - \frac{x^2}{a^2} = 1 \tag{7}$$

with $a = b$ is called an ***equilateral hyperbola***. Show that a hyperbola satisfying an equation of the form in (7) is equilateral if and only if its asymptotes are perpendicular to one another.

In Exercises 25–36, determine the type of conic section represented by the given equation. Determine the vertex of each parabola and the center of each ellipse or hyperbola.

25. $2x^2 - 4x - 3y + 3 = 0$
26. $x^2 + 3y^2 = 3$
27. $9x^2 - 4y^2 - 54x + 32y - 53 = 0$
28. $12y^2 - 12y + 8x + 15 = 0$
29. $x^2 + 6x - y + 11 = 0$
30. $9x^2 + 16y^2 + 72x - 96y + 287 = 0$
31. $9x^2 = 4y^2 - 8y + 3$
32. $x^2 + 6x + 9y^2 - 27 = 0$
33. $3y^2 + 6y - 2x = 3$
34. $25x^2 - 2y^2 = 100x + 8y - 142$
35. $9x^2 + 9y^2 + 6x - 24y = 28$
36. $9x^2 - 4y^2 - 54x + 32y - 19 = 0$

KEY TERMS

polynomial function
quadratic function
minimum value of a function
maximum value of a function
rational function
  vertical asymptote
  horizontal asymptote
  oblique asymptote
conic section
parabola
  axis
  vertex
  directrix
  focus

ellipse
  center
  vertex
  major axis
  minor axis
  focus
hyperbola
  center
  vertex
  transverse axis
  asymptote
  focus

REVIEW EXERCISES

In Exercises 1–4, sketch the graph of the given function, and identify the vertex, axis, and intercepts.

1. $f(x) = \frac{1}{4}x^2 - 4$
2. $f(x) = x^2 + 8x + 15$

3. $f(x) = -x^2 - 4x - 7$ 4. $f(x) = x^2 - 3x + 2$

In Exercises 5–16, sketch the graph of the given function, and identify all intercepts and asymptotes.

5. $f(x) = 2 - x^3$

6. $f(x) = 4x^3 - x$

7. $f(x) = x^4 + 2$

8. $g(x) = -\dfrac{1}{(x-3)^2}$

9. $f(x) = -x^2(x-2)$

10. $f(x) = (x+1)(x-2)(x-3)$

11. $f(x) = x^4 - 10x^2 + 9$

12. $g(x) = \dfrac{2x}{x+3}$

13. $g(x) = \dfrac{x+1}{x-1}$

14. $y = \dfrac{(x-1)(x-3)}{(x+2)(x-4)}$

15. $y = \dfrac{x^2 + 3x + 2}{x^3 + 5x^2 + 4x}$

16. $y = \dfrac{x^2}{4x^2 - 1}$

In Exercises 17–22, sketch the graph of the conic section, indicating all pertinent information (center, axes, vertices, asymptotes).

17. $x^2 + y^2 + 2x - 4y + 2 = 0$
18. $6x + y^2 + 6y = 0$
19. $x^2 - y^2 - x - y = 4$
20. $4x^2 + 25y^2 - 8x + 100y + 4 = 0$
21. $12x^2 + 72x + 72 = 9y^2 + 72y$
22. $18y^2 - 36y - 2x + 10 = 0$

23. Let $f(x) = 3x^2 + bx - 4$. One x intercept of the graph of f is 4. Determine the other x intercept.

24. Let $f(x) = 3x^2 - 2x - c$. Determine the value of c for which
a. the graph contains the point $(-1, 4)$.
b. the graph has exactly one x intercept.
c. the x intercepts are 6 units apart.

25. Find the values of h for which the point $(-5, 1)$ lies on the ellipse having equation

$$\frac{(x-h)^2}{25} + \frac{(y+5)^2}{100} = 1$$

26. Show that if u is any real number, then the point

$$\left(\frac{u^2 - 1}{u^2 + 1}, \frac{2u}{u^2 + 1}\right)$$

lies on the circle $x^2 + y^2 = 1$.

27. The Student Union plans to organize a group trip to Mexico during spring vacation. The cost per person is set at \$400 for 300 people, and it decreases \$1 for each additional person up to 100 additional people. What is the cost per person that brings the maximum revenue, and how many people must participate at that cost?

28. Find two numbers whose product is as large as possible if the sum of twice one number and 3 times the other is 36.

29. The height at time t of a cannon ball fired at a 30° angle with respect to the ground is given by

$$h = -16t^2 + \frac{1}{2} v_0 t$$

where v_0 is the initial, or muzzle, velocity. If the muzzle velocity is 1024 feet per second, determine the maximum height attained by the cannon ball.

30. If the muzzle velocity of the cannon in Exercise 29 is doubled, how is the maximum height affected?

31. A wire 2 feet long is to be cut into 2 pieces, one of which will be bent into the shape of a square and the other into the shape of a circle. Determine the lengths into which the wire must be cut if the sum of the areas of the square and circle is to be minimum.

32. What is the largest possible area of a rectangular tablecloth whose perimeter is 24 feet?

33. Let $f(x) = x^3 + x^2$ for $0 \le x \le 1$. Determine the dimensions of the rectangle of maximum area that has base along the x axis and can be inscribed in the shaded region in Figure 4.68. (Hint: The area can be expressed as a quadratic function of x^2. Find the value of x^2 that yields the maximum area.)

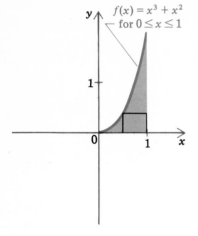

$f(x) = x^3 + x^2$
for $0 \le x \le 1$

FIGURE 4.68

EXPONENTIAL AND LOGARITHMIC FUNCTIONS

5

This chapter is devoted to two kinds of functions that cannot be formed by any of the elementary algebraic operations (addition, subtraction, multiplication, division, and roots). These are the exponential and logarithmic functions. They not only are extremely important in advanced mathematics but also occur with great frequency in the study of such varied topics as radioactivity, compound interest, noise level, and atmospheric pressure.

Numerical answers that involve exponential or logarithmic functions usually can only be given approximately in decimal form. For a high degree of accuracy one needs tables or a calculator. We will normally use a calculator and give our answers accurate to eight places.

5.1

EXPONENTIAL FUNCTIONS

In Section 1.5 we defined the expression a^r for any $a > 0$ and any *rational* number r. However, until now we have not defined expressions such as $2^{\sqrt{3}}$, where the exponent is irrational. In this section we will give a meaning to a^x, in which x can be any number, irrational or rational.

To see how a^x can be defined for an arbitrary irrational number x, let us select a specific number $a > 0$, say $a = 2$. By plotting points from the table

r	-3	-2	-1	0	1	2	3
2^r	$\frac{1}{8}$	$\frac{1}{4}$	$\frac{1}{2}$	1	2	4	8

we obtain Figure 5.1a. The more points $(r, 2^r)$ we plot for other rational values of r, the more the points seem to fall on a smooth curve (Figure 5.1b). In fact, there is exactly one smooth curve that contains all points of the form $(r, 2^r)$, where r is rational (Figure 5.1c). As a result, for any number x we define 2^x to be the y coordinate of the point that lies on the curve directly above x on the x axis. From Figure 5.1c we see that the curve contains the point $(0, 1)$ and slopes upward to the right, which means that

$$\text{if } x < z, \quad \text{then} \quad 2^x < 2^z$$

Notice that if r is close to x, then 2^r is close to 2^x, and moreover, the closer r is to x, the closer 2^r is to 2^x. Thus we can think of 2^x as the "limiting value" of 2^r, where r is rational and r gets closer and closer to x. For example, if $x = \sqrt{3}$, we could let r have the values

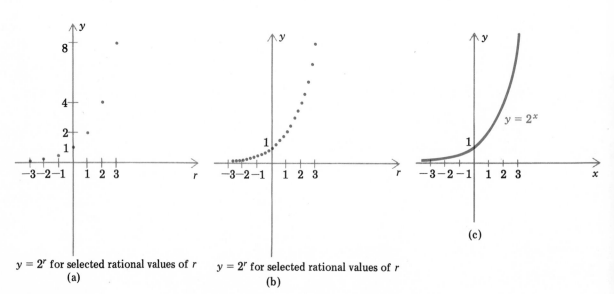

$y = 2^r$ for selected rational values of r
(a)

$y = 2^r$ for selected rational values of r
(b)

(c)

FIGURE 5.1

$$1, \quad 1.7, \quad 1.73, \quad 1.732, \quad 1.73205, \quad 1.7320508, \ldots$$

which are obtained by taking more and more of the decimal expansion of $\sqrt{3}$, so that the successive values of r approach $\sqrt{3}$. The corresponding values of 2^r are

$$2^1, \quad 2^{1.7}, \quad 2^{1.73}, \quad 2^{1.732}, \quad 2^{1.73205}, \quad 2^{1.7320508}, \ldots$$

Evaluating these numbers 2^r with a calculator, we obtain the following table:

r	1	1.7	1.73	1.732	1.73205	1.7320508
2^r	2	3.2490096	3.3172782	3.3218801	3.3219952	3.3219971

It appears from the table that the decimal expansion of $2^{\sqrt{3}}$ begins 3.32199 . . . .

The situation is analogous if a is any fixed number with $a > 1$. For rational values of r the points (r, a^r) lie on a unique smooth curve, and for any number x we define a^x to be the y coordinate of the point lying on that curve directly above the number x on the x axis (Figure 5.2). Notice also that the point $(0, 1)$ is on the curve, and the curve slopes upward to the right, which means that

$$\text{for } a > 1: \quad \text{if } x < z, \quad \text{then} \quad a^x < a^z \qquad (1)$$

Moreover, if x is negative with $-x$ very large, then a^x is close to 0, whereas if x is positive and very large, then a^x is very large.

Now if $0 < a < 1$, then the method of defining a^x is the same. The main difference is that the smooth curve connecting the points (r, a^r) for rational values of r slopes downward to the right, which means that

$$\text{for } 0 < a < 1: \quad \text{if } x < z, \quad \text{then} \quad a^x > a^z \qquad (2)$$

(Figure 5.3). Moreover, if x is positive and very large, then a^x is

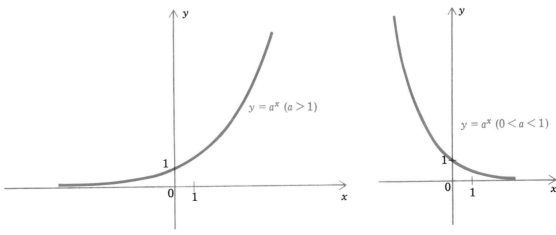

FIGURE 5.2

FIGURE 5.3

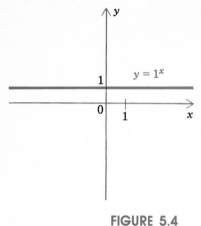

FIGURE 5.4

close to 0, whereas if x is negative and $-x$ is very large, then a^x is very large.

Finally, if $a = 1$, then $a^r = 1^r = 1$ for any rational value of r, so the points (r, a^r) all lie on the horizontal line $y = 1$ (Figure 5.4). By defining 1^x to be the y coordinate of the point on the curve above x on the x axis, we find that $1^x = 1$ for any real number x.

Led by our discussion above, we define a^x for any fixed $a > 0$ and any number x as follows:

> a^x = the y coordinate of the point that lies above the point x on the x axis and lies on the smooth curve that contains all points of the form (r, a^r), where r is a rational number

No matter what the values of a and x are (so long as $a > 0$), a^x is the "limiting value" of a^r, where r is rational and r gets closer and closer to x.

The "definition" we have just presented is intuitive, not a rigorous mathematical definition, which must be left to more advanced texts. Nevertheless, we will abide by our "definition," which is consistent with the definition of a^r given earlier for rational values of r.

In the expression a^x the number a is called the **base** and x the **exponent** or **power**. By holding the base fixed and allowing the exponent to vary, we obtain a function.

DEFINITION 5.1   Let $a > 0$. The function f defined by

$$f(x) = a^x$$

is called the **exponential function with base a**.

For any $a > 0$ with $a \neq 1$, the following are properties of the function a^x and its graph:

1. $a^x > 0$ for all x, and the range of a^x consists of *all* positive numbers.
2. The y intercept is 1, and there is no x intercept.
3. The x axis is a horizontal asymptote of the graph of a^x.
4. If $x < z$, then $\begin{cases} a^x < a^z & \text{for } a > 1 \\ a^x > a^z & \text{for } 0 < a < 1 \end{cases}$

We notice that if $a > 1$, then the graph of a^x slopes upward to the right, with the rate of slope becoming larger farther to the right. By contrast, if $0 < a < 1$, then the graph of a^x slopes downward to the right, with the rate of slope becoming smaller farther to the right.

As we did earlier when $a = 2$, we can sketch the graph of the exponential function a^x for any specific positive value of a by computing a few values of a^x for specially chosen rational values of x

and then connecting the corresponding points on the graph with a smooth curve. Of course, the more values we compute, the more accurate our sketch is likely to be.

Example 1. Let $f(x) = (\frac{3}{2})^x$. Sketch the graph of f.

Solution. Using the values in the table

x	-3	-2	-1	0	1	2	3
$f(x)$	$\frac{8}{27}$	$\frac{4}{9}$	$\frac{2}{3}$	1	$\frac{3}{2}$	$\frac{9}{4}$	$\frac{27}{8}$

we plot the corresponding points. Then we sketch a smooth curve through the points (Figure 5.5). $\square$

Example 2. Let $g(x) = (\frac{1}{3})^x$. Sketch the graph of g.

Solution. We make a short table of values:

x	-2	-1	0	1	2
$g(x)$	9	3	1	$\frac{1}{3}$	$\frac{1}{9}$

Then we plot the corresponding points and connect them with a smooth curve (Figure 5.6). $\square$

The various laws of exponents that apply to a^x for rational values of x remain valid for arbitrary real values of x. Although we will not prove the laws, we list them below:

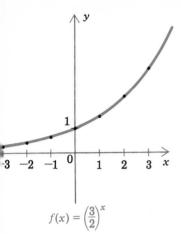

$f(x) = \left(\frac{3}{2}\right)^x$

FIGURE 5.5

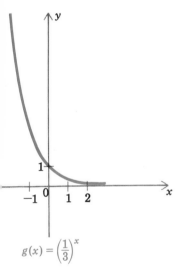

$g(x) = \left(\frac{1}{3}\right)^x$

FIGURE 5.6

> **LAWS OF EXPONENTS** Let a and b be fixed positive numbers. For any numbers x and y we have
>
> i. $a^x a^y = a^{x+y}$ v. $\left(\dfrac{a}{b}\right)^x = \dfrac{a^x}{b^x}$
>
> ii. $(a^x)^y = a^{xy}$ vi. $\dfrac{a^x}{a^y} = a^{x-y}$
>
> iii. $(ab)^x = a^x b^x$ vii. $a^{-x} = \dfrac{1}{a^x} = \left(\dfrac{1}{a}\right)^x$
>
> iv. $1^x = 1$ viii. $a^0 = 1$

Example 3. Simplify $\dfrac{4^{\sqrt{6}} \cdot 16^{\sqrt{6}}}{4^{3\sqrt{6}-1}}$.

Solution. By the Laws of Exponents,

$$\frac{4^{\sqrt{6}} \cdot 16^{\sqrt{6}}}{4^{3\sqrt{6}-1}} = \frac{4^{\sqrt{6}}(4^2)^{\sqrt{6}}}{4^{3\sqrt{6}-1}} = \frac{4^{\sqrt{6}} \cdot 4^{2\sqrt{6}}}{4^{3\sqrt{6}-1}} = \frac{4^{\sqrt{6}+2\sqrt{6}}}{4^{3\sqrt{6}-1}}$$

$$= \frac{4^{3\sqrt{6}}}{4^{3\sqrt{6}} \cdot 4^{-1}} = \frac{1}{4^{-1}} = 4 \quad \square$$

From Law (vii) of Exponents,

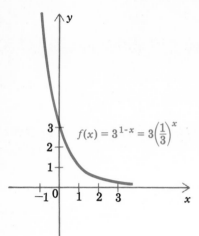

$$3^{-x} = \left(\frac{1}{3}\right)^x$$

so that the graph of 3^{-x} is the same as the graph of $(\frac{1}{3})^x$, which appears in Figure 5.6. We will use this fact in the solution of the following example.

Example 4. Let $f(x) = 3^{1-x}$. Sketch the graph of f.

Solution. By the Laws of Exponents and our remark above,

$$f(x) = 3^{1-x} = 3^1 3^{-x} = 3\left(\frac{1}{3}\right)^x$$

It follows that the graph of f can be obtained by multiplying the y coordinate of each point on the graph of $(\frac{1}{3})^x$ (shown in Figure 5.6) by 3. The resulting graph of f appears in Figure 5.7. ☐

FIGURE 5.7

The Exponential Function e^x

Of all the possible bases a for an exponential function, one outshines the others in importance. It is the one which was first called e by the Swiss mathematician Leonhard Euler (1707–1783). This number e is an irrational number whose decimal expansion begins

$$e = 2.71828182845904523536 \ldots$$

One mathematical way of obtaining e is as the limit of numbers of the form $(1 + 1/x)^x$ for positive values of x that become arbitrarily large. For a few integral values of x we have

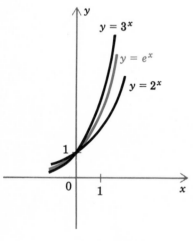

$$\left(1 + \frac{1}{10}\right)^{10} \approx 2.5937425 \qquad \left(1 + \frac{1}{1000}\right)^{1000} \approx 2.7169239$$

$$\left(1 + \frac{1}{100}\right)^{100} \approx 2.7048138 \qquad \left(1 + \frac{1}{10,000}\right)^{10,000} \approx 2.7181459$$

The function e^x is called the **natural exponential function**, or simply the **exponential function**. It is remarkable how often it arises both inside and outside of mathematics (see Section 5.6 for a few applications). Since $2 < e < 3$, the graph of e^x has the same general shape as the graphs of 2^x and 3^x and lies between them (Figure 5.8).

Since e^x occurs so frequently in applications, it is useful to be able to approximate its value for various values of x. A calculator can be used for this purpose. The method of obtaining e^x on a calculator varies, depending on the instrument. But the basic idea is the same. On one such calculator we would evaluate $e^{0.256}$ by pressing first the decimal key that registers a point, then the 2, 5, and 6 keys,

FIGURE 5.8

and finally the e^x key. Displayed on the calculator would be the approximate value 1.2917527, so that

$$e^{0.256} \approx 1.2917527$$

Next we will sketch the graph of a function related to the exponential function and basic to the study of probability.

Example 5. Let $f(x) = e^{-x^2}$. Sketch the graph of f.

Solution. We first make a table of approximate values, with the help of a calculator:

x	0	$\frac{1}{2}$	1	$\frac{3}{2}$	2
$f(x)$	1	0.77880078	0.36787944	0.10539922	0.018315639

Then we plot the corresponding points on the graph. Because

$$f(-x) = e^{-(-x)^2} = e^{-x^2} = f(x)$$

f is even, and thus its graph is symmetric with respect to the y axis. The graph of f is sketched in Figure 5.9. $\square$

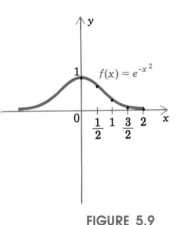

FIGURE 5.9

Exponential Equations

Let $a > 0$ with $a \neq 1$. From (1) and (2) it follows that if $x \neq z$, then $a^x \neq a^z$. This means that for $a > 0$ with $a \neq 1$,

$$\text{if } a^x = a^z, \quad \text{then} \quad x = z \tag{3}$$

Another way of expressing (3) is to say that such a function a^x is **one-to-one**. As a result, the function a^x has an inverse, a fact that will be of fundamental importance in the next section.

We can use (3) to solve certain equations involving exponential functions.

Example 6. Solve the equation $9^x = 3^{x+2}$ for x.

Solution. In order to use (3), we need the same base on both sides of the equation. Since $9 = 3^2$, the Laws of Exponents imply that

$$9^x = (3^2)^x = 3^{2x}$$

Thus the given equation is equivalent to

$$3^{2x} = 3^{x+2}$$

From (3) we find that

$$2x = x + 2 \quad \text{so that} \quad x = 2$$

Check: $9^2 = 81$ and $3^{2+2} = 3^4 = 81$

Therefore the solution of the given equation is 2. $\square$

EXERCISES 5.1

In Exercises 1–14, sketch the graph of the given function.

1. $f(x) = 4^x$

2. $f(x) = \left(\dfrac{1}{2}\right)^x$

3. $f(x) = -(3^x)$

4. $f(x) = -3(2^x)$

5. $f(x) = 1^{x+2}$

6. $f(t) = 1 + 2^t$

7. $g(t) = 3 - 2^{-t}$

8. $g(x) = e^{-x}$

9. $g(x) = 2^{-2x}$

10. $h(x) = 3^{x+1}$

11. $h(x) = \sqrt{2}(2^{x+1/2})$

12. $g(x) = (0.5)^{2x-1}$

13. $g(x) = \left(\dfrac{1}{2}\right)^{|x|}$

14. $g(x) = \left(\dfrac{1}{2}\right)^{-|x|}$

In Exercises 15–22, simplify the given expression.

15. $\dfrac{\pi^{\sqrt{5}}\pi^{3-\sqrt{5}}}{\pi}$

16. $\dfrac{12^{2e}\,12^{3-3e}}{12^{-e}\cdot 144}$

17. $\dfrac{5^{\sqrt{6}}25^{3\sqrt{3}}}{125^{\sqrt{12}}}$

18. $(\sqrt{2}^{\sqrt{2}})^2$

19. $(2^{\sqrt{2}})^{-\sqrt{2}}$

20. $\left(\dfrac{1}{7^{\sqrt{5}}}\right)^{2\sqrt{5}}$

21. $(\sqrt[3]{9})^\pi(\sqrt[3]{3})^\pi$

22. $\dfrac{(\sqrt{6})^{\sqrt{3}}(\sqrt{5})^{\sqrt{3}}}{(\sqrt{10})^{\sqrt{3}}}$

In Exercises 23–32, find the base a if the graph of $y = a^x$ contains the given point.

23. $(3, 8)$ 24. $(3, 27)$ 25. $(1, 7)$ 26. $(2, \frac{1}{4})$

27. $(2, \frac{1}{9})$ 28. $(\frac{1}{2}, \frac{1}{2})$ 29. $(-\frac{1}{2}, 2)$ 30. $(3, -\frac{1}{64})$

31. $(\frac{1}{3}, \frac{1}{4})$ 32. $(2, a)$

c In Exercises 33–38, use a calculator to approximate the given number.

33. $e^{0.34}$

34. $e^{0.94}$

35. $e^{-1.2}$

36. e^{-3}

37. $\sqrt{e}$

38. $\sqrt[4]{e}$

c In Exercises 39–42, use a calculator to determine the smallest integer x that satisfies the given inequality.

39. $2^{\sqrt{2x}} > 25$

40. $(1.1)^{\sqrt{2x}} > 38$

41. $x^{\sqrt{3}} > 200$

42. $x^{\sqrt{2/3}} > 13.3$

In Exercises 43–48, solve the given equation.

43. $2^{3x} = 2^{5x^2/4}$

44. $(5^x)^{x+1} = 5^{2x+12}$

45. $(\sqrt{3})^{x-1} = 3^{x+\sqrt{2}}$

46. $(6^{1/6})^{x^2} = (\sqrt{6})^{x\sqrt{5}}$

47. $(5^x)^{x+1} = (\sqrt{5})^{2x+10}$

48. $4^{\sqrt{x+1}} = 2^{3x-2}$

49. The **hyperbolic sine** function, whose value at x is denoted by $\sinh x$, is given by

$$\sinh x = \frac{e^x - e^{-x}}{2}$$

Sketch the graph of $\sinh x$. (This function is important in engineering.)

50. The **hyperbolic cosine** function, whose value at x is denoted by cosh x, is given by

$$\cosh x = \frac{e^x + e^{-x}}{2}$$

Sketch the graph of cosh x. (This function is also important in engineering.)

51. Use the formulas given in Exercises 49 and 50 to show that for every number x,

$$(\cosh x)^2 - (\sinh x)^2 = 1$$

This is the fundamental equation relating the hyperbolic sine and cosine functions.

52. Use a calculator to determine which is larger, 3^π or π^3.

53. Use a calculator to determine which is larger, $(e^e)^e$ or $e^{(e^e)}$.

54. Use a calculator to compute the value of $(1 + 1/x)^x$ for $x = 100{,}000{,}000 = 10^8$. The result should be the same as the value given by your calculator for e^x with $x = 1$, although the numbers $(1 + 1/10^8)^{10^8}$ and $e^1 = e$ are different.

55. In calculus it is shown that the sums of the form

$$1 + x + \frac{x^2}{2!} + \frac{x^3}{3!} + \cdots + \frac{x^n}{n!} \qquad \text{where } n \text{ is a positive integer}$$

approach e^x as n increases without bound.
   a. Using a calculator, show that if $x = 1$ and $n = 10$, then the sum agrees with $e^1 = e$ to eight places.
   b. Using a calculator, determine to how many places $e^{1.5}$ and the sum agree if $x = 1.5$ and $n = 10$.

56. Let $a > 0$. Use Law (i) of Exponents twice to show that

$$a^{x+y+z} = a^x a^y a^z$$

57. Use (1) of this section to prove (2). (*Hint:* If $0 < a < 1$, then $1/a > 1$.)

58. Let $f(x) = e^{-ae^{-bx}}$, where a and b are positive constants. The graph of f, which is referred to as the **Gompertz growth curve**, is used by actuaries in the preparation of life expectancy tables. If $x < z$, determine whether $f(x) < f(z)$ or $f(x) > f(z)$.

59. The air pressure $p(x)$ at a height of x feet above sea level is given approximately by

$$p(x) = e^{-(4.101 \times 10^{-4})x}$$

where $p(x)$ is given in "atmospheres." Determine the air pressure at a height of
   a. 29,028 feet (summit of Mt. Everest)
   b. 14,110 feet (summit of Pikes Peak)
   c. 1,250 feet (top of Empire State Building)
   d. −282 feet (bottom of Death Valley)

60. One test that measures the thyroid condition in a human uses a small dose D of radioactive iodine I^{131} injected into the blood stream. The amount A remaining in the blood after t days is related to D by the formula

$$A(t) = De^{-0.0086t}$$

Suppose a dose is injected into the bloodstream at noon. Determine the percent in the bloodstream

a. at 1 P.M. b. one half day later c. one day later

5.2

LOGARITHMIC FUNCTIONS

Let $a > 0$ with $a \neq 1$. As we observed in the preceding section, the range of the function a^x consists of all positive numbers. Consequently if $y > 0$, then there is a number x such that $a^x = y$, and in fact this number x is unique. (Indeed, if z also has the property that $a^z = y$, then $a^x = a^z$, so by (3) of Section 5.1, it follows that $x = z$.) A special name is given to x.

DEFINITION 5.2

Let $a > 0$ with $a \neq 1$. For any positive number y, the **logarithm of y to the base a** is the unique number x such that $a^x = y$, and is denoted by $\log_a y$.

According to the definition of $\log_a y$, for any x and any $y > 0$, we have

$$\log_a y = x \quad \text{if and only if} \quad a^x = y \tag{1}$$

The following examples illustrate the use of (1) in passing back and forth between equations involving exponents and equations involving logarithms.

Example 1. Convert the following equations involving exponents to equations involving logarithms.

a. $10^2 = 100$ b. $81^{1/4} = 3$ c. $16 = 3^c$

Solution. In each case we use (1) to transform a statement of the form $a^x = y$ to one of the form $\log_a y = x$:

a. From $10^2 = 100$ we obtain $\log_{10} 100 = 2$.
b. From $81^{1/4} = 3$ we obtain $\log_{81} 3 = \frac{1}{4}$.
c. From $16 = 3^c$ we obtain $\log_3 16 = c$. □

Example 2. Convert the following equations involving logarithms to equations involving exponents.

a. $\log_3 9 = 2$ b. $\log_2 32 = 5$ c. $\log_c 81 = 7$

Solution. In each part we use (1) to transform a statement of the form $\log_a y = x$ to one of the form $a^x = y$.

a. From $\log_3 9 = 2$ we obtain $3^2 = 9$.
b. From $\log_2 32 = 5$ we obtain $2^5 = 32$.
c. From $\log_c 81 = 7$ we obtain $c^7 = 81$. $\square$

We can sometimes recognize the numerical value of a logarithm more easily by converting to exponents, as we illustrate now.

Example 3. Evaluate the following expressions.

a. $\log_2 8$ b. $\log_3 3$ c. $\log_4 256$

Solution.

a. If $\log_2 8 = x$, then by (1), $2^x = 8$. But $2^3 = 8$, so that $x = 3$. Thus $\log_2 8 = 3$.
b. If $\log_3 3 = x$, then by (1), $3^x = 3$. But $3^1 = 3$, so that $x = 1$. Thus $\log_3 3 = 1$.
c. If $\log_4 256 = x$, then by (1), $4^x = 256$. But $4^4 = 256$, so that $x = 4$. Thus $\log_4 256 = 4$. $\square$

Example 4. Evaluate the following expressions.

a. $\log_{10} 1$ b. $\log_{25} \frac{1}{5}$ c. $\log_4 8$

Solution.

a. If $\log_{10} 1 = x$, then by (1), $10^x = 1$. But $10^0 = 1$, so that $x = 0$. Therefore $\log_{10} 1 = 0$.
b. If $\log_{25} \frac{1}{5} = x$, then by (1), $25^x = \frac{1}{5}$. But $25^{-1/2} = \frac{1}{5}$, so that $x = -\frac{1}{2}$. As a result, $\log_{25} \frac{1}{5} = -\frac{1}{2}$.
c. If $\log_4 8 = x$, then $4^x = 8$. To determine x, we need to write 8 as a power of 4. Since

$$8 = 4 \cdot 2 = 4^1 4^{1/2} = 4^{3/2}$$

it follows that $4^x = 8$ if $4^x = 4^{3/2}$, which means that $x = \frac{3}{2}$. Consequently $\log_4 8 = \frac{3}{2}$. $\square$

Next we solve equations involving logarithms.

Example 5. Solve the following equations for x.

a. $\log_{10} x = 0$ c. $\log_x \pi = -1$
b. $\log_e x = 1$ d. $\log_x 81 = 4$

Solution.

a. By (1) the equation $\log_{10} x = 0$ is equivalent to $10^0 = x$, so that $x = 10^0 = 1$. Thus the solution is 1.
b. By (1) the equation $\log_e x = 1$ is equivalent to $e^1 = x$, so that $x = e^1 = e$. Thus the solution is e.
c. By (1) the equation $\log_x \pi = -1$ is equivalent to $x^{-1} = \pi$, or $1/x = \pi$. Thus $x = 1/\pi$, so the solution is $1/\pi$.
d. By (1) the equation $\log_x 81 = 4$ is equivalent to $x^4 = 81$ with $x > 0$. But the only positive value of x for which $x^4 = 81$ is 3. Therefore 3 is the solution. $\square$

By parts (a) and (b) of Example 5, $\log_{10} 1 = 0$ and $\log_e e = 1$, respectively. These are special cases of the following general formulas that hold for any $a > 0$ with $a \neq 1$:

$$\log_a 1 = 0 \tag{2}$$

$$\log_a a = 1 \tag{3}$$

Example 6. Solve the following equations for x.

a. $\log_4 \dfrac{x}{5} = 3$ b. $\log_{10} (x - 6) = 2$

Solution.

a. By (1), the equation $\log_4 (x/5) = 3$ is equivalent to $4^3 = x/5$, so that

$$x = 5(4^3) = 5(64) = 320$$

Therefore the solution is 320.

b. By (1), the equation $\log_{10} (x - 6) = 2$ is equivalent to $10^2 = x - 6$, so that

$$x = 10^2 + 6 = 106$$

Therefore the solution is 106. □

Having defined $\log_a y$ for any positive number y, we can define the associated function (with x as variable).

DEFINITION 5.3   Let $a > 0$ with $a \neq 1$. The **logarithmic function to the base a** is the function f defined by

$$f(x) = \log_a x \quad \text{for } x > 0$$

Caution: Notice that the domain of $\log_a x$ consists of all $x > 0$ and no other numbers. As a result, we cannot take the logarithm of 0 or any negative number.

From (1) and Definition 5.3 we conclude that the functions $\log_a x$ and a^x are inverses of each other (see Definition 3.4, in Section 3.7). Thus the range of the function $\log_a x$ is identical to the domain of the function a^x, which consists of all real numbers.
The basic formulas

$$f(f^{-1}(x)) = x \quad \text{and} \quad f^{-1}(f(x)) = x$$

relating a function and its inverse, yield the formulas

$$a^{\log_a x} = x \quad \text{for } x > 0 \tag{4}$$

$$\log_a (a^x) = x \quad \text{for all } x \tag{5}$$

For example,

$$10^{\log_{10}\sqrt{7}} = \sqrt{7} \quad \text{and} \quad \log_e (e^{5.1}) = 5.1$$

Now suppose that x and z are positive numbers and

$$\log_a x = \log_a z$$

Then by (4),

$$x = a^{\log_a x} = a^{\log_a z} = z$$

Thus we have shown that if $\log_a x = \log_a z$, then $x = z$. The converse is obviously true, so we obtain the important result

$$\log_a x = \log_a z \quad \text{if and only if} \quad x = z \tag{6}$$

We can use (6) to solve certain equations involving logarithms.

Example 7. Solve the equation $\log_2 (x^2 - 7) = \log_2 6x$.

Solution. By (6) with $a = 2$, the given equation is equivalent to the following equations:

$$x^2 - 7 = 6x$$
$$x^2 - 6x - 7 = 0$$
$$(x - 7)(x + 1) = 0$$
$$x = 7 \quad \text{or} \quad x = -1$$

Check: $\log_2 (7^2 - 7) = \log_2 42$ and $\log_2 (6 \cdot 7) = \log_2 42$
$\log_2 [(-1)^2 - 7]$ is undefined, since $(-1)^2 - 7 = -6$ is negative.

Therefore -1 is an extraneous solution, and consequently the solution of the given equation is 7. $\square$

Graphs of Logarithmic Functions

Since every logarithmic function is the inverse of an exponential function, we can obtain the graph of a logarithmic function by reflecting the graph of the corresponding exponential function through the line $y = x$.

Example 8. Sketch the graphs of the following functions:
    a. $\log_2 x$ b. $\log_{1/3} x$

Solution. We sketched the graphs of 2^x and $(\frac{1}{3})^x$ in Figures 5.1c and 5.6, respectively. By reflecting these graphs through the line $y = x$, we obtain the graphs of $\log_2 x$ and $\log_{1/3} x$, respectively, in Figure 5.10a, b. $\square$

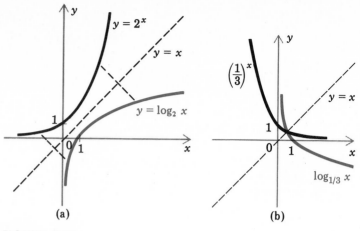

(a)   (b)

FIGURE 5.10

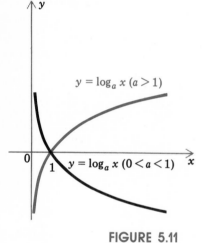

FIGURE 5.11

If $a > 0$ with $a \neq 1$, then the graph of the function $\log_a x$ is in Figure 5.11, depending on whether $a > 1$ or $0 < a < 1$. The following features of the function $\log_a x$ and its graph are inherited from the corresponding features of the function a^x and its graph.

i. The range of $\log_a x$ consists of all numbers.
ii. The x intercept is 1, and there is no y intercept.
iii. The y axis is a vertical asymptote of the graph.
iv. If $x < z$, then $\begin{cases} \log_a x < \log_a z & \text{for } a > 1 \\ \log_a x > \log_a z & \text{for } 0 < a < 1 \end{cases}$

We also observe that far to the right of the y axis the graph of the function $\log_a x$ becomes very flat, although the graph does not have a horizontal asymptote.

Special Bases

The base of a logarithmic function can be any positive number except 1, which means that $\sqrt{2}$ and $1/\pi$ are perfectly reasonable bases. However, from the viewpoint of mathematical theory and practical applications, the base is almost always greater than 1, and moreover, there are two bases that are far more widely used than any other: e and 10. In calculus and in other branches of mathematics, the logarithm to the base e plays a special role, and is called the *natural logarithmic function*. Its value at x is normally denoted by $\ln x$, so that we would write

$$\ln x \quad \text{instead of} \quad \log_e x$$

(The expression "ln" stands for the Latin phrase "logarithmus naturalis.") In contrast, in numerical calculation the logarithm to the base 10 has historically reigned supreme, and it is called the *common logarithmic function*. Its value at x is frequently denoted by log x, so that we would write

$$\log x \quad \text{instead of} \quad \log_{10} x$$

The advent of calculators as well as larger computers have made numerical calculations with common logarithms virtually obsolete.

EXERCISES 5.2

In Exercises 1–6, convert the exponential equation to a logarithmic equation.

1. $10^1 = 10$ 2. $2^4 = 16$ 3. $(\frac{1}{2})^{-5} = 32$
4. $343^{1/3} = 7$ 5. $x^{-4} = 16$ 6. $2^x = 9$

In Exercises 7–12, convert the logarithmic equation to an exponential equation.

7. $\log_{10} 1000 = 3$ 8. $\log_{12} 1 = 0$ 9. $\log_4 32 = 2.5$
10. $\log_{1.5} x = 3$ 11. $\ln x = 2$ 12. $\ln e^5 = x$

In Exercises 13–42, evaluate the given logarithm.

13. $\log_{10} 1000$ 14. $\log_2 4$ 15. $\log_2 16$
16. $\log_2 \frac{1}{4}$ 17. $\log_3 \frac{1}{81}$ 18. $\log_\pi \pi$
19. $\log_\pi 1$ 20. $\log_\pi (1/\pi)$ 21. $\log_\pi \pi^{3/2}$
22. $\log_{10} (10^e)$ 23. $\ln (e^{-1})$ 24. $\log_{10} \sqrt{10}$
25. $\log_8 2$ 26. $\log_{16} 2$ 27. $\log_{10} 0.01$
28. $\log_{10} 0.0001$ 29. $\ln (1/e)$ 30. $\log_{1/2} \frac{1}{8}$
31. $\log_{1/3} 81$ 32. $\log_4 8$ 33. $\log_4 32$
34. $\log_{27} 9$ 35. $\log_{27} 81$ 36. $\log_9 \frac{1}{27}$
37. $\log_{\sqrt{2}} 8$ 38. $\log_{\sqrt{2}} \frac{1}{32}$ 39. $\log_{2\sqrt{3}} 144$
40. $\log_2 (2^3 + 2^3)$ 41. $\log_4 (8^3 + 8^3)$ 42. $\log_{\sqrt{30}} (5^2 + 5)$

In Exercises 43–64, solve the given equation for x.

43. $\ln x = 0$ 44. $\log_{10} x = 1$
45. $\log_2 x = 4$ 46. $\log_\pi x = -1$
47. $\log_8 x = \frac{1}{3}$ 48. $\log_{1/8} x = \frac{1}{3}$
49. $\log_4 x = -2$ 50. $\log_{\sqrt{2}} x = 2$
51. $\log_{\sqrt{2}} x = 0$ 52. $\log_{1.5} x = -3$
53. $\log_2 (x/3) = 2$ 54. $\log_2 \sqrt{x} = 3$
55. $\log_4 (x - 3) = 3$ 56. $\log_2 (4x - 7) = 4$

57. $\ln (x^2 - 8) = \ln 2x$ 58. $\log_{10} \dfrac{x-1}{x-2} = \log_{10} \dfrac{x+1}{x+2}$

59. $\log_{10} \dfrac{x^2 + 4}{x} = \log_{10} \dfrac{x^2 + 4x + 4}{x + 1}$

60. $\log_3 |x - 1| = 2$

61. $\log_{10} |x^2 - 21| = 2$

62. $2^{\log_3 27} = x$

63. $3^{\log_2 x} = 27$

64. $4^x = (\frac{1}{4})^x$

In Exercises 65–70, solve the given equation for x.

65. $\log_x 16 = 2$

66. $\log_x 64 = 2$

67. $\log_x 125 = 3$

68. $\log_{x-1} 81 = 4$

69. $\log_{x^2} 16 = 2$

70. $\log_{x^2 - 2x} 64 = 2$

In Exercises 71–78, find the base a of the logarithmic function whose graph contains the given point.

71. $(4, 2)$

72. $(64, 3)$

73. $(3, \frac{1}{2})$

74. $(3, -\frac{1}{2})$

75. $(\frac{1}{27}, 3)$

76. $(\frac{1}{27}, -3)$

77. $(\frac{1}{2}, \frac{1}{2})$

78. $(\frac{1}{2}, -\frac{1}{2})$

In Exercises 79–86, sketch the graph of the given function.

79. $f(x) = 3 \log_2 x$

80. $f(x) = -\log_2 x$

81. $f(x) = \log_2 2x$

82. $f(x) = \log_2 |x|$

83. $f(x) = \log_4 (2 - x)$

84. $g(x) = \log_4 (x - 3)$

85. $f(x) = |\log_{1/3} x|$

86. $f(x) = -\log_{1/4} 3x$

87. Is $3 \log_{10} \frac{1}{6} > 2 \log_{10} \frac{1}{6}$? Explain your answer.

88. Show that if $\log_a b = 0$, then $b = 1$.

5.3

LAWS OF LOGARITHMS AND CHANGE OF BASE

Just as there are laws of exponents, there are also laws of logarithms. The most important ones are presented below. All of them follow from the Laws of Exponents given in Section 5.1 and from the definition of logarithms.

LAWS OF LOGARITHMS Let a be a fixed positive number with $a \neq 1$. For any positive numbers x and y,

i. $\log_a (xy) = \log_a x + \log_a y$

ii. $\log_a \dfrac{1}{x} = -\log_a x$

iii. $\log_a \dfrac{x}{y} = \log_a x - \log_a y$

iv. $\log_a (x^c) = c \log_a x$ for any number c

v. $\log_a 1 = 0$

vi. $\log_a a = 1$

To prove (i), let $r = \log_a x$ and $s = \log_a y$. It follows from the definition of logarithm that $a^r = x$ and $a^s = y$. Therefore $xy = a^r a^s$. By the Laws of Exponents (Section 5.1), $a^r a^s = a^{r+s}$, so that

$xy = a^{r+s}$. By another application of the definition of logarithm, it follows that $\log_a (xy) = r + s$. Thus

$$\log_a (xy) = r + s = \log_a x + \log_a y$$

which is (i). Hints for proving (ii)–(iv) are given in Exercises 72–74. Finally, we observe that (v) and (vi) are restatements, respectively, of (2) and (3) of Section 5.2.

The formula in (i) is frequently referred to as *the* Law of Logarithms. It contains the important feature of logarithms that the logarithm of a *product* of two numbers is the *sum* of the logarithms of the two numbers. In a similar vein, (iii) says that the logarithm of a *quotient* of two numbers is the *difference* of the logarithms of the two numbers.

Example 1. Using the facts that $\log_{10} 2 \approx 0.3010$ and $\log_{10} 3 \approx 0.4771$, approximate the numerical value of each of the following logarithms.

    a. $\log_{10} 6$ c. $\log_{10} 1.5$

    b. $\log_{10} \frac{1}{2}$ d. $\log_{10} \sqrt{2}$

Solution.

    a. Since $6 = 2 \cdot 3$, we use (i) to obtain

$$\log_{10} 6 = \log_{10} (2 \cdot 3) = \log_{10} 2 + \log_{10} 3$$
$$\approx 0.3010 + 0.4771 = 0.7781$$

    b. We use (ii) to deduce that

$$\log_{10} \frac{1}{2} = -\log_{10} 2 \approx -0.3010$$

    c. Since $1.5 = \frac{3}{2}$, we apply (iii), which yields

$$\log_{10} 1.5 = \log_{10} \frac{3}{2} = \log_{10} 3 - \log_{10} 2$$
$$\approx 0.4771 - 0.3010 = 0.1761$$

    d. Since $\sqrt{2} = 2^{1/2}$, we can use (iv) to conclude that

$$\log_{10} \sqrt{2} = \log_{10} (2^{1/2}) = \frac{1}{2} \log_{10} 2$$
$$\approx \frac{1}{2} (0.3010) = 0.1505 \quad \square$$

Example 2. Using the facts that $\log_{10} 3 \approx 0.4771$ and $\log_{10} 7 \approx 0.8451$, approximate the numerical value of $\log_{10} \frac{9}{49}$.

Solution. Since $9 = 3^2$ and $49 = 7^2$, we use (iii) and (iv) to conclude that

$$\log_{10} \frac{9}{49} = \log_{10} 9 - \log_{10} 49 = \log_{10} (3^2) - \log_{10} (7^2)$$

$$= 2 \log_{10} 3 - 2 \log_{10} 7 \approx 2(0.4771) - 2(0.8451)$$

$$= -0.7360 \quad \square$$

Caution: There is no law of logarithms that simplifies either $\log_a (x + y)$ or $(\log_a x)/(\log_a y)$. In particular, $\log_a (x + y)$ is *almost never* equal to $\log_a x + \log_a y$, and $(\log_a x)/(\log_a y)$ is *almost never* equal to $\log_a x - \log_a y$.

The Law of Logarithms (i) can be generalized to more than two variables. For three variables it reads as follows:

$$\log_a xyz = \log_a x + \log_a y + \log_a z \tag{1}$$

Example 3. Using the facts that $\log_{10} 3 \approx 0.4771$ and $\log_{10} 7 \approx 0.8451$, approximate the numerical value of $\log_{10} 210$.

Solution. Since $210 = 3 \cdot 7 \cdot 10$, we use (1) to find that

$$\log_{10} 210 = \log_{10} (3 \cdot 7 \cdot 10)$$

$$= \log_{10} 3 + \log_{10} 7 + \log_{10} 10$$

$$\approx 0.4771 + 0.8451 + 1$$

$$= 2.3222$$

Therefore $\log_{10} 210 \approx 2.3222$. $\quad \square$

By combining (i) and (iii) we can obtain

$$\log_a \frac{xy}{z} = \log_a x + \log_a y - \log_a z \tag{2}$$

Example 4. Express $\log_2 \frac{9}{2} + \log_2 \frac{8}{5} - \log_2 \frac{9}{8}$ as a single logarithm, and simplify your answer.

Solution. By (2),

$$\log_2 \frac{9}{2} + \log_2 \frac{8}{5} - \log_2 \frac{9}{8} = \log_2 \left(\frac{\frac{9}{2} \cdot \frac{8}{5}}{\frac{9}{8}} \right) = \log_2 \left(\frac{9}{2} \cdot \frac{8}{5} \cdot \frac{8}{9} \right)$$

$$= \log_2 \frac{32}{5} \quad \square$$

It is possible to simplify the answer to Example 4 further, although the simplified version is not a single logarithm:

$$\log_2 \frac{32}{5} = \log_2 32 - \log_2 5 = \log_2 (2^5) - \log_2 5 = 5 - \log_2 5$$

Solving Equations Involving Logarithms

In the next two examples we will return to solving equations involving logarithms.

Example 5. Solve the equation $3 \log_{10} x = \log_{10} 8$ for x.

Solution. We use (iv) above, and then (6) of Section 5.2 in the following set of equivalent equations:

$$3 \log_{10} x = \log_{10} 8$$
$$\log_{10} (x^3) = \log_{10} 8$$
$$x^3 = 8$$
$$x = 2$$

Check: $3 \log_{10} 2 = \log_{10} (2^3) = \log_{10} 8$

Consequently the solution is 2. □

Example 6. Solve the equation $\log_8 (x + 9) - \log_8 (x - 1) = \frac{1}{3}$ for x.

Solution. This time we use (iii) and the definition of the logarithm to the base 8:

$$\log_8 (x + 9) - \log_8 (x - 1) = \frac{1}{3}$$
$$\log_8 \frac{x + 9}{x - 1} = \frac{1}{3}$$
$$\frac{x + 9}{x - 1} = 8^{1/3} = 2$$
$$x + 9 = 2(x - 1)$$
$$x + 9 = 2x - 2$$
$$x = 11$$

Check: $\log_8 (11 + 9) - \log_8 (11 - 1) = \log_8 20 - \log_8 10$
$$= \log_8 \frac{20}{10} = \log_8 2$$
$$= \log_8 (8^{1/3}) = \frac{1}{3}$$

Thus the solution of the given equation is 11. □

Change of Base

It is sometimes necessary to convert from logarithms in one base to logarithms in another base, that is, to write $\log_b x$ in terms of $\log_a x$ (where values of $\log_a x$ are presumably known). Since calculators usually have keys only for natural logarithms and common logarithms, this conversion may even be necessary when one wishes to use a calculator to make computations involving logarithms.

Suppose a and b are positive numbers different from 1 and we wish to find the numerical value of $\log_b x$ in terms of $\log_a x$. Let

$$r = \log_b x \quad \text{so that} \quad x = b^r$$

Since $a \neq 1$ by hypothesis, we can take logarithms to the base a of both sides of the latter equation and obtain

$$\log_a x = \log_a (b^r)$$

But by (iv),

$$\log_a (b^r) = r \log_a b \tag{3}$$

Since $b \neq 1$ by hypothesis, it follows that $\log_a b \neq 0$. Therefore we may divide both sides of (3) by $\log_a b$ to obtain

$$r = \frac{\log_a x}{\log_a b}$$

Substituting $\log_b x$ for r, we obtain the following formula.

> **CHANGE OF BASE FORMULA**
>
> $$\log_b x = \frac{\log_a x}{\log_a b} \quad \text{for } x > 0 \tag{4}$$

As we observed earlier, most calculators have a key for the common logarithm of a given number. Now if $a = 10$, then (4) becomes

> $$\log_b x = \frac{\log_{10} x}{\log_{10} b} \quad \text{for } x > 0 \tag{5}$$

Since calculators can compute $\log_{10} x$ and $\log_{10} b$ quickly and accurately, a good approximate value for $\log_b x$ can be obtained by calculator.

Example 7. Approximate the value of $\log_2 5$.

Solution. By (5),

$$\log_2 5 = \frac{\log_{10} 5}{\log_{10} 2} \tag{6}$$

Using (6), we find by calculator that

$$\log_2 5 \approx 2.3219281 \quad \square$$

If $x = a$, then the change of base formula (4) becomes

$$\log_b a = \frac{\log_a a}{\log_a b} \tag{7}$$

Since $\log_a a = 1$, (7) simplifies to

$$\log_b a = \frac{1}{\log_a b} \tag{8}$$

a formula that is occasionally of use.

Example 8. Using the fact that $\log_{10} e \approx 0.4343$, approximate ln 10.

Solution. By (8),

$$\ln 10 = \log_e 10 = \frac{1}{\log_{10} e} \approx \frac{1}{0.4343} \approx 2.303 \quad \square$$

The invention of logarithms is generally credited to John Napier (1550–1617), a Scotsman of noble heritage for whom mathematics was a hobby. His invention arose through his desire to associate ratios of pairs of numbers in a geometric progression with ratios of pairs of numbers in an arithmetic progression. Thus the name *logarithmus*, which means "ratio number" and has been anglicized to "logarithm." What resulted from Napier's creation was an association between quotients and differences, namely Law of Logarithms (iii). The base of Napier's logarithms happened to be related to e, and this is the reason why $\log_e x$ (that is, ln x) is sometimes called a Napierian logarithm. Not only did Napier invent his special logarithms, but he made an extensive table of seven-place logarithms that was published in 1614.

EXERCISES 5.3

In Exercises 1–14, approximate the numerical value of the given logarithm. Use the facts that $\log_{10} 2 \approx 0.3010$, $\log_{10} 3 \approx 0.4771$, and $\log_{10} 5 \approx 0.6990$.

1. $\log_{10} 6$
2. $\log_{10} 30$
3. $\log_{10} 12$
4. $\log_{10} 75$
5. $\log_{10} \frac{1}{3}$
6. $\log_{10} \frac{1}{8}$
7. $\log_{10} \frac{10}{3}$
8. $\log_{10} \frac{2}{25}$
9. $\log_{10} \frac{9}{16}$
10. $\log_{10} 7.5$
11. $\log_{10} 0.2$
12. $\log_{10} \sqrt{5}$
13. $\log_{10} \sqrt{30}$
14. $\log_{10} \sqrt[3]{2}$

In Exercises 15–28, approximate the numerical value of the given logarithm. Use the facts that $\ln 2 \approx 0.6931$, $\ln 7 \approx 1.9459$, and $\ln 10 \approx 2.3026$.

15. $\ln 20$ 16. $\ln 70$ 17. $\ln \frac{1}{7}$ 18. $\ln 0.1$

19. $\ln 49$ 20. $\ln 7000$ 21. $\ln 5$ 22. $\ln 35$

23. $\ln 24.5$ 24. $\ln \sqrt{2}$ 25. $\ln \sqrt[3]{14}$ 26. $\ln \sqrt[4]{40}$

27. $\ln 14^{2/3}$ 28. $\ln 40^{0.8}$

In Exercises 29–32, find the numerical value of the given expression.

29. $\log_3 (27^5)$ 30. $\log_4 [(\frac{1}{32})^6]$

31. $\log_{16} (8^{7/9})$ 32. $\log_{27} (9^{0.2})$

In Exercises 33–36, approximate the numerical value of the given logarithm. Use the fact that $\log_{10} 4 \approx 0.6021$.

33. $\log_{10} 40$ 34. $\log_{10} 400{,}000{,}000$

35. $\log_{10} \frac{1}{4}$ 36. $\log_{10} 0.00000025$

In Exercises 37–42, approximate the numerical value of the given logarithm. Use the fact that $\log_{10} a = 2.7318$.

37. $\log_{10} (1000a)$ 38. $\log_{10} (a \times 10^6)$

39. $\log_{10} (a \times 10^{-9})$ 40. $\log_{10} \dfrac{a}{100}$

41. $\log_{10} (a^{10})$ 42. $\log_{10} (a^{-5})$

In Exercises 43–48, rewrite the given expression as a single logarithm.

43. $\log_2 3 + \log_2 \frac{4}{3} + \log_2 \frac{5}{4}$

44. $\ln (3/e) + \ln \frac{1}{6} + \ln 2$

45. $\log_2 (xy^2) + \log_2 (y^{-3}) + \log_2 (x^5)$

*46. $\ln 40 + \log_2 8$

*47. $\log_{10} 2 + 3 \log_{\sqrt{10}} x - \log_{10} (2y)$

*48. $\log_2 (9x) - \log_4 (7x^2) - \log_8 (3/x^2)$

In Exercises 49–54, express the given logarithm in terms of $\log_a x$, $\log_a y$, and $\log_a z$.

49. $\log_a \dfrac{xy}{z}$ 50. $\log_a (x^2yz^3)$ 51. $\log_a (z^3\sqrt{xy})$

52. $\log_a \dfrac{1}{xyz}$ 53. $\log_a \sqrt{\dfrac{xy^5}{z^3}}$ 54. $\log_a \sqrt[4]{\dfrac{x\sqrt{y}}{y^{1/3}}}$

In Exercises 55–60, solve the given equation for x.

55. $2 \log_{10} x = \log_{10} 16$

56. $4 \log_2 x = \log_2 81$

57. $-2 \log_{10} x = \log_{10} 4$

58. $\log_{10} (4x - 1) - \log_{10} \left(\dfrac{x}{5}\right) = 1$

59. $\log_2 (x + 1) + \log_2 (x - 1) = 3$

60. $\log_2 (2x - 1) - \log_2 (3x + 1) = -1$

In Exercises 61–66, use the change of base formulas to obtain the desired results.

61. Given that $\log_{10} 7 \approx 0.8450$ and $\log_{10} 2 \approx 0.3010$, approximate $\log_2 7$.

62. Given that $\log_{10} 5 \approx 0.6990$ and $\log_{10} e \approx 0.4343$, approximate $\ln 5$.

63. Given that $\log_2 6 \approx 2.5850$ and $\log_2 3 \approx 1.5850$, approximate $\log_3 6$.

64. Given that $\ln 7 \approx 1.9459$ and $\log_5 e \approx 0.6213$, approximate $\log_5 7$.

65. Given that $\log_2 3 \approx 1.5850$, approximate $\log_3 2$.

66. Given that $\log_{10} 2 \approx 0.3010$, approximate $\log_2 10$.

67. a. Show that $\log_2 x = \log_4 x^2$.
    b. Show that $\log_a x = \log_{a^2} x^2$ for any positive numbers a and x such that $a \neq 1$.

68. Prove that $\ln (x + \sqrt{x^2 - 1}) = -\ln (x - \sqrt{x^2 - 1})$.

69. Prove that $\log_a (1/x) = \log_{1/a} x$.

70. Show that if x satisfies $\dfrac{\ln x}{x} < \dfrac{\ln 2}{2}$, then $x^2 < 2^x$.

71. Let $b = e^{-1/10^7}$. Show that $\log_b x = -10^7 \ln x$ for any $x > 0$. (The "logarithm" that John Napier defined in the seventeenth century was actually $\log_b x + 10^7 \ln 10^7$. Thus, although his logarithm was not the natural logarithm, it was very closely related to it.)

72. Prove Law (ii) by using Law (i) with y replaced by $1/x$.

73. Prove Law (iii) by using Laws (i) and (ii).

74. Prove Law (iv). (*Hint:* Let $r = \log_a x$. Using first the definition of the logarithm to the base a, next a law of exponents, and then the definition of the logarithm to the base a again, show in turn that $x = a^r$, $x^c = a^{cr}$, and $cr = \log_a (x^c)$, and complete the proof.)

5.4

NUMERICAL COMPUTATIONS WITH COMMON LOGARITHMS

Before calculators were invented, computations were frequently performed either with a slide rule or with the aid of common logarithms and a table of (approximate) logarithmic values. Common logarithms were used because our number system is based on the number 10. Even though calculators have made these methods obsolete, we will discuss the use of the logarithmic table in computation. There are two reasons for this: First, the method illustrates the basic properties of logarithms. Second, some of the techniques we will use apply to other tables as well. Since all logarithms in this section have base 10, we will write $\log x$ instead of $\log_{10} x$ throughout the section.

Recall from Section 1.3 that any positive number x can be written in scientific notation as

$$x = b \times 10^n \tag{1}$$

where n is an integer and $1 \le b < 10$. Taking logarithms of both sides of (1) and then using the Laws of Logarithms, we find that

$$\log x = \log (b \times 10^n) = \log b + \log (10^n)$$

$$= (\log b) + n \log 10 = (\log b) + n$$

so that

$$\log x = n + \log b \qquad (2)$$

Thus the common logarithm of *any* positive number x is the sum of an integer n and the common logarithm $\log b$ of a number b between 1 and 10. The number n is called the **characteristic** of $\log x$, and the number $\log b$ is the **mantissa** of $\log x$.* Since $1 \le b < 10$, we have

$$0 = \log 1 \le \log b < \log 10 = 1$$

so the mantissa of $\log x$ is always between 0 and 1.

Table C in the Appendix gives the common logarithms of various numbers between 1 and 10 in steps of 0.01. A portion of one page of the table is reproduced here:

x	0	1	2	3	4	5	6	7	8	9
3.5	.5441	.5453	.5465	.5478	.5490	.5502	.5514	.5527	.5539	.5551
3.6	.5563	.5575	.5587	.5599	.5611	.5623	.5635	.5647	.5658	.5670
3.7	.5682	.5694	.5705	.5717	.5729	.5740	.5752	.5763	.5775	.5786
3.8	.5798	.5809	.5821	.5832	.5843	.5855	.5866	.5877	.5888	.5899
3.9	.5911	.5922	.5933	.5944	.5955	.5966	.5977	.5988	.5999	.6010

To use the table to find the logarithm of a number such as 3.81, which is between 1 and 10, first locate the row labeled 3.8 and then the column labeled 1 at the top. The entry belonging to both row and column is .5809 (without the customary 0 before the decimal in order to keep the table as compact as possible), so that

$$\log 3.81 \approx 0.5809$$

Caution: Most the the entries in Table C are not exact values but only approximations to the exact values, rounded off to four decimal places, which means that the maximum error in any entry is 0.00005. Thus the entry of .5809 for $\log 3.81$ implies that the true value of $\log 3.81$ lies between 0.58085 and 0.58095.

---

* The word "mantissa" is of Latin origin, meaning "an addition." The mantissa is "added on" to the characteristic, which is the integral, or main, part of the logarithm.

Before finding numerical values of common logarithms, we emphasize that there is no need for the common logarithm table to have logarithms of numbers that are not between 1 and 10, since the use of scientific notation reduces the problem of finding the logarithm of any positive number to that of finding the logarithm of a number between 1 and 10 (see (1) and (2)).

Example 1. Use Table C to approximate the following numbers.
   a. log 143   b. log 0.00756

Solution.
   a. First we write 143 in scientific notation as

$$143 = 1.43 \times 10^2$$

Next we use (2) to deduce that

$$\log 143 = 2 + \log 1.43$$

Finally we see from Table C that $\log 1.43 \approx 0.1553$, so that

$$\log 143 \approx 2 + 0.1553 = 2.1553$$

   b. Proceeding as in (a), we find that

$$0.00756 = 7.56 \times 10^{-3}$$

$$\log 0.00756 = -3 + \log 7.56$$

From Table C we notice that $\log 7.56 \approx 0.8785$, so that

$$\log 0.00756 \approx -3 + 0.8785 \quad \square$$

Notice that in the final answer of (b) we did not combine the characteristic -3 with the mantissa 0.8785. When the characteristic is negative, the logarithm is normally easier to use with the characteristic and mantissa kept separate.

Now we turn the process around and find a number whose common logarithm is given.

Example 2. Use Table C to find a number whose common logarithm is approximately 3.9253.

Solution. First observe that

$$3.9253 = 3 + 0.9253$$

This means that if we find a number b whose logarithm is given in Table C by .9253, then $b \times 10^3$ is the number we seek, since in that case

$$\log (b \times 10^3) = 3 + \log b \approx 3 + 0.9253 = 3.9253$$

Looking in Table C for the entry .9253, we find it in the row labeled 8.4 and the column labeled 2 at the top. It follows that

$$\log 8.42 \approx 0.9253$$

Therefore 8.42×10^3, which of course is 8420, is a number whose common logarithm is approximately 3.9253.   □

In the next example, where we again will find a number whose common logarithm is given, we will have to be careful to write the given number in the form $n + \log b$, where n is an integer and $0 \le \log b < 1$.

Example 3. Use Table C to find a number whose common logarithm is approximately -1.4123.

Solution. Although $-1.4123 = -1 - 0.4123$ the expression $-1 - 0.4123$ does not have the form $c + \log b$, where $0 \le \log b < 1$ (since -0.4123 is not between 0 and 1). But notice that

$$-1.4123 = -2 + 0.5877$$

and $-2 + 0.5877$ has the correct form because $0 < 0.5877 < 1$. From Table C we see that

$$\log 3.87 \approx 0.5877$$

Thus 3.87×10^{-2}, which is the same as 0.0387, is a number whose common logarithm is approximately -1.4123.   □

If x is any number, then the positive number y with $\log y = x$ is called the ***antilogarithm*** of x. For instance, the solution of Example 2 tells us that the antilogarithm of 3.9253 is approximately 8420, and the solution of Example 3 tells us that the antilogarithm of -1.4123 is approximately 0.0387.

To obtain an alternative description of an antilogarithm, we recall from (1) in Section 5.2 that

$$\log y = x \quad \text{if and only if} \quad 10^x = y$$

Since $\log y = x$ means that y is the antilogarithm of x, it follows that the antilograrithm of x is 10^x.

Linear Interpolation   By Table C,

$$\log 268 = \log (2.68 \times 10^2) = 2 + \log 2.68 \approx 2 + 0.4281 = 2.4281$$

and

$$\log 269 = \log (2.69 \times 10^2) = 2 + \log 2.69 \approx 2 + 0.4298 = 2.4298$$

But suppose that we desire to compute log 268.4, which lies between log 268 and log 269. The process starts in the same way:

$$\log 268.4 = \log (2.684 \times 10^2) = 2 + \log 2.684$$

But 2.684 has three digits to the right of the decimal, whereas Table C contains entries for numbers with only 1 or 2 digits to the right of the decimal. The way we go about finding a value for log 268.4 is as follows.

From Table C,

$$\log 2.68 \approx 0.4281$$

$$\log 2.684 = ??$$

$$\log 2.69 \approx 0.4298$$

Since $2.684 - 2.68 = 0.004$ and $2.69 - 2.68 = 0.01$,   evidently 2.684 is $0.004/0.01 = \frac{4}{10}$ of the distance from 2.68 to 2.69. Therefore we take log 2.684 to be $\frac{4}{10}$ of the distance from log 2.68 to log 2.69, that is, from 0.4281 to 0.4298. Since

$$0.4298 - 0.4281 = 0.0017 \quad \text{and} \quad \frac{4}{10} (0.0017) = 0.00068 \approx 0.0007$$

(rounded off to four decimal places, as are the entries in Table C), we take

$$\log 2.684 \approx (\log 2.68) + 0.0007 = 0.4281 + 0.0007 = 0.4288$$

Consequently

$$\log 268.4 = 2 + \log 2.684 \approx 2 + 0.4288 = 2.4288$$

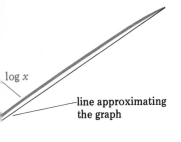

log x

line approximating the graph

2.69

FIGURE 5.12

This method of approximating a logarithm is called *linear interpolation*, because in effect the method treats the logarithmic function as a linear function on intervals of length 0.01, that is, changes in the mantissa are taken to be proportional to changes in the numbers (Figure 5.12). On such intervals the graph of log x is very nearly linear, so the estimate is reasonable.

Example 4. Use linear interpolation to find an approximate value of log 2817.

Solution. Notice that

$$\log 2817 = \log (2.817 \times 10^3) = 3 + \log 2.817$$

From Table C,

$$\log 2.81 \approx 0.4487$$

$$\log 2.817 = ??$$

$$\log 2.82 \approx 0.4502$$

Since $2.817 - 2.81 = 0.007$ and $2.82 - 2.81 = 0.01$, evidently 2.817 is $0.007/0.01 = \frac{7}{10}$ of the distance from 2.81 to 2.82. Therefore, we take log 2.817 to be $\frac{7}{10}$ of the distance from log 2.81 to log 2.82, that is, from 0.4487 to 0.4502. Since

$$0.4502 - 0.4487 = 0.0015 \quad \text{and} \quad \frac{7}{10}(0.0015) = 0.00105 \approx 0.0011$$

(rounded off to four decimal places), we take

$$\log 2.817 \approx (\log 2.81) + 0.0011 \approx 0.4487 + 0.0011 = 0.4498$$

Consequently

$$\log 2817 = 3 + \log 2.817 \approx 3 + 0.4498 = 3.4498 \quad \square$$

Linear interpolation can also be used to approximate antilogarithms of numbers not appearing in Table C. For example, suppose we wish to find the antilogarithm of 0.7961. Observe that by the table,

$$\log 6.25 \approx 0.7959$$

$$\log \text{ ??} = 0.7961$$

$$\log 6.26 \approx 0.7966$$

Since $0.7961 - 0.7959 = 0.0002$ and $0.7966 - 0.7959 = 0.0007$, it follows that 0.7961 is $0.0002/0.0007 = \frac{2}{7}$ of the distance from 0.7959 to 0.7966. Therefore we take the antilogarithm of 0.7961 to be $\frac{2}{7}$ of the distance from the antilogarithm of 0.7959 to the antilogarithm of 0.7966, that is, from 6.25 to 6.26. Now

$$6.26 - 6.25 = 0.01 \quad \text{and} \quad \frac{2}{7}(0.01) \approx 0.003$$

(rounded off to three decimal places), so the value of the antilogarithm of 0.7961 is approximately

$$6.25 + 0.003 = 6.253$$

Notice that logarithms are given in Table C for numbers with two decimal places, whereas the answer just given with the help of linear interpolation was rounded off to three decimal places. In general the third decimal place will be more reliable than a fourth or higher decimal place, and as a result, we will limit ourselves to three decimal places when we approximate antilogarithms with linear interpolation.

Example 5. Use linear interpolation to find an approximate value of the antilogarithm of 0.3764.

Solution. By Table C,

$$\log 2.37 \approx 0.3747$$

$$\log ?? = 0.3764$$

$$\log 2.38 \approx 0.3766$$

Since $0.3764 - 0.3747 = 0.0017$ and $0.3766 - 0.3747 = 0.0019$, it follows that 0.3764 is $0.0017/0.0019 = \frac{17}{19}$ of the distance from 0.3747 to 0.3766. Therefore we take the antilogarithm of 0.3764 to be $\frac{17}{19}$ of the distance from the antilogarithm of 0.3747 to the antilogarithm of 0.3766, that is, from 2.37 to 2.38. Since

$$\frac{17}{19}(2.38 - 2.37) = \frac{17}{19}(0.01) \approx 0.009$$

(rounded off to three decimal places), the antilogarithm we seek is approximately

$$2.37 + 0.009 = 2.379 \quad \square$$

We complete the section with an example that would have required the use of logarithm tables a few years ago.

Example 6. Use common logarithms to approximate $(8.53)^{7.27}$.

Solution. First we take the logarithm of the given expression, and then rearrange the result with the help of the Laws of Logarithms and Table C:

$$\log[(8.53)^{7.27}] = (7.27)\log 8.53 \approx (7.27)(0.9309) \approx 6.7676$$

In order to obtain the antilogarithm of $6 + 0.7676$, we observe from Table C that

$$\log 5.85 \approx 0.7672$$

$$\log ?? = 0.7676$$

$$\log 5.86 \approx 0.7679$$

Therefore 0.7676 is $0.0004/0.0007 = \frac{4}{7}$ of the distance from 0.7672 to 0.7679. Consequently we take the antilogarithm of 0.7676 to be $\frac{4}{7}$ of the distance from 5.85 to 5.86, that is, approximately 5.856. Thus the antilogarithm of $6 + 0.7676$ is approximately 5.856×10^6, or 5,856,000. We conclude that

$$(8.53)^{7.27} \approx 5,856,000 \quad \square$$

One calculator provides the value 5,861,374.3 for the answer to Example 6. Thus modern technology, in the form of calculators, can save much time and effort in computations and, moreover, can provide accuracy vastly superior to that available to people even as recently as 40 years ago.

EXERCISES 5.4

In Exercises 1–20, use Table C to approximate the common logarithm of the given number. Use linear interpolation if needed.

1. 1.01
2. 3.87
3. 520
4. 1,750,000
5. 0.0981
6. 0.00145
7. 5.23×10^4
8. $(7.7)^3$
9. $\sqrt{1.41}$
10. $\sqrt[3]{429}$
11. $\dfrac{1}{693}$
12. $\dfrac{1}{0.00872}$
13. 3.281
14. 29.35
15. 15,340,000
16. 0.1927
17. 0.001437
18. 4.793×10^{-2}
19. $(23.29)^3$
20. $\sqrt{49.75}$

In Exercises 21–36, use Table C to approximate the antilogarithm of the given number. Use linear interpolation if needed.

21. 0.8756
22. 0.5514
23. 4.9983
24. 5.0043
25. 8.8000
26. −1.5331
27. −2.2882
28. −6.3546
29. 0.9752
30. 1.6081
31. 2.1913
32. 6.5499
33. −1.6814
34. −2.6979
35. −0.3933
36. −3.7208

In Exercises 37–44, use common logarithms to approximate the given expression.

37. $\dfrac{2.7 \times 90.73}{3.48}$
38. $\dfrac{5.1 \times 1.021 \times 3.2}{1.015 \times 97.37}$
39. $(7.612)^{2/3}$
40. $\sqrt[4]{3} \cdot \sqrt[3]{4}$
41. $(4.512)^{1/2} \times (923.4)^3$
42. $\dfrac{(3.5)^{1/3}}{\sqrt{62,100}}$
43. $4^{-4.4}$
44. $\dfrac{(28100)^{3.2} \times (54,300)^{1.1}}{(0.00316)^2 \times \sqrt{2}}$

45. The equation

$$\log w = 1.9532 + 3.135 \log s$$

has been used to relate the weight w (in kilograms) to the sitting height s (in meters) of people. Using this equation and Table C or a calculator, approximate
  a. the weight of a person whose sitting height is 1 meter.
  b. the weight of a person whose sitting height is $\frac{1}{2}$ meter.
  c. the sitting height of a person whose weight is 36 kilograms.
  d. the sitting height of a person whose weight is 20 kilograms.

46. For small values of $|h|$, the number $\log (1 + h)$ may be approximated by using the formula

$$\log (1 + h) \approx 0.4343 \left(h - \frac{1}{2} h^2 \right)$$

Use this equation to approximate
a. $\log 1.1$ b. $\log 1.01$
(Compare your answers with the values given by a calculator or Table C.)

5.5

APPLICATIONS OF COMMON LOGARITHMS

This section is devoted to mathematical and scientific applications of common logarithms. The mathematical application concerns solutions of equations involving exponentials, and the scientific applications are represented by loudness of sound and magnitude of earthquakes, both of which employ common logarithms in their formulation. As before, we will write $\log x$ for $\log_{10} x$. Our computations will be made by calculator. They could also be made by using Table C.

Solutions of Equations

For the examples below, we will use the fact that

$$x = y \quad \text{if and only if} \quad \log_a x = \log_a y$$

When we pass from an equation of the form $x = y$ to the equivalent equation $\log_a x = \log_a y$, we say that we *take logarithms of both sides* of the equation $x = y$.

Example 1. Solve the equation $4^x = 17$ for x.

Solution. We take common logarithms of both side and then use Law of Logarithms (iv):

$$4^x = 17$$
$$\log (4^x) = \log 17$$
$$x \log 4 = \log 17$$
$$x = \frac{\log 17}{\log 4}$$

By calculator we find that

$$x \approx 2.0437314 \quad \square$$

Example 2. Solve the equation $2^{x+1} = 3^{x^2-1}$ for x.

Solution. Taking common logarithms of both sides and then using (iv) of the Laws of Logarithms, we obtain

$$\log (2^{x+1}) = \log (3^{x^2-1})$$

$$(x + 1) \log 2 = (x^2 - 1) \log 3$$

$$(x + 1) \log 2 = (x + 1)[(x - 1) \log 3]$$

From this last equation we conclude that either $x = -1$ or

$$\log 2 = (x - 1) \log 3$$

Solving for $x - 1$ in the latter equation, we obtain

$$x - 1 = \frac{\log 2}{\log 3} \quad \text{so that} \quad x = \frac{\log 2}{\log 3} + 1$$

Thus the solutions of the given equation are -1 and $[(\log 2)/(\log 3)] + 1$ (which by calculator we find is approximately 1.6309298). □

Loudness of Sound

We turn now to physical applications of common logarithms, represented by loudness of sound and magnitude of earthquakes. In each of these applications the primary reason for using common logarithms is to restrict the numbers involved to a more manageable range. For instance, if a certain quantity q can take any value from, say, 10^{-100} (which is miniscule) to 10^{100} (which is astronomical), then the value of $\log q$ lies between -100 and 100.

The *intensity* of a sound wave is the amount of energy that the wave carries through a unit area in a unit time, or equivalently, the amount of power per unit area. The unit we will use for the intensity of sound is one watt per square meter. The *threshold of audibility* of a sound wave is the minimal intensity the sound wave can have and still be audible to the normal human ear. The threshold of audibility of a sound wave varies according to the pitch, or frequency, of the wave. To simplify the discussion that follows, let us assume that each sound wave under consideration has the same frequency—100 hertz (cycles per second), producing a sound about an octave and a half below middle C on the piano.

Let I_0 be the intensity of such a sound wave at the threshold of audibility, approximately 10^{-12} watts per square meter. If x denotes the intensity of any sound wave of the same frequency, then the *noise level* (or *loudness*) $L(x)$ of the sound wave is defined by

$$L(x) = 10 \log \frac{x}{I_0} \tag{1}$$

The units for $L(x)$ are *decibels*, in honor of Alexander Graham Bell (1847–1922). Notice that if $x = I_0$, then (1) becomes

$$L(I_0) = 10 \log \frac{I_0}{I_0} = 10 \log 1 = 0$$

so that at the threshold of audibility the noise level is just 0.

Example 3. The noise level of a loud conversation is approximately 66 decibels, and the noise level of a whisper is approximately 22 decibels. Let x denote the intensity of a sound wave from a loud conversation and z the intensity of a sound wave for a whisper. Determine the ratio x/z of the intensities of the two sound waves.

Solution. By hypothesis, we may assume that

$$L(x) = 66 \quad \text{and} \quad L(z) = 22$$

so that

$$L(x) - L(z) = 66 - 22 = 44$$

In order to determine x/z, we first use (1) and the Laws of Logarithms to alter the expression $L(x) - L(z)$:

$$44 = L(x) - L(z) = 10 \log \frac{x}{I_0} - 10 \log \frac{z}{I_0}$$

$$= 10 \left(\log \frac{x}{I_0} - \log \frac{z}{I_0} \right) = 10 \log \frac{x/I_0}{z/I_0} = 10 \log \frac{x}{z}$$

Thus $\log x/z = 4.4$. It follows that

$$\frac{x}{z} = 10^{4.4}$$

Using a calculator, we find that

$$\frac{x}{z} \approx 25{,}118.864$$

This means that the intensity of a sound wave from a loud conversation is approximately 25,000 times as large as that of a whisper. □

Earthquakes   During the past three decades several formulas have been proposed for converting seismographic readings into a unified scale that would represent the magnitude of an earthquake. The scale most commonly used is the **Richter scale**, devised by Charles F. Richter, an American geologist. In order to describe measurement on the Richter scale, we first introduce a reference, or **zero-level**, earthquake, which by definition is any earthquake whose largest seismic wave would measure 0.001 millimeter (1 micron) on a standard seismograph located 100 kilometers from the epicenter of the earthquake. The ***magnitude*** *M* of a given earthquake can be obtained by the formula

$$M = \log \frac{a}{a_0} = \log a - \log a_0 \qquad (2)$$

where *a* is the amplitude measured on a standard seismograph of the largest seismic wave of the given earthquake, and a_0 is the amplitude on the same seismograph of the largest seismic wave of a zero-level earthquake with the same epicenter. Values of a_0 for various distances from the epicenter have been calculated, so one only needs to measure the number *a* in order to be able to assess the magnitude of a given earthquake.

Example 4. The Alaskan Good Friday earthquake of March 28, 1964, measured 8.5 on the Richter scale. Find the ratio a/a_0.

Solution. From (2) we have

$$\log \frac{a}{a_0} = M = 8.5$$

so that

$$\frac{a}{a_0} = 10^{8.5}$$

By calculator we find that

$$\frac{a}{a_0} \approx 316{,}227{,}770 \quad \square$$

The effect of the Good Friday earthquake on a main street in Anchorage, Alaska. Part of the street was left some 20 feet below the rest of it.

Theoretically, the magnitude of an earthquake can be any real number. In practice, however, a standard seismograph can only record those earthquakes whose magnitudes exceed 2. On the other end of the scale, no magnitude has ever exceeded 9.0, and only half a dozen times has one been 8.5 or higher. Furthermore, formula (2) really only applies for a given earthquake when the seismograph can handle the largest seismic wave of the earthquake; this means that accurate measurement of the magnitude of a strong earthquake can be accomplished only when the seismograph is several hundred kilometers away.

EXERCISES 5.5

In Exercises 1–12, solve the given equation. Leave your answer in terms of common logarithms.

1. $2^x = 5$
2. $3^{-x} = 17$
3. $5^{2x} = 51$
4. $7^{\sqrt{2x}} = 3$
5. $4^{x+2} = 5$
6. $(\frac{1}{4})^{x+2} = 5$
7. $5^{2x+3} = 3^{5x-2}$
8. $8^{4x-1} = 6^{1-x}$
9. $3^x = 2^{(x^2)}$
10. $3^x = 2^{(x^3)}$
11. $4^{x+2} = 5^{x^2-4}$
12. $2^{x-1} = 10^{x^2+x-2}$

13. Using (1), show that if $L(x_1)$ decibels corresponds to intensity x_1 and $L(x_2)$ corresponds to intensity x_2, then

$$L(x_2) - L(x_1) = 10 \log \frac{x_2}{x_1}$$

14. Determine the number of decibels that corresponds to each of the following intensities (in watts per square meter).
    a. 10^{-11} (refrigerator motor)
    b. 10^{-5} (conversation)
    c. 10^{-4} (electric typewriter)

15. Using the value of 10^{-12} for I_0, find the intensity of a sound at 50 decibels.

16. What is the ratio of the intensity of a given sound to that of one that is 100 decibels higher?

17. What is the difference in the noise levels of two sounds one of which is 100 times as intense as the other?

18. Suppose the cannon fired at the time of a football touchdown produces a sound intensity 1000 times as strong as a cheerleader's shouts. Determine how many more decibels the noise level from the cannon produces than the cheerleader's cry.

19. The human ear can just barely distinguish between two sounds if one is 0.6 decibels higher than the other. What is the ratio of the intensity of one sound to that of another sound which is lower than the first and is just barely distinguishable from the first sound?

20. Suppose soundproofing a room results in a noise reduction of 30 decibels in the noise from a busy highway. What percentage of the original intensity is the reduced intensity?

21. The strongest earthquakes ever recorded occurred off the coast of Ecuador and Columbia in 1906, and in Japan in 1933. Each had magnitude 8.9. Find the ratio of the amplitude of the largest wave of such a quake to the corresponding amplitude of a zero-level quake.

22. Determine the ratio a/a_0 for an earthquake whose magnitude is 6.3.

23. Suppose a seismograph is located exactly 100 kilometers from the epicenter of an earthquake. Determine the magnitude of the earthquake if the largest amplitude of the seismic waves registered on the seismograph is
    a. 1 micron   b. 1 millimeter   c. 1 centimeter

24. If an earthquake has magnitude 2, find the amplitude of its largest wave 100 kilometers from the epicenter.

25. Suppose the amplitude of the maximal seismic wave is doubled. By how much is the magnitude of the earthquake increased?

26. The magnitude of the Good Friday Alaskan earthquake of 1964 is sometimes given as 8.4. What is the ratio of the maximum amplitude of an earthquake of magnitude 8.4 to that of an earthquake of magnitude 8.5?

27. Of seven major earthquakes to hit Iran in recent times, two have measured approximately 6.9 on the Richter scale (those occurring in April 1972 and in March 1977). Find the ratio of the amplitudes of the largest waves of these earthquakes to that of the San Francisco earthquake of 1906, which measured 8.3 in magnitude.

28. In chemistry the *hydrogen potential* (denoted pH) of a solution is defined by

$$\text{pH} = \log \frac{1}{[H^+]} = -\log [H^+]$$

where H^+ is the concentration of hydrogen ions in the solution in moles per liter. The pH of a solution ranges from 0 to 14. If the pH of a solution is less than 7, the solution is acidic and turns litmus paper red. In contrast, if the pH is greater than 7, it is alkaline and turns litmus paper blue. Distilled water has a pH of 7 and is neither acid nor alkaline. Determine the hydrogen ion concentration of pure water.

29. Find the pH of a soft drink whose hydrogen ion concentration is roughly 0.004 mole per liter. Is the drink highly acidic, or highly alkaline?

30. Orange juice has a pH of approximately 3.5, and tomato juice has a pH of approximately 4.2. What would be the ratio of hydrogen ion concentration in orange juice to that in tomato juice?

31. Wheat grows best in soil that has a pH of between 6 and 7.5, whereas oats grow best in soil that has a pH of between 5 and 6.2. Suppose the hydrogen ion concentration in the soil in a field is found to be 4.47×10^{-7} mole per liter. Is the soil better suited to wheat or to oats?

5.6

EXPONENTIAL GROWTH AND DECAY

In physical and mathematical applications the exponential function that is most often used is the exponential function with base e, that is, the function e^x. In particular, many physical quantities, such as the amount of a specific radioactive substance, depend on time according to an equation of the form

$$f(t) = Ae^{ct} \tag{1}$$

where A and c are constants. If we substitute 0 for t in (1), we have

$$f(0) = Ae^{c \cdot 0} = Ae^0 = A \cdot 1 = A$$

so that $A = f(0)$. Thus (1) becomes

$$f(t) = f(0)e^{ct} \tag{2}$$

Since $e^{ct} > 0$ for all t, it follows from (2) that $f(t)$ has the same sign as $f(0)$. In applications $f(t)$ is almost always positive. As a result, we will assume throughout this section that $f(t) > 0$ for all t.
    If $c > 0$, then customarily one lets $c = k$, so that

$$f(t) = f(0)e^{kt} \quad \text{with} \quad k > 0 \tag{3}$$

In this case, $f(t)$ increases with t (Figure 5.13a), and we say that f **grows exponentially**. Analogously, if $c < 0$, then customarily one lets $c = -k$, so that

$$f(t) = f(0)e^{-kt} \quad \text{with } k > 0 \tag{4}$$

In this case, $f(t)$ decreases with t (Figure 5.13b), and we say that f **decays exponentially**. If f grows exponentially, then for any positive integer n, $f(t) > t^n$ for all sufficiently large t (Figure 5.14a), and we say that $f(t)$ grows faster than any function of the form t^n. Similarly, if f decays exponentially, then for any positive integer n, $f(t) < t^{-n}$ for all sufficiently large t (Figure 5.14b), and we say that $f(t)$ decreases faster than any function of the form t^{-n}.

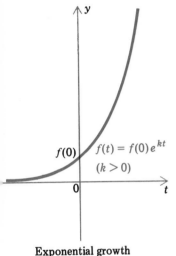

$f(0)$ $f(t) = f(0)e^{kt}$
$(k > 0)$

Exponential growth
(a)

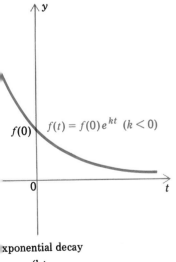

$f(0)$ $f(t) = f(0)e^{kt}$ $(k < 0)$

Exponential decay
(b)

FIGURE 5.13

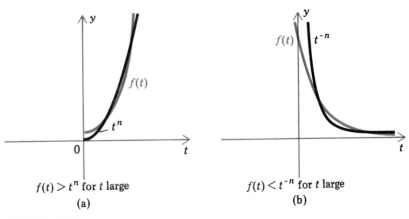

$f(t) > t^n$ for t large
(a)

$f(t) < t^{-n}$ for t large
(b)

FIGURE 5.14

Solving Exponential Equations by Natural Logarithms

In the examples of exponential growth and decay that we will consider, it will be necessary to solve equations containing expressions like e^{kt}. Since

$$\ln (e^{kt}) = kt \tag{5}$$

the use of natural logarithms will simplify the solutions. This is illustrated in the next two examples.

Example 1. Solve the equation $e^{5k} = 2$ for k.

Solution. We take the natural logarithm of each side, use (5), and then solve for k:

$$e^{5k} = 2$$
$$\ln (e^{5k}) = \ln 2$$
$$5k = \ln 2$$
$$k = \frac{1}{5} \ln 2$$

By calculator we find that $k \approx 0.13862944$. □

Example 2. Solve the equation $e^{(t/5) \ln 2} = 3$ for t.

Solution. We take the natural logarithm of each side, use (5), and solve for t:

$$e^{(t/5) \ln 2} = 3$$
$$\ln (e^{(t/5) \ln 2}) = \ln 3$$
$$\frac{t}{5} \ln 2 = \ln 3$$
$$t = 5 \frac{\ln 3}{\ln 2}$$

By calculator we find that $t \approx 7.9248125$. □

Population Growth

When the number of organisms in a group is very large, the graph of the function representing the number in the group at any given time is very nearly a smooth curve. If the group has plenty of food and space and no enemies (such as disease or predators), the curve can be essentially the graph of an exponential function of the form of (3). This means that the population of the group can grow exponentially. Statistics show that various types of bacteria, insects, and rodents, and even certain human populations, have in the past grown exponentially for awhile—until conditions such as lack of food, overcrowding, and war changed the growth pattern.

Example 3. Experiment shows that under optimal conditions and at a temperature of 29°C, the population of a colony of 1000 rice weevils can grow exponentially to about 2139 weevils in a week. Assuming that there are 1000 initially and that their population grows exponentially, find a formula for the number of weevils at time t.

Solution. Let $f(t)$ denote the number of weevils at time t, where t is measured in weeks. We take $t = 0$ to represent the initial time, which means that $f(0) = 1000$, the initial number of weevils. Then (3) becomes

$$f(t) = 1000e^{kt}$$

and we need to determine the numerical value of k. By assumption there are 2139 weevils after 1 week, so that $f(1) = 2139$. Thus

$$2139 = f(1) = 1000e^{k(1)} = 1000e^{k}$$

or equivalently,

$$e^{k} = \frac{2139}{1000} = 2.139$$

Taking natural logarithms of both sides yields

$$k = \ln 2.139$$

Therefore

$$f(t) = 1000e^{(\ln 2.139)t}$$

which can be written in the alternative form

$$f(t) = 1000(2.139^{t}) \quad \square$$

Example 4. From the data given in Example 3, determine the length of time in weeks for the population of weevils to double.

Solution. Since the initial population of weevils is 1000, what we seek is the value of t for which

$$f(t) = 2000$$

By the solution of Example 3, this means that we wish to find the value of t for which

$$2000 = 1000e^{(\ln 2.139)t}$$

or equivalently,

$$2 = e^{(\ln 2.139)t}$$

Taking natural logarithms of both sides, we obtain

$$\ln 2 = (\ln 2.139)t$$

so that

$$t = \frac{\ln 2}{\ln 2.139}$$

By calculator, we find that

$$t \approx 0.91162981$$

Consequently the population of weevils will double after approximately 0.91162981 weeks, which is just over 6 days and 9 hours. □

The length of time it takes a population growing exponentially to double is called the **doubling time** of the population. Thus the doubling time for rice weevils under optimal conditions is approximately 6 days and 9 hours.

Carbon Dating   Certain isotopes, such as carbon 14 (C^{14}) and uranium 238 (U^{238}), are radioactive, which means that they are unstable and in time decay, changing into other substances by the emission of various particles from their nuclei. For example, C^{14} decays into nitrogen 14 by emitting an electron (a beta particle), and U^{238} decays into thorium 234 by emitting a helium 4 nucleus (an alpha particle).

If $f(t)$ is the amount of a given radioactive element in a substance at time t, then by experiment it is known that f satisfies (4) for a suitable constant k depending on the isotope. Notice that the amount $f(t)$ decreases as t increases, and in fact, $f(t)$ eventually approaches 0, no matter how much of the element is initially present. The **half-life** of a radioactive element is the length of time necessary for half of any given amount of it to decay.

Example 5. The half-life of C^{14} is approximately 5730 years. Using this figure, show that the amount $f(t)$ of C^{14} remaining after an elapsed time of t years is given by

$$f(t) = f(0)e^{-(1/5730)(\ln 2)t} \tag{6}$$

Solution. We need to determine the constant k so that

$$f(t) = f(0)e^{-kt} \quad \text{for } t > 0 \tag{7}$$

By assumption, half the original C^{14} remains after 5730 years, so that

$$f(5730) = \frac{1}{2}f(0)$$

However, by substituting 5730 for t in (7), we obtain

$$f(5730) = f(0)e^{-5730k}$$

From the last two equations it follows that

$$\frac{1}{2} f(0) = f(0)e^{-5730k}$$

or equivalently,

$$\frac{1}{2} = e^{-5730k}$$

Taking natural logarithms of both sides, we obtain

$$\ln \frac{1}{2} = -5730k$$

Therefore

$$k = -\frac{1}{5730} \ln \frac{1}{2} = \left(-\frac{1}{5730}\right)(-\ln 2) = \frac{1}{5730} \ln 2$$

Consequently the formula we seek is

$$f(t) = f(0)e^{-(1/5730)(\ln 2)t} \qquad \square$$

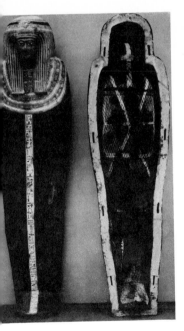

ancient Egyptian mummy in
wood coffin. Carbon dating is
ten used in determining the age
mummies.

Every living object contains C^{14}. Because of the radioactivity of C^{14}, you might think that the amount of C^{14} in living objects would be ever decreasing. But to offset the decay of C^{14} in any living object, there is an increase of C^{14} in the atmosphere due to cosmic radiation. This C^{14} enters the carbon cycle of living plants and animals, with the effect that the amount per gram of C^{14} in living organisms remains constant. However, when an organism dies, it abruptly ceases to assimilate any more C^{14}, and consequently the amount of C^{14} per gram in the organism starts decreasing. This decrease in C^{14} after death is the basis for the process known as *carbon dating* of once-living organisms.

Example 6. A mummy, known as Whiskey Lil, was discovered in 1955 in Chimney Cave, Lake Winnemucca, Nevada. By carbon dating it was learned that approximately 0.739 of the original C^{14} per gram was still present in 1955. When did Whiskey Lil die?

Solution. We let $t = 0$ at the time of death and let t_0 be the length of time between the death of Whiskey Lil and 1955. First we will determine the value of t_0, and then we will calculate when she died. By hypothesis,

$$f(t_0) = 0.739f(0)$$

and by (6) with t replaced by t_0,

$$f(t_0) = f(0)e^{-(1/5730)(\ln 2)t_0}$$

Therefore

$$0.739 f(0) = f(0)e^{-(1/5730)(\ln 2)t_0}$$

or equivalently,

$$0.739 = e^{-(1/5730)(\ln 2)t_0}$$

Taking logarithms of both sides, we find that

$$\ln 0.739 = -\left(\frac{1}{5730}\ln 2\right)t_0$$

Solving for t_0, we obtain

$$t_0 = -\frac{5730 \ln 0.739}{\ln 2}$$

By calculator, we conclude that

$$t_0 \approx 2500.3069$$

Therefore Whiskey Lil died about 2500 years before 1955 A.D., which means about 545 B.C. □

Since the mid-1950s, when Willard Libby proposed dating objects by analysis of electron emissions of C^{14}, carbon dating has been intimately connected with the questions of when the Americas were settled and what kinds of tools early cultures included. Bones, skulls, artifacts—archaeological finds of all kinds—were subjected to carbon dating analysis. Until recently the best estimates pointed to the first American settlements as having taken place around 11,000 B.C., but new evidence suggests, again with the help of carbon dating, that the first settlements were perhaps as much as 16,000 years earlier. (See Exercise 25.)

Yet there are limitations to the use of carbon dating. First, the outer limits of dates that can be ascertained by carbon dating are approximately 1500 A.D. and 50,000 B.C. Second, carbon dating gives approximate, rather than precise, dates; the error can be more than 10%.

EXERCISES 5.6

In Exercises 1–10, solve the given equation for k or t, whichever is appropriate. Leave your answer in terms of natural logarithms.

1. $e^{3k} = 4$
2. $e^{-4k} = \frac{1}{5}$
3. $7e^{-1.2k} = 3.5$
4. $3e^{(t/2)\ln 2} = 4$
5. $1.7e^{-(t/3)\ln 2} = 0.85$
6. $0.005e^{-(t/1.5)\ln 3} = 0.0004$
7. $e^{2t} - 3e^t + 2 = 0$ (*Hint:* This is a quadratic equation in e^t.)
8. $e^{4t} - e^{2t} - 6 = 0$
9. $e^t + e^{-t} = 2$
10. $2^{kt} = e^{-t}$

11. Suppose the number of organisms of a species is given by (3). Show that the doubling time is $(\ln 2)/k$. (*Hint:* Compute $f((\ln 2)/k)$.)

12. With the data given in Example 3, determine how long it would take for the population of 1000 rice weevils to reach
    a. 3000   b. 4000

13. If the temperature is lowered to 23°C, then an initial population of 1000 rice weevils could ideally grow to approximately 1537 during a week's time.
    a. Determine a formula for the number of rice weevils at a future time t.
    b. Determine the doubling time of the weevils.

14. If the temperature is increased to 33.5°C, then the doubling time of the rice beetle expands to approximately 5.78 weeks (about 40.5 days).
    a. Find the numerical value of k in the population formula.
    b. Determine how long it would take for a population of 217 to increase to 864.

15. Experiment has shown that under ideal conditions and a constant temperature of 28.5°C the population of a certain type of flour beetle grows according to the formula

    $$f(t) = f(0)e^{0.71t}$$

    where t is measured in weeks. Does it take more than a week for the population to double? Explain your answer.

16. Using the information from Exercise 15, determine the length of time it would take for a population of 5000 flour beetles to grow to 17,000.

17. It is known that a population of one type of grain beetle can grow from 1000 to 1994 in a week if the temperature is set at 32.3°C, whereas it can grow from 2500 to 4465 in a week at a temperature of 29.1°C. Determine at which temperature the doubling time is less.

18. Suppose the number of bacteria in a culture grows exponentially and increases from 1 million to 3 million in 6 hours.
    a. Find a formula for the number $f(t)$ of bacteria at time t.
    b. How long would it take for the number of bacteria to quadruple?

19. The world population in 1962 was approximately 3.15 billion, and in 1971 it was approximately 3.706 billion. Assuming that the world population was growing exponentially during the interim and continued to do so, determine what the (approximate) world population would be in 1984.

20. In 1930 the world's population was 2 billion, and in 1976 the population passed the 4 billion mark. Assuming that the population grew exponentially during the interim, determine the constant k.

21. Suppose a radioactive substance decays according to (4). Show that the half-life of the substance is $(\ln 2)/k$. (See Exercise 11.)

22. What percent of the original C^{14} content in a skull would remain now if the animal lived around
    a. 1500 A.D.?   b. 50,000 B.C.?

23. In the late 1960s a wrench made from bone was unearthed in southeast Arizona. The date of the bone was calculated to be approximately 11,200 B.C. What percentage of the original C^{14} remained?

24. About 1960, awls used in basketmaking and formed from antlers were found in an archaeological site in western Tennessee. At that time approximately 32.86% of the original C^{14} remained. How old was the antler?

25. During the beginning of the 1970s a toothed bone scraper found in northern Yukon, Canada, gave new evidence of human settlements in North America more ancient than had been supposed. The C^{14} content remaining in the scraper was calculated to be approximately 3% of the original C^{14}. What would the minimum age of the scraper have been?

26. Some trees that were destroyed during the Fourth Ice Age now contain only 27% of the amount of C^{14} per gram that living trees of the same type now contain. When did the Fourth Ice Age occur?

27. The recently discovered Dead Sea Scrolls were determined by carbon dating to have been written about 33 A.D. What percentage of the original amount of C^{14} presently remains in the Scrolls?

28. Suppose you are told that a bone found in an archaelogical site is more than 17,000 years old. When you submit the bone to carbon dating analysis, you determine that approximately 20% of the original C^{14} remains. Are your findings compatible with the claim of the bone's age? Explain why or why not.

29. Determine the percentage of C^{14} in a skull that decays during a century. (Your answer should show that the same percentage of C^{14} is lost, no matter what century is chosen.)

30. The half-life of strontium 90 (Sr^{90}) is 19.9 years. How long is required for a given amount of Sr^{90} to decay to
a. 25% of its original amount?
b. 20% of its original amount?

31. The half-life of uranium 238 (U^{238}) is approximately 4.5 billion years. How long is required for 1% of a given quantity of U^{238} to decay?

32. Suppose the half-life of a radioactive isotope is h years. Show that the amount $f(t)$ of the isotope after t years is given by

$$f(t) = f(0)2^{-t/h}$$

33. Halley's Law states that the barometric pressure $p(x)$ in inches of mercury at x miles above sea level is given by

$$p(x) = 29.92e^{-0.2x} \quad \text{for } x \geq 0$$

Find the barometric pressure
a. at sea level
b. 1 mile above sea level
c. atop Mt. Whitney (14,495 feet above sea level)

34. The Bouguer—Lambert Law states that as a light beam passes through water, the intensity of the beam x meters from the surface of the water is given by

$$f(x) = ce^{-1.4x}$$

where c is the intensity at the surface. Use a calculator to determine the ratio of the intensity 10 meters below the surface to the intensity at the surface—that is, find $f(10)/c$. (From your answer it will be apparent why most flora cannot survive at depths greater than 10 meters.)

35. Let $A(t)$ denote the area in square centimeters of the unhealed portion of a wound t days after it is sustained, and let $A(0)$ represent the area of the original wound. Then according to the Law of Healing, $A(t)$ is given by

$$A(t) = A(0)e^{-0.15t} \quad \text{for } t \geq 0$$

Suppose that just after a collision, a basketball player is told he cannot play until 90% of a leg wound is healed. Would he be able to play in a tournament in two weeks?

5.7

COMPOUND AND CONTINUOUS INTEREST

Recall from Section 2.2 that if P dollars are deposited into an account earning simple interest at an annual rate r, then the amount of interest after n years is Prn, and therefore the amount A_n in the account after n years is given by the following formula:

SIMPLE INTEREST

$$A_n = P + Prn = P(1 + rn) \tag{1}$$

In particular, after 1 year the amount A_1 in the account is given by

$$A_1 = P(1 + r) \tag{2}$$

For example, if the rate were 6%, then $A_1 = P(1 + 0.06) = 1.06\,P$.

If instead of using simple interest, the bank compounds interest annually, then the amount in the account after 1 year is A_1, as before. But during the second year interest is earned on A_1 dollars (rather than on the original principal P). Thus after 2 years the amount A_2 in the account is given by

$$A_2 = A_1 + A_1 r = A_1(1 + r)$$

Substituting for A_1 from (2), we have

$$A_2 = P(1 + r)^2$$

During the third year interest is earned on A_2 dollars, and in the same way as above, we find that the amount A_3 in the account after 3 years is given by the formula

$$A_3 = A_2 + A_2 r = A_2(1 + r) = P(1 + r)^3$$

More generally, the formula for the amount A_n in the account after n years of compounding interest annually is given by the following formula:

INTEREST COMPOUNDED ANNUALLY

$$A_n = P(1 + r)^n \tag{3}$$

Nowadays most banks go a step farther and compound interest not annually, but quarterly, monthly, or even daily (so the interest period becomes 3 months, 1 month, or 1 day, respectively). If a bank offers 6% interest compounded monthly, then since there are 12 months in a year, the interest rate for each month is 6% divided by 12, or $\frac{1}{2}$%, and if 6% interest is compounded daily, the rate for each day is $\frac{6}{365}$% (except in leap years). In general, if a bank offers compound interest at an annual rate r compounded k times a year, then the interest rate for each interest period is r/k, so that after the first interest period a deposit of P dollars is worth A_1 dollars, where

$$A_1 = P + P \cdot \frac{r}{k} = P\left(1 + \frac{r}{k}\right)$$

If A_n represents the amount in the account after n interest periods, if r is the annual interest rate, and if the interest is compounded k times a year, then A_n can be shown to be given by the following formula:

INTEREST COMPOUNDED k TIMES ANNUALLY

$$A_n = P\left(1 + \frac{r}{k}\right)^n \tag{4}$$

Notice that (4) agrees with (3) when $k = 1$, that is, when the interest is compounded once a year (annually).

Example 1. Suppose that $1000 is deposited into an account. How much money is in the account after 5 years if the account earns 6% interest compounded

    a. annually?
    b. monthly?
    c. daily?

Solution.

    a. We use (4) with $P = 1000$, $r = 0.06$, $k = 1$, and $n = 5$:

$$A_5 = 1000(1 + 0.06)^5 = 1000(1.06)^5 \approx 1338.23$$

where the final computation can be accomplished either by hand or by calculator. Thus $1338.23 is in the account after 5 years.

b. Here we use (4) with $P = 1000$, $r = 0.06$, $k = 12$, and $n = 5(12) = 60$:

$$A_{60} = 1000 \left(1 + \frac{0.06}{12}\right)^{60} = 1000(1.005)^{60}$$

Using a calculator, we find that $A_{60} \approx 1348.85$. Consequently the account contains approximately $1348.85 after 5 years.

c. Again we use (4), but this time with $P = 1000$, $r = 0.06$, $k = 365$, and $n = 5(365) = 1825$:

$$A_{1825} = 1000 \left(1 + \frac{0.06}{365}\right)^{1825}$$

Using a calculator, we find that $A_{1825} \approx 1349.83$. This means that after 5 years the account contains $1349.83.   □

Observe from the results of Example 1 that the shorter the interest period is (that is, the more times per year interest is compounded), the more money is in the account after 5 years. More generally, the shorter the interest period is, the more money there would be in the account after any given length of time.

Continuously Compounded Interest

As we mentioned above, the shorter the interest period is, the more money there is in the account after a fixed period of time. A natural question to ask is: What would happen to the money in the account in, say, one year, if the interest period shrinks to 0? Since (4) tells us that with k interest periods a year the amount in the account at the end of one year would be given by

$$A_k = P \left(1 + \frac{r}{k}\right)^k$$

the question just posed is equivalent to the following question: What happens to A_k as k grows without bound? Would A_k also grow without bound? Although we will not prove it, the answer is that A_k approaches a limiting value A as k increases without bound, and moreover,

$$A = Pe^r \tag{5}$$

When the interest for one year is computed by (5), we say that interest has been *compounded continuously*. Some banks actually do offer interest compounded continuously on savings accounts.

Now let us obtain a formula for the amount in an account after t

years when interest is compounded continuously at an annual rate r. By (5) the amount after one year is Pe^r. Therefore after 2 years the amount is $(Pe^r)e^r = Pe^{2r}$, and in general, the amount $A(t)$ in the account after t years is given by the following formula:

INTEREST COMPOUNDED CONTINUOUSLY

$$A(t) = Pe^{rt} \qquad\qquad (6)$$

Formula (6) is valid for any positive value of t, integer or not.

Example 2. Suppose that $1000 is deposited into an account that pays 6% interest compounded continuously. How much is in the account after 5 years?

Solution. We use (6) with $P = 1000$, $r = 0.06$, and $t = 5$:

$$A(5) = 1000e^{(0.06)5} = 1000e^{0.3}$$

Using a calculator, we find that $A(5) \approx 1349.86$. This means that the account contains $1349.86 after 5 years. ☐

Compounding continuously can be regarded as a kind of ultimate—it produces more than compounding annually, monthly, or even daily. However, it is worthwhile noticing that the amount attained by compounding $1000 continuously for 5 years is a mere 3¢ more than that attained by compounding daily, and only $1.01 more than that attained by compounding monthly.

EXERCISES 5.7

1. Suppose $1000 is deposited into an account. How much money will be in the account after 5 years if the account earns 5% interest compounded
   a. annually?   b. monthly?   c. daily?

2. Suppose the interest in Exercise 1 were compounded continuously. How much would be in the account after 5 years? Compare your answer with those given in Exercise 1.

3. Suppose $10,000 is deposited into an account. How much money will be in the account after 10 years if the account earns 9% interest compounded
   a. annually?   b. monthly?   c. daily?

4. Suppose the interest in Exercise 3 were compounded continuously. How much would be in the account after 10 years? Compare your answer with those given in Exercise 3.

5. How much money would you have to deposit into an account that earns 7% interest compounded semiannually if you wished the account to contain $1000 after 4 years?

6. How much money would you have to deposit into an account that earns 7% interest compounded continuously if you wished the account to contain $1000 after 4 years? Compare your answer with that given in Exercise 5.

7. Suppose a benefactor plans to open an account that will be yours in 5 years. Which would you prefer—a $2000 initial deposit that earns 6% simple interest, or a $1500 initial deposit that earns 12% interest compounded monthly? Justify your answer.

8. Approximately how many months does it take for $1000 to double in an account that earns
   a. 6% interest compounded monthly?
   b. 9% interest compounded monthly?

9. Suppose $5000 is put into a savings account compounded monthly, and after 3 years the account is worth $7000. What is the interest rate?

10. It has been estimated that in the past, productivity in the United States has increased at a rate of 4% annually. Assuming that this is true and productivity is compounded annually, determine how much productivity increased during a 25-year span.

11. If inflation is 9% per year over a 6-year period and is compounded annually, what is the overall inflation during the period?

12. Suppose that on January 1, $800,000 is deposited in an account earning 6% interest compounded daily. Determine the difference, if any, between the amounts in the account at the end of a regular year (with 365 days) and a leap year (with 366 days).

13. Approximately how long does it take for $1000 to double in an account that earns interest compounded continuously at a rate of
   a. 6%?   b. 9%?

14. You have $1000 with which to open a savings account. Which would be worth more at the end of a 10-year period—an account that draws 6% simple interest, or an account that earns 5% interest compounded continuously? Justify your answer.

15. In 1626, Manhattan was purchased from the Indians for $24. What would the value be in 1986 if the Indians had been able to invest the $24 at 5% interest compounded
   a. annually?   b. continuously?

16. If the $24 the Indians received were compounded continuously at a rate of 5%, during what year would it have been worth 1 billion dollars?

17. The rate r at which a savings account draws compound interest is sometimes called the **nominal rate** of the account. The **effective rate E** (or effective annual interest rate) of the account is the simple interest that would yield the same interest during the course of one year.
   a. Assume that interest is compounded k times a year. Show that E is given by

$$E = \left(1 + \frac{r}{k}\right)^k - 1$$

b. What is the effective rate for an account that draws interest at a rate of 7.5% compounded monthly?

c. Find a formula for the nominal interest rate r in terms of E and k.

18. A loan worth P dollars is to be paid off in n equal installments of A dollars each, and the interest rate is r per pay period. The **installment payment** A is given by the formula

$$A = P \left(\frac{r(1 + r)^n}{(1 + r)^n - 1} \right)$$

Solve for n in terms of P, r, and A.

The formula for A given in Exercise 18 applies to Exercises 19–22.

19. Pat receives a loan of $30,000, with an annual interest rate of 9% compounded annually, and the loan is to be paid off in 20 equal yearly installments.

a. Determine the annual payment.

b. Determine the total to be paid over the life of the loan.

20. Marian receives a loan of $6000, with an annual interest of 11.5% compounded monthly, and is to pay off the loan in equal monthly installments over a 10-year period.

a. Determine the monthly payment.

b. Determine the total to be paid back over the life of the loan.

21. Suppose Marian can get a 30-year loan at 9% per year, and can afford to pay up to $263 a month. Determine the maximum loan under these conditions.

22. Jack and Jill each take out a 5-year loan of $8000, with an annual interest rate of 10%. Jack pays the loan back in equal annual installments, whereas Jill pays the loan back in equal monthly installments. Who pays back more over the life of the loan, Jack or Jill? What is the difference?

*23. Suppose you negotiate a 3-year contract according to which you are paid $20,000 the first year and receive a total salary of $70,000 over the 3-year-period. If the salary is treated as being compounded annually, what is the rate at which it is compounded?

KEY TERMS

exponential function
  base
  exponent
  natural exponential function
logarithmic function
  base
  natural logarithm
  common logarithm
  characteristic
  mantissa

antilogarithm
linear interpolation
exponential growth
  doubling time
exponential decay
  half-life
compound interest
continuously compounded
  interest

KEY FORMULAS

Laws of Exponents	Laws of Logarithms

$a^x a^y = a^{x+y}$ $\qquad a^{\log_a x} = x$

$(a^x)^y = a^{xy}$ $\qquad \log_a (a^x) = x$

$(ab)^x = a^x b^x$ $\qquad \log_a (xy) = \log_a x + \log_a y$

$\left(\dfrac{a}{b}\right)^x = \dfrac{a^x}{b^x}$ $\qquad \log_a \dfrac{1}{x} = -\log_a x$

$\dfrac{a^x}{a^y} = a^{x-y}$ $\qquad \log_a \dfrac{x}{y} = \log_a x - \log_a y$

$a^{-x} = \dfrac{1}{a^x} = \left(\dfrac{1}{a}\right)^x$ $\qquad \log_a (x^c) = c \log_a x$

$a^0 = 1$ $\qquad \log_b x = \dfrac{\log_a x}{\log_a b}$

$1^x = 1$

$\log_a 1 = 0$

$\log_a a = 1$

REVIEW EXERCISES

In Exercises 1–4, sketch the graph of the given function.

1. $f(x) = 3^x$
2. $f(x) = (\frac{1}{2})^{x-1}$
3. $f(x) = \frac{1}{2} \log_2 x$
4. $f(x) = |\log_2 x|$

In Exercises 5–8, simplify the given expression.

5. $(\sqrt{2}^{\sqrt{2}})^{\sqrt{2}}$
6. $\dfrac{5^{1-\sqrt{2}} \cdot 25^{1/\sqrt{2}}}{5^{2+\sqrt{2}}}$

7. $\ln 12 - \ln 2 + 2 \ln e^3$
8. $\log_{10} 5 - \log_{10} \dfrac{50}{7}$

In Exercises 9–14, solve the given equation.

9. $7^{x-1} = (\sqrt{7})^{-2x^2}$
10. $2^{x^3} = 4^x$
11. $\log_2 x = -3$
12. $\log_x 5 = 2$
13. $\log_{10} (3 - 5x) - \log_{10} (1 - x) = 1$
14. $3 \log_2 \sqrt{x} = 2 \log_2 8$

In Exercises 15–20, solve the given equation. Leave your answer in terms of logarithms.

15. $2^{-x} = 7$
16. $4^{x-4} = 5^{3x+5}$
17. $2^x = 3^{(x^1)}$
18. $e^{2x} = \dfrac{1}{e^{(x^2)}}$
19. $e^{5k} = 4.37$
20. $e^{-3t \ln 2} = 0.37$

In Exercises 21–24, evaluate the given expression.

21. $\log_2 \frac{1}{16}$

22. $\log_{10} (10^{-5})$

23. $e^{\ln 5}$

24. $10^{\log_{10} 1.23}$

In Exercises 25–28, approximate the numerical value of the given logarithms. Use the approximations $\log_{10} 6 \approx 0.7782$, $\log_{10} 7 \approx 0.8451$, and $\log_{10} 11 \approx 1.0414$.

25. $\log_{10} 420$

26. $\log_{10} \frac{77}{36}$

27. $\log_{10} 0.0077$

28. $\log_{10} (6^{-1} \times 11^{10})$

In Exercises 29–32, use Table C to approximate the common logarithm of the given number. Use linear interpolation if needed.

29. 4.13

30. 0.00956

31. 6.384×10^8

32. 0.1977

In Exercises 33–36, use Table C to approximate the antilogarithm of the given number. Use linear interpolation if needed.

33. 0.5514   34. -3.7372 35. 0.6752   36. -1.1300

37. Given that $\log_{10} 8 \approx 0.9031$ and $\log_{10} e \approx 0.4343$, approximate $\ln 8$.

38. Given that $\ln 5 \approx 1.6094$ and $\ln 2 \approx 0.6931$, approximate $\log_2 5$.

39. Given that $\ln 10 \approx 2.3026$, approximate $\log_{10} e$.

40. Solve the equation $y = \ln (x - \sqrt{x^2 - 1})$ for x.

41. Suppose $y = a + b \ln x$, where a and b are constants. Express x in terms of y.

42. Show that $\log_2 x = 2 \log_4 x$.

43. Show that $\log_2 x = 3 \log_8 x$.

44. Let $f(x) = e^{cx}$, where c is a constant. Suppose that the graph of f passes through the point $(\frac{1}{3}, e^2)$. Determine c.

45. Let $g(x) = ae^{cx}$, where a and c are constants. Suppose that the graph of g passes through the points $(0, \frac{1}{5})$ and $(-1, \frac{1}{5}e^2)$. Determine a and c.

46. The noise level of one sound is 20 decibels higher than that of a second sound. Find the ratio of the intensity of the first sound to that of the second.

47. If an earthquake measures 6.1 on the Richter scale, what will one that is 100 times as intense measure?

48. It has been shown experimentally that under optimal conditions, the population of a colony of brown rats could grow exponentially from 50 rats to approximately 11,070 during the course of a year.
    a. Determine the doubling time.
    b. How long would it take for the population to increase from 120 to 340 rats?

49. The total amount of timber in a young forest increases exponentially.
    a. If the annual growth rate k is 3%, how long does it take for the total amount of timber to double?
    b. How much sooner would the total amount of timber double if the rate of growth were 4%?

50. The radioactive isotope potassium 40 (K^{40}) has a half-life of approximately 91.3 billion years. If a rock formed 4 billion years ago (shortly

after the earth came into existence) initially contained 1 gram of K^{40}, how much of it would the rock contain today?

51. Radioactive argon 39 (Ar^{39}) has a half-life of 4 minutes. How long would it take for 99% of a given amount of Ar^{39} to decay?

52. Until about 200 years ago, charcoal, which is derived from wood, was used in the smelting of iron, and thus it is possible to use carbon dating to determine the age of objects made long ago out of an iron alloy (such as steel). Fragments of a steel sword unearthed in Yugoslavia in 1970 were found to contain only 77.286% of the C^{14} the sword would have contained at the time it was produced. Determine the age of the sword.

53. Nails buried long ago at a Roman Legionary fortress in Scotland were found in 1960 to contain 79.948% of the C^{14} they would have contained when produced. Determine when the nails were produced. (According to historical accounts, the fort was built in 83 A.D. and dismantled in 87 A.D.)

54. Suppose $2000 is deposited into a savings account. How much money is in the account after 10 years if the account earns 8% compounded
    a. annually?
    b. monthly?
    c. daily?

55. Suppose $2000 is deposited into a savings account. How much money is in the account after 10 years if the account earns 8% compounded continuously?

56. How much money must be deposited into a savings account that earns 6% compounded continuously if the account is to contain $5000 after 8 years?

SYSTEMS OF EQUATIONS AND INEQUALITIES

6

In Chapter 2 we studied methods of solving equations having a single unknown or variable (usually denoted by x). However, some problems involve more than one unknown. Suppose, for example, that we wish to find two numbers x and y whose sum is 10 and whose product is 21. Then there are two unknowns, denoted by x and y. Moreover, since the sum of the two numbers is 10 and their product is 21, it follows that x and y are related by the two equations

$$x + y = 10$$
$$xy = 21$$

Thus our problem is to find numbers x and y that satisfy both of the above equations. In other problems there may be more than two unknowns and more than two equations relating them. In each such case we would wish to find values of the unknowns that satisfy all the equations relating the unknowns. The collection of equations relating the unknowns is called a *system of equations* and will be a central topic of this chapter.

The outline of the chapter is as follows: In Sections 6.1 through 6.3 we examine several methods of solving systems of equations. Sections 6.4 through 6.7 involve a discussion of matrices and determinants and their use in solving systems of equations. Section 6.8 is devoted to systems of inequalities and their solutions, and the final section of the chapter concerns linear programming, which utilizes both systems of equations and systems of inequalities.

6.1

SYSTEMS OF EQUATIONS TWO VARIABLES

By a system of equations in two variables x and y we mean a collection of equations that contain the variables x and y and no others. Examples are

$$\begin{cases} x + y = 10 \\ xy = 21 \end{cases} \tag{1}$$

and

$$\begin{cases} 2x + y = 1 \\ -3x + 2y = 9 \end{cases} \tag{2}$$

As is done above, we will use a brace to group together the equations in a given system.

A *solution* of a system of equations in two variables x and y is an ordered pair (a, b) such that if x is replaced by a and y is replaced by b in all equations, then all the equations are valid. For example, $(3, 7)$ is a solution of (1) because if we replace x by 3 and y by 7, then (1) becomes

$$\begin{cases} 3 + 7 = 10 \\ 3 \cdot 7 = 21 \end{cases}$$

which is a system of valid equations. Likewise $(7, 3)$ is a solution of (1), as you can check. What may be less obvious, but will be shown a little later, is the fact that $(3, 7)$ and $(7, 3)$ are the only solutions of (1). The process of finding all solutions of a system of equations is called *solving* the system.

Caution: It is crucial to note that *each* equation in a system must be satisfied by a pair (a, b) for (a, b) to be a solution of the system. Indeed, (2, 8) is a solution of $x + y = 10$ in (1) because $2 + 8 = 10$, but since $2 \cdot 8 = 16$, (2, 8) is *not* a solution of $xy = 21$, and therefore (2, 8) is *not* a solution of the system in (1).

One can interpret the solutions of a system of equations in two variables graphically. For a specific example, let us consider (1) again. The graphs of the two equations are shown in Figure 6.1. Notice that the solutions (3, 7) and (7, 3) are the points at which the graphs of the two equations in (1) intersect. More generally, we observe that a point (a, b) is on the graph of each equation in a system if and only if each equation in the system is valid when x is replaced by a and y is replaced by b. Consequently (a, b) is a solution of the system if and only if (a, b) lies on the graph of each equation in the system. When the graphs of the equations in a system are easy to draw and happen to intersect at obvious points in the plane, it is possible to use graphs to detect solutions of the system. However, generally other methods are preferable because they are more likely to lead to precise solutions and because they do not rely on precise graphs.

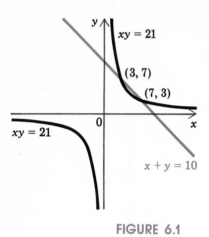

FIGURE 6.1

The Method of Substitution

Let us return to the system

$$\begin{cases} x + y = 10 \\ \quad xy = 21 \end{cases} \tag{3}$$

and determine all solutions of (3). Solving for y in terms of x in the first equation of (3), we obtain

$$y = -x + 10 \tag{4}$$

Next we substitute $-x + 10$ for y in the second equation of (3), which leads to the following chain of equivalent equations:

$$x(-x + 10) = 21$$
$$-x^2 + 10x - 21 = 0$$
$$x^2 - 10x + 21 = 0$$

The left side of the last equation factors to give us

$$(x - 3)(x - 7) = 0$$

It follows that $x = 3$ or $x = 7$. If $x = 3$, then by (4),

$$y = -3 + 10 = 7$$

If $x = 7$, then again by (4),

$$y = -7 + 10 = 3$$

Thus the only possible solutions of (3) are (3, 7) and (7, 3). We noted earlier that they are indeed solutions, so (3) is solved.

The method we have just used is called the **substitution method** of solving a system of equations, because after solving for one variable in one equation, we substituted for that variable in the other equation. To guide us in using the method, we list the steps to be followed:

SUBSTITUTION METHOD

   i. Solve one of the equations for y in terms of x (or for x in terms of y).

  ii. Substitute for y (or x) in the other equation in order to obtain an equation in the single variable x (or y).

 iii. Find all solutions of the equation obtained in (ii).

 iv. Use the equation obtained in (i) and the values for x (or y) obtained in (iii) in order to find the corresponding values of y (or x).

  v. Check every proposed solution by substituting in *each* equation of the given system.

Example 1. Use the substitution method to solve the system

$$\begin{cases} 2x + y = 1 \\ -3x + 2y = 9 \end{cases}$$

Solution. Solving for y in the first equation, we obtain

$$y = -2x + 1 \tag{5}$$

Substituting $-2x + 1$ for y in the second equation yields

$$-3x + 2(-2x + 1) = 9$$

$$-3x - 4x + 2 = 9$$

$$-7x = 7$$

$$x = -1$$

From (5) we see that if $x = -1$, then

$$y = -2(-1) + 1 = 2 + 1 = 3$$

Thus the proposed solution is $(-1, 3)$.

Check: $\begin{cases} 2(-1) + 3 = -2 + 3 = 1 \\ -3(-1) + 2(3) = 3 + 6 = 9 \end{cases}$

Thus the only solution of the given system is $(-1, 3)$. $\quad\square$

Example 2. Use the substitution method to solve the system

$$\begin{cases} xy + 2y^2 = 8 \\ x - 2y = 4 \end{cases}$$

Solution. Solving for x in the second equation yields

$$x = 2y + 4 \tag{6}$$

By substituting $2y + 4$ for x in the first equation, we obtain

$$(2y + 4)y + 2y^2 = 8$$
$$2y^2 + 4y + 2y^2 = 8$$
$$4y^2 + 4y - 8 = 0$$
$$y^2 + y - 2 = 0$$
$$(y + 2)(y - 1) = 0$$

Therefore $y = -2$ or $y = 1$. If $y = -2$, then by (6),

$$x = 2(-2) + 4 = 0$$

which means that $(0, -2)$ is a proposed solution. If $y = 1$, then again by (6),

$$x = 2(1) + 4 = 6$$

so that $(6, 1)$ is a proposed solution.

Check: $\begin{cases} (0)(-2) + 2(-2)^2 = 8 \\ 0 - 2(-2) = 4 \end{cases}$ and $\begin{cases} (6)(1) + 2(1)^2 = 8 \\ 6 - 2(1) = 4 \end{cases}$

Consequently the solutions of the given system are $(0, -2)$ and $(6, 1)$. $\square$

Example 3. Use the substitution method to solve the system

$$\begin{cases} \dfrac{y}{x - 1} = x + 1 \\ x^2 + y^2 = 1 \end{cases}$$

Solution. Multiplying both sides of the first equation by $x - 1$, we have

$$y = (x - 1)(x + 1) = x^2 - 1 \tag{7}$$

By substituting $x^2 - 1$ for y in the second equation, we find that

$$x^2 + (x^2 - 1)^2 = 1$$

$$x^2 + (x^4 - 2x^2 + 1) = 1$$

$$x^4 - x^2 = 0$$

$$x^2(x - 1)(x + 1) = 0$$

Thus $x = 0$, $x = 1$, or $x = -1$. From (7) we see that $y = -1$ if $x = 0$, $y = 0$ if $x = 1$, and $y = 0$ if $x = -1$. Therefore there are three proposed solutions: $(0, -1)$, $(1, 0)$, and $(-1, 0)$. You can check that $(0, -1)$ and $(-1, 0)$ actually are solutions of the given system. However, for the proposed solution $(1, 0)$, when we attempt to substitute 1 for x and 0 for y in the left side of the first equation, we obtain the meaningless expression $\dfrac{0}{1 - 1}$. We conclude that there are only two solutions: $(0, -1)$ and $(-1, 0)$. $\square$

A proposed solution of a system that is not a solution is called an **_extraneous solution_**. Thus $(1, 0)$ is an extraneous solution of the system in Example 3. Such solutions often arise when we square both sides of an equation or multiply both sides by an expression that may be 0.

So far we have solved systems that had one or two solutions. In general, there can be any number of solutions: none, any finite number, or infinitely many. The following example illustrates a system with no solutions.

Example 4. Show that the system

$$\begin{cases} x - 2y = 3 \\ 2x - 4y = 7 \end{cases}$$

has no solutions.

Solution. If (x, y) is a solution of the system, then from the first equation we have

$$x = 2y + 3$$

Substituting $2y + 3$ for x in the second equation yields

$$2(2y + 3) - 4y = 7$$

$$4y + 6 - 4y = 7$$

$$6 = 7$$

which is impossible. Thus no solution exists. $\square$

A system of equations that has no solutions is called **_inconsistent_**. Thus the system in Example 4 is inconsistent. Any system that has solutions is by contrast **consistent**. The systems given in Examples 1–3 exemplify consistent systems.

It is also possible to prove geometrically that the system in Example 4 is inconsistent. Observe that $x - 2y = 3$ and $2x - 4y = 7$ are equations of lines. The slope–intercept form of $x - 2y = 3$ is

$$y = \frac{1}{2} x - \frac{3}{2}$$

and the slope–intercept form of $2x - 4y = 7$ is

$$y = \frac{1}{2} x - \frac{7}{4}$$

It follows that the lines have the same slope, $\frac{1}{2}$, but different y intercepts. Thus the two lines are parallel and do not intersect. Since any solution of the system must be on both lines, there are no solutions of the system in Example 4.

Linear Systems

In general, any equation of the form $ax + by = c$, where a, b, and c are constants with either $a \neq 0$ or $b \neq 0$, is called a **linear equation**, since its graph is a line. A system containing only linear equations is called a **linear system**.

Now consider a linear system containing two equations in two variables. The graph of each equation is a straight line. Since two lines are identical, intersect at a single point, or do not intersect at all, it follows that a linear system has either infinitely many solutions, one solution, or no solutions. Because the nonlinear system appearing in Example 2 has two solutions, we know that nonlinear systems are not limited to infinitely many, one, or no solutions.

Our next example involves a linear system with infinitely many solutions.

Example 5. Show that the system

$$\begin{cases} 2x - 5y = 1 \\ 8x - 20y = 4 \end{cases} \tag{8}$$

has infinitely many solutions.

Solution. If we multiply both sides of the first equation by 4, it becomes

$$8x - 20y = 4$$

which is the same as the second equation. Solving for y, we obtain

$$y = \frac{1}{20} (8x - 4) = \frac{1}{5} (2x - 1)$$

Thus (a, b) is a proposed solution of the given system if and only if

$$b = \frac{1}{5}(2a - 1)$$

This means that any ordered pair of the form $(a, \frac{1}{5}(2a - 1))$ is a proposed solution, where a can be any number.

Check: $\begin{cases} 2a - 5\left[\dfrac{1}{5}(2a - 1)\right] = 2a - 2a + 1 = 1 \\[3mm] 8a - 20\left[\dfrac{1}{5}(2a - 1)\right] = 8a - 8a + 4 = 4 \end{cases}$

Therefore the given system has infinitely many solutions. □

Actual solutions of the system in Example 5 can be found by assigning particular values to a. For example, the solutions arising from the values 0, 1, 3, and -7 for a are $(0, -\frac{1}{5})$, $(1, \frac{1}{5})$, $(3, 1)$, and $(-7, -3)$, respectively.

An Applied Problem

We conclude this section with an applied problem that involves the mixing of two liquids.

Example 6. Suppose that in the process of developing photographs, one needs 2 gallons of water at 70°F. If the temperature of tap water is 54°F and that of water heated on a stove is 182°F, determine how much water at each temperature must be used.

Solution. Let

x = the amount in gallons of tap water at 54°F to be used
y = the amount in gallons of stove water at 182°F to be used

By assumption, 2 gallons are needed, so that one equation relating x and y is

$$x + y = 2 \tag{9}$$

Next, since the proportion of tap water in the 2-gallon mixture is $x/2$ and the proportion of stove water is $y/2$, the temperature of the mixture is $(x/2)54 + (y/2)182$. Thus the assumption that the temperature of the mixture is to be 70°F means that

$$\left(\frac{x}{2}\right)54 + \left(\frac{y}{2}\right)182 = 70$$

which simplifies to

$$27x + 91y = 70 \tag{10}$$

Equations (9) and (10) form the system

$$\begin{cases} x + y = 2 \\ 27x + 91y = 70 \end{cases} \tag{11}$$

Solving for x in the first equation of (11), we have

$$x = 2 - y \tag{12}$$

Substituting $2 - y$ for x in the second equation of (11) and solving for y, we obtain

$$27(2 - y) + 91y = 70$$
$$54 - 27y + 91y = 70$$
$$64y = 16$$
$$y = \frac{1}{4}$$

From (12) we conclude that

$$x = 2 - \frac{1}{4} = \frac{7}{4}$$

Consequently $\frac{7}{4}$ gallons of tap water and $\frac{1}{4}$ gallon of stove water are needed. $\square$

     In closing we mention that systems of equations in two variables have a long history. In fact, systems such as the one in (3) have appeared in one guise or another since antiquity. For example, a tablet dating from around 300 B.C. shows problems related to finding the dimensions of a rectangle with unit area and given semiperimeter s. If the sides of the rectangle are denoted by x and y, then these conditions on the area and semiperimeter give rise to the system

$$\begin{cases} xy = 1 \\ x + y = s \end{cases}$$

This system is a close relative of the system in (3) and can be solved the same way (see Exercise 31).

EXERCISES 6.1

In Exercises 1–4, verify that the given ordered pair is a solution of the given system.

1. $\begin{cases} 2x + 3y = 8 \\ -5x + \dfrac{1}{2}y = -4 \end{cases}$ $(1, 2)$

2. $\begin{cases} -\dfrac{2}{3}x + 3y = 1 \\ 17x - 6y = -2 \end{cases}$ $(0, \frac{1}{3})$

3. $\begin{cases} 2xy = 12 \\ \dfrac{1}{y} - \dfrac{1}{x} = \dfrac{1}{6} \end{cases}$ $(-2, -3)$

4. $\begin{cases} \dfrac{y}{x-1} = x + 1 \\ x^2 + y^2 = 1 \end{cases}$ $(0, -1)$

In Exercises 5–30, find all solutions of the given system of equations.

5. $\begin{cases} 3x - 5y = 13 \\ y = -2 \end{cases}$

6. $\begin{cases} x - y = 0 \\ 2x - 5y = 9 \end{cases}$

7. $\begin{cases} 2y - 3z = 3 \\ 3y + z = 10 \end{cases}$

8. $\begin{cases} z - 3y = 11 \\ 3y - 7z = -5 \end{cases}$

9. $\begin{cases} 7x - 5y = -6 \\ 15y - 21x = 18 \end{cases}$

10. $\begin{cases} 2x - 3y = 5 \\ 4x - 6y = 10 \end{cases}$

11. $\begin{cases} x = 15 - 4y \\ x = -13 + 3y \end{cases}$

12. $\begin{cases} x - 3y = 9 \\ \frac{1}{3}x - y = 27 \end{cases}$

13. $\begin{cases} 4y - z = 4.7 \\ 3y + 2z = 9.3 \end{cases}$

14. $\begin{cases} 7y + 5z = -3 \\ y + 2z = -2.1 \end{cases}$

15. $\begin{cases} \frac{1}{2}x - \frac{1}{3}y = 0 \\ 3x - 2y = 1 \end{cases}$

16. $\begin{cases} 2x + 3y - 5 = 0 \\ x - 2y - 4 = 0 \end{cases}$

17. $\begin{cases} 3x - 2y + 9 = 0 \\ y - 1.5x - 6 = 0 \end{cases}$

18. $\begin{cases} 5x - y - 7 = 0 \\ 15x = 3y + 21 \end{cases}$

19. $\begin{cases} 3x - y = 0 \\ xy = 3 \end{cases}$

20. $\begin{cases} x + 2y = 8 \\ xy = 6 \end{cases}$

21. $\begin{cases} x^2 + 4y^2 = 32 \\ x + 2y = 8 \end{cases}$

22. $\begin{cases} y^2 + z^2 = 25 \\ y - 2z = 0 \end{cases}$

23. $\begin{cases} x^2 - 3xy + 2y^2 = 27 \\ 2x + y = 0 \end{cases}$

24. $\begin{cases} x + y = 1 \\ \dfrac{1}{x} + \dfrac{1}{y} = 4 \end{cases}$

25. $\begin{cases} 6y - x - 1 = 0 \\ y = x^2 \end{cases}$

26. $\begin{cases} y = e^x \\ 0 = e^{2x} - 2y - 8 \end{cases}$

27. $\begin{cases} y = 2^x \\ 6 = 4^x - y \end{cases}$

28. $\begin{cases} x^2 + 2y^2 = 24 \\ \log_2 x + \log_2 y = 3 \end{cases}$

29. $\begin{cases} x + y = a - b \\ x - y = a + b \end{cases}$ where a and b are constants

30. $\begin{cases} x + 2y = a + b \\ 2x - 3y = 4a - 2b \end{cases}$ where a and b are constants

31. Let s be a given positive number, and consider the system

$$\begin{cases} xy = 1 \\ x + y = s \end{cases}$$

a. Solve the system under the condition that $s \geq 2$.
b. Show that the system has no solution if $0 < s < 2$.

32. Solve the system

$$\begin{cases} \dfrac{1}{x^2} - \dfrac{1}{y^2} = 20 \\ \dfrac{1}{x} - \dfrac{1}{y} = 2 \end{cases}$$

(*Hint:* Let $u = 1/x$ and $v = 1/y$, solve the resulting system in u and v, and then find the corresponding values of x and y.)

33. Consider the system

$$\begin{cases} x^2 + 3y^2 = a \\ 1 + x = y \end{cases}$$

Determine the values of a for which the system has
a. exactly two solutions
b. exactly one solution
c. no solutions

34. Determine the values of a and b for which the line with equation $ax + by = 1$ passes through the points $(2, -1)$ and $(-3, 3)$.

35. Determine the values of a and b for which the line with equation $ax + by = 1$ has y intercept $\frac{7}{5}$ and passes through the point $(-4, -1)$.

36. The sum of two numbers is 11, and their product is 28. Find the numbers.

37. The difference of two numbers is 6, and their product is 216. Find the numbers.

38. The difference of two numbers is 4, and the sum of their squares is 136. Find the numbers.

39. The area of a right triangle is 20, and the length of the hypotenuse is $4\sqrt{5}$. Find the lengths of the legs of the triangle.

40. The lengths of the sides of a triangle are 4, 8, and 10. Find the length of the altitude directed toward the longest side.

41. The perimeter of a rectangular room is 82 feet, and the area is 408 square feet. Find the dimensions of the room.

42. A Metroliner train travels the 216 miles from New York to Washington, D.C., at an average speed of 24 miles per hour faster than a local train. If the local train takes 90 minutes longer than the Metroliner, find the average speed of the Metroliner.

43. In making yogurt one places the ingredients in water at 110° Fahrenheit. How much tap water at 60° Fahrenheit and how much stove water at 140° Fahrenheit should be mixed to obtain 3 quarts of water at 110° Fahrenheit?

44. A person has a total of $800 in two savings accounts, one yielding 6% simple interest per year and the other $7\frac{1}{2}$% simple interest per year. If the total interest earned during a single year is $51.75, how much is invested in the 6% savings account?

45. In 8 years a girl will be $\frac{5}{8}$ as old as her father was 8 years ago, whereas in 4 years she will be $\frac{4}{9}$ as old as he was 4 years ago. What is the present age of each?

46. When a 4-foot wide sidewalk is built along one end and one side of a rectangular plot of ground, the ground area is reduced from 7200 square feet to 6536 square feet. Find the original dimensions of the plot.

47. A smaller cube sits on top of a larger cube, which sits on the ground. If the combined height is 6 centimeters and the uncovered portions have a combined surface area of 96 square centimeters, find the lengths of the sides of the cubes.

6.2

SOLUTIONS OF SYSTEMS BY THE ELIMINATION METHOD

In addition to the substitution method discussed in Section 6.1, there is another important method for solving systems of equations: the elimination method. It is especially effective for solving linear systems with more than two equations and two variables, and for understanding methods we will discuss later in the chapter.

The elimination method involves manipulating the equations in a system in ways that do not affect the solutions of the system. Two systems that have the same solutions are called *equivalent*, and the strategy will be to replace a given system by an equivalent system whose solutions are more easily identified.

One way to manipulate a system is to multiply an equation in the system by a nonzero constant, by which we mean to multiply each side of the equation by that constant. For example, multiplying the second equation of the system

$$\begin{cases} -2x + 3y = 10 \\ x - y = 6 \end{cases}$$

by 2 yields the equivalent system

$$\begin{cases} -2x + 3y = 10 \\ 2x - 2y = 12 \end{cases} \tag{1}$$

Since multiplying an equation by a nonzero number does not alter the solutions of the equation, this manipulation leads to an equivalent system.

A second manipulation of a system involves replacing one of the equations in the system by the sum (or difference) of the equations in the system. For example, taking the sum of the equations in (1) gives us

$$(-2x + 3y) + (2x - 2y) = 10 + 12$$

or more simply,

$$y = 22$$

Replacing the second equation in (1) by the simplified form $y = 22$ of the sum yields

$$\begin{cases} -2x + 3y = 10 \\ \qquad\quad y = 22 \end{cases} \tag{2}$$

The system in (2), obtained from the system in (1) by the addition manipulation, has the same solution as does the system in (1). Thus the two systems are equivalent. Observe that x does not appear in the second equation of (2). Thus in passing from (1) to (2) we have eliminated x from the second equation.

The method of solving a system by utilizing the two manipulations—multiplying an equation by a constant, and substituting the sum (or difference) of two equations for one of the equations—in order to eliminate one of the variables is called the *elimination method.*

Example 1. Use the elimination method to solve the system

$$\begin{cases} 2x + 3y = -4 \\ x - 2y = \quad 5 \end{cases}$$

Solution. First we multiply the second equation by -2:

$$\begin{cases} \quad 2x + 3y = -4 \\ -2x + 4y = -10 \qquad 2 \times \text{second equation} \end{cases}$$

Next we replace the second equation by its sum with the first equation:

$$\begin{cases} 2x + 3y = -4 \\ \qquad 7y = -14 \qquad \text{first equation} + \text{second equation} \end{cases}$$

This last system can be solved easily, since x has been eliminated from its second equation. Indeed, the second equation implies that $y = -2$, so by substituting -2 for y in the first equation we obtain

$$2x + 3(-2) = -4$$
$$2x = -4 + 6 = 2$$
$$x = 1$$

Thus the proposed solution is $(1, -2)$.

Check: $\begin{cases} 2(1) + 3(-2) = -4 \\ \quad 1 - 2(-2) = \quad 5 \end{cases}$

We conclude that $(1, -2)$ is the solution of the given system. □

Let us observe that if the variables appearing in the system are changed but the coefficients of the variables are unaltered, then the solutions of the system remain the same. Thus the solution of the system

$$\begin{cases} 2y + 3z = -4 \\ \ y - 2z = \quad 5 \end{cases} \tag{3}$$

obtained from the system in Example 1 by replacing x by y and y by z, is the ordered pair $(1, -2)$. We will use this fact later.

In attempting to use the elimination method to solve a linear system, it is possible to obtain a new system that has two identical equations. In that case there are infinitely many solutions, as the following example shows.

Example 2. Use the elimination method to solve the system

$$\begin{cases} \quad 2x - \ y = \quad 4 \\ -6x + 3y = -12 \end{cases}$$

Solution. If we multiply the second equation by $-\frac{1}{3}$, we obtain the system

$$\begin{cases} 2x - y = 4 \\ 2x - y = 4 \qquad -\frac{1}{3} \times \text{second equation} \end{cases}$$

It follows that

$$y = 2x - 4$$

Thus (a, b) is a proposed solution of the given system if and only if

$$b = 2a - 4$$

Check: $\begin{cases} \quad 2a - (2a - 4) = 2a - 2a + 4 = 4 \\ -6a + 3(2a - 4) = -6a + 6a - 12 = -12 \end{cases}$

We conclude that for any number a, $(a, 2a - 4)$ is a solution of the given system. □

The elimination method can also identify inconsistent systems, that is, systems with no solutions.

Example 3. Use the elimination method to show that the system

$$\begin{cases} 4x - 2y = 8 \\ -6x + 3y = 3 \end{cases}$$

is inconsistent.

Solution. Multiplying the first equation by 3 and the second by -2, we obtain the system

$$\begin{cases} 12x - 6y = 24 & 3 \times \text{first equation} \\ 12x - 6y = -6 & -2 \times \text{second equation} \end{cases}$$

If we subtract the second equation from the first, we obtain $0 = 30$, which is impossible. Consequently the system has no solutions and hence is inconsistent. □

EXERCISES 6.2

In Exercises 1–14, use the elimination method to find all solutions of the given system of equations.

1. $\begin{cases} 2x + 3y = 3 \\ x + y = 2 \end{cases}$

2. $\begin{cases} 2x + y = 4 \\ 3x + y = 9 \end{cases}$

3. $\begin{cases} 7x - 5y = -6 \\ x + 5y = 22 \end{cases}$

4. $\begin{cases} 5x + 7y = 2 \\ 3x + 2y = -1 \end{cases}$

5. $\begin{cases} -5u + 3v = 1 \\ 9u - 10v = 12 \end{cases}$

6. $\begin{cases} \frac{2}{3}u - \frac{1}{4}v = 0 \\ \frac{3}{5}u + \frac{14}{5}v = 4 \end{cases}$

7. $\begin{cases} \frac{1}{2}x - 5y = 5 \\ -3x + 30y = -30 \end{cases}$

8. $\begin{cases} 5x - 2y = 1.4 \\ 2x + 3y = 9.3 \end{cases}$

9. $\begin{cases} 4x - 3y = 6 \\ -2x + 1.5y = 5 \end{cases}$

10. $\begin{cases} \dfrac{5x - 3y}{2} - 6 = 0 \\ \dfrac{x + y}{4} - 1 = 0 \end{cases}$

11. $\begin{cases} 6x - 4y = 7 \\ -4x + \frac{8}{3}y = -\frac{14}{3} \end{cases}$

12. $\begin{cases} 14x + 10y - 18 = 0 \\ 42 - 35x - 25y = 0 \end{cases}$

13. $\begin{cases} \dfrac{x}{a} + \dfrac{y}{b} = 6 \\ \dfrac{x}{a} - \dfrac{y}{b} = 2 \end{cases}$ where a and b are constants

14. $\begin{cases} ax + by = a^2 + b^2 \\ bx - ay = a^2 + b^2 \end{cases}$ where a and b are constants

15. Show that for each real number a, the system

$$\begin{cases} 3x + 2y = 7 \\ 4x - 2y = a \end{cases}$$

has exactly one solution.

16. Consider the system

$$\begin{cases} 6x - 4y = 8 \\ -9x + 6y = a \end{cases}$$

Find the values of a for which the system has
a. an infinite number of solutions
b. no solutions

17. Find the values of a and b for which the line with equation $ax + by = 6$ passes through the points $(10, 6)$ and $(-2, -3)$.
18. Let $f(x) = ax^2 + bx + 4$. Find the values of a and b for which $(1, 6)$ and $(-2, 18)$ lie on the graph of f.
19. Two students have $50 together. One student loses a $5 bet to the other, and then both have the same amount of money. How much money did each student have before the bet?
20. When a passenger jet travels with a constant air speed (the speed of the plane with respect to the air) from Seattle to Chicago, it has a 50 mph tailwind and makes the trip in 3 hours and 36 minutes. On the return flight it flies with the same air speed, has a headwind of 50 mph, and makes the trip in $4\frac{1}{2}$ hours. Find the air speed of the jet and the distance flown from Seattle to Chicago.
21. How many ounces of 24-carat (pure) gold and how many ounces of 14-carat gold must be fused together to obtain a 5-ounce bar of 18-carat gold?
22. An investor splits $1000 between two savings accounts that earn simple interest at the same rate. At the end of 3 years one account contains $472, and after 5 years the other contains $780. Find the interest rate and the amount invested in each account.
23. A theater that has a seating capacity of 200 charges $3 for adults and $1.50 for children. If the receipts for a given sold-out performance are $510, how many children's tickets were sold?

6.3

LINEAR SYSTEMS WITH MORE THAN TWO VARIABLES

Until now we have discussed and solved systems of equations with two variables. However, a system can have any number of variables. For example, the system

$$\begin{cases} x - 2y + z = -1 \\ 2x - 3y + 4z = -5 \\ 3x - 4y + 2z = 1 \end{cases} \tag{1}$$

has three variables, and the system

$$\begin{cases} w + 2x - y + 3z = 0 \\ 2w + y - z = 3 \\ -w + x - y + \frac{1}{2}z = -1 \\ 9w + 3x - 4y - z = 2 \end{cases} \tag{2}$$

has four variables. In this section we will restrict our attention to systems having three variables. However, the methods developed here can be extended to systems having four or more variables.

By a **solution** of a system of equations in three variables x, y, and z we mean an ordered triple (a, b, c) of numbers such that if x is replaced by a, y by b, and z by c in all equations of the system, all the resulting equations are valid. Thus $(3, 1, -2)$ is a solution of (1) because when we let $x = 3$, $y = 1$, and $z = -2$ in (1) we obtain the system

$$\begin{cases} 3 - 2(1) + (-2) = 3 - 2 - 2 = -1 \\ 2(3) - 3(1) + 4(-2) = 6 - 3 - 8 = -5 \\ 3(3) - 4(1) + 2(-2) = 9 - 4 - 4 = 1 \end{cases} \tag{3}$$

all of whose equations are valid.

An equation in three variables of the form

$$ax + by + cz = d$$

where a, b, c, and d are constants and either $a \neq 0$, $b \neq 0$, or $c \neq 0$, is called a **linear equation**, and a system of linear equations is called a **linear system**. The system in (1) is an example of a linear system; in this section we will solve only linear systems with three equations and three variables. One feature of linear systems with three equations and three variables is that there are either infinitely many solutions, one solution, or no solutions. We have **solved** the system once we have determined all its solutions.

The methods of substitution and elimination that we presented for systems having two variables have counterparts for systems with three variables.

The Substitution Method

Although the method of substitution is not as conveniently applied for systems with three variables as it is for systems of two variables, the basic idea is the same. We use one of the equations to express one of the variables in terms of the other two variables. Then we substitute for the chosen variable in the other equations. That yields a system of two equations and two variables, which we solve by the methods of the preceding two sections.

Example 1. Use the substitution method to solve the system

$$\begin{cases} x - 2y + z = -1 \\ 2x - 3y + 4z = -5 \\ 3x - 4y + 2z = 1 \end{cases} \quad (4)$$

Solution. We use the first equation to express x in terms of y and z:

$$x = 2y - z - 1 \quad (5)$$

Now we substitute $2y - z - 1$ for x in the second and third equations of (4), obtaining the system

$$\begin{cases} 2(2y - z - 1) - 3y + 4z = -5 \\ 3(2y - z - 1) - 4y + 2z = 1 \end{cases}$$

which condenses to

$$\begin{cases} y + 2z = -3 \\ 2y - z = 4 \end{cases} \quad (6)$$

Next we solve the system in (6) by substitution. We use the first equation in (6) to express y in terms of z:

$$y = -2z - 3 \quad (7)$$

Substituting $-2z - 3$ for y in the second equation of (6) gives us

$$2(-2z - 3) - z = 4$$
$$-5z = 10$$
$$z = -2$$

To find y we use (7) with the known value -2 of z:

$$y = -2(-2) - 3 = 1$$

Then to find x we use (5), substituting 1 for y and -2 for z:

$$x = 2(1) - (-2) - 1 = 3$$

Thus the only possible solution is $(3, 1, -2)$. That it is in fact a solution was shown in (3). □

The Elimination Method

Consider the system

$$\begin{cases} 3x + 5y + 4z = 9 \\ 3x + 3y + z = 13 \\ y - 2z = 5 \end{cases} \quad (8)$$

One way of solving this system involves subtracting the second equation from the first and substituting the result for the second equation to obtain

$$\begin{cases} 3x + 5y + 4z = 9 \\ 2y + 3z = -4 \\ y - 2z = 5 \end{cases} \quad \text{first equation} - \text{second equation} \qquad (9)$$

The latter two equations of (9) form a system in the two variables y and z. By the elimination method the system

$$\begin{cases} 2y + 3z = -4 \\ y - 2z = 5 \end{cases}$$

is equivalent to the system

$$\begin{cases} 2y + 3z = -4 \\ 7z = -14 \end{cases}$$

(See Example 1 and the system in (3), both in Section 6.2.) Therefore (9) is equivalent to the system

$$\begin{cases} 3x + 5y + 4z = 9 \\ 2y + 3z = -4 \\ 7z = -14 \end{cases}$$

This system is easy to solve. First we find that $z = -2$ from the last equation. Then, substituting -2 for z in the middle equation, we find y:

$$2y + 3(-2) = -4$$
$$2y = -4 + 6 = 2$$
$$y = 1$$

Finally, we substitute 1 for y and -2 for z in the first equation to obtain x:

$$3x + 5(1) + 4(-2) = 9$$
$$3x = 12$$
$$x = 4$$

The proposed solution of the system in (8) is therefore $(4, 1, -2)$. A routine check bears out the validity of the solution. (See Exercise 1.)

Our normal strategy for solving systems of three equations in three variables by the elimination method involves replacing a given system by an equivalent one in which one equation has one variable, a second equation has two variables, and the remaining equation has three variables. The manipulations that we use in passing from one system to an equivalent one are similar to those used for two variables:

i. Multiplying an equation in the system by a nonzero constant.

ii. Replacing an equation in the system by the result of adding the equation to, or subtracting it from, another equation in the system.

As you will see below, we frequently combine (i) and (ii).

Example 2. Use the elimination method to solve the system

$$\begin{cases} x + 3y + 4z = 2 \\ 2x - y - 5z = 3 \\ 3x + 2y - 2z = -1 \end{cases} \tag{10}$$

Solution. We begin by multiplying the first equation by 2 and then subtracting the second equation from the result:

$$\begin{array}{ll} 2x + 6y + 8z = 4 & 2 \times \text{first equation} \\ \underline{2x - y - 5z = 3} & \text{second equation} \\ 7y + 13z = 1 \end{array}$$

Replacing the second equation in (10) by $7y + 13z = 1$ yields the new system

$$\begin{cases} x + 3y + 4z = 2 \\ 7y + 13z = 1 \\ 3x + 2y - 2z = -1 \end{cases} \tag{11}$$

Notice that x has been eliminated from the second equation. Next we multiply the first equation in (11) by 3 and subtract the third equation from the result:

$$\begin{array}{ll} 3x + 9y + 12z = 6 & 3 \times \text{first equation} \\ \underline{3x + 2y - 2z = -1} & \text{third equation} \\ 7y + 14z = 7 \end{array}$$

Replacing the third equation in (11) by $7y + 14z = 7$, we obtain the system

$$\begin{cases} x + 3y + 4z = 2 \\ 7y + 13z = 1 \\ 7y + 14z = 7 \end{cases} \tag{12}$$

Subtracting the second equation in (12) from the third, and substituting the result for the third equation in (12), we obtain the system

$$\begin{cases} x + 3y + 4z = 2 \\ 7y + 13z = 1 \\ z = 6 \end{cases} \qquad (13)$$

This system has one equation in one variable, one in two variables, and one in three variables, so it is easy to substitute in the second equation to find y, and then to substitute in the first equation to find x. Indeed, substituting 6 for z in the second equation of (13), we find that

$$7y + 13(6) = 1$$

$$7y = 1 - 78 = -77$$

$$y = -11$$

Now we substitute -11 for y and 6 for z in the first equation of (13) and solve for x:

$$x + 3(-11) + 4(6) = 2$$

$$x = 11$$

Therefore $(11, -11, 6)$ is the proposed solution of (10).

$$\text{Check:} \begin{cases} 11 + 3(-11) + 4(6) = 2 \\ 2(11) - (-11) - 5(6) = 3 \\ 3(11) + 2(-11) - 2(6) = -1 \end{cases}$$

We conclude that the solution of the given system is $(11, -11, 6)$. □

In the next example we illustrate how the method of elimination can be used to find all solutions of a system with infinitely many solutions.

Example 3. Use the method of elimination to solve the system

$$\begin{cases} x - 2y - 3z = 1 \\ 2x + y + 4z = 2 \\ 8x - 6y - 4z = 8 \end{cases} \qquad (14)$$

Solution. We begin by multiplying the first equation by 2 and subtracting the resulting equation from the second equation:

$$\begin{array}{ll} 2x + y + 4z = 2 & \text{second equation} \\ \underline{2x - 4y - 6z = 2} & 2 \times \text{first equation} \\ 5y + 10z = 0 \end{array}$$

Replacing the second equation in (14) by $5y + 10z = 0$, we obtain the system

$$\begin{cases} x - 2y - 3z = 1 \\ 5y + 10z = 0 \\ 8x - 6y - 4z = 8 \end{cases} \quad (15)$$

Next we multiply the first equation in (15) by 8 and subtract the resulting equation from the third equation:

$$\begin{array}{ll} 8x - 6y - 4z = 8 & \text{third equation} \\ \underline{8x - 16y - 24z = 8} & 8 \times \text{first equation} \\ 10y + 20z = 0 & \end{array}$$

Replacing the third equation in (15) by $10y + 20z = 0$ yields the system

$$\begin{cases} x - 2y - 3z = 1 \\ 5y + 10z = 0 \\ 10y + 20z = 0 \end{cases} \quad (16)$$

Multiplying the third equation in (16) by $\frac{1}{2}$ brings us to the system

$$\begin{cases} x - 2y - 3z = 1 \\ 5y + 10z = 0 \\ 5y + 10z = 0 \end{cases} \quad (17)$$

Subtracting the third equation in (17) from the second and substituting the result for the third equation, we obtain the system

$$\begin{cases} x - 2y - 3z = 1 \\ 5y + 10z = 0 \\ 0 = 0 \end{cases} \quad (18)$$

The third equation yields no value for x, y, or z. However, if we solve for y in the second equation in (18), we obtain

$$\begin{array}{l} 5y = -10z \\ y = -2z \end{array} \quad (19)$$

Now we substitute $-2z$ for y in the first equation in (18) and solve for x:

$$\begin{array}{l} x - 2(-2z) - 3z = 1 \\ x + z = 1 \\ x = 1 - z \end{array} \quad (20)$$

It follows that (a, b, c) is a proposed solution of the given system if and only if $b = -2c$ (from (19)) and $a = 1 - c$ (from (20)). Consequently there are infinitely many proposed solutions, given by $(1 - c, -2c, c)$ for any number c. In Exercise 4 you are asked to check that for every value of c, $(1 - c, -2c, c)$ satisfies all three equations in the given system. $\square$

EXERCISES 6.3

In Exercises 1–4, verify that the given ordered triple is a solution of the given system.

1.
$$\begin{cases} 3x + 5y + 4z = 9 \\ 3x + 3y + z = 13 \\ y - 2z = 5 \end{cases} \quad (4, 1, -2)$$

2.
$$\begin{cases} 4x - 2y + 3z = 2 \\ -2x + \tfrac{1}{2}y - \tfrac{1}{2}z = 1 \\ 7x - 13y + 2z = -3 \end{cases} \quad (-1, 0, 2)$$

3.
$$\begin{cases} 2x + y - 6z = 1 \\ x + y - 4z = 1 \\ -4x - 3y + z = -\tfrac{5}{6} \end{cases} \quad (-\tfrac{1}{3}, \tfrac{2}{3}, -\tfrac{1}{6})$$

4.
$$\begin{cases} x - 2y - 3z = 1 \\ 2x + y + 4z = 2 \\ 8x - 6y - 4z = 8 \end{cases} \quad (1 - c, -2c, c) \text{ for any number } c$$

In Exercises 5–24, find all solutions of the given system.

5.
$$\begin{cases} 3x - 2y + z = 10 \\ 5y - z = 2 \\ 3z = -6 \end{cases}$$

6.
$$\begin{cases} -2x + 3y + 5z = -1 \\ 6y - z = -11 \\ -2y + 4z = 0 \end{cases}$$

7.
$$\begin{cases} x - y + z = 5 \\ 2x + 3y - z = -4 \\ x + y + 2z = 6 \end{cases}$$

8.
$$\begin{cases} 2x - y + 3z = 1 \\ -4x + y - 2z = 4 \\ x - y - 2z = -2 \end{cases}$$

9.
$$\begin{cases} x - y - z = 0 \\ x + 3y - z = 0 \\ x + y - z = 0 \end{cases}$$

10.
$$\begin{cases} 4u - 3v - 3w = 22 \\ 2u - v - 6w = 15 \\ u + 3v - 2w = 3 \end{cases}$$

11.
$$\begin{cases} u + 2v + 3w = 5 \\ 2u - 8v - 14w = 5 \\ 2u - 2v - 4w = 7 \end{cases}$$

12.
$$\begin{cases} 3x - 4y + z = 0 \\ 7x - z = 0 \\ -9x + 12y - 3z = 0 \end{cases}$$

13. $\begin{cases} 6x - 3y + 7z = 0 \\ 4x + 5y + 9z = 0 \\ 13x - 11y - 4z = 0 \end{cases}$

14. $\begin{cases} 2x - 3y + 4z + 12 = 0 \\ -4x + y - z - 7 = 0 \\ 3x + y - z = 0 \end{cases}$

15. $\begin{cases} 4x - 5y = 11 \\ 2x + z = 7 \\ 2y + z = 1 \end{cases}$

16. $\begin{cases} 2x - 3y + 5z = 4 \\ x + 5y - 2z = 5 \\ 5x - y + 8z = 6 \end{cases}$

17. $\begin{cases} 2x - y - 7 = 0 \\ 3y - 4z - 5 = 0 \\ -x + 2z + 7 = 0 \end{cases}$

18. $\begin{cases} x + 5y - z = 11 \\ 2x - y = 7 \\ 3x + y + z = 11 \end{cases}$

19. $\begin{cases} x + y = 0 \\ 2x - 4z = 3 \\ x - y + z = -2 \end{cases}$

20. $\begin{cases} x + 2y - z = -9 \\ -x + y + z = 3 \\ 2x + 3y + z = 8 \end{cases}$

21. $\begin{cases} x + y + z = 0 \\ x - y - z = 1 \\ 2x - 2y + 4z = 3 \end{cases}$

22. $\begin{cases} \frac{1}{2}x - \frac{1}{3}y - \frac{1}{6}z = \frac{5}{6} \\ \frac{1}{3}x + \frac{1}{6}y + z = -\frac{3}{2} \\ -\frac{2}{3}x - \frac{1}{2}y + \frac{1}{3}z = -\frac{1}{3} \end{cases}$

23. $\begin{cases} 2.4x - 1.6y - 0.8z = 4.8 \\ 1.2x + 1.4y - 2z = -0.6 \\ -x + 0.8y - 1.8z = -3.2 \end{cases}$

24. $\begin{cases} 40x - 50y - 10z = -21 \\ -5x - 3y + 20z = 13 \\ -10x + 20y + 30z = 31 \end{cases}$

25. Let $f(x) = ax^2 + bx + c$. Find the values of $a, b,$ and c for which the graph of f contains the points $(-1, 10), (1, 6),$ and $(2, 13)$.

26. Let $f(x) = ae^x + be^{-x} + c$. Find the values of $a, b,$ and c for which the graph of f contains the points $(0, 1), (\ln 2, 6),$ and $(\ln 4, 13)$.

27. The sum of three numbers is 58. The sum of the smaller two numbers is two more than the largest number, and the largest number is double the smallest number. Find the three numbers.

28. One angle in a certain triangle is $\frac{1}{3}$ a second angle in the triangle, and the third angle in the triangle is twice the sum of the other two. Find the angles. (*Hint:* The sum of the angles in any triangle is equal to 180°.)

29. A collection of quarters, dimes, and nickels is worth $4.55 and has 44 coins in it. If there are 6 more nickels than dimes in the collection, find the number of quarters in the collection.

30. Peter, Paul, and Mary have a combined total of $200. If Peter gives $10 to Paul, they will have the same amount of money, and Mary will have twice as much as either Peter or Paul. How much money does each have before the transaction?

31. A university theater charges $2 for adults other than students, $0.75 for students, and $0.25 for children. Suppose that 580 people attend a performance of *My Fair Lady,* including 8 times as many students as children. If the receipts total $690, determine how many students attend the performance.

32. A concert hall has 2100 seats. The orchestra seats cost $8 each, whereas the seats in the first and second balcony cost $6 and $3, respectively. For a full house the hall grosses $13,800. If all the seats in both balconies were sold for $5 each, then the gross for a full house would be $14,100. Find the total number of balcony seats.

6.4

SOLUTIONS OF SYSTEMS BY MATRICES

The elimination method of solving systems of equations can be re-formulated so as to reduce the amount of writing necessary. The re-formulation involves a bracketed rectangular array of numbers such as

$$\begin{bmatrix} 2 & -1 & 4 \\ 3 & 1 & -14 \end{bmatrix} \tag{1}$$

or

$$\begin{bmatrix} 3 & -1 & 2 \\ 1 & 0 & 3 \\ 6 & -4 & 2 \end{bmatrix} \tag{2}$$

called a matrix.

DEFINITION 6.1   A *matrix* is a rectangular array

$$\begin{bmatrix} a_{11} & a_{12} & \cdots & a_{1n} \\ a_{21} & a_{22} & \cdots & a_{2n} \\ & & & \\ \cdot & \cdot & & \cdot \\ \cdot & \cdot & & \cdot \\ \cdot & \cdot & & \cdot \\ a_{m1} & a_{m2} & \cdots & a_{mn} \end{bmatrix} \tag{3}$$

where $a_{11}, a_{12}, \ldots, a_{1n}; a_{21}, a_{22}, \ldots, a_{2n}; \ldots; a_{m1}, a_{m2}, \ldots, a_{mn}$ are numbers, called the *entries* of the matrix.

The matrix in (3) is called an $m \times n$ matrix (read "m by n matrix") to emphasize the fact that there are m rows and n columns of numbers in the matrix. The matrix in (1) is a 2×3 matrix, because there are 2 rows and 3 columns. For that matrix we have

$$a_{11} = 2, \quad a_{12} = -1, \quad a_{13} = 4, \quad a_{21} = 3, \quad a_{22} = 1, \quad a_{23} = -14$$

Similarly, the matrix in (2) is a 3×3 matrix, because there are 3 rows and 3 columns. A 1×1 matrix has the form $[a]$, where a is a real number. Thus a 1×1 matrix is essentially the same as a number. An $n \times n$ matrix is called a *square matrix of order n*. The matrix in (2) is a square matrix of order 3.

Now consider the system

$$\begin{cases} 2x - y = 4 \\ 3x + y = -14 \end{cases} \tag{4}$$

There are two matrices associated with the system. First there is the *coefficient matrix*

$$\begin{bmatrix} 2 & -1 \\ 3 & 1 \end{bmatrix}$$

consisting of the coefficients of x and y. Notice that the coefficients 2 and 3 of x occupy the first column, and the coefficients -1 and 1 of y the second column. The rows correspond to the equations in the system, in their given order.

The second matrix related to a system of equations is the *augmented matrix*, which for (4) is

$$\begin{bmatrix} 2 & -1 & \vdots & 4 \\ 3 & 1 & \vdots & -14 \end{bmatrix}$$

The augmented matrix is the matrix obtained by attaching to the coefficient matrix the constants to the right of the equality signs in the system. The vertical dashed line serves to separate the coefficient matrix from the column of constants. For the system

$$\begin{cases} x - y - 2z = -2 \\ 4x + 5z = \tfrac{1}{2} \\ 7x + 15y - z = 0 \end{cases} \tag{5}$$

the coefficient matrix and the augmented matrix are

$$\begin{bmatrix} 1 & -1 & -2 \\ 4 & 0 & 5 \\ 7 & 15 & -1 \end{bmatrix} \quad \text{and} \quad \begin{bmatrix} 1 & -1 & -2 & \vdots & -2 \\ 4 & 0 & 5 & \vdots & \tfrac{1}{2} \\ 7 & 15 & -1 & \vdots & 0 \end{bmatrix} \tag{6}$$

respectively. Notice that the second equation in the system (5) is equivalent to

$$4x + 0y + 5z = \frac{1}{2}$$

and as a result, 0 represents the coefficient of y in the second row of the matrices in (6).

Equivalent Matrices

The solutions of a given system of equations are unaffected by the order in which the equations appear. Similarly, the solutions of the given system are unaffected by multiplying an equation by a constant or by replacing an equation by the sum of it and another equation. These alterations in a system of equations correspond to the first three of the following alterations in the rows of the associated augmented matrix:

> i. Interchanging two rows
> ii. Replacing a row by a nonzero multiple of that row
> iii. Replacing a row by the sum of it and another row
> iv. Replacing a row by the sum of it and a nonzero multiple of another row

Operations (i)–(iv) are sometimes called *elementary row operations* on a matrix. Frequently two of the four operations are performed in a single step. In fact, (iv) is a combination of (ii) and (iii), as is the operation of substituting for one row the difference of it and another row.

We say that two matrices are *equivalent* if one of them can be derived from the other by one or more of the operations (i)–(iv). We will use the symbol $\leftrightarrow$ between two matrices to indicate that the matrices are equivalent. From the definition of equivalence, if two matrices are equivalent, then they have the same number of rows and the same number of columns. Moreover, if matrices A, B, and C have the property that $A \leftrightarrow B$ and $B \leftrightarrow C$, then $A \leftrightarrow C$.

In applying elementary row operations to a matrix, we will denote the top row by R_1, the next row by R_2, and so forth. If we write, say, $R_2 \rightarrow R_1 + (-3)R_2$, we mean that we are replacing the second row by the new row obtained by multiplying the second row by -3 and adding the result to the first row.

Example 1. Show that

$$
\begin{bmatrix} 3 & -1 & 0 & -3 \\ 1 & -\frac{1}{3} & 3 & -2 \\ -6 & -4 & 2 & 0 \end{bmatrix}
\longleftrightarrow
\begin{bmatrix} 3 & -1 & 0 & -3 \\ 0 & -3 & 1 & -3 \\ 0 & 0 & -9 & 3 \end{bmatrix}
$$

Solution. We proceed as follows:

$$
\begin{bmatrix} 3 & -1 & 0 & -3 \\ 1 & -\frac{1}{3} & 3 & -2 \\ -6 & -4 & 2 & 0 \end{bmatrix}
\xleftarrow{R_2 \rightarrow R_1 + (-3)R_2}
\begin{bmatrix} 3 & -1 & 0 & -3 \\ 0 & 0 & -9 & 3 \\ -6 & -4 & 2 & 0 \end{bmatrix}
\xleftarrow{R_3 \rightarrow R_1 + \frac{1}{2}R_3}
$$

$$
\begin{bmatrix} 3 & -1 & 0 & -3 \\ 0 & 0 & -9 & 3 \\ 0 & -3 & 1 & -3 \end{bmatrix}
\xleftarrow[R_3 \rightarrow R_2]{R_2 \rightarrow R_3}
\begin{bmatrix} 3 & -1 & 0 & -3 \\ 0 & -3 & 1 & -3 \\ 0 & 0 & -9 & 3 \end{bmatrix}
$$

Thus the two given matrices are equivalent. $\square$

Recall that every system of equations has an associated augmented matrix. Each elementary row operation performed on the augmented matrix corresponds to an operation on the given system that yields a system with the same solutions. As a result, two systems of equations that have equivalent augmented matrices have the same solutions.

Example 2. Use augmented matrices to solve the system

$$\begin{cases} 3x - y & = -3 \\ x - \tfrac{1}{3}y + 3z = -2 \\ -6x - 4y + 2z = 0 \end{cases} \qquad (7)$$

Solution. The associated augmented matrix is

$$\begin{bmatrix} 3 & -1 & 0 & \vdots & -3 \\ 1 & -\tfrac{1}{3} & 3 & \vdots & -2 \\ -6 & -4 & 2 & \vdots & 0 \end{bmatrix}$$

By Example 1,

$$\begin{bmatrix} 3 & -1 & 0 & \vdots & -3 \\ 1 & -\tfrac{1}{3} & 3 & \vdots & -2 \\ -6 & -4 & 2 & \vdots & 0 \end{bmatrix} \longleftrightarrow \begin{bmatrix} 3 & -1 & 0 & \vdots & -3 \\ 0 & -3 & 1 & \vdots & -3 \\ 0 & 0 & -9 & \vdots & 3 \end{bmatrix}$$

But the latter matrix is the augmented matrix for the system

$$\begin{cases} 3x - y & = -3 \\ -3y + z = -3 \\ -9z = 3 \end{cases} \qquad (8)$$

By the comments preceding the example, the solutions of (8) are exactly the solutions of (7). From (8),

$$z = \frac{3}{-9} = -\frac{1}{3}$$

To find y we substitute $-\tfrac{1}{3}$ for z in the second equation of (8) and solve for y:

$$-3y + \left(-\frac{1}{3}\right) = -3$$

$$-3y = -3 + \frac{1}{3} = -\frac{8}{3}$$

$$y = \frac{8}{9}$$

To find x we substitute $\tfrac{8}{9}$ for y in the first equation of (8) and solve for x:

$$3x = -3 + \frac{8}{9} = -\frac{19}{9}$$

$$x = -\frac{19}{27}$$

Thus $(-\frac{19}{27}, \frac{8}{9}, -\frac{1}{3})$ is a proposed solution of (8) and hence of (7).

Check:
$$\begin{cases} 3\left(-\frac{19}{27}\right) - \frac{8}{9} = -3 \\ -\frac{19}{27} - \frac{1}{3}\left(\frac{8}{9}\right) + 3\left(-\frac{1}{3}\right) = -2 \\ -6\left(-\frac{19}{27}\right) - 4\left(\frac{8}{9}\right) + 2\left(-\frac{1}{3}\right) = 0 \end{cases}$$

Consequently $(-\frac{19}{27}, \frac{8}{9}, -\frac{1}{3})$ is the solution of the given system.   □

We remark that the third augmented matrix appearing in the solution of Example 2 has zeros everywhere below the diagonal 3, -3, -9 starting at the upper left corner. For this reason, the final equation of the associated system in (8) has the lone variable z. In finding solutions of systems by means of augmented matrices, the goal is to replace the augmented matrix of a given system by an equivalent matrix with zeros below the designated diagonal, whether the augmented matrix is a 3×4 or a 2×3 matrix.

Example 3. Use augmented matrices to show that the system

$$\begin{cases} x + y + z = -1 \\ x + 2y - z = -6 \\ 2x + y + 4z = 3 \end{cases} \tag{9}$$

has infinitely many solutions.

Solution. The augmented matrix is

$$\left[\begin{array}{ccc|c} 1 & 1 & 1 & -1 \\ 1 & 2 & -1 & -6 \\ 2 & 1 & 4 & 3 \end{array}\right]$$

We apply elementary row operations as follows:

$$\left[\begin{array}{ccc|c} 1 & 1 & 1 & -1 \\ 1 & 2 & -1 & -6 \\ 2 & 1 & 4 & 3 \end{array}\right] \xrightarrow{R_2 \to R_2 + (-1)R_1} \left[\begin{array}{ccc|c} 1 & 1 & 1 & -1 \\ 0 & 1 & -2 & -5 \\ 2 & 1 & 4 & 3 \end{array}\right] \xrightarrow{R_3 \to R_3 + (-2)R_1}$$

$$\left[\begin{array}{ccc|c} 1 & 1 & 1 & -1 \\ 0 & 1 & -2 & -5 \\ 0 & -1 & 2 & 5 \end{array}\right] \xrightarrow{R_3 \to R_3 + R_2} \left[\begin{array}{ccc|c} 1 & 1 & 1 & -1 \\ 0 & 1 & -2 & -5 \\ 0 & 0 & 0 & 0 \end{array}\right]$$

The system associated with the last matrix above is

$$\begin{cases} x + y + \ z = -1 \\ \qquad y - 2z = -5 \\ \qquad\qquad 0 = \ \ 0 \end{cases} \tag{10}$$

From the second equation in (10) it follows that

$$y = 2z - 5$$

Substituting $2z - 5$ for y in the first equation in (10), we obtain

$$x + (2z - 5) + z = -1$$

or equivalently,

$$x = -3z + 4$$

Consequently (a, b, c) is a solution of (10), and hence of (9), only if $b = 2c - 5$ and $a = -3c + 4$. Thus the solutions of (9) are $(-3c + 4, 2c - 5, c)$ for any number c. You can check that each triple of that form actually satisfies the given system. $\square$

EXERCISES 6.4

In Exercises 1–6, write the coefficient matrix and the augmented matrix of the given system.

1. $\begin{cases} 2x - 4y = 7 \\ -5x + \ y = 6 \end{cases}$

2. $\begin{cases} -x - \ y = 0 \\ 4x + \frac{1}{2}y = 3 \end{cases}$

3. $\begin{cases} x + y = 2 \\ \qquad y = 1 \end{cases}$

4. $\begin{cases} x - \ y - \ z = \ \ 1 \\ 2x + 4y - 5z = -3 \\ -7x + \ y + \ z = -2 \end{cases}$

5. $\begin{cases} \frac{1}{3}x - \frac{1}{4}y + \frac{1}{2}z = 1 \\ \frac{2}{5}x + z = 3 \\ x - \frac{1}{2}y = 4 \end{cases}$

6. $\begin{cases} x - y = 0 \\ x + z = 1 \\ 2y - z = 3 \end{cases}$

In Exercises 7–12, show that the given matrices are equivalent, giving all reasons.

7. $\begin{bmatrix} 1 & 3 \\ 4 & 2 \end{bmatrix}$ and $\begin{bmatrix} 4 & 2 \\ 1 & 3 \end{bmatrix}$

8. $\begin{bmatrix} 2 & -6 & 4 \\ 6 & -3 & 7 \end{bmatrix}$ and $\begin{bmatrix} 14 & -12 & 18 \\ 6 & -3 & 7 \end{bmatrix}$

9. $\begin{bmatrix} 1 & 2 & 3 \\ -2 & 1 & 4 \\ 3 & -1 & 0 \end{bmatrix}$ and $\begin{bmatrix} 1 & 2 & 3 \\ 0 & 1 & 2 \\ 0 & 0 & 5 \end{bmatrix}$

10. $\begin{bmatrix} -1 & 2 & 0 & 4 \\ 3 & 1 & -2 & -1 \\ 1 & 5 & -3 & 0 \end{bmatrix}$ and $\begin{bmatrix} -1 & 2 & 0 & 4 \\ 0 & 7 & -2 & 11 \\ 0 & 0 & 1 & 7 \end{bmatrix}$

11. $\begin{bmatrix} -1 & 1 & -1 & 1 \\ 2 & 3 & 0 & -1 \\ 1 & 1 & -2 & 4 \end{bmatrix}$ and $\begin{bmatrix} -1 & 1 & -1 & 1 \\ 0 & 5 & -2 & 1 \\ 0 & 0 & -11 & 23 \end{bmatrix}$

12. $\begin{bmatrix} 1 & -1 & 5 & 11 \\ 2 & 0 & -1 & 7 \\ 3 & 1 & 1 & 11 \end{bmatrix}$ and $\begin{bmatrix} 1 & -1 & 5 & 11 \\ 0 & 2 & -11 & -15 \\ 0 & 0 & 8 & 8 \end{bmatrix}$

In Exercises 13–24, use augmented matrices to solve the given system.

13. $\begin{cases} x + y = 0 \\ 2x - 3y = 5 \end{cases}$

14. $\begin{cases} -3x - 4y = 13 \\ -x + 4y = -1 \end{cases}$

15. $\begin{cases} 4x + 10y = 19 \\ -6x + 12y = -15 \end{cases}$

16. $\begin{cases} x - \frac{1}{2}y = 3 \\ \frac{1}{4}x + y = \frac{3}{4} \end{cases}$

17. $\begin{cases} 0.1x + 0.3y = 0.18 \\ 1.7x + 2.4y = 1.98 \end{cases}$

18. $\begin{cases} x - y - z = 2 \\ 2x + y - z = 3 \\ x + y + z = 0 \end{cases}$

19. $\begin{cases} x + 3y - 5z = -20 \\ 2x - 5y + 3z = 10 \\ -x + 6y - 5z = -19 \end{cases}$

20. $\begin{cases} x - 2y + 3z = 0 \\ 4x - 9y + z = 0 \\ 5x + 2y - z = 0 \end{cases}$

21. $\begin{cases} x - y = -3 \\ 2x + z = 5 \\ y - 2z = 3 \end{cases}$

22. $\begin{cases} x + y - z = 12 \\ y - 2z = 8 \\ 3y + z = 3 \end{cases}$

23. $\begin{cases} x + y - 4z = 0 \\ 2x + y - 3z = 2 \\ -3x - y + 2z = -4 \end{cases}$

24. $\begin{cases} x - y - z = -3 \\ 3x - 2y - 6z = -9 \\ -5x + 7y - z = 15 \end{cases}$

In Exercises 25–26, use augmented matrices to show that the given system is inconsistent.

25. $\begin{cases} 4x - y - 2z = 4 \\ x - y - \frac{1}{2}z = 1 \\ 2x - y - z = 8 \end{cases}$

26. $\begin{cases} 6x - 12y + z = 5 \\ 9x - 18y + 2z = 3 \\ 2x - 4y + 3z = 2 \end{cases}$

6.5

DETERMINANTS

In the preceding section we discussed solutions of systems of linear equations by augmented matrices. If the given system happens to have the same number of equations as variables, so that the system's coefficient matrix is a square matrix, then the system can frequently be solved by a second method that utilizes a special number associated with the coefficient matrix, called the determinant of the matrix. Since determinants are of interest in other contexts as well, we devote the present section to determinants and their properties. In the following section we will see how to solve systems by determinants.

In general, the determinant of a square matrix

$$\begin{bmatrix} a_{11} & a_{12} & \cdots & a_{1n} \\ a_{21} & a_{22} & \cdots & a_{2n} \\ \cdot & \cdot & & \cdot \\ \cdot & \cdot & & \cdot \\ \cdot & \cdot & & \cdot \\ a_{n1} & a_{n2} & \cdots & a_{nn} \end{bmatrix}$$

is a number denoted by

$$\begin{vmatrix} a_{11} & a_{12} & \cdots & a_{1n} \\ a_{21} & a_{22} & \cdots & a_{2n} \\ \cdot & \cdot & & \cdot \\ \cdot & \cdot & & \cdot \\ \cdot & \cdot & & \cdot \\ a_{n1} & a_{n2} & \cdots & a_{nn} \end{vmatrix}$$

Normally, computation of the determinant varies according to the order of the matrix.

For a 1×1 matrix $[a_{11}]$, the **determinant** is defined by

$$|a_{11}| = a_{11}$$

Thus the determinant of a 1×1 matrix is just the number in the matrix. For example, $|-2|$ is -2.

Caution: The symbol $|a_{11}|$ can denote either the determinant of the 1×1 matrix $[a_{11}]$ or the absolute value of the number a_{11}. In any given context it will be clear which is meant, so no confusion should arise.

For a 2×2 matrix

$$\begin{bmatrix} a_{11} & a_{12} \\ a_{21} & a_{22} \end{bmatrix}$$

the *determinant* is defined by

$$\begin{vmatrix} a_{11} & a_{12} \\ a_{21} & a_{22} \end{vmatrix} = a_{11}a_{22} - a_{21}a_{12} \tag{1}$$

which is the difference of the products of the numbers in the two diagonals.

Example 1. Evaluate $\begin{vmatrix} 2 & -4 \\ -5 & 3 \end{vmatrix}$.

Solution. By (1),

$$\begin{vmatrix} 2 & -4 \\ -5 & 3 \end{vmatrix} = (2)(3) - (-5)(-4) = 6 - 20 = -14 \quad \square$$

For a 3×3 matrix A, where

$$A = \begin{bmatrix} a_{11} & a_{12} & a_{13} \\ a_{21} & a_{22} & a_{23} \\ a_{31} & a_{32} & a_{33} \end{bmatrix}$$

the *determinant* of A, denoted by $\det A$, is defined as follows:

$$\det A = \begin{vmatrix} a_{11} & a_{12} & a_{13} \\ a_{21} & a_{22} & a_{23} \\ a_{31} & a_{32} & a_{33} \end{vmatrix} = a_{11}a_{22}a_{33} + a_{12}a_{23}a_{31} + a_{13}a_{21}a_{32} \\ - a_{31}a_{22}a_{13} - a_{32}a_{23}a_{11} - a_{33}a_{21}a_{12} \tag{2}$$

A way to remember the determinant given in (2) is by the diagram

Example 2. Evaluate $\begin{vmatrix} 1 & 2 & -1 \\ 4 & 3 & -4 \\ -2 & 0 & -3 \end{vmatrix}$.

Solution. By (2),

$$\begin{vmatrix} 1 & 2 & -1 \\ 4 & 3 & -4 \\ -2 & 0 & -3 \end{vmatrix} = (1)(3)(-3) + (2)(-4)(-2) + (-1)(4)(0)$$
$$- (-2)(3)(-1) - (0)(-4)(1) - (-3)(4)(2)$$
$$= -9 + 16 + 0 - 6 - 0 + 24$$
$$= 25 \quad \square$$

For square matrices of order 4 and above, the definition of the determinant is more complicated. As a prelude to the definition, we introduce the notion of a minor of a matrix.

DEFINITION 6.2   Let a_{ij} be the entry in the ith row and jth column of a matrix. The *minor M_{ij}* of the matrix is the determinant of the matrix formed by deleting the ith row and jth column (that is, the row and column containing a_{ij}).

Example 3. Determine the minor M_{23} of

$$\begin{bmatrix} 1 & 2 & -1 \\ 4 & 3 & -4 \\ -2 & 0 & -3 \end{bmatrix}$$

Solution. By definition, to obtain M_{23} we first delete the second row and third column:

$$\begin{bmatrix} 1 & 2 & -1 \\ 4 & 3 & -4 \\ -2 & 0 & -3 \end{bmatrix}$$

which yields the matrix

$$\begin{bmatrix} 1 & 2 \\ -2 & 0 \end{bmatrix}$$

Then M_{23} is the determinant of this matrix, which means that

$$M_{23} = \begin{vmatrix} 1 & 2 \\ -2 & 0 \end{vmatrix} = 0 - (-4) = 4 \quad \square$$

Example 4. Find the minor M_{43} of

$$\begin{bmatrix} 2 & 3 & 0 & 1 \\ 0 & -1 & 2 & 4 \\ -2 & 2 & 3 & 0 \\ 3 & 5 & 1 & -2 \end{bmatrix}$$

Solution. By definition, for M_{43} we delete the fourth row and third column to obtain

$$M_{43} = \begin{vmatrix} 2 & 3 & 1 \\ 0 & -1 & 4 \\ -2 & 2 & 0 \end{vmatrix} = 0 - 24 + 0 - 2 - 16 - 0 = -42 \quad \square$$

The definition in (2) of the determinant of a 3×3 matrix can be reformulated in terms of minors. More precisely,

$$\begin{vmatrix} a_{11} & a_{12} & a_{13} \\ a_{21} & a_{22} & a_{23} \\ a_{31} & a_{32} & a_{33} \end{vmatrix} = a_{11}M_{11} - a_{12}M_{12} + a_{13}M_{13} \tag{3}$$

Notice that the sum in (3) contains the term $-a_{12}M_{12}$, with the minus sign. If we let the *cofactor* A_{ij} of a_{ij} be defined by

$$A_{ij} = (-1)^{i+j}M_{ij}$$

then we find, for instance, that

$$A_{11} = (-1)^{1+1}M_{11} = (-1)^2 M_{11} = \quad M_{11}$$
$$A_{12} = (-1)^{1+2}M_{12} = (-1)^3 M_{12} = \quad -M_{12}$$
$$A_{13} = (-1)^{1+3}M_{13} = (-1)^4 M_{13} = \quad M_{13}$$

Therefore by (3) the determinant can be written in terms of cofactors:

$$\begin{vmatrix} a_{11} & a_{12} & a_{13} \\ a_{21} & a_{22} & a_{23} \\ a_{31} & a_{32} & a_{33} \end{vmatrix} = a_{11}A_{11} + a_{12}A_{12} + a_{13}A_{13} \tag{4}$$

Formula (4) tells us that the determinant of a 3×3 matrix is obtained by multiplying the entries in the first row of the matrix by their respective cofactors and taking the sum of the resulting products. This leads us to the formal definition of the determinant for any square matrix of order greater than or equal to 3.

DEFINITION 6.3   The *determinant* of a given square matrix of order greater than or equal to 3 is the number obtained as follows: Select a row or column of the matrix. Multiply each entry in the selected row or column by its cofactor and take the sum of the resulting products.

Notice that because of (4), the definition of determinant given in Definition 6.3 is consistent with the definition of the determinant of a 3×3 matrix given in (2).

According to Definition 6.3, the evaluation of the determinant of a square matrix of order 4 is based on calculating four cofactors, each of which involves the determinant of a square matrix of order 3. Computation of the determinant of a square matrix of order 5 involves evaluating five cofactors, each associated with a square matrix of order 4. As you can well imagine, as the order of the square matrix increases from 4 to 5 and even higher, the computations needed in order to evaluate the determinant soon become unwieldy.

In using Definition 6.3 to calculate the determinant of a square matrix, one begins with the selection of a row or column. If the pth row of the nth-order square matrix is chosen, then

$$\begin{vmatrix} a_{11} & a_{12} & \cdots & a_{1n} \\ a_{21} & a_{22} & \cdots & a_{2n} \\ a_{n1} & a_{n2} & \cdots & a_{nn} \end{vmatrix} = a_{p1}A_{p1} + a_{p2}A_{p2} + \cdots + a_{pn}A_{pn}$$

whereas if the pth column of the matrix is chosen, then

$$\begin{vmatrix} a_{11} & a_{12} & \cdots & a_{1n} \\ a_{21} & a_{22} & \cdots & a_{2n} \\ a_{n1} & a_{n2} & \cdots & a_{nn} \end{vmatrix} = a_{1p}A_{1p} + a_{2p}A_{2p} + \cdots + a_{np}A_{np}$$

It can be proved that the same number is obtained for the value of the determinant, no matter which row or column is used.

Example 5. Use Definition 6.3 to evaluate the determinant

$$\begin{vmatrix} -3 & 2 & 1 \\ -5 & 0 & 6 \\ -2 & -1 & 3 \end{vmatrix}$$

Solution. Because of the presence of 0 in the second row, we select the second row. We find that

$$\begin{vmatrix} -3 & 2 & 1 \\ -5 & 0 & 6 \\ -2 & -1 & 3 \end{vmatrix} = (-5)A_{21} + (0)A_{22} + 6A_{23}$$

$$= (-5)(-1)^{2+1}\begin{vmatrix} 2 & 1 \\ -1 & 3 \end{vmatrix} + 0 + (6)(-1)^{2+3}\begin{vmatrix} -3 & 2 \\ -2 & -1 \end{vmatrix}$$

$$= 5[6 - (-1)] - 6[3 - (-4)]$$

$$= 35 - 42 = -7 \quad \square$$

As the solution of Example 5 indicates, if an entry in a row or column is 0 (such as a_{22} in Example 5), the product of the entry and its cofactor is 0. Thus when we compute a determinant by means of cofactors and in so doing select a row (or column), there is no need to include a term for any entry in the row (or column) that is 0. Therefore it can help to select a row or column containing zero entries.

The following properties of determinants aid in their evaluation.

> i. If all entries of a single row or column are 0, then the determinant is 0.
> ii. If two rows or two columns are identical, then the determinant is 0.
> iii. Interchanging two rows or two columns alters the determinant by a factor of -1.
> iv. If a nonzero multiple of one row (or column) is added to a second row (or column), then the determinant is unaltered.
> v. If all the entries of a single row (or column) are multiplied by a constant c, then the determinant of the new matrix is c times the determinant of the original matrix.

Example 6. Evaluate

$$\begin{vmatrix} 3 & 8 & 5 & 8 \\ 1 & 6 & -4 & 7 \\ 0 & 3 & 9 & -2 \\ 1 & 6 & -4 & 7 \end{vmatrix}$$

Solution. Since the second and fourth rows are identical, the determinant is 0 by (ii). $\square$

One procedure for evaluating a determinant is to use (iv) in order to make all, or all but one, of the entries in a particular row or column equal to 0, and then to evaluate the determinant by using the cofactors of that row or column.

Example 7. Evaluate

$$\begin{vmatrix} 2 & -1 & 0 & 3 \\ 1 & 2 & 5 & -4 \\ 3 & -1 & 2 & 0 \\ -1 & 3 & 2 & 5 \end{vmatrix}$$

Solution. If we multiply the second column by 2 and add the resulting column to the first column, we obtain the determinant

$$\begin{vmatrix} 0 & -1 & 0 & 3 \\ 5 & 2 & 5 & -4 \\ 1 & -1 & 2 & 0 \\ 5 & 3 & 2 & 5 \end{vmatrix}$$

Next we multiply the second column by 3 and add the resulting column to the fourth column, obtaining

$$\begin{vmatrix} 0 & -1 & 0 & 0 \\ 5 & 2 & 5 & 2 \\ 1 & -1 & 2 & -3 \\ 5 & 3 & 2 & 14 \end{vmatrix} \tag{5}$$

By (iv) neither of the two operations we have performed changes the value of the determinant. However, since all but one entry in the first row of the determinant in (5) are 0, we can readily evaluate the determinant by using cofactors of the first row:

$$\begin{vmatrix} 0 & -1 & 0 & 0 \\ 5 & 2 & 5 & 2 \\ 1 & -1 & 2 & -3 \\ 5 & 3 & 2 & 14 \end{vmatrix} = (0)A_{11} + (-1)A_{12} + (0)A_{13} + (0)A_{14}$$

$$= (-1)(-1)^{1+2} \begin{vmatrix} 5 & 5 & 2 \\ 1 & 2 & -3 \\ 5 & 2 & 14 \end{vmatrix}$$

$$= 140 - 75 + 4 - 20 - (-30) - 70$$

$$= 9 \quad \square$$

EXERCISES 6.5

In Exercises 1–2, find the minors and cofactors of the entries of the second row of the given matrix.

1. $\begin{bmatrix} -4 & 0 & 2 \\ 3 & 6 & -1 \\ 2 & -5 & 7 \end{bmatrix}$

2. $\begin{bmatrix} -1 & -3 & 1 & 2 \\ -2 & 0 & -1 & 1 \\ 3 & 2 & 0 & 4 \\ 0 & -3 & 1 & -2 \end{bmatrix}$

In Exercises 3–18, evaluate the given determinant.

3. $\begin{vmatrix} 2 & 0 \\ 3 & 4 \end{vmatrix}$

4. $\begin{vmatrix} 4 & -1 \\ -2 & -3 \end{vmatrix}$

5. $\begin{vmatrix} 6 & 0 \\ 3 & 0 \end{vmatrix}$

6. $\begin{vmatrix} a & b \\ -b & a \end{vmatrix}$

7. $\begin{vmatrix} 3 & -1 & 0 \\ 2 & 4 & 1 \\ -2 & 5 & -3 \end{vmatrix}$

8. $\begin{vmatrix} 1 & -1 & 1 \\ 2 & 2 & -2 \\ 3 & 3 & 3 \end{vmatrix}$

9. $\begin{vmatrix} -5 & -4 & 2 \\ -3 & 6 & 4 \\ -2 & -3 & 1 \end{vmatrix}$

10. $\begin{vmatrix} 6 & 9 & -1 \\ 4 & 6 & 4 \\ 2 & 3 & 7 \end{vmatrix}$

11. $\begin{vmatrix} 3 & 2 & -5 \\ 4 & -1 & -2 \\ 5 & -3 & 1 \end{vmatrix}$

12. $\begin{vmatrix} 3 & 0 & 0 \\ 0 & -2 & 0 \\ 0 & 0 & 5 \end{vmatrix}$

13. $\begin{vmatrix} -1 & 1 & 1 \\ 1 & -1 & 1 \\ 1 & 1 & -1 \end{vmatrix}$

14. $\begin{vmatrix} a & r & s \\ 0 & b & t \\ 0 & 0 & c \end{vmatrix}$

15. $\begin{vmatrix} 0 & 1 & 0 & 2 \\ 4 & 3 & -1 & 2 \\ 6 & 2 & 0 & 3 \\ 1 & -1 & 4 & -5 \end{vmatrix}$

16. $\begin{vmatrix} 1 & -2 & -4 & 5 \\ 2 & -4 & -6 & 10 \\ 1 & 3 & 0 & 2 \\ -1 & 7 & 1 & -3 \end{vmatrix}$

17. $\begin{vmatrix} 4 & 1 & -1 & -2 \\ -1 & 1 & 1 & 2 \\ -2 & 0 & 1 & 2 \\ -2 & 1 & 1 & 3 \end{vmatrix}$

18. $\begin{vmatrix} 1 & -2 & -3 & -4 \\ 0 & 2 & -3 & -4 \\ 0 & 0 & 3 & -4 \\ 0 & 0 & 0 & 4 \end{vmatrix}$

In Exercises 19–22, use properties (i)–(v) to explain why each equation is valid.

19. $\begin{vmatrix} 5 & -4 & 3 \\ 6 & 2 & 4 \\ 8 & 1 & -1 \end{vmatrix} = -\begin{vmatrix} 8 & 1 & -1 \\ 6 & 2 & 4 \\ 5 & -4 & 3 \end{vmatrix}$

20. $\begin{vmatrix} 2 & -4 & 7 \\ 5 & 1 & 0 \\ 3 & -6 & 5 \end{vmatrix} = \begin{vmatrix} 5 & 1 & 0 \\ 3 & -6 & 5 \\ 2 & -4 & 7 \end{vmatrix}$

21. $\begin{vmatrix} 3 & 5 & 8 \\ -2 & -3 & 4 \\ 1 & 7 & -2 \end{vmatrix} = -\frac{1}{4} \begin{vmatrix} 3 & 5 & 8 \\ 8 & 12 & -16 \\ 1 & 7 & -2 \end{vmatrix}$

22. $\begin{vmatrix} 2 & 3 & -1 \\ 5 & 1 & 4 \\ -6 & -9 & 3 \end{vmatrix} = 0$

23. Find all values of x satisfying

$$\begin{vmatrix} x & 1 & 0 \\ -1 & 3 & 4 \\ 3 & 1 & x \end{vmatrix} = 30$$

24. Show that

$$\begin{vmatrix} 1-x & 0 & 0 \\ 17 & 2-x & 0 \\ 31 & 13 & 3-x \end{vmatrix} = 0$$

if and only if $x = 1$, $x = 2$, or $x = 3$.

25. Show that

$$\begin{vmatrix} a & b \\ c & d \end{vmatrix} = - \begin{vmatrix} c & d \\ a & b \end{vmatrix}$$

26. Show that an equation of the straight line passing through the distinct points (a_1, b_1) and (a_2, b_2) is given by

$$\begin{vmatrix} 1 & x & y \\ 1 & a_1 & b_1 \\ 1 & a_2 & b_2 \end{vmatrix} = 0$$

27. Show that

$$\begin{vmatrix} 1 & a & a^2 \\ 1 & b & b^2 \\ 1 & c & c^2 \end{vmatrix} = (a-b)(b-c)(c-a)$$

28. It can be shown that the area $\mathcal{A}$ of the triangle in the plane with vertices $(0, 0)$, (a, b), and (c, d) is given by

$$\mathcal{A} = \left| \frac{1}{2} \begin{vmatrix} a & b \\ c & d \end{vmatrix} \right|$$

Find the area of the triangle whose vertices are
a. $(0, 0), (2, 1), (-1, 3)$ b. $(0, 0), (-2, -2), (5, 7)$

29. Suppose that three sides of a parallelepiped no two of which are parallel are represented by the vectors $\mathbf{a} = (a_1, a_2, a_3)$, $\mathbf{b} = (b_1, b_2, b_3)$, and $\mathbf{c} = (c_1, c_2, c_3)$. Then it can be shown that the volume V of the parallelepiped is given by

$$V = \left\| \begin{vmatrix} a_1 & a_2 & a_3 \\ b_1 & b_2 & b_3 \\ c_1 & c_2 & c_3 \end{vmatrix} \right\|$$

Find the volume of the parallelepiped determined by the vectors
a. $\mathbf{a} = (2, 1, 0)$, $\mathbf{b} = (-1, -4, 3)$, $\mathbf{c} = (0, -2, 1)$
b. $\mathbf{a} = (1, 3, 4)$, $\mathbf{b} = (2, 2, -1)$, $\mathbf{c} = (4, 0, -1)$

6.6

SOLUTIONS OF SYSTEMS BY DETERMINANTS: CRAMER'S RULE

Having learned to evaluate determinants, we are ready to use them in solving linear systems of equations. In order to do so, we will confine our attention to systems that have the same number of equations and variables.

Let us begin by solving the linear system

$$\begin{cases} a_1x + b_1y = c_1 \\ a_2x + b_2y = c_2 \end{cases} \tag{1}$$

for x and y, with the assumption that the determinant D of the coefficient matrix is not 0, that is, that

$$D = \begin{vmatrix} a_1 & b_1 \\ a_2 & b_2 \end{vmatrix} = a_1b_2 - a_2b_1 \neq 0 \tag{2}$$

To solve for x, we multiply the first equation in (1) by b_2 and the second equation by b_1, so the coefficient of y in each will be the same:

$$a_1b_2x + b_1b_2y = c_1b_2$$
$$a_2b_1x + b_1b_2y = c_2b_1 \tag{3}$$

Subtracting the second equation in (3) from the first yields

$$(a_1b_2 - a_2b_1)x = c_1b_2 - c_2b_1$$

Using (2), we find that

$$x = \frac{c_1b_2 - c_2b_1}{a_1b_2 - a_2b_1}$$

or in determinant form,

$$x = \frac{\begin{vmatrix} c_1 & b_1 \\ c_2 & b_2 \\ \hline a_1 & b_1 \\ a_2 & b_2 \end{vmatrix}} = \frac{\begin{vmatrix} c_1 & b_1 \\ c_2 & b_2 \end{vmatrix}}{D} \tag{4}$$

Similarly, we find that

$$y = \frac{\begin{vmatrix} a_1 & c_1 \\ a_2 & c_2 \\ \hline a_1 & b_1 \\ a_2 & b_2 \end{vmatrix}} = \frac{\begin{vmatrix} a_1 & c_1 \\ a_2 & c_2 \end{vmatrix}}{D} \tag{5}$$

(see Exercise 19). If we let the numerator of (4) be D_x and the numerator of (5) be D_y, so that

$$D_x = \begin{vmatrix} c_1 & b_1 \\ c_2 & b_2 \end{vmatrix} \quad \text{and} \quad D_y = \begin{vmatrix} a_1 & c_1 \\ a_2 & c_2 \end{vmatrix}$$

then (4) and (5) can be rewritten in the condensed form

$$x = \frac{D_x}{D} \quad \text{and} \quad y = \frac{D_y}{D} \tag{6}$$

Notice that D_x is obtained from D by replacing the first column with the entries c_1 and c_2, and D_y comes from D by replacing the second column with c_1, c_2.

Formulas (4) and (5) give the unique solution of the system in (1), provided that $D \neq 0$. Together, the formulas are known as **Cramer's Rule** after the Swiss mathematician Gabriel Cramer (1704–1752).

Although Cramer's Rule always provides us with a solution for the system of equations, we might make errors in the computations leading to the solution. As a result, it is important to check the proposed solution, as we have done until now in the chapter. However, to save space we will omit the checks from now on.

Example 1. Solve the linear system

$$\begin{cases} 3x - 5y = -2 \\ x + 2y = 4 \end{cases}$$

by means of determinants.

Solution. Since

$$D = \begin{vmatrix} 3 & -5 \\ 1 & 2 \end{vmatrix} = 6 - (-5) = 11$$

(2) is satisfied, and Cramer's Rule applies. Thus by (4),

$$x = \frac{\begin{vmatrix} -2 & -5 \\ 4 & 2 \end{vmatrix}}{D} = \frac{-4 - (-20)}{11} = \frac{16}{11}$$

and by (5),

$$y = \frac{\begin{vmatrix} 3 & -2 \\ 1 & 4 \end{vmatrix}}{D} = \frac{12 - (-2)}{11} = \frac{14}{11}$$

Consequently the solution of the given system is $(\frac{16}{11}, \frac{14}{11})$.   □

Next we present the three-variable version of Cramer's Rule. Consider the system

$$\begin{cases} a_1 x + b_1 y + c_1 z = d_1 \\ a_2 x + b_2 y + c_2 z = d_2 \\ a_3 x + b_3 y + c_3 z = d_3 \end{cases} \qquad (7)$$

and assume that the determinant D of the coefficient matrix is nonzero, that is,

$$D = \begin{vmatrix} a_1 & b_1 & c_1 \\ a_2 & b_2 & c_2 \\ a_3 & b_3 & c_3 \end{vmatrix} \neq 0$$

Then there is a unique solution of the system in (7) given by

$$x = \frac{\begin{vmatrix} d_1 & b_1 & c_1 \\ d_2 & b_2 & c_2 \\ d_3 & b_3 & c_3 \end{vmatrix}}{D}, \quad y = \frac{\begin{vmatrix} a_1 & d_1 & c_1 \\ a_2 & d_2 & c_2 \\ a_3 & d_3 & c_3 \end{vmatrix}}{D}, \quad z = \frac{\begin{vmatrix} a_1 & b_1 & d_1 \\ a_2 & b_2 & d_2 \\ a_3 & b_3 & d_3 \end{vmatrix}}{D} \qquad (8)$$

This result is also known as Cramer's Rule.

If we let

$$D_x = \begin{vmatrix} d_1 & b_1 & c_1 \\ d_2 & b_2 & c_2 \\ d_3 & b_3 & c_3 \end{vmatrix}, \quad D_y = \begin{vmatrix} a_1 & d_1 & c_1 \\ a_2 & d_2 & c_2 \\ a_3 & d_3 & c_3 \end{vmatrix}, \quad D_z = \begin{vmatrix} a_1 & b_1 & d_1 \\ a_2 & b_2 & d_2 \\ a_3 & b_3 & d_3 \end{vmatrix}$$

obtained from D by replacing the respective columns of D by the column consisting of d_1, d_2, d_3, then the formulas in (8) can be written in the condensed form

$$x = \frac{D_x}{D}, \quad y = \frac{D_y}{D}, \quad \text{and} \quad z = \frac{D_z}{D} \tag{9}$$

This corresponds to the simplified formulas for x and y in (6).

Example 2. Solve the linear system

$$\begin{cases} 2x - 3y - z = 0 \\ -5x + y - 4z = 1 \\ 3x + z = -2 \end{cases}$$

by means of determinants.

Solution. Since

$$D = \begin{vmatrix} 2 & -3 & -1 \\ -5 & 1 & -4 \\ 3 & 0 & 1 \end{vmatrix} = 2 + 36 + 0 - (-3) - 0 - 15 = 26$$

Cramer's Rule applies, so that by (8),

$$x = \frac{\begin{vmatrix} 0 & -3 & -1 \\ 1 & 1 & -4 \\ -2 & 0 & 1 \end{vmatrix}}{D} = \frac{0 - 24 + 0 - 2 - 0 - (-3)}{26} = -\frac{23}{26}$$

$$y = \frac{\begin{vmatrix} 2 & 0 & -1 \\ -5 & 1 & -4 \\ 3 & -2 & 1 \end{vmatrix}}{D} = \frac{2 + 0 - 10 - (-3) - 16 - 0}{26} = -\frac{21}{26}$$

$$z = \frac{\begin{vmatrix} 2 & -3 & 0 \\ -5 & 1 & 1 \\ 3 & 0 & -2 \end{vmatrix}}{D} = \frac{-4 - 9 + 0 - 0 - 0 - (-30)}{26} = \frac{17}{26}$$

Thus $(-\frac{23}{26}, -\frac{21}{26}, \frac{17}{26})$ is the solution of the given system. $\square$

An interesting consequence of Cramer's Rule concerns systems all of whose equations have 0 on the right side. Such a system is called ***homogeneous***. If a homogeneous system is in the form of (7) with $D \neq 0$, then the unique solution is $(0, 0, 0)$, since $d_1 = d_2 = d_3 = 0$ in (8). For example, for the solutions of the linear system

$$\begin{cases} 6x - 4y + 3z = 0 \\ -5x + y - 3z = 0 \\ 7x - y - 8z = 0 \end{cases}$$

we can first calculate that $D \neq 0$ and then conclude that the unique solution of the system is $(0, 0, 0)$. An analogous result holds for homogeneous systems of two equations.

If $D = 0$ for a given linear system of equations, then Cramer's Rule does not apply. In that case either the system is inconsistent or there are infinitely many solutions, which are best found by augmented matrices.

Finally, we mention that for a linear system of n equations with variables $x_1, x_2, \ldots, x_n$, if the determinant D of the coefficient matrix is not 0, then the analogue of Cramer's Rule in (9) remains valid. In particular, $(x_1, x_2, \ldots, x_n)$ is a solution of the system if and only if

$$x_1 = \frac{D_{x_1}}{D}, \ x_2 = \frac{D_{x_2}}{D}, \ \ldots, \ x_n = \frac{D_{x_n}}{D}$$

But the solutions are increasingly complicated to evaluate as n increases. In fact, if $n \geq 4$, then solving such a system by Cramer's Rule is more complicated than solving by augmented matrices, and Cramer's Rule is generally not used.

EXERCISES 6.6

1–4. Use Cramer's Rule to solve the systems in Exercises 5–8 of Section 6.1 (page 315).

5–9. Use Cramer's Rule to solve the systems in Exercises 1–5 of Section 6.2 (page 320).

10–14. Use Cramer's Rule to solve the systems in Exercises 17–21 of Section 6.3 (page 329).

In Exercises 15–18, use Cramer's Rule to solve the given system.

15.
$$\begin{cases} 2w + x - y + 2z = -3 \\ 3x + 2y + z = 0 \\ -w + x + y - 2z = 0 \\ 3w + 2x - y + 4z = -4 \end{cases}$$

16.
$$\begin{cases} 2w + 2x + y - z = 0 \\ w + 4x + y + 2z = 3 \\ -w + 2x + 2y - z = -2 \\ 3w - 2x + 4y + 2z = 1 \end{cases}$$

17.
$$\begin{cases} w - x + y - 2z = 0 \\ 2w + 2x + 2y - z = 5 \\ -3w + x - z = -4 \\ x + y + z = 7 \end{cases}$$

18.
$$\begin{cases} 2w + x + y - z = -4 \\ 4w - x + 5y + 2z = -6 \\ -2w + 2x - z = 2 \\ -6w - 3x + 3y + 2z = 4 \end{cases}$$

19. Prove (5). (*Hint:* Use the same procedure we used to prove (4).)

6.7

PRODUCTS OF MATRICES

This section is devoted mainly to products of matrices and the relation of such products to systems of equations. Before we study multiplication of matrices, let us make the following definitions.

(a) Let A and B be $m \times n$ matrices. Then $A = B$ if and only if the corresponding entries of A and B are equal. Thus

$$\begin{bmatrix} a & b \\ c & d \end{bmatrix} = \begin{bmatrix} e & f \\ g & h \end{bmatrix} \quad \text{if and only if} \quad a = e, b = f, c = g, \text{ and } d = h$$

(b) Let A and B be $m \times n$ matrices. Then the **sum** $A + B$ is the $m \times n$ matrix formed by adding the corresponding entries of A and B together. Thus

$$\begin{bmatrix} 2 & 3 & 1 \\ 4 & -5 & 0 \end{bmatrix} + \begin{bmatrix} 7 & -3 & -6 \\ 1 & 2 & 8 \end{bmatrix} = \begin{bmatrix} 2 + 7 & 3 + (-3) & 1 + (-6) \\ 4 + 1 & -5 + 2 & 0 + 8 \end{bmatrix}$$

$$= \begin{bmatrix} 9 & 0 & -5 \\ 5 & -3 & 8 \end{bmatrix}$$

In contrast, the sum of

$$\begin{bmatrix} 2 & 3 & 1 \\ 4 & -5 & 0 \end{bmatrix} \quad \text{and} \quad \begin{bmatrix} 7 & 1 \\ -3 & 2 \\ -6 & 8 \end{bmatrix}$$

is undefined because the matrices do not have the same number of rows (nor the same number of columns, for that matter).

(c) Let A be a matrix and c a number. Then the **constant multiple** cA is the matrix each of whose entries is c times the corresponding entry in A. Thus

$$(-3) \begin{bmatrix} 2 & \pi & -4 \\ 6 & -2 & 1 \\ -1 & 5 & -\sqrt{2} \end{bmatrix} = \begin{bmatrix} (-3)2 & (-3)\pi & (-3)(-4) \\ (-3)6 & (-3)(-2) & (-3)(1) \\ (-3)(-1) & (-3)(5) & (-3)(-\sqrt{2}) \end{bmatrix}$$

$$= \begin{bmatrix} -6 & -3\pi & 12 \\ -18 & 6 & -3 \\ 3 & -15 & 3\sqrt{2} \end{bmatrix}$$

You might well anticipate that the product AB of two matrices A and B would involve products of corresponding entries, to reflect the way we add matrices. However, from the standpoint of applications of matrices, another definition of the product AB is more useful. It is restricted to those matrices A and B such that A has the same number of columns as B has rows.

To define the product, let A be an $m \times n$ matrix and B an $n \times p$ matrix. Then we define the **product** AB to be the $m \times p$ matrix C whose entry c_{ij} in the ith row and jth column is given by

$$c_{ij} = a_{i1}b_{1j} + a_{i2}b_{2j} + \cdots + a_{in}b_{nj}$$

In other words, c_{ij} is obtained by multiplying the ith row of A entrywise with the jth column of B and then adding.

Example 1. Let $A = \begin{vmatrix} 3 & -1 & 4 \\ 2 & 0 & 1 \end{vmatrix}$ and

$$B = \begin{bmatrix} 1 & 0 & -3 & -2 \\ -2 & 4 & 5 & -1 \\ 3 & -1 & 0 & 6 \end{bmatrix}.$$ Calculate AB.

Solution. Since A is a 2×3 matrix and B a 3×4 matrix, the product AB is defined and is a 2×4 matrix, which we designate by C. Now

$$AB = \begin{bmatrix} 3 & -1 & 4 \\ 2 & 0 & 1 \end{bmatrix} \begin{bmatrix} 1 & 0 & -3 & -2 \\ -2 & 4 & 5 & -1 \\ 3 & -1 & 0 & 6 \end{bmatrix} = C = \begin{bmatrix} c_{11} & c_{12} & c_{13} & c_{14} \\ c_{21} & c_{22} & c_{23} & c_{24} \end{bmatrix}$$

where

$$c_{11} = (3)(1) + (-1)(-2) + (4)(3) = 17$$
$$c_{12} = (3)(0) + (-1)(4) + (4)(-1) = -8$$
$$c_{13} = (3)(-3) + (-1)(5) + (4)(0) = -14$$
$$c_{14} = (3)(-2) + (-1)(-1) + (4)(6) = 19$$
$$c_{21} = (2)(1) + (0)(-2) + (1)(3) = 5$$
$$c_{22} = (2)(0) + (0)(4) + (1)(-1) = -1$$
$$c_{23} = (2)(-3) + (0)(5) + (1)(0) = -6$$
$$c_{24} = (2)(-2) + (0)(-1) + (1)(6) = 2$$

Thus

$$AB = \begin{bmatrix} 17 & -8 & -14 & 19 \\ 5 & -1 & -6 & 2 \end{bmatrix} \quad \square$$

Caution: Notice that although the product AB is defined for the matrices A and B of Example 1, the product BA is *not* defined. The reason is that the number of columns of B (which is 4) is not equal to the number of rows of A (which is 2). In general, even if both the products AB and BA of two matrices A and B are defined, AB and BA need not be equal. (See Exercises 19–22.)

In general, if the product AB of two matrices A and B is defined, then the relationship among the numbers of columns and rows of A, B, and the product $C = AB$ is as follows:

$$\begin{array}{ccccc} A & \times & B & = & C \\ {\scriptstyle m \times n} & & {\scriptstyle n \times p} & & {\scriptstyle m \times p} \end{array}$$

Example 2. Let $A = \begin{bmatrix} 1 & 4 \\ -3 & -2 \\ 5 & 0 \end{bmatrix}$ and $X = \begin{bmatrix} -1 \\ 6 \end{bmatrix}$. Find AX.

Solution. Since A is a 3×2 matrix and X is a 2×1 matrix, AX is a 3×1 matrix, whose entries are computed as follows:

$$AX = \begin{bmatrix} 1 & 4 \\ -3 & -2 \\ 5 & 0 \end{bmatrix} \begin{bmatrix} -1 \\ 6 \end{bmatrix} = \begin{bmatrix} (1)(-1) + (4)(6) \\ (-3)(-1) + (-2)(6) \\ (5)(-1) + (0)(6) \end{bmatrix} = \begin{bmatrix} 23 \\ -9 \\ -5 \end{bmatrix} \quad \square$$

If A, B, and C are $m \times n$, $n \times p$, and $p \times r$ matrices, respectively, then AB is an $m \times p$ matrix, so $(AB)C$ is defined and is an $m \times r$ matrix. Similarly, BC is an $n \times r$ matrix, so $A(BC)$ is defined and is an $m \times r$ matrix. What may not be so evident, but is nevertheless true, is that

$$(AB)C = A(BC) \tag{1}$$

so it does not matter whether we first multiply A with B, or B with C. (See Exercises 23 and 24.) We will use this fact later.

Identity Matrices

For any positive integer n, let I_n be the square matrix of order n which has 1's down the diagonal starting at the upper left corner, and all other entries 0. This matrix is called the ***nth-order identity matrix***. For example,

$$I_2 = \begin{bmatrix} 1 & 0 \\ 0 & 1 \end{bmatrix} \quad \text{and} \quad I_3 = \begin{bmatrix} 1 & 0 & 0 \\ 0 & 1 & 0 \\ 0 & 0 & 1 \end{bmatrix}$$

Example 3. Let $A = \begin{bmatrix} -1 & 4 \\ 2 & -3 \end{bmatrix}$. Show that $AI_2 = I_2A = A$.

Solution. By definition,

$$AI_2 = \begin{bmatrix} -1 & 4 \\ 2 & -3 \end{bmatrix} \begin{bmatrix} 1 & 0 \\ 0 & 1 \end{bmatrix} = \begin{bmatrix} (-1)(1) + (4)(0) & (-1)(0) + (4)(1) \\ (2)(1) + (-3)(0) & (2)(0) + (-3)(1) \end{bmatrix}$$

$$= \begin{bmatrix} -1 & 4 \\ 2 & -3 \end{bmatrix} = A$$

and

$$I_2A = \begin{bmatrix} 1 & 0 \\ 0 & 1 \end{bmatrix} \begin{bmatrix} -1 & 4 \\ 2 & -3 \end{bmatrix} = \begin{bmatrix} (1)(-1) + (0)(2) & (1)(4) + (0)(-3) \\ (0)(-1) + (1)(2) & (0)(4) + (1)(-3) \end{bmatrix}$$

$$= \begin{bmatrix} -1 & 4 \\ 2 & -3 \end{bmatrix} = A \quad \square$$

If A is any square matrix of order n, then it can be shown that

$$AI_n = I_nA = A \qquad (2)$$

Example 3 lends credence to this fact, which says that the nth-order identity matrix has the property that the product of it and any square matrix A of order n is A. In this way I_n resembles the number 1 in multiplication, because for any real number a, we have $a \cdot 1 = 1 \cdot a = a$.

Inverse Matrices

If A is a square matrix of order n, then A *has an inverse* if and only if there is a square matrix B of order n such that

$$AB = I_n$$

In that case B is unique, and we also have $BA = I_n$. The matrix B is called the *inverse* of A and is written A^{-1}. Thus

$$AA^{-1} = I_n \quad \text{and} \quad A^{-1}A = I_n$$

Two questions immediately arise:

1. Which square matrices have inverses?
2. Assuming that a given square matrix has an inverse, how does one determine the entries of the inverse?

The answer to the first question is contained in the following theorem, given without proof.

THEOREM 6.4

Let A be a square matrix. Then A has an inverse if and only if the determinant of A is not 0.

To simplify the notation below, we will denote the determinant of a square matrix A by $\det A$.

Example 4. Let $A = \begin{bmatrix} 3 & 0 \\ 1 & 2 \end{bmatrix}$. Show that A has an inverse.

Solution. Since

$$\det A = \begin{vmatrix} 3 & 0 \\ 1 & 2 \end{vmatrix} = 6 - 0 = 6$$

Theorem 6.4 assures us that A has an inverse. $\square$

It is possible to give a formula for the inverse of a matrix. As you might imagine, the formula is simple for 2×2 matrices. If

$$A = \begin{bmatrix} a & b \\ c & d \end{bmatrix}$$

and $\det A \neq 0$, then

$$A^{-1} = \frac{1}{\det A}\begin{bmatrix} d & -b \\ -c & a \end{bmatrix} \tag{3}$$

as you can check by showing that $AA^{-1} = I_2$.

Example 5. Let $A = \begin{bmatrix} 3 & 0 \\ 1 & 2 \end{bmatrix}$. Find A^{-1}.

Solution. By Example 4 we know that $\det A = 6$, and thus A^{-1} exists. Then (3) with $a = 3$, $b = 0$, $c = 1$, and $d = 2$, implies that A^{-1} is given by

$$A^{-1} = \frac{1}{6}\begin{bmatrix} 2 & 0 \\ -1 & 3 \end{bmatrix} \quad \square$$

If $n \geq 3$ and A is an $n \times n$ matrix that has an inverse, then the formula that yields A^{-1} is more cumbersome to apply. Instead of giving such a formula, we will present a procedure that not only tells whether the inverse exists but also yields the inverse when it does exist. The procedure is as follows:

i. Write the $n \times 2n$ matrix formed by adjoining I_n to the right of A:

$$\begin{bmatrix} a_{11} & a_{12} & \cdots & a_{1n} & 1 & 0 & 0 & \cdots & 0 \\ a_{21} & a_{22} & \cdots & a_{2n} & 0 & 1 & 0 & \cdots & 0 \\ \cdot & \cdot & & \cdot & & \cdot & & & \cdot \\ \cdot & \cdot & & \cdot & & \cdot & \cdot & & \cdot \\ \cdot & \cdot & & \cdot & & \cdot & \cdot & \cdot & \cdot \\ a_{n1} & a_{n2} & \cdots & a_{nn} & 0 & 0 & 0 & \cdots & 1 \end{bmatrix}$$

ii. Perform elementary row operations on the matrix in (i), with the goal of bringing the matrix into the form

$$\begin{bmatrix} 1 & 0 & 0 & \cdots & 0 & b_{11} & b_{12} & \cdots & b_{1n} \\ 0 & 1 & 0 & \cdots & 0 & b_{21} & b_{22} & \cdots & b_{2n} \\ \cdot & \cdot & \cdot & & \cdot & \cdot & \cdot & & \cdot \\ \cdot & \cdot & \cdot & & \cdot & \cdot & \cdot & & \cdot \\ \cdot & \cdot & \cdot & & \cdot & \cdot & \cdot & & \cdot \\ 0 & 0 & 0 & \cdots & 1 & b_{n1} & b_{n2} & \cdots & b_{nn} \end{bmatrix}$$

Notice that this matrix has the entries of I_n on the left, and the entries of a matrix that we will call B on the right.

iii. If it is possible to obtain the matrix appearing in step (ii), then not only does A have an inverse, but also $A^{-1} = B$. Conversely, if it is not possible to attain the matrix appearing in step (ii), then A has no inverse.

Example 6. Let

$$A = \begin{bmatrix} 1 & 2 & -1 \\ 4 & 3 & -4 \\ -2 & 0 & -3 \end{bmatrix}$$

Show that A has an inverse, and find A^{-1}.

Solution. Using the procedure (i)–(iii) given above, we first form the matrix

$$\begin{bmatrix} 1 & 2 & -1 & 1 & 0 & 0 \\ 4 & 3 & -4 & 0 & 1 & 0 \\ -2 & 0 & -3 & 0 & 0 & 1 \end{bmatrix}$$

Next we perform elementary row operations to bring the left side of the matrix into the form given in step (ii):

$$\begin{bmatrix} 1 & 2 & -1 & 1 & 0 & 0 \\ 4 & 3 & -4 & 0 & 1 & 0 \\ -2 & 0 & -3 & 0 & 0 & 1 \end{bmatrix} \xleftarrow{R_2 \to R_2 + (-4)R_1} \begin{bmatrix} 1 & 2 & -1 & 1 & 0 & 0 \\ 0 & -5 & 0 & -4 & 1 & 0 \\ -2 & 0 & -3 & 0 & 0 & 1 \end{bmatrix} \xleftarrow{R_3 \to R_3 + 2R_1}$$

$$\begin{bmatrix} 1 & 2 & -1 & 1 & 0 & 0 \\ 0 & -5 & 0 & -4 & 1 & 0 \\ 0 & 4 & -5 & 2 & 0 & 1 \end{bmatrix} \xleftarrow{R_2 \to (-\frac{1}{5})R_2} \begin{bmatrix} 1 & 2 & -1 & 1 & 0 & 0 \\ 0 & 1 & 0 & \frac{4}{5} & -\frac{1}{5} & 0 \\ 0 & 4 & -5 & 2 & 0 & 1 \end{bmatrix} \xleftarrow{R_1 \to R_1 + (-2)R_2}$$

$$\begin{bmatrix} 1 & 0 & -1 & -\frac{3}{5} & \frac{2}{5} & 0 \\ 0 & 1 & 0 & \frac{4}{5} & -\frac{1}{5} & 0 \\ 0 & 4 & -5 & 2 & 0 & 1 \end{bmatrix} \xleftarrow{R_3 \to R_3 + (-4)R_2} \begin{bmatrix} 1 & 0 & -1 & -\frac{3}{5} & \frac{2}{5} & 0 \\ 0 & 1 & 0 & \frac{4}{5} & -\frac{1}{5} & 0 \\ 0 & 0 & -5 & -\frac{6}{5} & \frac{4}{5} & 1 \end{bmatrix} \xleftarrow{R_3 \to (-\frac{1}{5})R_3}$$

$$\begin{bmatrix} 1 & 0 & -1 & -\frac{3}{5} & \frac{2}{5} & 0 \\ 0 & 1 & 0 & \frac{4}{5} & -\frac{1}{5} & 0 \\ 0 & 0 & 1 & \frac{6}{25} & -\frac{4}{25} & -\frac{1}{5} \end{bmatrix} \xleftarrow{R_1 \to R_1 + R_3} \begin{bmatrix} 1 & 0 & 0 & -\frac{9}{25} & \frac{6}{25} & -\frac{1}{5} \\ 0 & 1 & 0 & \frac{4}{5} & -\frac{1}{5} & 0 \\ 0 & 0 & 1 & \frac{6}{25} & -\frac{4}{25} & -\frac{1}{5} \end{bmatrix}$$

Therefore by (iii), A has an inverse, and

$$A^{-1} = \begin{bmatrix} -\frac{9}{25} & \frac{6}{25} & -\frac{1}{5} \\ \frac{4}{5} & -\frac{1}{5} & 0 \\ \frac{6}{25} & -\frac{4}{25} & -\frac{1}{5} \end{bmatrix} = \frac{1}{25} \begin{bmatrix} -9 & 6 & -5 \\ 20 & -5 & 0 \\ 6 & -4 & -5 \end{bmatrix}$$

You can check this result by showing that $AA^{-1} = I_3$. □

Solutions of Linear Systems Using Inverses

We return to linear systems of equations and present one final method of solving them. The method works when the system has the same number of equations as variables, provided that the coefficient matrix has an inverse. For simplicity our discussion will involve systems consisting of three equations in three variables.

Consider the system

$$\begin{cases} a_1x + b_1y + c_1z = k_1 \\ a_2x + b_2y + c_2z = k_2 \\ a_3x + b_3y + c_3z = k_3 \end{cases} \tag{4}$$

Let

$$A = \begin{bmatrix} a_1 & b_1 & c_1 \\ a_2 & b_2 & c_2 \\ a_3 & b_3 & c_3 \end{bmatrix}, \quad X = \begin{bmatrix} x \\ y \\ z \end{bmatrix}, \quad \text{and} \quad K = \begin{bmatrix} k_1 \\ k_2 \\ k_3 \end{bmatrix}$$

Then

$$AX = \begin{bmatrix} a_1 & b_1 & c_1 \\ a_2 & b_2 & c_2 \\ a_3 & b_3 & c_3 \end{bmatrix} \begin{bmatrix} x \\ y \\ z \end{bmatrix} = \begin{bmatrix} a_1x + b_1y + c_1z \\ a_2x + b_2y + c_2z \\ a_3x + b_3y + c_3z \end{bmatrix}$$

By the definition of the equality of two matrices,

$$AX = K \tag{5}$$

if and only if

$$\begin{bmatrix} a_1x + b_1y + c_1z \\ a_2x + b_2y + c_2z \\ a_3x + b_3y + c_3z \end{bmatrix} = \begin{bmatrix} k_1 \\ k_2 \\ k_3 \end{bmatrix}$$

which means that the corresponding entries of AX and K are the same, that is, the equations in (4) are satisfied. Therefore (4) and (5) are equivalent. This means that determining values of x, y, and z such that (4) is satisfied is equivalent to finding a matrix X such that (5) is satisfied. In other words, solving the system of equations in (4) is equivalent to solving the matrix equation in (5) for the matrix X.

How can we solve (5) for X? Suppose that the determinant of the coefficient matrix A is nonzero. Then A^{-1} exists by Theorem 6.4, so that (5) is equivalent to

$$A^{-1}(AX) = A^{-1}K \tag{6}$$

Using (1), (2), and the definition of inverse, we conclude that

$$A^{-1}(AX) = (A^{-1}A)X = I_3X = X$$

Therefore (6) becomes

$$X = A^{-1}K \tag{7}$$

In other words, if det $A \neq 0$, then in order to solve (5) for X, we need only compute the product $A^{-1}K$. And since

$$X = \begin{bmatrix} x \\ y \\ z \end{bmatrix}$$

once we know X, we know the solution of the system (4).

Example 7. Use (7) to solve the linear system of equations

$$\begin{cases} x + 2y - z = 6 \\ 4x + 3y - 4z = -1 \\ -2x \qquad - 3z = 3 \end{cases}$$

Solution. As our discussion preceding this example indicated, the given system of equations is equivalent to the matrix equation

$$AX = K \tag{8}$$

where

$$A = \begin{bmatrix} 1 & 2 & -1 \\ 4 & 3 & -4 \\ -2 & 0 & -3 \end{bmatrix}, \quad X = \begin{bmatrix} x \\ y \\ z \end{bmatrix}, \quad \text{and} \quad K = \begin{bmatrix} 6 \\ -1 \\ 3 \end{bmatrix}$$

By Example 6 we know that A^{-1} exists and that

$$A^{-1} = \frac{1}{25} \begin{bmatrix} -9 & 6 & -5 \\ 20 & -5 & 0 \\ 6 & -4 & -5 \end{bmatrix}$$

It then follows from (7) that the solution X of (8) is given by

$$X = A^{-1}K$$

We find that

$$\begin{bmatrix} x \\ y \\ z \end{bmatrix} = X = A^{-1}K = \frac{1}{25} \begin{bmatrix} -9 & 6 & -5 \\ 20 & -5 & 0 \\ 6 & -4 & -5 \end{bmatrix} \begin{bmatrix} 6 \\ -1 \\ 3 \end{bmatrix}$$

$$= \frac{1}{25} \begin{bmatrix} (-9)(6) + (6)(-1) + (-5)(3) \\ (20)(6) + (-5)(-1) + (0)(3) \\ (6)(6) + (-4)(-1) + (-5)(3) \end{bmatrix} = \frac{1}{25} \begin{bmatrix} -75 \\ 125 \\ 25 \end{bmatrix} = \begin{bmatrix} -3 \\ 5 \\ 1 \end{bmatrix}$$

By the definition of equality of matrices, this means that $x = -3$, $y = 5$, and $z = 1$. Consequently $(-3, 5, 1)$ is the proposed solution of the given system of equations.

Check:
$$\begin{cases} -3 + 2(5) - 1 = & 6 \\ 4(-3) + 3(5) - 4(1) = & -1 \\ -2(-3) - 3(1) = & 3 \end{cases}$$

We conclude that $(-3, 5, 1)$ is the solution of the given system. $\quad\square$

EXERCISES 6.7

In Exercises 1–4, perform the indicated operation.

1. $\begin{bmatrix} 2 & 1 & 3 \\ -4 & 1 & 5 \end{bmatrix} + \begin{bmatrix} 3 & -3 & 0 \\ 1 & 4 & 2 \end{bmatrix}$

2. $\begin{bmatrix} -1 & 3 & 5 \\ 2 & 6 & -4 \\ -3 & 1 & 2 \end{bmatrix} + \begin{bmatrix} 6 & -1 & 5 \\ 1 & -3 & 2 \\ -4 & 5 & 1 \end{bmatrix}$

3. $-3 \begin{bmatrix} 2 & -4 \\ -5 & 12 \end{bmatrix}$

4. $\dfrac{1}{2} \begin{bmatrix} 2 & 4 & 0 \\ -3 & 6 & -2 \\ 14 & -8 & 10 \end{bmatrix}$

In Exercises 5–18, calculate the given product.

5. $\begin{bmatrix} 1 & 2 \\ 4 & 3 \end{bmatrix} \begin{bmatrix} -2 & -1 \\ 0 & 3 \end{bmatrix}$

6. $\begin{bmatrix} 0 & 1 \\ -5 & 6 \end{bmatrix} \begin{bmatrix} 3 & -4 \\ 4 & 7 \end{bmatrix}$

7. $\begin{bmatrix} \frac{1}{2} & 2 \\ -1 & 0 \end{bmatrix} \begin{bmatrix} 4 & -1 \\ \frac{1}{2} & 1 \end{bmatrix}$

8. $\begin{bmatrix} 1 & 2 \\ -3 & -4 \end{bmatrix} \begin{bmatrix} 3 \\ -1 \end{bmatrix}$

9. $\begin{bmatrix} -4 & 1 \\ 3 & 6 \end{bmatrix} \begin{bmatrix} 1 \\ -1 \end{bmatrix}$

10. $[2 \;\; -5] \begin{bmatrix} 6 \\ -1 \end{bmatrix}$

11. $\begin{bmatrix} 2 \\ -1 \end{bmatrix} [3 \;\; -4]$

12. $\begin{bmatrix} 3 & 1 & -1 \\ 0 & 2 & -3 \\ 4 & 0 & -2 \end{bmatrix} \begin{bmatrix} 0 & -1 & 2 \\ 2 & -3 & 4 \\ 3 & 0 & 2 \end{bmatrix}$

13. $\begin{bmatrix} 1 & 0 & 0 \\ 0 & 2 & 0 \\ 0 & 0 & 3 \end{bmatrix} \begin{bmatrix} 4 & -2 & 5 \\ 9 & -1 & -7 \\ -3 & 6 & 3 \end{bmatrix}$

14. $\begin{bmatrix} -1 & 1 & 2 \\ 3 & 0 & -2 \\ -4 & \frac{1}{2} & 0 \end{bmatrix} \begin{bmatrix} -1 & 2 & 0 \\ -2 & 4 & 6 \\ 7 & -6 & -\frac{1}{2} \end{bmatrix}$

15. $\begin{bmatrix} 2 & -3 & 1 \\ -6 & 9 & -2 \\ -3 & 5 & -1 \end{bmatrix} \begin{bmatrix} -1 & -2 & 3 \\ 0 & -1 & 2 \\ 3 & 1 & 0 \end{bmatrix}$

16. $\begin{bmatrix} 3 & -4 \\ 2 & 5 \\ 1 & 6 \end{bmatrix} \begin{bmatrix} 1 \\ -2 \end{bmatrix}$

17. $[-3 \quad -6 \quad 2] \begin{bmatrix} -1 \\ 3 \\ 4 \end{bmatrix}$

18. $\begin{bmatrix} -5 \\ 1 \\ -3 \end{bmatrix} [2 \quad -1]$

In Exercises 19–22, calculate AB and BA.

19. $A = \begin{bmatrix} -2 & 0 \\ -1 & 3 \end{bmatrix}$, $B = \begin{bmatrix} 4 & 3 \\ -1 & 0 \end{bmatrix}$

20. $A = \begin{bmatrix} -7 & 3 \\ 3 & 1 \end{bmatrix}$, $B = \begin{bmatrix} -1 & 0 \\ 0 & -1 \end{bmatrix}$

21. $A = \begin{bmatrix} 2 & 3 & -1 \\ 4 & -3 & 5 \\ 1 & -1 & 0 \end{bmatrix}$, $B = \begin{bmatrix} 1 & 2 & 0 \\ -1 & 0 & 3 \\ 4 & -1 & 2 \end{bmatrix}$

22. $A = \begin{bmatrix} \frac{1}{2} & -\frac{1}{3} & \frac{1}{4} \\ 2 & \frac{1}{2} & \frac{1}{6} \\ 1 & -\frac{1}{4} & \frac{2}{9} \end{bmatrix}$, $B = \begin{bmatrix} 4 & 0 & 0 \\ 0 & 4 & 0 \\ 0 & 0 & 4 \end{bmatrix}$

In Exercises 23–24, calculate $(AB)C$ and $A(BC)$, and thereby support (1).

23. $A = \begin{bmatrix} 0 & -2 \\ 3 & -1 \end{bmatrix}$, $B = \begin{bmatrix} 3 & -1 \\ 1 & 0 \end{bmatrix}$, $C = \begin{bmatrix} 4 & 0 \\ -1 & -3 \end{bmatrix}$

24. $A = \begin{bmatrix} 3 & 0 & 1 \\ 2 & -1 & -3 \\ 0 & 4 & -1 \end{bmatrix}$, $B = \begin{bmatrix} 1 & -1 & 0 \\ 0 & -1 & 2 \\ 0 & 1 & 0 \end{bmatrix}$, $C = \begin{bmatrix} 1 & 2 & -3 \\ -2 & 0 & 1 \\ -1 & 2 & 0 \end{bmatrix}$

In Exercises 25–40, determine whether the given matrix has an inverse. If it does, compute the inverse.

25. $\begin{bmatrix} 0 & 1 \\ 1 & 0 \end{bmatrix}$

26. $\begin{bmatrix} 3 & -2 \\ 1 & 0 \end{bmatrix}$

27. $\begin{bmatrix} 4 & -6 \\ 2 & -3 \end{bmatrix}$

28. $\begin{bmatrix} -2 & 2 \\ 1 & 1 \end{bmatrix}$

29. $\begin{bmatrix} 6 & 1 \\ -3 & 2 \end{bmatrix}$

30. $\begin{bmatrix} -6 & 9 \\ 1 & -\frac{3}{2} \end{bmatrix}$

31. $\begin{bmatrix} a & 0 \\ c & b \end{bmatrix}$ where $ab \neq 0$

32. $\begin{bmatrix} 1 & -1 & 0 \\ 0 & -1 & 1 \\ 1 & 2 & -1 \end{bmatrix}$

33. $\begin{bmatrix} 1 & 0 & 0 \\ 0 & 1 & 0 \\ 0 & 0 & 1 \end{bmatrix}$

34. $\begin{bmatrix} 0 & 0 & 1 \\ 0 & 1 & 0 \\ 1 & 0 & 0 \end{bmatrix}$

35. $\begin{bmatrix} 2 & 0 & 4 \\ 3 & -1 & 4 \\ 2 & 1 & 6 \end{bmatrix}$

36. $\begin{bmatrix} 2 & -3 & 1 \\ 4 & -1 & -2 \\ -1 & 0 & 2 \end{bmatrix}$

37. $\begin{bmatrix} -1 & -2 & 3 \\ 0 & -1 & 2 \\ 3 & 1 & 0 \end{bmatrix}$

38. $\begin{bmatrix} 2 & -1 & -2 \\ 0 & -3 & 1 \\ 4 & -14 & 0 \end{bmatrix}$

39. $\begin{bmatrix} 0 & 4 & 2 \\ 1 & -3 & -1 \\ -4 & 5 & -2 \end{bmatrix}$

40. $\begin{bmatrix} a & 0 & 0 \\ d & b & 0 \\ e & f & c \end{bmatrix}$ where $abc \neq 0$

In Exercises 41–48, solve the given system by using the inverse of the coefficient matrix.

41. $\begin{cases} 3x + 4y = -11 \\ 2x - y = 0 \end{cases}$

42. $\begin{cases} -2x + 3y = 13 \\ x + 5y = 39 \end{cases}$

43. $\begin{cases} -4x - 2y = 22 \\ 3x - 5y = -23 \end{cases}$

44. $\begin{cases} 12x + y = 0 \\ 9x - 8y = 0 \end{cases}$

45. $\begin{cases} 2x + 4y - z = -4 \\ x - y + 3z = 8 \\ 3x + 3y - 4z = -8 \end{cases}$

46. $\begin{cases} 2x - 4y + z = -9 \\ -3x - y - 2z = 0 \\ 4x + 3y + z = 5 \end{cases}$

47. $\begin{cases} 3x - 7y + 5z = 7 \\ 2x + y - 5z = -3 \\ -x - y + z = \frac{1}{2} \end{cases}$

48. $\begin{cases} 9x + 2y - 7z = 0 \\ -3x + 2y + 6z = 0 \\ 5x - 3y - 4z = 0 \end{cases}$

49. Show that $\begin{bmatrix} 0 & 1 \\ 0 & 1 \end{bmatrix}\begin{bmatrix} 1 & 0 \\ 0 & 0 \end{bmatrix} = \begin{bmatrix} 0 & 0 \\ 0 & 0 \end{bmatrix}$.

50. Let A and B be 2×2 matrices. Show det $(AB) = (\det A)(\det B)$.

6.8
SYSTEMS OF INEQUALITIES

In Sections 2.6 through 2.8 and 3.9 we solved single inequalities. Now we return to inequalities and solve systems of two or more inequalities, such as

$$\begin{cases} x + y \leq 1 \\ 2x - y + 1 \geq 0 \end{cases} \tag{1}$$

and

$$\begin{cases} x^2 + y^2 \leq 4 \\ \qquad x^2 \geq 3y \\ \qquad x \geq 1 \end{cases} \qquad (2)$$

An ordered pair (x, y) is a **solution** of a given system of inequalities if and only if (x, y) satisfies each inequality in the system individually. Thus $(1, -2)$ is a solution of (1) because

$$1 + (-2) \leq 1 \quad \text{and} \quad 2(1) - (-2) + 1 \geq 0$$

With a system of two or more inequalities, the set of solutions is most easily presented on a graph, and this is the way we will give the solutions. The solutions normally form a region whose boundary is composed of line segments or curves. An essential part of graphing the set of solutions will involve determining the points at which such pieces of the boundary meet.

Example 1. Sketch the solutions of the system

$$\begin{cases} y \leq -x + 1 \\ y \leq \ 2x + 1 \end{cases}$$

Solution. A pair (x, y) is a solution of the first inequality if (x, y) lies below or on the line $y = -x + 1$ (with slope -1 and y intercept 1). Likewise, (x, y) is a solution of the second inequality if (x, y) lies below or on the line $y = 2x + 1$ (with slope 2 and y intercept 1). The solutions of the system consist of those (x, y) satisfying both conditions, and such (x, y) are shaded in Figure 6.2. Because the y intercepts of the lines forming the boundary are both 1, evidently the lines meet at $(0, 1)$. □

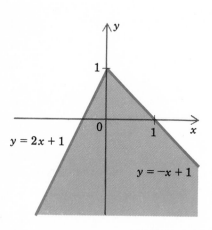

FIGURE 6.2

To aid in solving other systems of inequalities, we will say that two inequalities are **equivalent** if they have the same solutions. For example, the inequalities

$$y \leq -x + 1 \quad \text{and} \quad x + y \leq 1$$

are equivalent, because the second can be obtained by adding x to each side of the first. Similarly,

$$y \leq 2x + 1 \quad \text{and} \quad 2x - y + 1 \geq 0$$

are equivalent because the second can be obtained by subtracting y from each side of the first. In addition, two systems of inequalities are **equivalent** if they have the same solutions. For example, the systems

$$\begin{cases} y \leq -x + 1 \\ y \leq \ 2x + 1 \end{cases} \quad \text{and} \quad \begin{cases} x + y \qquad \leq 1 \\ 2x - y + 1 \geq 0 \end{cases}$$

are equivalent, by our preceding comments.

If the boundary of the set of solutions of a system of inequalities consists of straight lines, it is usually convenient, when possible, to write the inequalities in the system in the equivalent standard form

$$y \leq ax + b \quad \text{or} \quad y \geq ax + b \tag{3}$$

because of their close relationship with the slope–intercept form of lines. We will use this procedure in the following example.

Example 2. Sketch the solutions of the system

$$\begin{cases} x \geq 0 \\ y \geq 0 \\ 2y + x \leq 8 \\ x + y \leq 5 \\ 2x + y \leq 8 \end{cases}$$

and determine the points of intersection of the boundary lines.

Solution. The given system is equivalent to the system

$$\begin{cases} x \geq 0 \\ y \geq 0 \\ y \leq -\dfrac{1}{2} x + 4 \\ y \leq -x + 5 \\ y \leq -2x + 8 \end{cases}$$

Observe that (x, y) satisfies $x \geq 0$ and $y \geq 0$ if and only if (x, y) lies in the first quadrant. For that reason, we can restrict our attention to the first quadrant when graphing the remaining inequalities. The graph of $y \leq -\frac{1}{2}x + 4$ consists of all points below or on the line $y = -\frac{1}{2}x + 4$ (with slope $-\frac{1}{2}$ and y intercept 4). The graph of $y \leq -x + 5$ consists of all points below or on the line $y = -x + 5$ (with slope -1 and y intercept 5). Finally, the graph of $y \leq -2x + 8$ consists of all points below or on the line $y = -2x + 8$ (with slope -2 and y intercept 8). Thus the solutions of the given system consist of all points in the first quadrant that lie on or below each of the three above-mentioned lines (Figure 6.3). Finally we determine the coordinates of the points O, A, B, C, and D shown in Figure 6.3. We know that O is the origin, $(0, 0)$, and D is the point $(0, 4)$ at which the line $y = -\frac{1}{2}x + 4$ intersects the y axis. Similarly, A is the point at which the line $y = -2x + 8$ intersects the x axis. We find the coordinates of A by setting y equal to 0 in the equation $y = -2x + 8$:

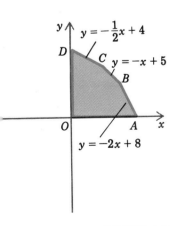

FIGURE 6.3

$$0 = -2x + 8$$

$$2x = 8$$

$$x = 4$$

Thus A is the point $(4, 0)$. Next, B is the point of intersection of the lines $y = -x + 5$ and $y = -2x + 8$, which (as you can check) is the point $(3, 2)$. Finally, C is the point of intersection of the lines $y = -\frac{1}{2}x + 4$ and $y = -x + 5$, which you can check is the point $(2, 3)$. This completes the solution of the example. □

Example 3. Sketch the solutions of the system

$$\begin{cases} x^2 + y^2 \leq 4 \\ (x - 2)^2 + (y + 2)^2 \geq 4 \end{cases}$$

Solution. The graph of the equation $x^2 + y^2 = 4$ is the circle centered at the origin with radius 2, so the solutions of $x^2 + y^2 \leq 4$ consist of all points (x, y) inside or on the circle. Similarly, the graph of the equation $(x - 2)^2 + (y + 2)^2 = 4$ is the circle centered at the point $(2, -2)$ with radius 2, so the solutions of $(x - 2)^2 + (y + 2)^2 \geq 4$ consist of all points (x, y) outside or on the circle. The solutions of the given system consist of those (x, y) satisfying both inequalities and form the region shaded in Figure 6.4. As indicated in Figure 6.4, the two circles meet at the points $(2, 0)$ and $(0, -2)$. You can check that this is true by solving the system consisting of the equations of the two circles. □

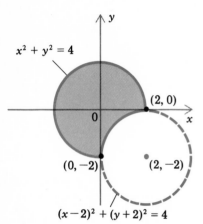

FIGURE 6.4

EXERCISES 6.8

In Exercises 1–6, determine whether the given ordered pair is a solution of the given system of inequalities.

1. $\begin{cases} 2x + y \leq 3 \\ 3x - 4y \leq 0 \end{cases}$ $(-1, 2)$

2. $\begin{cases} 4x - 3y \geq -1 \\ 2x - 3y \leq 20 \end{cases}$ $(3, -4)$

3. $\begin{cases} 2x - 9y \leq 0 \\ 3x - 2y \leq 0 \end{cases}$ $(\frac{1}{2}, \frac{1}{3})$

4. $\begin{cases} x \geq 0 \\ y \geq 0 \\ 2x - y \geq 0 \\ 2x + 3y \leq 50 \end{cases}$ $(6, 11)$

5. $\begin{cases} x \geq 0 \\ y \leq 5 \\ 2x - y \leq 10 \\ 5x + 6y \leq 96 \end{cases}$ $(3, 8)$

6. $\begin{cases} x \geq 0 \\ y \leq 5 \\ 7x + 8y \leq 25 \\ 9x + 10y \leq 30 \end{cases}$ $(0, 3)$

In Exercises 7–10, sketch the solutions of the given system of inequalities without calculating the points of intersection of the boundary lines.

7. $\begin{cases} y \le -2x + 4 \\ y \ge x + 2 \end{cases}$

8. $\begin{cases} 4x + 2y \le 5 \\ -3x + 2y \le 6 \end{cases}$

9. $\begin{cases} x \ge 0 \\ y \ge 0 \\ 2x + y \le 6 \\ 4x + y \le 8 \end{cases}$

10. $\begin{cases} x \ge 0 \\ y \ge 0 \\ y - \frac{1}{2}x \le 3 \\ 2x - y \le 4 \end{cases}$

In Exercises 11–22, sketch the solutions of the given system of inequalities, and find the points of intersection of the boundary lines.

11. $\begin{cases} x \ge 0 \\ x \le 3 \\ y \ge 0 \\ x + y \le 5 \end{cases}$

12. $\begin{cases} y \ge 0 \\ y - 2x \le 3 \\ x + y \le 6 \end{cases}$

13. $\begin{cases} x \ge 0 \\ y \ge 0 \\ y \le 7 \\ 2y \ge x - 4 \end{cases}$

14. $\begin{cases} y \ge 0 \\ x \le 3 \\ y - x \le 1 \\ x \ge 2y - 3 \end{cases}$

15. $\begin{cases} y \le 2x + 3 \\ x \le 2y - 2 \\ y + 2x \le 11 \end{cases}$

16. $\begin{cases} y \ge 0 \\ x - y \ge 1 \\ 2y \le x \\ 2x + y \le 10 \end{cases}$

17. $\begin{cases} x \ge 0 \\ y \ge 0 \\ 2y + x \le 20 \\ x \le -y + 12 \\ 2x + y \le 20 \end{cases}$

18. $\begin{cases} x \ge 1 \\ x \le 15 \\ y \ge 2 \\ y \le 7 \\ x + 3y \le 27 \end{cases}$

19. $\begin{cases} x^2 + y^2 \le 9 \\ (x + 3)^2 + (y + 3)^2 \ge 9 \end{cases}$

20. $\begin{cases} x^2 + y^2 \le 16 \\ x^2 + (y - 2)^2 \ge 1 \end{cases}$

21. $\begin{cases} y \ge x^2 \\ y \le 1 - x^2 \end{cases}$

22. $\begin{cases} y \le x + 2 \\ y \le 4x \\ y \ge x^2 \end{cases}$

6.9
LINEAR PROGRAMMING

In the preceding sections of the chapter we have studied both systems of equations and systems of inequalities. We conclude the chapter with a discussion of linear programming, a topic that centers around both kinds of systems. Basically linear programming involves finding the maximum or minimum value of a linear expression, which in two variables has the form

$$ax + by$$

where the variables x and y are subject to a set of conditions or constraints given in the form of linear inequalities. The linear expression $ax + by$ might represent the profit resulting from allocating resources so that x units of one product and y units of another product are manufactured. In that case the linear inequalities might represent constraints on the resources used in production. The problem would be to allocate or "program" resources in such a way as to maximize the profit. This is the origin of the term "linear programming."

For example, consider the problem of maximizing the expression $3x + y$ subject to the linear inequalities

$$\begin{cases} x \geq 0 \\ y \geq 0 \\ 2x + y \leq 4 \\ x + 2y \leq 5 \end{cases} \tag{1}$$

Perhaps x represents the amount in thousands of gallons of yogurt and y the amount in thousands of gallons of ice cream that a company sells daily. If the profit in thousands of dollars per thousand gallons of yogurt is 3 and the profit in thousands of dollars per thousand gallons of ice cream is 1, then $3x + y$ would represent the daily profit in thousands of dollars on the sale of x thousand gallons of yogurt and y thousand gallons of ice cream. The first two constraints in (1), which are $x \geq 0$ and $y \geq 0$, say that only a nonnegative amount of the foods can be sold, which is reasonable. The next constraint, $2x + y \leq 4$, might represent restrictions on the available labor, and the final constraint, $x + 2y \leq 5$, might represent restrictions on plant equipment. After all this information is provided, the basic problem is to determine the values of x and y that maximize the profit $3x + y$ subject to the constraints in (1). After we develop the necessary tools, we will solve this problem in Example 1.

In the last 30 years linear programming has found widespread application in science and industry. Normally there are far more than two variables involved in linear programming problems that arise. However, in order to keep the calculations manageable, we will restrict our attention to those linear programming problems that involve only two variables.

Assume now that we have a linear programming problem such as the one discussed earlier, so that we wish to find the maximum or minimum value of a linear expression $ax + by$ subject to a set of linear inequalities (the constraints). Any ordered pair (x, y) that satisfies all the constraints is called a *feasible solution* of the problem. The set of all feasible solutions is called the *feasible set.* A feasible solution that yields the desired maximum or minimum value of

$ax + by$ is called an **optimal solution** of the problem. For the problem of maximizing $3x + y$ subject to the constraints in (1), the points $(1, \frac{1}{2})$ and $(1, 2)$ are feasible solutions, since they each satisfy all constraints:

$$\begin{cases} 1 \geq 0 \\ \dfrac{1}{2} \geq 0 \\ 2(1) + \dfrac{1}{2} \leq 4 \\ 1 + 2\left(\dfrac{1}{2}\right) \leq 5 \end{cases} \quad \text{and} \quad \begin{cases} 1 \geq 0 \\ 2 \geq 0 \\ 2(1) + 2 \leq 4 \\ 1 + 2(2) \leq 5 \end{cases}$$

For the linear expression $3x + y$ the points $(1, \frac{1}{2})$ and $(1, 2)$ yield

$$3(1) + \frac{1}{2} = \frac{7}{2} \quad \text{and} \quad 3(1) + 2 = 5$$

respectively. We conclude that $(1, \frac{1}{2})$ does not yield the maximum value of $3x + y$ and hence is *not* an optimal solution. It will follow from the solution of Example 1 that $(1, 2)$ is also not an optimal solution.

We can sketch the feasible set for a given linear programming problem by sketching the solutions of the system of linear inequalities that form the constraints. We do this by the methods of Section 6.8. Since all of the constraints are linear inequalities, the boundaries of the feasible set will be lines or line segments (Figure 6.5). The point at which two boundary lines intersect is called a *vertex* of the feasible set. The following theorem, which we will not prove, assures us that if optimal solutions exist, then at least one of them can be found among the vertices of the feasible set.

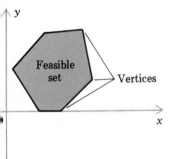

FIGURE 6.5

THEOREM 6.5   *Fundamental Theorem of Linear Programming.* Assume that a linear programming problem has an optimal solution. Then at least one vertex of the feasible set is an optimal solution.

Since the feasible set may well have only a few vertices, the Fundamental Theorem of Linear Programming can reduce the work needed to find an optimal solution of a linear programming problem. The procedure is as follows:

i. Determine the feasible set by graphing the linear inequalities that form the constraints.
ii. Find the vertices of the feasible set by solving two at a time the equations of the boundary lines of the feasible set.
iii. Compute the value of $ax + by$ at each of the vertices.
iv. Select the largest value computed in step (iii) if $ax + by$ is to be maximized, and select the smallest value if $ax + by$ is to be minimized.

Before presenting examples and problems, we mention that a given linear programming problem may or may not have an optimal solution. However, all the problems in this book actually do have optimal solutions.

Example 1. Find the maximum and minimum values of $3x + y$ subject to the constraints

$$\begin{cases} x \geq 0 \\ y \geq 0 \\ 2x + y \leq 4 \\ x + 2y \leq 5 \end{cases} \tag{2}$$

Solution. First we will determine the feasible set by graphing the inequalities in (2). Notice that (2) is equivalent to the system

$$\begin{cases} x \geq 0 \\ y \geq 0 \\ y \leq -2x + 4 \\ y \leq -\dfrac{1}{2}x + \dfrac{5}{2} \end{cases}$$

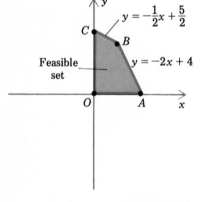

FIGURE 6.6

The inequalities $x \geq 0$ and $y \geq 0$ restrict the feasible set to the first quadrant. Thus the feasible set is the set of all points in the first quadrant that lie on or below each of the lines $y = -2x + 4$ (with slope -2 and y intercept 4) and $y = -\frac{1}{2}x + \frac{5}{2}$ (with slope $-\frac{1}{2}$ and y intercept $\frac{5}{2}$). The feasible set is sketched in Figure 6.6. Next we will determine the vertices O, A, B, and C of the feasible set. We know that O is the origin, and C is the point $(0, \frac{5}{2})$ at which the line $y = -\frac{1}{2}x + \frac{5}{2}$ crosses the y axis. Similarly, A is the point at which the line $y = -2x + 4$ crosses the x axis. To find the coordinates of A we substitute 0 for y in the equation $y = -2x + 4$:

$$0 = -2x + 4$$
$$2x = 4$$
$$x = 2$$

Thus A is the point $(2, 0)$. Since B is the point of intersection of the lines $y = -2x + 4$ and $y = -\frac{1}{2}x + \frac{5}{2}$, we find the coordinates of B by solving the system

$$\begin{cases} y = -2x + 4 \\ y = -\dfrac{1}{2}x + \dfrac{5}{2} \end{cases} \tag{3}$$

Substituting $-\frac{1}{2}x + \frac{5}{2}$ for y in the first equation of (3) yields

$$-\frac{1}{2}x + \frac{5}{2} = -2x + 4$$

$$\frac{3}{2}x = \frac{3}{2}$$

$$x = 1$$

Substituting 1 for x in the first equation of (3), we find that

$$y = -2(1) + 4 = 2$$

Therefore the solution of (3) is $(1, 2)$, which tells us that B is the point $(1, 2)$. In order to find the maximum value of $3x + y$ among the feasible solutions, we compute the values of $3x + y$ at the vertices O, A, B, and C:

(x, y)	$O = (0, 0)$	$A = (2, 0)$	$B = (1, 2)$	$C = (0, \frac{5}{2})$
$3x + y$	0	6	5	$\frac{5}{2}$

From the table we see that the maximum value of $3x + y$ at the vertices is 6, corresponding to $(2, 0)$. We conclude from the Fundamental Theorem of Linear Programming that $(2, 0)$ is an optimal solution and that 6 is the maximum value of $3x + y$ subject to the constraints listed in (2). Likewise, we see from the table that the minimum value of $3x + y$ at the vertices is 0, corresponding to $(0, 0)$. We conclude that $(0, 0)$ is an optimal solution and that 0 is the minimum value of $3x + y$ subject to the constraints listed in (2). $\square$

Example 2. A small business produces tables and chairs. Suppose the material for each table costs \$75 and the material for each chair costs \$50, and that the business can spend a maximum of \$1200 per day for material. Suppose also that 3 hours of labor are required to assemble each table and 5 hours to assemble each chair, and that up to 75 hours of labor are available daily at a cost of \$5 per hour. Finally, let us assume that the delivery truck can deliver at most 18 items (tables or chairs) per day. If each table is sold for \$150 and each chair for \$125, determine the maximum daily profit and how many tables and chairs should be produced daily for maximum profit.

Solution. We will translate the given problem into a linear programming problem. Ultimately we wish to determine how many tables and how many chairs to produce daily, so let x be the number of tables and y the number of chairs to be produced daily. Now the material for each table costs \$75, and the labor costs \$15 (3 hours at \$5 per hour). Thus the total cost of producing each table is \$90. Likewise, the mateial for each chair costs \$50, and the labor costs \$25 (5 hours at \$5 per hour). Thus the total cost of producing each chair is \$75. Therefore the profit from each table sold for \$150 is \$60 (because $150 - 90 = 60$), and the profit for each chair sold for $125 is $50 (because $125 - 75 = 50$). Consequently if x tables and y chairs are produced and sold daily, then the daily profit is

$$60x + 50y$$

dollars. Since the business wishes to maximize its daily profit, we have now identified the linear expression to be maximized. Next we will determine the constraints. Recall first that the material for each table costs $75 and the material for each chair costs $50. Thus if x tables and y chairs are produced daily, then the daily cost of material is $75x + 50y$ dollars. Since by hypothesis the business can pay a maximum of $1200 per day for material, we must have

$$75x + 50y \le 1200$$

which reduces to

$$3x + 2y \le 48$$

Next recall that each table requires 3 hours to assemble and each chair 5 hours. Therefore with x tables and y chairs produced daily, the number of hours required daily is $3x + 5y$. Since the number of hours available daily is by hypothesis limited to 75, we have

$$3x + 5y \le 75$$

Recall also from the hypothesis that at most 18 items (tables or chairs) can be delivered a day. Thus

$$x + y \le 18$$

Finally, we have

$$x \ge 0 \quad \text{and} \quad y \ge 0$$

since the number of items produced cannot be negative.

Now we can restate the problem as a linear programming problem: Find the maximum value of $60x + 50y$ subject to the constraints

$$\begin{cases} x \ge 0 \\ y \ge 0 \\ 3x + 2y \le 48 \\ x + y \le 18 \\ 3x + 5y \le 75 \end{cases} \tag{4}$$

The first thing we do as we begin to solve the problem is to rewrite the system in (4) in the equivalent form

$$\begin{cases} x \geq 0 \\ y \geq 0 \\ y \leq -\frac{3}{2}x + 24 \\ y \leq -x + 18 \\ y \leq -\frac{3}{5}x + 15 \end{cases}$$

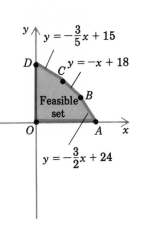

FIGURE 6.7

The feasible set for this problem is shown in Figure 6.7. There are 5 vertices: O, A, B, C, and D. To determine their coordinates, we notice that 0 is the origin, A is the point at which the line $y = -\frac{3}{2}x + 24$ intersects the x axis, B is the point of intersection of the lines $y = -x + 18$ and $y = -\frac{3}{2}x + 24$, C is the point of intersection of the lines $y = -\frac{3}{5}x + 15$ and $y = -x + 18$, and D is the point at which the line $y = -\frac{3}{5}x + 15$ intersects the y axis. You can verify that

$$O = (0, 0), \quad A = (16, 0), \quad B = (12, 6), \quad C = (7.5, 10.5), \quad D = (0, 15)$$

To complete the solution of the problem we evaluate $60x + 50y$ at the 5 vertices:

(x, y)	$O = (0, 0)$	$A = (16, 0)$	$B = (12, 6)$	$C = (7.5, 10.5)$	$D = (0, 15)$
$60x + 50y$	0	960	1020	975	750

From the table we see that the maximum value of $60x + 50y$ at the vertices is 1020, and occurs at $(12, 6)$. It follows from the Fundamental Theorem of Linear Programming that the maximum value of $60x + 50y$ subject to the constraints in (4) is 1020, corresponding to the optimal solution $(12, 6)$. We conclude that 12 tables and 6 chairs should be produced daily to maximize profit, and that the maximum daily profit is $1020. $\quad \square$

EXERCISES 6.9

In Exercises 1–10, find the maximum and minimum values of the given linear expression subject to the given constraints, and find points at which those values are assumed.

1. $-4x + y$ subject to

$$\begin{cases} x \geq -1 \\ x \leq 3 \\ y \geq -2 \\ y \leq 1 \end{cases}$$

2. $3x - 2y$ subject to

$$\begin{cases} x \geq 0 \\ y \geq 0 \\ x \leq 5 \\ y \leq x \end{cases}$$

3. $\frac{1}{2}x + \frac{1}{3}y$ subject to

$$\begin{cases} x \geq 0 \\ y \geq 0 \\ y \leq 6 \\ 3x + y \leq 18 \end{cases}$$

4. $0.01x + 0.25y$ subject to

$$\begin{cases} x \geq 0 \\ y \geq 0 \\ 2x + 3y \leq 24 \\ 2x + y \leq 12 \end{cases}$$

5. $4y - 5x$ subject to

$$\begin{cases} y \geq 2 \\ 2x - y \leq 2 \\ 2y - x \leq 8 \end{cases}$$

6. $4y - 5x$ subject to

$$\begin{cases} x \leq 8 \\ y \geq 0 \\ 2x - y \geq 2 \\ 2y - x \leq 8 \end{cases}$$

7. $2x + \frac{1}{3}y$ subject to

$$\begin{cases} x \leq 6 \\ 3x + 2y \geq 18 \\ y \leq 2x + 9 \end{cases}$$

8. $\frac{1}{6}x + \frac{1}{4}y$ subject to

$$\begin{cases} x \geq 0 \\ y \geq 0 \\ y \leq 4 \\ x + 2y \geq 2 \\ 2x + 3y \leq 22 \end{cases}$$

9. $2x + 4y$ subject to

$$\begin{cases} x \geq 0 \\ y \geq 0 \\ 2x + 5y \leq 50 \\ 2x + 3y \leq 34 \\ 3x + y \leq 30 \end{cases}$$

10. $4x + 5y$ subject to

$$\begin{cases} y \geq 0 \\ y \leq 6 \\ x + 3y \geq 10 \\ 3y - x \leq 8 \\ 2x + y \leq 36 \end{cases}$$

11. A sidewalk vendor sells submarines and soft drinks from a cart. Each carton of submarines weighs 5 pounds, takes up 1 cubic foot of space in the cart, and yields a profit of $8 to the vendor. Each case of soft drinks weighs 20 pounds, takes up 0.3 cubic foot of space in the cart and yields a profit of $3 to the vendor. The cart can carry up to 225 pounds and has a capacity of 8 cubic feet. Assuming that the vendor can sell whatever the cart can hold, find the maximum profit that the vendor can reap from one cartload.

12. A grocer plans to buy up to $1000 worth of peanuts and raisins and sell the mixture for $1 per pound. Peanuts cost 25¢ per pound, and raisins cost 50¢ per pound. The weight of the peanuts in the mixture should be at least three times and at most six times the weight of the raisins. How many pounds of peanuts and how many pounds of raisins should the grocer buy in order to maximize the profit?

13. A corporate executive plans to invest up to $60,000 in municipal and corporate bonds for the next year. The municipal bonds yield 6% interest per year, and the corporate bonds 8%. If the executive wishes to invest at least $10,000 in corporate bonds and at

least twice as much money in municipal bonds as in corporate bonds, find the maximum amount of interest that can be earned during the year.

14. A builder can acquire up to 125 lots on which to build either modern or colonial houses. Each modern house requires an investment of $30,000 and yields a profit of $10,000, whereas each colonial house requires an investment of $20,000 and yields a profit of $8000. If the builder can invest up to $3,000,000 and wishes the number of colonial houses built to be at most 4 times the number of modern houses, determine how many modern houses should be built to yield the maximum profit.

15. Smith's orchard produces peaches and apples. On the average, one bushel of peaches can be picked in 20 minutes and one bushel of apples in 30 minutes. Suppose that the labor force can pick up to 240 hours each day, that Smith's delivery truck can deliver at most 500 bushels daily to market, and that each day Smith must retain 40 bushels of fruit for sales on the premises. If there is a $2 profit on each bushel of peaches and $2.50 on each bushel of apples, determine the number of bushels of peaches that must be harvested daily to achieve the maximum daily profit, and compute that profit.

16. Suppose the price per bushel of apples drops to $2 in Exercise 15. Show that there would be multiple ways of obtaining a maximum profit.

17. A garden shop has 1100 pounds of Kentucky Bluegrass seed and 900 pounds of Kentucky Tall Fescue seed. In each bag of Premium seed the shop puts 4 pounds of Bluegrass and no Fescue seed, whereas in each bag of Shady seed it puts one pound of Bluegrass and 3 pounds of Fescue seed. If the profit on each bag of Premium seed is $3 and on each bag of Shady seed is $2, determine the maximum profit the shop can earn from the sale of Premium and Shady seed.

18. A sporting goods manufacturer produces sleeping bags and tents. Each sleeping bag requires $\frac{2}{3}$ hour of cutting, 1 hour of machine sewing, and $\frac{1}{2}$ hour of hand sewing, and brings a profit of $15. Each tent requires 1 hour of cutting, 3 hours of machine sewing, and $\frac{1}{2}$ hour of hand sewing, and brings a profit of $25. Each day the shop can provide 36 hours of labor for cutting, 90 hours for machine sewing, and 24 hours for hand sewing. Determine the number of sleeping bags that should be produced each day to bring in the maximum daily profit.

KEY TERMS

system of equations	matrix
solution of a system of equations	square matrix of order n
method of substitution	coefficient matrix
method of elimination	augmented matrix
linear system	equivalent matrices

elementary row operation
determinant
minor
cofactor
identity matrix
inverse of a matrix
system of inequalities

solution of a system of inequalities
equivalent systems of inequalities
linear programming
feasible solution
optimal solution

KEY THEOREMS

Cramer's Rule
Fundamental Theorem of Linear Programming

REVIEW EXERCISES

In Exercises 1–14, find all solutions of the given system of equations.

1.
$$\begin{cases} x + y = 7 \\ -2x - 7y = -4 \end{cases}$$

2.
$$\begin{cases} 2x + 3y = 0 \\ -4x + 6y = -4 \end{cases}$$

3.
$$\begin{cases} 6x - 12y = 1 \\ -9x + 18y = 7 \end{cases}$$

4.
$$\begin{cases} x - 3y = -5 \\ -2x + 6y = 10 \end{cases}$$

5.
$$\begin{cases} x - 2y = 0 \\ -3x + 6y = 0 \end{cases}$$

6.
$$\begin{cases} x^2 + y^2 = 10 \\ 9x^2 + 2y^2 = 27 \end{cases}$$

7.
$$\begin{cases} 2xy = -1 \\ 4x - y = 3 \end{cases}$$

8.
$$\begin{cases} e^{x+y} = 1 \\ 2x - y = 3 \end{cases}$$

9.
$$\begin{cases} 2x - xy = 0 \\ 3x + xy = 5 \end{cases}$$

10.
$$\begin{cases} rs - 5s + r = 5 \\ 2s - r = 3 \end{cases}$$

11.
$$\begin{cases} 3x - 2y - z = 3 \\ 2x + 5y + 2z = 0 \\ 5x + 7y + 2z = -1 \end{cases}$$

12.
$$\begin{cases} 4x - 7y + 11z = 5 \\ 6x + 5y - 9z = -21 \\ 5x - y + z = 0 \end{cases}$$

13.
$$\begin{cases} x - y - z = 1 \\ 2x + y - 8z = -1 \\ -3x + 2y + 5z = -2 \end{cases}$$

14.
$$\begin{cases} 3x - 5y + 6z = 1 \\ 6x + 7y - 12z = 6 \\ -12x + 2y - 6z = -7 \end{cases}$$

In Exercises 15–18, solve the given system by using augmented matrices.

15.
$$\begin{cases} x - 2y + 2z = 1 \\ 3x + 2y - z = -6 \\ 4x - 3y + 3z = -6 \end{cases}$$

16.
$$\begin{cases} 2x - 3y - z = 0 \\ 7x - 3y + z = -2 \\ 4x - y + z = -1 \end{cases}$$

17. $\begin{cases} 2x + 3y + 2z = 0 \\ 3x + 5y - 5z = 5 \\ -2x + y - z = 1 \end{cases}$

18. $\begin{cases} x - y + z = 3 \\ -3x + 2y + 3z = -7 \\ -5x + 3y + 7z = -11 \end{cases}$

In Exercises 19–24, evaluate the given determinant.

19. $\begin{vmatrix} 3 & -5 \\ 2 & 4 \end{vmatrix}$

20. $\begin{vmatrix} -4 & -5 \\ -7 & -13 \end{vmatrix}$

21. $\begin{vmatrix} -3 & 4 & 2 \\ 1 & 0 & 6 \\ 3 & 2 & -5 \end{vmatrix}$

22. $\begin{vmatrix} 2 & -4 & -5 \\ 0 & 1 & 1 \\ -3 & 5 & 6 \end{vmatrix}$

23. $\begin{vmatrix} 2 & -1 & 0 & -2 \\ 1 & 0 & -3 & -1 \\ 4 & -1 & 2 & 1 \\ -2 & 2 & 1 & 0 \end{vmatrix}$

24. $\begin{vmatrix} -2 & -1 & -1 & -3 \\ 1 & 0 & -2 & 0 \\ 3 & 2 & -6 & 4 \\ 3 & 2 & 1 & 5 \end{vmatrix}$

In Exercises 25–28, use Cramer's Rule to solve the given system.

25. $\begin{cases} 3x - 2y = 1 \\ -5x + 4y = 3 \end{cases}$

26. $\begin{cases} 2x + 16y = 1 \\ 3x + 26y = 2 \end{cases}$

27. $\begin{cases} 3x + 2y + 2z = 9 \\ 2x - 2y + 3z = 1 \\ 4x - 2y + 4z = 0 \end{cases}$

28. $\begin{cases} 3x + 2y + 2z = 2 \\ -4x - 3y - 2z = 4 \\ -2x + 3y + 3z = 3 \end{cases}$

In Exercises 29–30, perform the indicated operations.

29. $\begin{bmatrix} 3 & -1 & 5 \\ 2 & 1 & 4 \\ -7 & 2 & 0 \end{bmatrix} + 3\begin{bmatrix} -1 & -4 & 2 \\ 0 & -5 & 8 \\ 7 & 3 & 0 \end{bmatrix}$

30. $\begin{bmatrix} 4 & 3 & -2 \\ 6 & -5 & 7 \\ 0 & -4 & \frac{1}{2} \end{bmatrix}\begin{bmatrix} -4 & 5 & 6 \\ 2 & 4 & -3 \\ 0 & -2 & 2 \end{bmatrix}$

In Exercises 31–36, determine whether the given matrix has an inverse. If it does, compute the inverse.

31. $\begin{bmatrix} 2 & -1 \\ 0 & 4 \end{bmatrix}$

32. $\begin{bmatrix} 4 & 3 \\ 8 & 6 \end{bmatrix}$

33. $\begin{bmatrix} 1 & -1 \\ 1 & 1 \end{bmatrix}$

34. $\begin{bmatrix} 2 & -3 & -1 \\ 4 & 0 & 5 \\ -8 & 6 & -3 \end{bmatrix}$

35. $\begin{bmatrix} 1 & -1 & 2 \\ 2 & 1 & 3 \\ 2 & -2 & 5 \end{bmatrix}$

36. $\begin{bmatrix} 5 & -2 & 4 \\ 8 & -1 & 7 \\ 6 & -2 & 5 \end{bmatrix}$

In Exercises 37–40, solve the given system by using the inverse of the coefficient matrix

37. $\begin{cases} 3x - 2y = -1 \\ 2x - 4y = 26 \end{cases}$ 38. $\begin{cases} 6x + 3y = 1 \\ 4x + 5y = \frac{8}{3} \end{cases}$

39. $\begin{cases} 2x + 5y + z = 9 \\ -3x + 3y - 4z = -3 \\ -8x + 6y - 10z = -10 \end{cases}$ 40. $\begin{cases} x + 2y + 3z = -3 \\ 2x + 7y + 5z = -\frac{17}{2} \\ -x - 5y - z = 5 \end{cases}$

In Exercises 41–44, sketch the solutions of the given system of inequalities, and find the points of intersection of the boundary lines.

41. $\begin{cases} x \geq 0 \\ y \geq 0 \\ y + \frac{1}{4}x \leq 1 \\ y + \frac{2}{3}x \leq \frac{2}{3} \end{cases}$ 42. $\begin{cases} x \leq 5 \\ y \geq 0 \\ 2y - x \leq 7 \\ 3x \leq 19 - 2y \end{cases}$

43. $\begin{cases} x \geq 0 \\ y \geq 0 \\ x + y \leq 10 \\ y \leq -2(x - 6) \\ 4x \leq 20 - y \end{cases}$ 44. $\begin{cases} y \geq 2x \\ 4x + y \geq 0 \\ 3x \leq 5 - y \\ y \leq x + 5 \end{cases}$

In Exercises 45–48, find the maximum and minimum values of the given linear expression subject to the given constraints, and find the points at which those values are assumed.

45. $2x + 4y$; the region in Exercise 41.

46. $-x + 2y$; the region in Exercise 42.

47. $3x + y$; the region in Exercise 43.

48. $4y - 5x$; the region in Exercise 44.

49. Let $A = \begin{bmatrix} 0 & 1 \\ 0 & 1 \end{bmatrix}$. Show that $A^2 = A$.

50. Let $A = \begin{bmatrix} 1 & 1 \\ 0 & 1 \end{bmatrix}$. Find a formula for A^n, for any positive integer n.

51. For which values of λ does the system

$$\begin{cases} \lambda x + 3y = 0 \\ 27x + \lambda y = 0 \end{cases}$$

have a solution other than the trivial solution $(0, 0)$?

52. For what values of a and b does the line $ax + by = 5$ contain the points $(15, 10)$ and $(-5, -5)$?

53. If 6 oranges and 12 lemons together cost $3.30, and 12 oranges and 6 lemons together cost $3.00, how much does a single orange cost?

54. If Joe were 4 inches taller, he would be $\frac{2}{3}$ as tall as his mother. If he were 7 inches shorter, he would be only half as tall as his mother. How tall is his mother?

55. A swimming club deposited a $20,000 surplus into two separate savings accounts earning simple interest for a period for one year. One account earned 8% interest and the other 10%. If the total interest earned during the year was $1920, how much was deposited in each account?

56. An advertising company must have at least 1000 letters addressed within four hours. Its own staff can address 100 letters per hour at a cost of $40 per hour, and a professional typing company can address 200 letters per hour at a cost of $120 per hour. For how many hours should the advertising company employ the typing company if it wishes to minimize its total cost for addressing letters?

57. A company plans to introduce a new line of storm doors and storm windows made out of glass and aluminum. Because of limitations on experienced labor available at its glass-cutting and aluminum-cutting buildings, the company must limit production. The labor required for each door and window and the profit from each are shown in Table 6.1. Determine how many storm doors and how many storm windows should be produced daily in order to maximize the total profit.

TABLE 6.1

	Storm door	Storm window	Available daily
Time for cutting glass	2 minutes	1 minute	1000 minutes
Time for cutting aluminum	4 minutes	1 minute	1600 minutes
Profit	$7	$2	

COMPLEX

NUMBERS

7

For many centuries mathematicians have studied solutions of polynomial equations, that is, equations of the form

$$a_n x^n + a_{n-1} x^{n-1} + \cdots + a_1 x + a_0 = 0 \qquad (1)$$

where n is a positive integer and a_0, a_1, . . . , a_{n-1}, a_n are real numbers. However, even if $n = 2$, there are equations in the form of (1) that have no solutions among the real numbers. The most famous such equation is

$$x^2 + 1 = 0 \qquad (2)$$

If a number i were a solution of (2), then of course $i^2 + 1 = 0$, so that i would have the property that

$$i^2 = -1$$

But the square of any *real* number is nonnegative. Thus equation (2) has no solution among the real numbers.

The great desire to have solutions of equations such as (2) led mathematicians to invent new numbers like the number i. But the thought of having numbers whose squares were negative was so foreign, and even repugnant to some, that these new numbers were referred to as "complex" or "imaginary" numbers. However, complex numbers have proved to be extremely useful, not only in mathematics but also in physics and electrical engineering. Today complex numbers are accepted and used throughout the scientific community.

7.1

INTRODUCTION TO COMPLEX NUMBERS

We begin our study of complex numbers by introducing a new number i with the property that

$$i^2 = -1 \tag{1}$$

It follows that i is a solution of the equation

$$x^2 = -1, \quad \text{or equivalently,} \quad x^2 + 1 = 0$$

In order to have solutions of other equations such as

$$x^2 + 9 = 0 \quad \text{and} \quad x^2 + 6x + 13 = 0 \tag{2}$$

we introduce numbers of the form $a + bi$, where a and b are any real numbers, and call $a + bi$ a **complex number**.* Later we will show that every quadratic equation (in particular, either of those in (2)) has a solution among the complex numbers.

The real number a is called the **real part** of $a + bi$, and the real number b is called the **imaginary part** of $a + bi$. Thus the real part of $-2 + 3i$ is -2 and the imaginary part is 3. Two complex numbers $a + bi$ and $c + di$ are said to be **equal** if $a = c$ and $b = d$, that is, if their real parts are equal and their imaginary parts are equal. If a is a real number, then $a + 0i$ is identified with the real number a, and we write a in place of $a + 0i$; in particular, $0 + 0i$ is written 0. Similarly, we write bi instead of $0 + bi$, and we call bi a **pure imaginary number**, or just an imaginary number. Thus $\frac{1}{4}i$ and $-i$ are pure imaginary numbers. Sometimes, to avoid confusion, we write $a + bi$ as $a + ib$. Thus we would write $2 - i \cos \pi^2$ instead of $2 - (\cos \pi^2)i$. Finally, by convention we write $a - bi$ rather than $a + (-b)i$. For example, we write $2 - 3i$ instead of $2 + (-3)i$.

---

* In some other fields, especially electrical engineering, j is used for i, since i refers to electric current.

The value of complex numbers to mathematicians and scientists is greatly enhanced by the fact that it is possible to define an addition and multiplication for them that retains the usual laws of arithmetic (such as commutativity and associativity). For the **sum** of two complex numbers $a + bi$ and $c + di$ we have

$$(a + bi) + (c + di) = (a + c) + (b + d)i \qquad (3)$$

In other words, we add two complex numbers by adding their real parts and their imaginary parts separately and joining the results as indicated.

Example 1. Write the following in the form $a + bi$.
a. $(4 + 2i) + (-3 + 5i)$ b. $(3 - 6i) + (-1 + 6i)$

Solution.

    a. By (3),

$$(4 + 2i) + (-3 + 5i) = (4 - 3) + (2 + 5)i = 1 + 7i$$

    b. By (3),

$$(3 - 6i) + (-1 + 6i) = (3 - 1) + (-6 + 6)i$$
$$= 2 + 0i = 2 \quad \square$$

From (3) it follows that

$$(a + bi) + (-a - bi) = (a - a) + (b - b)i = 0 + 0i = 0$$

so $-a - bi$ is the additive inverse of $a + bi$ and is denoted $-(a + bi)$. Thus

$$-(a + bi) = -a - bi$$

For the **difference** of two complex numbers $a + bi$ and $c + di$ we have

$$(a + bi) - (c + di) = (a - c) + (b - d)i \qquad (4)$$

Example 2. Write the difference $(5 + 3i) - (-2 + 4i)$ in the form $a + bi$.

Solution. By (4),

$$(5 + 3i) - (-2 + 4i) = [5 - (-2)] + (3 - 4)i = 7 - i \quad \square$$

To introduce the product of two complex numbers $a + bi$ and $c + di$, we multiply as if the numbers were polynomials in i and use the fact that $i^2 = -1$:

$$(a + bi)(c + di) = ac + adi + bci + bdi^2$$
$$= ac + (ad + bc)i - bd$$

By rearranging the terms on the right side of the last line, we obtain the **product** of $a + bi$ and $c + di$:

$$(a + bi)(c + di) = (ac - bd) + (ad + bc)i \qquad (5)$$

It is not necessary to memorize formula (5), because to find the product $(a + bi)(c + di)$ we can simply multiply the individual numbers in the parentheses, as we did above, and use the fact that $i^2 = -1$.

Example 3. Write the product of $4 - 5i$ and $-2 - i$ in the form $a + bi$.

Solution. Multiplying out, we find that

$$(4 - 5i)(-2 - i) = (4)(-2) + (4)(-i) + (-5i)(-2) + (-5i)(-i)$$
$$= -8 - 4i + 10i + 5i^2$$
$$= -8 - 4i + 10i - 5$$
$$= -13 + 6i \quad \square$$

Among products of complex numbers are powers of i. We list some of these powers below.

$$i^1 = i \qquad\qquad i^5 = i^4 i = (1)i = i$$
$$i^2 = -1 \qquad\qquad i^6 = i^4 i^2 = (1)i^2 = -1$$
$$i^3 = i^2 i = (-1)i = -i \qquad i^7 = i^4 i^3 = (1)i^3 = -i$$
$$i^4 = i^2 i^2 = (-1)(-1) = 1 \qquad i^8 = i^4 i^4 = (1)(1) = 1$$

As this list suggests, the powers of i repeat every 4 powers, and because $i^4 = 1$, we can compute any integral power of i. For example,

$$i^{15} = i^4 \cdot i^4 \cdot i^4 \cdot i^3 = (1)(1)(1)(-i) = -i$$

Although we will not do so, it is possible to prove that all the algebraic properties (such as commutativity and associativity) of real numbers remain true for addition and multiplication of complex numbers. Thus all the algebraic formulas we derived for real numbers hold for complex numbers. For example,

$$(v + w)^3 = v^3 + 3v^2 w + 3vw^2 + w^3 \qquad (6)$$

for any complex numbers v and w.

Example 4. Let $z = 2 - 3i$. Write z^3 in the form $a + bi$.

Solution. We use (6) with $v = 2$ and $w = -3i$:

$$z^3 = (2 - 3i)^3$$
$$= 2^3 + 3(2^2)(-3i) + 3(2)(-3i)^2 + (-3i)^3$$
$$= 8 - 36i + 54i^2 - 27i^3$$
$$= 8 - 36i - 54 + 27i$$
$$= -46 - 9i \quad \square$$

Square Roots of Negative Numbers

Since $i^2 = -1$, we may think of i as the square root of -1. But observe that

$$(-i)^2 = i^2 = -1$$

so that we could just as well think of $-i$ as the square root of -1. Similarly, if a is any positive real number, then

$$(\sqrt{a}\,i)^2 = -a \quad \text{and} \quad (-\sqrt{a}\,i)^2 = -a \tag{7}$$

We call $\sqrt{a}\,i$ the *principal square root* of $-a$ and write

$$\sqrt{-a} = \sqrt{a}\,i \tag{8}$$

Thus

$$\sqrt{-1} = i, \quad \sqrt{-4} = \sqrt{4}\,i = 2i, \quad \text{and} \quad \sqrt{-27} = \sqrt{27}\,i = 3\sqrt{3}\,i$$

It follows from (8) that negative numbers as well as nonnegative numbers have square roots when complex numbers are admitted.

Caution: If a and b are positive real numbers, then

$$\sqrt{ab} = \sqrt{a}\,\sqrt{b}$$

However, this equation is *not* valid if a and b are negative numbers. Indeed,

$$\sqrt{(-4)(-9)} = \sqrt{36} = 6$$

whereas by (8)

$$\sqrt{-4}\,\sqrt{-9} = (\sqrt{4}\,i)(\sqrt{9}\,i) = (2i)(3i) = 6i^2 = -6$$

Thus

$$\sqrt{(-4)(-9)} \neq \sqrt{(-4)}\,\sqrt{(-9)}$$

**Solutions of
a Quadratic Equation
with Real Coefficients**

Complex numbers were introduced to afford solutions of polynomial equations. We begin with equations of the form

$$x^2 + c = 0 \quad \text{with } c > 0$$

which are related to $x^2 + 1 = 0$ mentioned at the outset of this section and whose solutions are obtained by the use of (7).

Example 5. Find the solutions of $x^2 + 9 = 0$.

Solution. The given equation is equivalent to

$$x^2 = -9$$

By (7) with $a = 9$, the solutions of the given equation are $3i$ and $-3i$. □

Next we turn to quadratic equations of the form

$$ax^2 + bx + c = 0$$

where a, b, and c are real numbers and $a \neq 0$. The solutions can be obtained by the quadratic formula

$$x = \frac{-b \pm \sqrt{b^2 - 4ac}}{2a} \tag{9}$$

If $b^2 - 4ac \geq 0$, this is just the formula we had in Section 2.3, and the solutions are real numbers. However, if $b^2 - 4ac < 0$, then $\sqrt{b^2 - 4ac}$ is the square root of the negative number $b^2 - 4ac$, and the solutions are complex numbers.

Example 6. Find the solutions of $x^2 + 6x + 13 = 0$.

Solution. The solutions are given by (9):

$$x = \frac{-6 \pm \sqrt{(6)^2 - 4(1)(13)}}{2(1)} = \frac{-6 \pm \sqrt{-16}}{2}$$

By (8) with $a = 16$, we have $\sqrt{-16} = 4i$, so that

$$x = \frac{-6 \pm 4i}{2} = -3 \pm 2i$$

Thus there are two solutions, $-3 + 2i$ and $-3 - 2i$. □

Example 7. Find the solutions of $3x^2 - 4x + 6 = 0$.

Solution. By (9) and (8), the solutions are given by

$$x = \frac{-(-4) \pm \sqrt{(-4)^2 - 4(3)(6)}}{2(3)}$$

$$= \frac{4 \pm \sqrt{-56}}{6} = \frac{4 \pm \sqrt{56}i}{6} = \frac{4 \pm \sqrt{4 \cdot 14}i}{6}$$

$$= \frac{4 \pm 2\sqrt{14}i}{6} = \frac{2}{3} \pm \frac{1}{3}\sqrt{14}i$$

Therefore there are two solutions, $\frac{2}{3} + \frac{1}{3}\sqrt{14}i$ and $\frac{2}{3} - \frac{1}{3}\sqrt{14}i$.   □

As before, you should check proposed solutions to validate them. For the solutions obtained in Example 6, the check would be as follows:

$$(-3 + 2i)^2 + 6(-3 + 2i) + 13$$
$$= (9 - 12i - 4) + (-18 + 12i) + 13 = 0$$
$$(-3 - 2i)^2 + 6(-3 - 2i) + 13$$
$$= (9 + 12i - 4) + (-18 - 12i) + 13 = 0$$

EXERCISES 7.1

In Exercises 1–18, perform the indicated operation. Express your answer in the form $a + bi$.

1. $(2 + 3i) + (4 + 6i)$
2. $(5 - 7i) + (-1 + 6i)$
3. $(\frac{1}{2} + \frac{2}{3}i) + (\frac{1}{4} - \frac{1}{3}i)$
4. $2i + (-8 - 7i)$
5. $5 + (-1 + 3i)$
6. $(9 - 7i) + (9 + 7i)$
7. $(-3 + \frac{1}{4}i) + (3 + \frac{1}{2}i)$
8. $(2 - 3i) - (4 + 5i)$
9. $(4 - 5i) - (6 - 5i)$
10. $2i - (3 - 9i)$
11. $-3i(\frac{1}{2} - 4i)$
12. $6(-\frac{5}{4} + \frac{2}{3}i)$
13. $i(-3i)(-\frac{5}{8}i)$
14. $(1 + i)(1 - i)$
15. $(-3 - 2i)(5 - 4i)$
16. $(\frac{2}{3} - \frac{1}{6}i)(3 - i)$
17. $(\sqrt{2} + i)(1 + \sqrt{2}i)$
18. $\left(c + \frac{1}{c}i\right)\left(\frac{1}{c} + ci\right)$, where c is a real number

In Exercises 19–32, perform the indicated operations. Express your answer in the form $a + bi$.

19. i^{17}
20. $(-i)^{23}$
21. $(-i)^{-54}$
22. $(1 + i)^4$
23. $(-1 + 3i)^3$
24. $(\frac{1}{2} - \frac{1}{3}i)^3$
25. $\left(\frac{\sqrt{2}}{2} + \frac{\sqrt{2}}{2}i\right)^2$
26. $\left(\frac{\sqrt{2}}{2} - \frac{\sqrt{2}}{2}i\right)^2$
27. $(1 + i)(1 - i)^3$
28. $(1 + i)^2 + (1 - i)^2$

29. $(3 - 2i)(1 - 4i) - 2i(1 - 5i)^2$ 30. $i^6(3 - 2i)$

31. $(i^{25} + i^7)i^3$ 32. $(i^{-15} + 2i^{-9} - 3i^{10})i^3$

In Exercises 33–40, evaluate the given expression. Express your answer in the form $a + bi$.

33. $\sqrt{-16}$ 34. $\sqrt{-49}$

35. $\sqrt{-15}$ 36. $\sqrt{(1 + i)(i - 1)}$

37. $\sqrt{i^{30}}$ 38. $\sqrt{-49} - \sqrt{-64}$

39. $(3 - \sqrt{-4})(2 + \sqrt{-9})$ 40. $(4 + \sqrt{-3})(2 + \sqrt{-12})$

In Exercises 41–52, find all solutions of the given equation.

41. $x^2 + 25 = 0$ 42. $x^2 + 100 = 0$

43. $x^2 + 48 = 0$ 44. $2x^2 + 2x + 5 = 0$

45. $-\frac{1}{2}x^2 + 5x - 25 = 0$ 46. $x^2 - 3x + 6 = 0$

47. $x^2 + 13 = 4x$ 48. $x^2 + x + 1 = 0$

49. $x + \dfrac{2}{x} = 2$ 50. $2x(1 - x) = 3$

51. $x^3 - 8 = 0$ (*Hint:* $x^3 - 8 = (x - 2)(x^2 + 2x + 4)$.)

52. $x^3 + 8 = 0$ (*Hint:* $x^3 + 8 = (x + 2)(x^2 - 2x + 4)$.)

In Exercises 53–55, find all complex numbers $x + yi$ that satisfy the given equation.

53. $-1 + (x + 2)i = y + 2 + 5i$

54. $2 + (x + y)i = y + 2xi$

55. $2 + 3yi = x^2 + x + (x + y)i$

56. By first expressing $(1 + i)^2$ in the form $a + bi$, calculate
    a. $(1 + i)^6$ b. $(1 + i)^9$ c. $(1 + i)^{10}$

57. Find all complex numbers z that satisfy $z^2 = i$. (*Hint:* Let $z = a + bi$ and determine a and b.)

58. Let a and b be real numbers.
    a. Show that $(a + bi)(a - bi) = a^2 + b^2$.
    b. Suppose that $a^2 + b^2 \neq 0$. Using part (a), find a complex number $c + di$ such that $(a + bi)(c + di) = 1$.

59. Show that if $(a + bi)(c + di) = 0$, then either $a + bi = 0$ or $c + di = 0$.   (*Hint:*   If $(a + bi)(c + di) = 0$,   then $(a + bi)(a - bi)(c + di)(c - di) = 0$. Show that the left side is equal to $(a^2 + b^2)(c^2 + d^2)$. Then conclude that either $a = b = 0$ or $c = d = 0$.)

60. a. Simplify $i + i^2 + i^3 + i^4 + i^5 + i^6 + i^7$.
    b. Simplify $i + i^2 + i^3 + \cdots + i^{41}$.

61. Let $a, b,$ and c be real numbers with $a \neq 0$.
    a. Show that the sum of the solutions of the equation $ax^2 + bx + c = 0$ is $-b/a$.
    b. Show that the product of the solutions of the equation $ax^2 + bx + c = 0$ is c/a.

62. Determine the complex numbers z for which z^2 is
    a. real
    b. pure imaginary

7.2

CONJUGATES AND DIVISION OF COMPLEX NUMBERS

Closely related to any complex number $a + bi$ is its ***conjugate*** $a - bi$, often denoted by $\overline{a + bi}$. For example,

$$\overline{2 + 7i} = 2 - 7i \quad \text{and} \quad \overline{4 - \frac{1}{3}i} = 4 + \frac{1}{3}i$$

It follows from the definition of the conjugate that a real number is equal to its conjugate. Thus

$$\overline{-5} = \overline{-5 + 0i} = -5 - 0i = -5$$

It also follows that the conjugate of a pure imaginary number is its negative. For example,

$$\overline{-6i} = 6i$$

The conjugate is important because of its role in the definition of the multiplicative inverse and the quotient of complex numbers. To define the multiplicative inverse, we observe first that

$$(a + bi)(a - bi) = [a \cdot a - b(-b)] + [a(-b) + ba]i$$

which condenses to

$$(a + bi)(a - bi) = a^2 + b^2 \tag{1}$$

For example,

$$(6 + i)(6 - i) = 6^2 + 1^2 = 37$$

Since a and b are real numbers, (1) tells us that the product of any complex number and its conjugate is a real number (and is even nonnegative).

Suppose $a + bi \neq 0$, so that $a^2 + b^2 \neq 0$. Then by (1),

$$(a + bi)\frac{1}{a^2 + b^2}(a - bi) = 1$$

This means that every nonzero complex number $a + bi$ has a multiplicative inverse. We will denote the inverse by $1/(a + bi)$ (or $(a + bi)^{-1}$). Thus

$$\frac{1}{a + bi} = \frac{1}{a^2 + b^2}(a - bi) \tag{2}$$

If $c + di \neq 0$, we define the ***quotient*** $\dfrac{a + bi}{c + di}$ by the equation

$$\frac{a + bi}{c + di} = (a + bi)\frac{1}{c + di} \tag{3}$$

It is not necessary to memorize formulas (2) or (3). In fact, since

$$\frac{1}{a + bi} = \frac{1}{a + bi} \cdot \frac{a - bi}{a - bi} \tag{4}$$

we can find $1/(a + bi)$ by multiplying both numerator and denominator by the conjugate $a - bi$ and then simplifying. Similarly,

$$\frac{a + bi}{c + di} = \frac{a + bi}{c + di} \cdot \frac{c - di}{c - di} \tag{5}$$

which means that to evaluate a quotient, we need only multiply both numerator and denominator of the quotient by the conjugate of the denominator and then simplify.

Example 1. Write the following expressions in the form $a + bi$.

   a. $\dfrac{1}{4 + i}$ b. $(1 - i)^{-1}$

Solution. We solve both parts by (4):

   a. $\dfrac{1}{4 + i} = \dfrac{1}{4 + i} \cdot \dfrac{4 - i}{4 - i} = \dfrac{1}{4^2 + 1^2}(4 - i)$

          $= \dfrac{1}{17}(4 - i) = \dfrac{4}{17} - \dfrac{1}{17}i$

   b. $(1 - i)^{-1} = \dfrac{1}{1 - i} = \dfrac{1}{1 - i} \cdot \dfrac{1 + i}{1 + i}$

          $= \dfrac{1}{1^2 + 1^2}(1 + i) = \dfrac{1}{2}(1 + i)$

          $= \dfrac{1}{2} + \dfrac{1}{2}i$ $\square$

Example 2. Write $\dfrac{5 + i}{-2 - 3i}$ in the form $a + bi$.

Solution. By (5),

$$\frac{5 + i}{-2 - 3i} = \frac{5 + i}{-2 - 3i} \cdot \frac{-2 + 3i}{-2 + 3i} = \frac{-10 + 15i - 2i - 3}{4 + 9}$$

$$= \frac{-13 + 13i}{13} = -1 + i \quad \square$$

If a complex number is denoted by z, then the conjugate is denoted by $\bar{z}$. For later reference we list some of the properties of conjugation, most of which can be proved directly from the definition of conjugate.

> i. $\overline{\overline{z}} = z$
> ii. $\overline{z + w} = \overline{z} + \overline{w}$
> iii. $\overline{zw} = \overline{z}\,\overline{w}$
> iv. $\overline{z^n} = \overline{z}^n$ for any positive integer n.
> v. If $z = a + bi$, then $a = \dfrac{1}{2}(z + \overline{z})$ and $b = \dfrac{1}{2i}(z - \overline{z})$.
> vi. $\overline{z} = z$ if and only if z is a real number.

It follows from (i) that z is the conjugate of $\overline{z}$, so that z and $\overline{z}$ are conjugates of one another. As a consequence of (v),

> $z + \overline{z}$ is real for any complex number z (6)

It is easy to see that this is true by looking at an example. Suppose that

$$z = 2 + 3i$$

Then

$$z + \overline{z} = (2 + 3i) + (2 - 3i) = (2 + 2) + (3 - 3)i = 4$$

which indeed is a real number. As you can see, the imaginary part 3 of z is canceled by the imaginary part -3 of $\overline{z}$.

We close the section with an observation concerning conjugates. By (9) of Section 7.1, the solutions x_1 and x_2 of a quadratic equation $ax^2 + bx + c = 0$, where a, b, and c are real numbers and $a \neq 0$, are given by

$$x_1 = \frac{-b + \sqrt{b^2 - 4ac}}{2a} = -\frac{b}{2a} + \frac{1}{2a}\sqrt{b^2 - 4ac}$$

and

$$x_2 = \frac{-b - \sqrt{b^2 - 4ac}}{2a} = -\frac{b}{2a} - \frac{1}{2a}\sqrt{b^2 - 4ac}$$

If $b^2 - 4ac < 0$, then $\dfrac{1}{2a}\sqrt{b^2 - 4ac}$ is the pure imaginary number $\dfrac{1}{2a}\sqrt{-(b^2 - 4ac)}\,i$, so that if

$$A = -\frac{b}{2a} \quad \text{and} \quad B = \frac{1}{2a}\sqrt{-(b^2 - 4ac)}$$

then

$$x_1 = A + Bi \quad \text{and} \quad x_2 = A - Bi$$

with A and B real numbers. It is apparent that x_2 is the conjugate of x_1. Thus if $b^2 - 4ac < 0$, then the solutions of $ax^2 + bx + c = 0$ are conjugates of one another. For example, for the equation $x^2 + 6x + 13 = 0$ we have $b^2 - 4ac = 6^2 - 4(1)(13) < 0$, and the solutions are the conjugates $-3 + 2i$ and $-3 - 2i$, as we found in Example 6 of Section 7.1.

EXERCISES 7.2

In Exercises 1–6, find the conjugate of the given complex number.

1. $7 + 5i$
2. $-3 - \frac{1}{2}i$
3. $6 - 19i$
4. $-13i$
5. 23
6. -23

In Exercises 7–26, write the given expression in the form $a + bi$.

7. $\overline{6 - 2i}$

8. $\overline{-5i}$

9. $\overline{5 - \frac{1}{2}i}$

10. $\overline{\frac{1}{2} + \sqrt{2}\,i}$

11. $(1 + i)^{-1}$

12. $\dfrac{1}{4 - 5i}$

13. i^{-1}

14. $(-2 - i)^{-1}$

15. $\dfrac{1}{3 - 6i}$

16. $\dfrac{5}{3 - i}$

17. $\dfrac{1 + i}{1 - i}$

18. $\dfrac{2i}{3 + 4i}$

19. $\dfrac{2 - 3i}{-5i}$

20. $\dfrac{-1 + 3i}{5 - 2i}$

21. $\dfrac{(-1 - 2i)^2}{4 - 3i}$

22. $\dfrac{1 - i}{(1 - 2i)^2}$

23. $(5 + 12i)\overline{(5 + 12i)}$

24. $\overline{(3 - 2i)(-1 + 4i)}$

25. $2 + i + \dfrac{1}{2 + i}$

26. $\overline{\left(\dfrac{2 - 3i}{1 - \sqrt{2}\,i} \right)}$

27. Suppose x is a real number. Show that

$$\frac{1 - ix}{1 + ix} = \frac{1 - x^2}{1 + x^2} - \left(\frac{2x}{1 + x^2} \right)i$$

28. Write the following in the form $a + bi$.

   a. $\dfrac{(1 + i)^5}{(1 - i)^3}$

   b. $\dfrac{(1 - i)^3 - 1}{(1 + i)^3 + 1}$

29. a. Show that $3 - 4i$ is a solution of the equation $x^2 - 6x + 25 = 0$.

   b. Without using the quadratic formula, determine the other solution of the equation in part (a).

30. a. Show that $-2 - 5i$ is a solution of the equation $x^2 + 4x + 29 = 0$.

   b. Without using the quadratic formula, determine the other solution of the equation in part (a).

31. Show that $\overline{z + w} = \overline{z} + \overline{w}$.

32. Show that if $z = a + bi$, then

$$a = \frac{1}{2}(z + \overline{z}) \quad \text{and} \quad b = \frac{1}{2i}(z - \overline{z})$$

33. Prove that $\overline{z} = z$ if and only if z is a real number.

*34. Find all complex numbers z such that $\overline{z} = z^2$. (*Hint:* Write $z = a + bi$ and determine the possible values for a and b.)

*35. Show that if both $z + w$ and zw are real, then either $w = \overline{z}$ or both z and w are real numbers.

7.3

THE COMPLEX PLANE

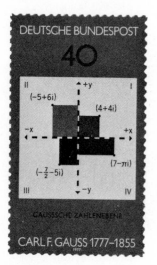

The complex number $a + bi$ can be identified with the ordered pair (a, b) of real numbers, which itself is identified with a point in the (Cartesian) plane. It follows that any complex number can be represented by a point in the plane (Figure 7.1). In this representation, the first coordinate of the point is the real part of the complex number, and the second coordinate is the imaginary part of the complex number. We use this identification to represent several complex numbers in Figure 7.2. The horizontal axis, which contains the complex numbers that are real, is called the ***real axis***. Analogously, the vertical axis contains the pure imaginary numbers and is called the ***imaginary axis***. When the plane is identified with complex numbers, it is referred to as the ***complex plane***. The complex plane is also called "the Argand diagram," after the French–Swiss mathematician Jean Robert Argand (1768–1822), and "the Gaussian plane," after the German mathematician Carl Friedrich Gauss (1777–1855). (In a postage stamp commemorating the two-hundredth year since the birth of Gauss, reproduced on this page, West Germany included the complex plane.)

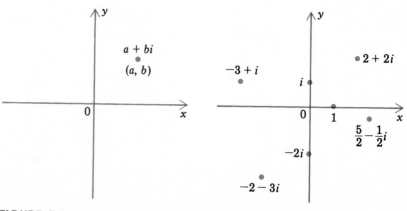

FIGURE 7.1   FIGURE 7.2

Absolute Value of a Complex Number

Let $a + bi$ be any complex number. Its **absolute value**, or **modulus**, written $|a + bi|$, is defined as the distance between the point (a, b) and the origin, namely, $\sqrt{a^2 + b^2}$:

$$|a + bi| = \sqrt{a^2 + b^2} \qquad (1)$$

(Figure 7.3). Thus if z is any complex number, then $|z|$ is the distance between z and the origin.

Example 1. Compute the absolute values of the following complex numbers.

a. i b. $2 - 5i$

Solution.

a. Since $i = 0 + 1 \cdot i$, (1) implies that

$$|i| = \sqrt{0^2 + 1^2} = \sqrt{1^2} = 1 \qquad (2)$$

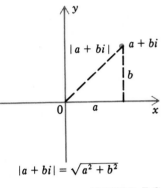

$$|a + bi| = \sqrt{a^2 + b^2}$$

FIGURE 7.3

b. By (1) we have

$$|2 - 5i| = \sqrt{2^2 + (-5)^2} = \sqrt{4 + 25} = \sqrt{29} \quad \square$$

The notion of absolute value of a complex number is consistent with absolute value as defined for real numbers in Section 1.2. Indeed, if $a + bi$ is real, so that $b = 0$, then

$$|a + bi| = |a + 0 \cdot i| = \sqrt{a^2 + 0^2} = \sqrt{a^2} = |a|$$

This means, for example, that

$$|-9| = 9$$

whether we think of -9 as a real number or as a complex number.

The following properties of absolute value hold for any complex numbers z and w and can be proved from the definition of absolute value:

$$|z| \geq 0 \qquad\qquad |zw| = |z||w|$$

$$|z| = 0 \quad \text{if and only if } z = 0 \qquad \left|\frac{z}{w}\right| = \frac{|z|}{|w|}$$

$$|z| = |-z| \qquad\qquad |z + w| \leq |z| + |w|$$

$$|z - w| = |w - z|$$

Example 2. Sketch the graph of $|z| = 2$.

Solution. Because $|z|$ is the distance between z and the origin, $|z| = 2$ means that the distance between z and the origin is 2. Therefore

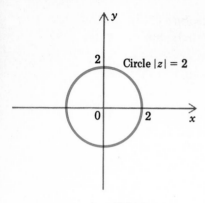

Circle $|z| = 2$

FIGURE 7.4

the collection of all such complex numbers z forms the circle of radius 2 centered at the origin. It is sketched in Figure 7.4. □

Now let $z = x + yi$ and $z_0 = a + bi$ be two complex numbers. Then

$$|z - z_0| = |(x + yi) - (a + bi)|$$
$$= |(x - a) + (y - b)i| = \sqrt{(x - a)^2 + (y - b)^2}$$

Therefore $|z - z_0|$ is the distance between the points (x, y) and (a, b) that represent z and z_0, respectively. It follows that if r is a positive real number and z_0 a fixed complex number, then the graph of the equation

$$|z - z_0| = r$$

is the circle of radius r with center at (a, b).

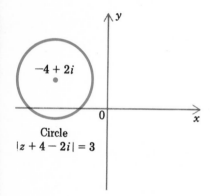

$-4 + 2i$

Circle
$|z + 4 - 2i| = 3$

FIGURE 7.5

Example 3. Sketch the graph of $|z + 4 - 2i| = 3$.

Solution. The graph is a circle of radius 3. Since

$$z + 4 - 2i = z - (-4 + 2i)$$

we find that the center of the circle is $-4 + 2i$. Now we are in a position to draw the graph (Figure 7.5). □

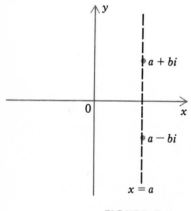

$a + bi$

$a - bi$

$x = a$

FIGURE 7.6

Conjugates have a geometric interpretation. Indeed, the real parts of $a + bi$ and $a - bi$ are the same (namely, a), so the two complex numbers both lie on the vertical line $x = a$. Since their imaginary parts are negatives of one another, the two conjugates are symmetric to one another with respect to the real axis (Figure 7.6). Therefore

$$\boxed{|\bar{z}| = |z|}$$

a fact that can be easily verified by algebraic considerations (see Exercise 25). Another important connection between the conjugate and the absolute value is the formula

$$\boxed{|z|^2 = z\bar{z}} \qquad (3)$$

To prove this formula, let $z = a + bi$. Then

$$z\bar{z} = (a + bi)(a - bi) = a^2 + b^2 = |z|^2$$

EXERCISES 7.3

In Exercises 1–13, compute the absolute value of the given number.

1. $-i$

2. -12

3. i^{13}

4. $3 + 4i$

5. $3 - 4i$

6. $-3 + 4i$

7. $1 - i$

8. $-5 - 12i$

9. $(1 + i)(-3 - 4i)$

10. $\dfrac{1}{2 - i}$

11. $\dfrac{5 - i}{-2 + 4i}$

12. $-3i(4 + i)(5 - 2i)^2$

13. $\dfrac{(1 - 2i)(3 - i)}{4i(3 + 4i)}$

14. Plot the points in the complex plane that correspond to the following complex numbers.

    a. $1 + i$ b. $3 - 4i$ c. $-2 + i$

    d. $-\dfrac{1}{2} - \dfrac{3}{2}i$ e. -3 f. $2i$

    g. $-3i$ h. 0

In Exercises 15–22, sketch the graph of the given equation.

15. $|z| = 1$

16. $|z| = 3$

17. $|z| = \sqrt{3}$

18. $|z| = \frac{1}{2}$

19. $|z - 1| = 1$

20. $|z - 5 - i| = 2$

21. $|z + 2 - 4i| = \frac{3}{2}$

22. $|z + 3 + 5i| = 4$

23. Show that $|z| = |-z|$.

24. Show that $|z - w| = |w - z|$.

25. Show that $|\bar{z}| = |z|$ by writing $z = a + bi$ and computing $|z|$ and $|\bar{z}|$.

26. If $\bar{z} = 1/z$, what can we say about $|z|$?

27. Let z and w be any complex numbers. Show that

$$|z - w|^2 + |z + w|^2 = 2(|z|^2 + |w|^2)$$

(*Hint:* Let $z = a + bi$ and $w = c + di$. Express each side of the given equation in terms of a, b, c, and d.)

*28. Suppose that z and w are distinct complex numbers with $|z| = 1$. Show that

$$\left| \frac{z - w}{1 - \bar{z}w} \right| = 1$$

(*Hint:* Multiply both numerator and denominator of the fraction by z.)

KEY TERMS

complex number
real part
imaginary part
conjugate

complex plane
absolute value (modulus) of a
   complex number

KEY FORMULAS

$i^2 = -1$

$(a + bi) + (c + di) = (a + c) + (b + d)i$

$(a + bi)(c + di) = (ac - bd) + (ad + bc)i$

$\overline{a + bi} = a - bi$

$|a + bi| = \sqrt{a^2 + b^2}$

REVIEW EXERCISES

In Exercises 1–19, perform the indicated operation, and express your answer in the form $a + bi$.

1. $(2 + 6i) + (3 - 3i)$

2. $(\sqrt{3} - i)(1 - \sqrt{3}i)$

3. $(4 - \frac{1}{2}i)(4 + \frac{1}{2}i)$

4. i^{-29}

5. $(2 - 3i)^4 \, i^3$

6. $(-3 + 3i)^6$

7. $(\sqrt{2} - \sqrt{6}i)^{-5}$

8. $(1 + i)^3 + (1 - i)^3$

9. $\overline{-3 + 4i}$

10. $\overline{-0.4 - 0.01i}$

11. $\overline{(1 - i)^5}$

12. $\dfrac{1}{-5i - 1}$

13. $\dfrac{-i}{2i - 3}$

14. $\dfrac{4 - 2i}{1 + 5i}$

15. $(\sqrt{2} - \sqrt{3}i)\dfrac{1}{\overline{\sqrt{2} + \sqrt{3}i}}$

16. $\overline{\left(\dfrac{-2 - i}{7 + 3i}\right)}$

17. $|8 - 6i|$

18. $\left|\dfrac{i}{3i - 1}\right|$

19. $\left|\dfrac{(-1 - 2i)(3 + i)}{2i(3 - i)}\right|$

20. Evaluate the given square root.
   a. $\sqrt{-36}$ b. $\sqrt{-(4 - 2i)(4 + 2i)}$ c. $\sqrt{i^{-14}}$

In Exercises 21–30, find all solutions (if any) of the given equation.

21. $x^2 + 49 = 0$

22. $x^2 + 32 = 0$

23. $x^2 - x + 1 = 0$

24. $3x^2 + 2 = 4x$

25. $\dfrac{1}{2} - \dfrac{1}{x} + \dfrac{3}{x^2} = 0$

26. $x + i = \dfrac{1}{x - i}$

27. $x = \dfrac{1}{x + i}$

28. $(x + 1)(x^2 + 1) = 0$

29. $\dfrac{1}{x + i} - \dfrac{1}{x - i} = 0$

30. $\dfrac{x}{x^2 + ix} = \dfrac{x}{x^2 + 1}$

In Exercises 31–34, sketch the graph of the given equation.

31. $|z| = \frac{2}{3}$

32. $|z - 2| = \frac{1}{2}$

33. $|z - 1 - i| = 3$

34. $|z + 4 + 6i| = \frac{5}{2}$

35. Describe geometrically the complex numbers $z = x + iy$ that satisfy the equation $\bar{z} = -iz$.

36. Find all complex numbers $z = x + iy$ that satisfy the equation

$$iz + 1 = z - 1$$

ROOTS OF POLYNOMIALS

8

In the preceding chapter we introduced the imaginary number i, which is a solution of the equation

$$x^2 + 1 = 0$$

More generally, we saw that any quadratic equation

$$ax^2 + bx + c = 0$$

has (possibly complex) solutions if a, b, and c are real numbers with $a \neq 0$. Now we turn to general polynomial equations, that is, equations of the form

$$a_n x^n + a_{n-1} x^{n-1} + \cdots + a_1 x + a_0 = 0 \tag{1}$$

where the coefficients a_n, a_{n-1}, . . . , a_1, a_0 are constants, real or complex. Any solution of (1) is called a *root* (or *zero*) of the associated polynomial $a_n x^n + a_{n-1} x^{n-1} + \cdots + a_1 x + a_0$. The results appearing in this chapter can be given either in terms of solutions of polynomial equations or in terms of roots of polynomials. Because it is simpler to give the results in terms of roots of polynomials, the main subject of this chapter is roots of polynomials.

We will see that every nonconstant polynomial has a (possibly complex) root, even if we allow the coefficients to be complex numbers. This result, known as the Fundamental Theorem of Algebra, is one of the deepest and most remarkable theorems in mathematics. We will also devise methods for actually finding roots in case the coefficients of the polynomial are integers.

8.1

DIVISION OF POLYNOMIALS

In Section 4.4 we studied rational functions, which are quotients of polynomials. Now we will see what the result is when we actually "divide" one polynomial by another.

Recall that a polynomial has the form

$$a_n x^n + a_{n-1} x^{n-1} + \cdots + a_1 x + a_0 \qquad (1)$$

where a_n, a_{n-1}, . . . , a_1, and a_0 are constants that henceforth are allowed to be either real or complex numbers. As before, if $a_n \neq 0$, then n is the degree of the polynomial.

The Division Algorithm

Let $f(x)$ and $g(x)$ be polynomials. We say that $g(x)$ *divides* $f(x)$ (or that $g(x)$ is a *factor* of $f(x)$, or that $f(x)$ is *divisible* by $g(x)$) if there is a polynomial $q(x)$ such that

$$f(x) = g(x)q(x)$$

For example, since

$$x^2 - 5x + 4 = (x - 1)(x - 4) \qquad (2)$$

it follows that $x^2 - 5x + 4$ is divisible both by $x - 1$ and by $x - 4$. Equation (2) is equivalent to the equation

$$\frac{x^2 - 5x + 4}{x - 1} = x - 4$$

and the division of $x^2 - 5x + 4$ by $x - 1$ can be carried out much as one divides integers by long division:

$$
\begin{array}{r}
x - 4 \\
x - 1\overline{\smash{\big)}\,x^2 - 5x + 4} \\
\underline{x^2 - x} \\
-4x + 4 \\
\underline{-4x + 4} \\
0
\end{array}
$$

This method of dividing one polynomial by another applies to other polynomials as well.

Now assume that $f(x)$ and $g(x)$ are polynomials such that $g(x) \neq 0$, where $f(x)$ is not necessarily divisible by $g(x)$. We can perform a long division patterned after our division above. We illustrate by dividing $2x^4 - 5x^3 + 5x^2 - 3x + 4$ by $x^2 - x$:

$$
\begin{array}{r}
2x^2 - 3x + 2 \\
x^2 - x\overline{\smash{\big)}\,2x^4 - 5x^3 + 5x^2 - 3x + 4} \\
\underline{2x^4 - 2x^3} \\
-3x^3 + 5x^2 \\
\underline{-3x^3 + 3x^2} \\
2x^2 - 3x \\
\underline{2x^2 - 2x} \\
-x + 4
\end{array}
$$

The result of this division can be expressed by the equation

$$2x^4 - 5x^3 + 5x^2 - 3x + 4 = (x^2 - x)(2x^2 - 3x + 2) + (-x + 4) \quad (3)$$

The polynomial $2x^2 - 3x + 2$ is called the quotient of $2x^4 - 5x^3 + 5x^2 - 3x + 4$ by the divisor $x^2 - x$, and $-x + 4$ is called the remainder. Observe that the degree of the remainder $-x + 4$ is less than the degree of the divisor $x^2 - x$. In general, when we use long division to divide a polynomial $f(x)$ by a nonzero polynomial $g(x)$, we continue the procedure until the remainder is either 0 or has degree less than the degree of the divisor $g(x)$. It is possible to prove that such a division is always possible. We now state the result formally, but without proof.

THEOREM 8.1   *The Division Algorithm.* Let $f(x)$ and $g(x)$ be polynomials with $g(x) \neq 0$. Then there are unique polynomials $q(x)$ and $r(x)$ with either $r(x) = 0$ or deg $r(x) <$ deg $g(x)$ such that

$$f(x) = g(x)q(x) + r(x)$$

The polynomial $q(x)$ is called the **quotient**, and the polynomial $r(x)$ is called the **remainder**. When we find $q(x)$ and $r(x)$, we say that we **divide $f(x)$ by $g(x)$**. If $f(x)$ and $g(x)$ have real coefficients, then so do $q(x)$ and $r(x)$.

Example 1. Find the quotient and remainder when $2x^4 - 5x^3 + 5x^2 - 3x + 4$ is divided by $x^2 - x$.

Solution. Let

$$f(x) = 2x^4 - 5x^3 + 5x^2 - 3x + 4 \quad \text{and} \quad g(x) = x^2 - x$$

From (3) we see that

$$q(x) = 2x^2 - 3x + 2 \quad \text{and} \quad r(x) = -x + 4 \quad \square$$

It can happen that either $q(x)$ or $r(x)$ is 0. For example, if $f(x) = 2x$ and $g(x) = x^2 - 1$, then

$$\overbrace{2x}^{f(x)} = \overbrace{(x^2 - 1)}^{g(x)}\overbrace{0}^{q(x)} + \overbrace{2x}^{r(x)}$$

so in this case, $q(x) = 0$. In general, if the degree of $f(x)$ is less than the degree of $g(x)$, then dividing $f(x)$ by $g(x)$ yields the quotient $q(x) = 0$. In contrast, if $f(x) = x^3 - 1$ and $g(x) = x - 1$, then

$$\overbrace{x^3 - 1}^{f(x)} = \overbrace{(x - 1)}^{g(x)}\overbrace{(x^2 + x + 1)}^{q(x)}$$

It follows that $r(x) = 0$. The remainder is 0 because $g(x)$ is a factor of $f(x)$.

**Synthetic
Division**

As we will see later, one frequently needs to divide a polynomial by a second polynomial of the special form $x - c$. A method called synthetic division normally reduces the writing involved in the long division in such a case. Consider dividing $2x^4 - 10x^3 + 38x - 10$ by $x - 3$, with the long division performed as above:

$$
\begin{array}{r}
2x^3 - 4x^2 - 12x + 2 \\
x - 3 \overline{\smash{)}2x^4 - 10x^3 + 38x - 10} \\
\underline{2x^4 - 6x^3} \\
-4x^3 \\
\underline{-4x^3 + 12x^2} \\
-12x^2 + 38x \\
\underline{-12x^2 + 36x} \\
2x - 10 \\
\underline{2x - 6} \\
-4
\end{array}
$$

The result is

$$2x^4 - 10x^3 + 38x - 10 = (x - 3)\overbrace{(2x^3 - 4x^2 - 12x + 2)}^{\text{quotient}} \overbrace{- 4}^{\text{remainder}}$$

Observe first of all that the numbers in color in each vertical column are duplicates of one another. Notice next that from the dividend downward, like powers of the variable x appear in vertical columns. Because of these two observations, we can condense the division as follows by eliminating the repetitions, collapsing the lower lines all onto a second line, and putting the quotient at the bottom along with the remainder:

$$
\begin{array}{r|rrrrr}
-3\big| & 2 & -10 & 0 & 38 & -10 \\
& & -6 & 12 & 36 & -6 \\
\hline
& 2 & -4 & -12 & 2 & \big|-4
\end{array}
$$

In the display we have separated the divisor number, -3, from the dividend numbers by the symbol $\lrcorner$, and the remainder -4 from the quotient numbers by the symbol $\llcorner$. Moreover, we placed a 0 in the first row because the polynomial $2x^4 - 10x^3 + 38x - 10$, with no x^2 term, is equal to the polynomial $2x^4 - 10x^3 + 0x^2 + 38x - 10$. Notice also that the coefficients in the bottom row of the display correspond to powers of x that are one less than the powers to which the coefficients in the top row of the display correspond.

There is one further simplification we can make. Notice that each number (except the first) in the third row of the condensed display is obtained from the numbers directly above it by *subtracting* the number in the second row from the number in the first row. We can achieve the same third row by replacing each number in the second row by its negative and *adding* the resulting number to the number directly above it in the first row. For later convenience we also replace -3 by its negative, the number 3. (More generally, if the divisor is $x - c$, we would put c at the left of the display.) When we make these replacements, our condensed display becomes

$$
\begin{array}{r|rrrrr}
3\big| & 2 & -10 & 0 & 38 & -10 \\
& & 6 & -12 & -36 & 6 \\
\hline
& 2 & -4 & -12 & 2 & \big|-4
\end{array}
$$

This final display is an example of **synthetic division**, wherein powers of the variable have disappeared and the entire division appears in three rows.

On the facing page the method of synthetic division for dividing a polynomial $f(x)$ by $x - c$ is outlined and illustrated with the division of $2x^4 - 2x^3 - 7x^2 - 57$ by $x - 3$. The solutions of Examples 2–4 are based on the outline.

Example 2. Use synthetic division to find the quotient and remainder when $3x^4 - 8x^3 + x^2 - 5x + 9$ is divided by $x - 2$.

Solution. Following the procedure outlined, we have:

$$
\begin{array}{r|rrrrr}
2\big| & 3 & -8 & 1 & -5 & 9 \\
& & 6 & -4 & -6 & -22 \\
\hline
& 3 & -2 & -3 & -11 & \big|-13
\end{array}
$$

Therefore the quotient is $3x^3 - 2x^2 - 3x - 11$, and the remainder is -13. $\square$

$$2 \quad -2 \quad -7 \quad 0 \quad -57$$

1. Write the coefficients of $f(x)$ in a horizontal line. Include a 0 for each missing power of x.

$$3 \underline{\rfloor \quad 2 \quad -2 \quad -7 \quad 0 \quad -57}$$

2. Write the number c to the left of the coefficients of $f(x)$. Separate c from the coefficients by the symbol $\rfloor$, and draw a horizontal line as indicated, well below the coefficients.

$$3 \underline{\rfloor \quad 2 \quad -2 \quad -7 \quad 0 \quad -57}$$
$$\quad \quad \downarrow$$
$$\quad \quad 2$$

3. Bring down the first coefficient of $f(x)$ as indicated.

$$3 \underline{\rfloor \quad 2 \quad -2 \quad -7 \quad 0 \quad -57}$$
$$\quad \quad \quad \quad 6$$
$$\quad \quad 2$$

4. Multiply the number below the line by c, and place the result under the second coefficient of $f(x)$.

$$3 \underline{\rfloor \quad 2 \quad -2 \quad -7 \quad 0 \quad -57}$$
$$\quad \quad \quad \quad 6$$
$$\quad \quad 2 \quad 4$$

5. Add the numbers in the second column and place the result under the line.

$$3 \underline{\rfloor \quad 2 \quad -2 \quad -7 \quad 0 \quad -57}$$
$$\quad \quad \quad \quad 6 \quad 12$$
$$\quad \quad 2 \quad 4$$

6. Multiply the number below the line in the second column by c and place the result below the third coefficient of $f(x)$.

$$3 \underline{\rfloor \quad 2 \quad -2 \quad -7 \quad 0 \quad -57}$$
$$\quad \quad \quad \quad 6 \quad 12$$
$$\quad \quad 2 \quad 4 \quad 5$$

7. Add the numbers in the third column and place the result below the line.

$$3 \underline{\rfloor \quad 2 \quad -2 \quad -7 \quad 0 \quad -57}$$
$$\quad \quad \quad \quad 6 \quad 12 \quad 15 \quad 45$$
$$\quad \quad 2 \quad 4 \quad 5 \quad 15 \underline{\lfloor -12}$$

8. Continue in this fashion until the last column has been completed. Place the symbol $\lfloor$ around the last number of the bottom row to isolate the remainder.

9. Read off the result. The last number on the right below the line is the remainder. The other numbers are the coefficients of the quotient. The degree of the quotient is one less than the degree of $f(x)$.

Quotient:
$$2x^3 + 4x^2 + 5x + 15$$
Remainder: -12
Result of Synthetic Division:
$$2x^4 - 2x^3 - 7x^2 - 57$$
$$= (x - 3)(2x^3 + 4x^2 + 5x + 15) - 12$$

Example 3.   Use synthetic division to show that $-x^3 - x^2 + 7x - 20$ is divisible by $x + 4$.

Solution.   In performing the synthetic division we use the fact that $x + 4 = x - (-4)$. Then we have

$$
\begin{array}{r|rrrr}
-4 & -1 & -1 & 7 & -20 \\
   &   & 4 & -12 & 20 \\
\hline
   & -1 & 3 & -5 & \underline{0}
\end{array}
$$

From the division we see that

$$
\begin{array}{l}
-x^3 - x^2 + 7x - 20 \qquad \overbrace{}^{\text{quotient}} \quad \overbrace{}^{\text{remainder}} \\
\quad = (x + 4)\overbrace{(-x^2 + 3x - 5)}^{} + \overbrace{0}^{} = (x + 4)(-x^2 + 3x - 5)
\end{array}
$$

so that $-x^3 - x^2 + 7x - 20$ is divisible by $x + 4$. $\square$

> **Caution:** Long division can be used with a divisor of any degree. However, synthetic division can only be used when the divisor is a polynomial of the form $x - c$.

In our final example, we use synthetic division with polynomials whose coefficients include complex as well as real numbers.

Example 4. Use synthetic division to divide $ix^3 + 3x^2 + (-4 + i)x + 1 + 9i$ by $x - i$.

Solution. Following the same procedure as above, we have

$$
\begin{array}{r|rrrr}
i & i & 3 & -4 + i & 1 + 9i \\
  &   & -1 & 2i & -3 - 4i \\
\hline
  & i & 2 & -4 + 3i & \underline{-2 + 5i}
\end{array}
$$

Therefore the quotient is $ix^2 + 2x - 4 + 3i$, and the remainder is $-2 + 5i$. $\square$

EXERCISES 8.1

In Exercises 1–6, perform the indicated long division.

1. $x^2 + 1 \overline{\smash{\big)}\, 2x^4 + 3x^3 - 4x + 7}$
2. $3x - 2 \overline{\smash{\big)}\, -9x^3 + 12x^2 + 2x + 12}$
3. $x^2 + 1 \overline{\smash{\big)}\, x^4 + 1}$
4. $x^2 + \sqrt{2}x + 1 \overline{\smash{\big)}\, x^4 + 1}$
5. $x^3 + x + 1 \overline{\smash{\big)}\, 3x^4 - 2x^3 + 3x^2 + x - 1}$
6. $x^2 + 3 \overline{\smash{\big)}\, x^3 + 4x^2 - 7}$

In Exercises 7–18, use synthetic division to find the quotient and the remainder when the first polynomial is divided by the second.

7. $3x^2 - 2x + 7$; $x + 1$
8. $-x^3 + 3x^2 - 2x$; $x - 2$
9. $2x^3 + 3x^2 + 5$; $x - \frac{1}{2}$
10. $x^3 + 3x^2 - 9x + 5$; $x + 5$
11. $6x^3 - x^2 + 2x + 1$; $x + \frac{1}{3}$
12. $2x^4 - 3x^2 + 2x - 5$; $x - 3$
13. $5x^5 + 16x^4 - 15x^3 + x^2 + 19x$; $x + 4$
14. $x^7 - x^3 + x$; $x - 1$
15. $x^7 - 1$; $x - 1$
16. $x^{10} - 1$; $x - 1$
17. $x^5 - ix + 1$; $x - i$
18. $x^4 + (1 - 3i)x^3 + (-5 - i)x^2 + (-6 + 7i)x - 1 + 8i$; $x + 1 - i$

In Exercises 19–30, use synthetic division to show that the first polynomial is a factor of the second.

19. $x - 3$; $2x^4 - 6x^3 - 3x^2 + 13x - 12$
20. $x + 2$; $x^6 + 6x^3 + 7x - 2$
21. $x - 1$; $3x^5 - 4x^3 - 6x + 7$
22. $x + 1$; $4x^4 + 3x^3 - 2x^2 + 2x + 3$
23. $x - 5$; $x^3 - 8x^2 + 17x - 10$
24. $x - \frac{1}{2}$; $8x^3 - 4x^2 - 2x + 1$
25. $x + 6$; $2x^3 + 9x^2 - 17x + 6$
26. $x + 2$; $x^5 + 2x^4 - x^3 - 2x^2 + 2x + 4$
27. $x + i$; $x^3 + 2x^2 + x + 2$
28. $x - i$; $x^4 + 3x^3 + 3x^2 + 3x + 2$
29. $x - i$; $x^5 + x^3$
30. $x - \frac{\sqrt{2}}{2} - \frac{\sqrt{2}}{2}i$; $x^2 - \sqrt{2}x + 1$

31. Use synthetic division to find the values of k for which $x - 2$ is a factor of $x^3 + k^2x^2 - 3kx + k^2 - 7$.
32. Use synthetic division to find the values of k for which $x + 2$ is a factor of $x^4 + k^2x + k - 10$.

8.2
THE REMAINDER AND FACTOR THEOREMS

When a polynomial $f(x)$ is divided by a first-degree polynomial $x - c$, the Division Algorithm yields

$$f(x) = (x - c)q(x) + r(x) \tag{1}$$

where $q(x)$ and $r(x)$ are polynomials and where either $r(x) = 0$ or $\deg r(x) < 1 = \deg (x - c)$. The fact that $r(x) = 0$ or $\deg r(x) < 1$ implies that $r(x)$ is a constant polynomial. It follows that for some constant b, $r(x) = b$, so that (1) becomes

$$f(x) = (x - c)q(x) + b \qquad (2)$$

Replacing x by c in (2), we obtain

$$f(c) = (c - c)q(c) + b = b$$

so that we can substitute $f(c)$ for b in (2) to obtain

$$f(x) = (x - c)q(x) + f(c) \qquad (3)$$

This proves the following theorem.

THEOREM 8.2   *The Remainder Theorem.* If a polynomial $f(x)$ is divided by $x - c$, then the remainder is $f(c)$.

Example 1. Let $f(x) = 3x^4 - 8x^3 + x^2 - 5x + 9$. Find the remainder when $f(x)$ is divided by $x - 2$.

Solution. Taking $c = 2$ in the Remainder Theorem, we find that the remainder is $f(2)$. But

$$f(2) = 3(2)^4 - 8(2)^3 + (2)^2 - 5(2) + 9 = -13$$

Therefore the remainder is -13.   □

Now we have found the remainder when $3x^4 - 8x^3 + x^2 - 5x + 9$ is divided by $x - 2$ in two different ways: by synthetic division (Example 2 in Section 8.1) and by the Remainder Theorem (Example 1 above). In each case the remainder is -13.

Example 2. Let $f(x) = ix^3 + (2 - i)x^2 - 3i + 2$. Find the remainder when $f(x)$ is divided by $x - 1 + i$.

Solution. We have $x - 1 + i = x - (1 - i)$. By the Remainder Theorem with $c = 1 - i$, the remainder is $f(1 - i)$. Since

$$f(1 - i) = i(1 - i)^3 + (2 - i)(1 - i)^2 - 3i + 2$$
$$= i(1 - 3i + 3i^2 - i^3) + (2 - i)(1 - 2i + i^2) - 3i + 2$$
$$= (i + 3 - 3i - 1) + (2 - 4i - 2 - i - 2 + i) - 3i + 2$$
$$= 2 - 9i$$

the remainder is $2 - 9i$.   □

The most important consequence of the Remainder Theorem is the Factor Theorem.

THEOREM 8.3

The Factor Theorem. Let $f(x)$ be any polynomial. Then $x - c$ is a factor of $f(x)$ if and only if $f(c) = 0$.

Proof

Suppose that $x - c$ is a factor of $f(x)$. Then

$$f(x) = (x - c)g(x)$$

for some polynomial $g(x)$. It follows that

$$f(c) = (c - c)g(c) = 0 \cdot g(c) = 0$$

Thus if $x - c$ is a factor of $f(x)$, then $f(c) = 0$. Conversely, assume that $f(c) = 0$. By (3),

$$f(x) = (x - c)q(x) + f(c) \tag{4}$$

for some polynomial $q(x)$. Since $f(c) = 0$ by assumption, (4) becomes

$$f(x) = (x - c)q(x)$$

so that $x - c$ is a factor of $f(x)$. ∎

Example 3. Show that $x - 1$ is a factor of $x^3 + x^2 - x - 1$, and then factor $x^3 + x^2 - x - 1$ into linear factors.

Solution. Let

$$f(x) = x^3 + x^2 - x - 1$$

Since

$$f(1) = 1^3 + 1^2 - 1 - 1 = 0$$

the Factor Theorem implies that $x - 1$ is a factor of $f(x)$. We use synthetic division to divide $f(x)$, that is, $x^3 + x^2 - x - 1$, by $x - 1$.

$$
\begin{array}{r|rrrr}
1 & 1 & 1 & -1 & -1 \\
  &   & 1 & 2 & 1 \\
\hline
  & 1 & 2 & 1 & \underline{0}
\end{array}
$$

Thus

$$x^3 + x^2 - x - 1 = (x - 1)(x^2 + 2x + 1)$$
$$= (x - 1)(x + 1)^2 \quad \square$$

Example 4. Show that $x + 2$ is not a factor of $-3x^3 + 2x + 4$.

Solution. Let

$$f(x) = -3x^3 + 2x + 4$$

By the Factor Theorem, $x + 2$, which is the same as $x - (-2)$, is a factor of $f(x)$ if and only if $f(-2) = 0$. But

$$f(-2) = (-3)(-2)^3 + 2(-2) + 4 = 24$$

so that $f(-2) \neq 0$. Therefore $x + 2$ is not a factor of $-3x^3 + 2x + 4$. □

Recall that a number c is a root (or zero) of a polynomial $f(x)$ if $f(c) = 0$.

Example 5. Show that 1 is a root of $x^5 - x^4 + x^3 - x^2 + x - 1$.

Solution. We need only show that if $x = 1$, then the value of the polynomial is 0. Indeed,

$$(1)^5 - (1)^4 + (1)^3 - (1)^2 + 1 - 1 = 0$$

Therefore 1 is a root of the given polynomial. □

The Factor Theorem implies that c is a root of $f(x)$ if and only if $x - c$ is a factor of $f(x)$.

Example 6. Find a polynomial of degree 3 that has roots 1, 2, and -3.

Solution. By the Factor Theorem, any polynomial that has roots 1, 2, and -3 must be divisible by $x - 1$, $x - 2$, and $x - (-3) = x + 3$. Such a polynomial is $(x - 1)(x - 2)(x + 3)$. Since

$$(x - 1)(x - 2)(x + 3) = (x^2 - 3x + 2)(x + 3) = x^3 - 7x + 6$$

it follows that the polynomial $x^3 - 7x + 6$ has the required properties. □

Factors
of $x^2 + a^2$

Let a be a nonzero real number, and let

$$f(x) = x^2 + a^2$$

If c is any real number, then

$$f(c) = c^2 + a^2 \neq 0$$

so that by the Factor Theorem, $f(x)$ has no factor of the form $x - c$. However, $f(x)$ does have the factors $x - ai$ and $x + ai$, since

$$f(ai) = (ai)^2 + a^2 = -a^2 + a^2 = 0$$

and

$$f(-ai) = (-ai)^2 + a^2 = -a^2 + a^2 = 0$$

Consequently for any nonzero real number a,

$$x^2 + a^2 = (x - ai)(x + ai)$$

For example,

$$x^2 + 1\ \ = (x - i)(x + i)$$
$$x^2 + 16 = (x - 4i)(x + 4i)$$
$$x^2 + 5\ \ = (x - \sqrt{5}\,i)(x + \sqrt{5}\,i)$$

EXERCISES 8.2

In Exercises 1–10, use the Remainder Theorem to find the remainder when the first polynomial is divided by the second.

1. $x^3 - 3x^2 + 2x + 1; x - 1$
2. $3x^4 - 6x^2 + 7; x + 1$
3. $\frac{1}{2}x^6 - 2x^3 - x + 7; x - 2$
4. $-x^5 + 7; x + 2$
5. $3x^4 - 4x^2 + 3x - 6; x - \sqrt{2}$
6. $-17x^3 + \pi x^2 - \sqrt{2}x + \frac{17}{32}; x$
7. $x^4 + a^2; x + a$
8. $x^4 - 2x^2 + 1 + i; x - i$
9. $x^2 + x + 1 - 3i; x - 1 + i$
10. $x^6 - 5x^3 + 1; x + i$

In Exercises 11–20, verify that the linear polynomial is a factor of the other polynomial, and then factor the other polynomial into linear and constant factors.

11. $x^3 - 6x^2 + 12x - 8; x - 2$
12. $x^3 + 2x^2 - 13x + 10; x - 1$
13. $x^3 + 3x^2 - 4; x + 2$
14. $x^3 + 4x^2 - 7x - 10; x + 5$
15. $2x^3 + 3x^2 - 8x + 3; x - \frac{1}{2}$
16. $x^3 - 3x^2 + 4x - 12; x - 3$
17. $x^4 - 2x^3 + 9x^2 - 18x; x - 2$
18. $x^3 + 3ix^2 - 3x - i; x + i$
19. $2x^4 - 7ix^3 - 7x^2 + 2ix; x - \frac{1}{2}i$
20. $x^3 + (1 + i)x^2 + (2 + i)x + 2; x - i$

In Exercises 21–28, show that the given number is a root of the given polynomial.

21. $2x^3 - 4x^2 - 3x + 3; -1$
22. $x^4 - 2x^3 + x^2 - 10x - 6; 3$
23. $x^3 + x^2 + x + \frac{3}{8}; -\frac{1}{2}$
24. $x^7 + 3x^4 - x; 0$
25. $\sqrt{2}x^5 - \dfrac{\sqrt{2}}{2}x^3 - x^2 - 4; \sqrt{2}$
26. $2x^3 - 5x^2 + 8x - 20; 2i$

27. $x^4 + 3ix^3 - 2x^2 - 6ix; -3i$

28. $2x^3 + 3x^2 - 2ix - 8i + 2; i + 1$

In Exercises 29–40, find a polynomial having the given roots and no others.

29. 3, -2

30. -1, 6

31. 0, 13, -1

32. -1, 1, 2

33. -2, 1, 4

34. 0, $\frac{1}{2}$, 1, -2

35. 1, -1, 2, -2

36. 1, 2, 3, 4

37. $5i$, $-5i$

38. $3 + 4i$, $3 - 4i$

39. $\sqrt{2} + 2i$, $\sqrt{2} - 2i$

40. $1 + \sqrt{3}i$, $1 - \sqrt{3}i$

In Exercises 41–48, factor the given polynomial into linear and constant factors.

41. $x^2 + 4$

42. $x^2 + 25$

43. $x^2 + 7$

44. $x^2 + 17$

45. $2x^2 + 2$

46. $3x^2 + 5$

47. $x^3 + x$

48. $-3x^3 - 21x$

49. For what value of k is $kx^3 + 3x^2 - 2x + 4$ divisible by $x - 2$?

50. For what values of k is $kx^2 + k^2x + 12$ divisible by $x + 3$?

51. Let n be a positive integer and a a real number.
   a. Show that $x - a$ is a factor of $x^n - a^n$. (*Hint:* Use the Factor Theorem.)
   b. Show that if n is even, then $x + a$ is not a factor of $x^n + a^n$.
   c. Show that if n is odd, then $x + a$ is a factor of $x^n + a^n$.

*52. Suppose w is a root of the polynomial $x^3 - x + 1$. Show that w is a root of the polynomial $-x^6 + 2x^4 - x^2 + 1$. (*Hint:* From the fact that $w^3 = w - 1$, find expressions for w^6 and w^4.)

8.3

THE FUNDAMENTAL THEOREM OF ALGEBRA

The Factor Theorem provides a simple test for determining whether a polynomial $f(x)$ is divisible by $x - c$: Just find out whether $f(c) = 0$, that is, whether c is a root of $f(x)$. However, the Factor Theorem does not indicate which polynomials actually have roots, nor does it give any clue to finding roots when they exist. A deep theorem called the Fundamental Theorem of Algebra solves all mystery concerning which polynomials have roots—it says that *every* nonconstant polynomial with real or complex coefficients has a (possibly complex) root. The theorem was first proved rigorously by Carl Friedrich Gauss at the age of 20. Since all known proofs of the Fundamental Theorem of Algebra involve concepts not covered in this book, we will state the theorem but not prove it.

THEOREM 8.4   *Fundamental Theorem of Algebra.* Every nonconstant polynomial has at least one complex root.

From the Fundamental Theorem of Algebra we know that even a complicated polynomial such as

$$x^8 - ix^5 - \sqrt{2}x^3 + (3i - 5)x + 1 \qquad (1)$$

has a root, although the theorem does not suggest how to find a root. Using the Fundamental Theorem, we can even conclude that in a certain sense the polynomial in (1) has 8 roots, and more generally, any polynomial of positive degree n has n roots.

In order to prove this last result, we will need the fact that for any nonzero polynomial $g(x)$,

$$\text{degree of } (x - c)g(x) = 1 + \text{degree of } g(x) \qquad (2)$$

After all, if

$$g(x) = a_m x^m + a_{m-1} x^{m-1} + \cdots + a_1 x + a_0 \quad \text{with } a_m \neq 0$$

then

$$(x - c)g(x)$$
$$= (x - c)(a_m x^m + a_{m-1} x^{m-1} + \cdots + a_1 x + a_0)$$
$$= a_m x^{m+1} + (a_{m-1} - ca_m)x^m + \cdots + (a_0 - ca_1)x + (-ca)$$

so that the degree of $g(x)$ is m, whereas the degree of $(x - c)g(x)$ is $m + 1$.

Now we are ready to show that if $f(x)$ is a nonconstant polynomial of degree n, then $f(x)$ can be written as a product of a constant and n linear factors. By the Fundamental Theorem of Algebra, $f(x)$ has a root c_1, so the Factor Theorem tells us that $x - c_1$ is a factor of $f(x)$, that is, there is a polynomial $g_1(x)$ such that

$$f(x) = (x - c_1)g_1(x) \qquad (3)$$

Since $f(x)$ is not the zero polynomial, neither is $g_1(x)$. By (3),

$$\text{degree of } (x - c_1)g_1(x) = \text{degree of } f(x) = n$$

so that by (2),

$$\text{degree of } g_1(x) = n - 1$$

If $n - 1 = 0$, then $g_1(x)$ is a constant polynomial. But if $n - 1 > 0$, we can apply the same procedure to $g_1(x)$ as we did to $f(x)$, and conclude that there is a nonzero polynomial $g_2(x)$ such that

$$g_1(x) = (x - c_2)g_2(x) \qquad (4)$$

where c_2 is a root of $g_1(x)$ and where $g_2(x)$ has degree $n - 2$. Combining (3) and (4), we have

$$f(x) = (x - c_1)(x - c_2)g_2(x)$$

If we continue this process, we find that there are complex numbers $c_1, c_2, \ldots, c_n$ and a nonzero polynomial $g_n(x)$ such that

$$f(x) = (x - c_1)(x - c_2) \cdots (x - c_n)g_n(x)$$

where the degree of $g_n(x)$ is $n - n = 0$. But then

$$g_n(x) = b$$

for some nonzero constant b, which leads us to the following corollary of the Fundamental Theorem of Algebra.

COROLLARY 8.5   Let $f(x)$ be a nonconstant polynomial of degree n. Then there are complex numbers $b, c_1, c_2, \ldots, c_n$ such that $b \neq 0$ and such that

$$f(x) = b(x - c_1)(x - c_2) \cdots (x - c_n) \tag{5}$$

The numbers $c_1, c_2, \ldots, c_n$ are the roots of $f(x)$, and b is the leading coefficient of $f(x)$.

> **Caution:** The numbers $c_1, c_2, \ldots, c_n$ in (5) need not be distinct from one another. Indeed, if
>
> $$f(x) = x^2 - 2ix - 1$$
>
> then
>
> $$f(x) = (x - i)^2$$
>
> which means that $c_1 = i = c_2$.

The import of Corollary 8.5 is that any nonconstant polynomial $f(x)$ can be written as a product of a constant and linear polynomials. This fact is expressed by saying that the polynomial $f(x)$ *factors completely*.

The roots mentioned in Corollary 8.5 are the *only* roots of $f(x)$. The reason is that if c is a number different from each of the numbers $c_1, c_2, \ldots, c_n$, then because $b, c - c_1, c - c_2, \ldots, c - c_n$ are all different from 0, we have

$$f(c) = b(c - c_1)(c - c_2) \cdots (c - c_n) \neq 0$$

Therefore c cannot be a root of $f(x)$. This yields the next theorem.

THEOREM 8.6   A polynomial of degree n has at most n distinct roots.

One consequence of Theorem 8.6 is that the complete factorization of $f(x)$ in (5) is unique except for the order in which we write the roots $c_1, c_2, \ldots, c_n$ of $f(x)$.

Example 1. Factor the polynomial $3x^5 + 2x^4 + x^3$ completely, and show that it has no nonzero real roots.

Solution. Notice that

$$3x^5 + 2x^4 + x^3 = x^3(3x^2 + 2x + 1)$$

so 0 is a root of the given polynomial. Next we will find the roots of $3x^2 + 2x + 1$, which means finding the solutions of the equation

$$3x^2 + 2x + 1 = 0 \qquad (6)$$

By the quadratic formula, the solutions of (6) are given by

$$x = \frac{-2 \pm \sqrt{2^2 - 4(3)(1)}}{2(3)} = \frac{-2 \pm \sqrt{-8}}{6}$$

$$= \frac{-2 \pm \sqrt{8}i}{6} = \frac{-2 \pm 2\sqrt{2}i}{6} = -\frac{1}{3} \pm \frac{1}{3}\sqrt{2}i$$

Therefore the roots of $3x^2 + 2x + 1$ are $-\frac{1}{3} + \frac{1}{3}\sqrt{2}i$ and $-\frac{1}{3} - \frac{1}{3}\sqrt{2}i$. Since the leading coefficient is 3, it follows from Corollary 8.5 that

$$3x^5 + 2x^4 + x^3$$

$$= 3(x - 0)^3 \left[x - \left(-\frac{1}{3} + \frac{1}{3}\sqrt{2}i \right) \right] \left[x - \left(-\frac{1}{3} - \frac{1}{3}\sqrt{2}i \right) \right]$$

$$= 3x^3 \left(x + \frac{1}{3} - \frac{1}{3}\sqrt{2}i \right) \left(x + \frac{1}{3} + \frac{1}{3}\sqrt{2}i \right)$$

It is now clear that the given polynomial has no nonzero real roots. $\square$

In general, the roots $c_1, c_2, \ldots, c_n$ of a nonconstant polynomial $f(x)$ need not be distinct from one another. If at least two of the roots are the same, then of course $f(x)$ has fewer than n distinct roots. A root c_k appearing m times in (5) is called a root of $f(x)$ of **multiplicity m**. For example, if

$$f(x) = -\frac{3}{4}(x - 9)^3(x - i)^7(x - 6)$$

then $f(x)$ has degree 11 (because $3 + 7 + 1 = 11$) and has roots 9, i, and 6. The root 9 has multiplicity 3, the root i has multiplicity 7, and the root 6 has multiplicity 1. If we count each root m times, where m is its multiplicity, then f has 11 roots, and in this sense every polynomial of positive degree n has n roots.

Example 2. Let $f(x) = x^4 - 2x^3 + 5x^2 - 8x + 4$. Given that 1 is a root of multiplicity 2 of $f(x)$, factor $f(x)$ completely.

Solution. Since 1 is a root of $f(x)$ of multiplicity 2, we know by Corollary 8.5 that $(x - 1)^2$ divides $f(x)$. Next we use synthetic division to divide $f(x)$ by $x - 1$:

$$
\begin{array}{r|rrrrr}
1 & 1 & -2 & 5 & -8 & 4 \\
  &   & 1 & -1 & 4 & -4 \\
\hline
  & 1 & -1 & 4 & -4 & \,\underline{|\,0} \\
\end{array}
$$

Thus

$$x^4 - 2x^3 + 5x^2 - 8x + 4 = (x - 1)(x^3 - x^2 + 4x - 4)$$

Now we use synthetic division again, this time to divide $x^3 - x^2 + 4x - 4$ by $x - 1$:

$$
\begin{array}{r|rrrr}
1 & 1 & -1 & 4 & -4 \\
  &   & 1 & 0 & 4 \\
\hline
  & 1 & 0 & 4 & \,\underline{|\,0} \\
\end{array}
$$

Therefore

$$x^3 - x^2 + 4x - 4 = (x - 1)(x^2 + 4)$$

Consequently

$$
\begin{aligned}
x^4 - 2x^3 + 5x^2 - 8x + 4 &= (x - 1)(x - 1)(x^2 + 4) \\
&= (x - 1)^2(x^2 + 4)
\end{aligned}
$$

Since

$$x^2 + 4 = (x + 2i)(x - 2i)$$

we conclude that

$$x^4 - 2x^3 + 5x^2 - 8x + 4 = (x - 1)^2(x + 2i)(x - 2i)$$

which is the complete factorization of $f(x)$. $\square$

We close by mentioning that one interpretation of the Fundamental Theorem of Algebra is that any equation of the form

$$a_n x^n + a_{n-1} x^{n-1} + \cdots + a_1 x + a_0 = 0$$

where $n \geq 1$, has a solution among the complex numbers. The corresponding interpretation of Theorem 8.6 is that such an equation has at most n distinct solutions.

EXERCISES 8.3

In Exercises 1–8, find the roots of the polynomial, along with their multiplicities.

1. $(x - 1)^2(2x + 1)^3$
2. $(x + \frac{1}{2})(x - \frac{4}{3})(x + \pi)^4$
3. $-3x^2(2x^2 + 6)^2$
4. $(3x - 5)(2x^2 + 5x - 3)$
5. $(x^2 - 9)(x^2 + 3)$
6. $(6x^2 + 7x - 3)^2$
7. $-\frac{1}{2}x^2(2x + 1)^2(3x^2 + 1)^3$
8. $17i[x(7x - 3)^2]^3(4x^2 + 1)$

In Exercises 9–16, factor the polynomial completely.

9. $x^3 - 3x^2 + 2x$
10. $x^3 + 3x^2 + 3x + 1$
11. $x^4 - 10x^2 + 24$
12. $2x^5 - 7x^4 - 4x^3$
13. $(x^2 + 9)^2$
14. $x^4 + 5x^2 - 36$
15. $-x^2 + 4x - 13$
16. $x^3 - ix^2 + 5x - 5i$

17. Show that $5x^2 - 2x + 9$ has no real roots.
18. Show that $(x^2 + 4)(3x^2 - x + 1)$ has no real roots.
19. Show that $-2x^3 + 2x^2 - 3x$ has exactly one real root, and find it.
20. Show that -1 is a root of multiplicity 2 of the polynomial $2x^3 + 5x^2 + 4x + 1$, and then factor the polynomial completely.
21. Show that 2 is a root of multiplicity 3 of the polynomial $x^5 - 6x^4 + 13x^3 - 14x^2 + 12x - 8$, and then factor the polynomial completely.
22. Show that $1 + i$ is a root of multiplicity 2 of the polynomial $x^3 - (1 + 2i)x^2 - 2x + 2i$, and then factor the polynomial completely.
23. Show that $-i$ is a root of multiplicity 4 of the polynomial $x^5 + 2ix^4 + 2x^3 + 8ix^2 - 7x - 2i$, and then factor the polynomial completely.

In Exercises 24–28, write a polynomial with the given roots and multiplicities, and no other roots. Leave the polynomial in factored form.

24. $2 + 3i$ with multiplicity 1, $3 - i$ with multiplicity 1, -4 with multiplicity 1.
25. 3 with multiplicity 4, $1 - i$ with multiplicity 1.
26. $\sqrt{2}$ with multiplicity 1, -4 with multiplicity 2, $3i$ with multiplicity 3.
27. $1 + i$ with multiplicity 2, $1 - i$ with multiplicity 2, -6 with multiplicity 3.
28. $2 - 4i$ with multiplicity 3, $-i$ with multiplicity 2, 0 with multiplicity 5.
29. Show that -3 is a root of multiplicity 2 of the polynomial $x^4 + 2x^3 - 11x^2 - 12x + 36$. Find the other root.

30. Show that 1 is a root of multiplicity 4 of the polynomial $x^5 - x^4 - 6x^3 + 14x^2 - 11x + 3$. Find the other root.

31. Show that -1 is a root of multiplicity 5 of the polynomial $x^6 + 3x^5 - 10x^3 - 15x^2 - 9x - 2$. Find the other root.

8.4

POLYNOMIALS WITH REAL COEFFICIENTS

Any finite set of complex numbers can serve as the roots of a polynomial with complex coefficients. For example, a polynomial whose roots are -3, $2i$, and $\frac{1}{2} - 4i$ is given by

$$f(x) = (x + 3)(x - 2i)\left[x - \left(\frac{1}{2} - 4i\right)\right]$$

However, if we insist that each coefficient $a_n, a_{n-1}, \ldots, a_1, a_0$ of the polynomial

$$f(x) = a_n x^n + a_{n-1} x^{n-1} + \cdots + a_1 x + a_0$$

be a *real* number, then it turns out that the conjugate of any nonreal root of $f(x)$ must also be a root of the polynomial. This fact was already observed in Section 7.2 for quadratic polynomials. We will now state and prove the result for polynomials of arbitrary degree.

THEOREM 8.7   Let $f(x)$ be a polynomial with real coefficients. If c is a root of $f(x)$, then so is the conjugate $\bar{c}$.

Proof   Let

$$f(x) = a_n x^n + a_{n-1} x^{n-1} + \cdots + a_1 x + a_0$$

where $a_n, a_{n-1}, \ldots, a_1, a_0$ are real numbers. Because the coefficients are real, each is equal to its conjugate. Thus for any complex number c,

$$
\begin{aligned}
f(\bar{c}) &= a_n(\bar{c})^n + a_{n-1}(\bar{c})^{n-1} + \cdots + a_1\bar{c} + a_0 \\
&= a_n(\overline{c^n}) + a_{n-1}(\overline{c^{n-1}}) + \cdots + a_1\bar{c} + a_0 \\
&= \overline{a_n}(\overline{c^n}) + \overline{a_{n-1}}(\overline{c^{n-1}}) + \cdots + \overline{a_1}\bar{c} + \overline{a_0} \\
&= \overline{a_n c^n} + \overline{a_{n-1} c^{n-1}} + \cdots + \overline{a_1 c} + \overline{a_0} \\
&= \overline{a_n c^n + a_{n-1} c^{n-1} + \cdots + a_1 c + a_0} \\
&= \overline{f(c)}
\end{aligned}
$$

Now suppose that c is a root of $f(x)$, so that $f(c) = 0$. By our preceding calculations, this means that

$$f(\bar{c}) = \overline{f(c)} = \bar{0} = 0$$

so that $\bar{c}$ is also a root of $f(x)$, which is what we wished to prove. ∎

Caution: We must emphasize that the conjugate of a root of a given polynomial need not also be a root if one or more of the coefficients are not real (see Exercise 24).

Let $f(x)$ be a polynomial with real coefficients, and let c be a root of $f(x)$. If c is a real number, then $\bar{c} = c$, so that Theorem 8.7 tells us nothing new about $f(x)$. In contrast, if c is a root of $f(x)$ and is not a real number, then Theorem 8.7 tells us that the conjugate $\bar{c}$ is a root of $f(x)$, and of course in this case $\bar{c} \neq c$. Thus nonreal roots of any polynomial with real coefficients come in conjugate pairs.

Example 1. Find a polynomial that has real coefficients, has roots 5 and $1 - 3i$, and has the value 100 for $x = 0$.

Solution. If $1 - 3i$ is a root of a polynomial $f(x)$ with real coefficients, then by Theorem 8.7 so is its conjugate $1 + 3i$. Since the remaining prescribed root is 5, we will let $f(x)$ be given by

$$f(x) = b(x - 5)[x - (1 - 3i)][x - (1 + 3i)]$$

where b is to be determined. Since

$$
\begin{aligned}
[x - (1 - 3i)][x - (1 + 3i)] &= [(x - 1) + 3i][(x - 1) - 3i] \\
&= (x - 1)^2 - (3i)^2 \\
&= (x^2 - 2x + 1) + 9 \\
&= x^2 - 2x + 10
\end{aligned}
$$

$f(x)$ is given by

$$f(x) = b(x - 5)(x^2 - 2x + 10) = b(x^3 - 7x^2 + 20x - 50) \quad (1)$$

The number b must be chosen so that $f(0) = 100$. By (1), it follows that

$$f(0) = -50b$$

Therefore $100 = -50b$, so that $b = -2$. Consequently

$$
\begin{aligned}
f(x) &= -2(x^3 - 7x^2 + 20x - 50) \\
&= -2x^3 + 14x^2 - 40x + 100
\end{aligned}
$$

is a polynomial having the required properties. $\quad\square$

Example 2. Let $f(x) = x^4 + 3x^3 + 3x^2 + 3x + 2$. Given that i is a root of $f(x)$, factor $f(x)$ completely.

Solution. Since all the coefficients of $f(x)$ are real, Theorem 8.7 applies. Thus the fact that i is a root of $f(x)$ implies that its conjugate

$-i$ is also a root. Now Corollary 8.5 tells us that $f(x)$ is divisible by $x - i$ and by $x - (-i) = x + i$. To complete the factorization of $f(x)$, we use synthetic division to divide first by $x - i$ and then by $x + i$:

$$
\begin{array}{r|ccccc}
i & 1 & 3 & 3 & 3 & 2 \\
 & & i & -1 + 3i & -3 + 2i & -2 \\
\hline
 & 1 & 3 + i & 2 + 3i & 2i & \boxed{0}
\end{array}
$$

Therefore

$$x^4 + 3x^3 + 3x^2 + 3x + 2 = (x - i)[x^3 + (3 + i)x^2 + (2 + 3i)x + 2i]$$

Dividing the bracketed term by $x + i$ yields

$$
\begin{array}{r|cccc}
-i & 1 & 3 + i & 2 + 3i & 2i \\
 & & -i & -3i & -2i \\
\hline
 & 1 & 3 & 2 & \boxed{0}
\end{array}
$$

Consequently

$$x^3 + (3 + i)x^2 + (2 + 3i)x + 2i = (x + i)(x^2 + 3x + 2)$$

so that

$$x^4 + 3x^3 + 3x^2 + 3x + 2 = (x - i)(x + i)(x^2 + 3x + 2)$$

Since

$$x^2 + 3x + 2 = (x + 1)(x + 2)$$

$f(x)$ is completely factored as follows:

$$x^4 + 3x^3 + 3x^2 + 3x + 2 = (x - i)(x + i)(x + 1)(x + 2) \quad \square$$

It is sometimes desirable to factor a polynomial with real coefficients so that the factors are also polynomials with real coefficients. Although it is not always possible to accomplish this if we insist that the factors have degree 1 (see the result of Example 2), it is possible if we allow the factors to have degree 1 or 2.

THEOREM 8.8   Every nonconstant polynomial with real coefficients can be expressed as a product of linear and quadratic polynomials with real coefficients.

Proof   Let $f(x)$ be a nonconstant polynomial of degree n. By Corollary 8.5,

$$f(x) = b(x - c_1)(x - c_2) \ldots (x - c_n) \tag{2}$$

where $c_1, c_2, \ldots, c_n$ are the roots of f (repeated with multiplicity), and where b is the leading coefficient of $f(x)$ and hence is real. Consider $x - c_1$. If c_1 is real, then $x - c_1$ is a linear polynomial with real coefficients. However, if c_1 is not real, then by Theorem 8.7 we know that $\bar{c}_1$ is also a root of $f(x)$, and hence $\bar{c}_1$ is one of the numbers $c_2, c_3, \ldots, c_n$. In this case $x - \bar{c}_1$ is a factor of $f(x)$. Multiplying $x - c_1$ and $x - \bar{c}_1$ together, we obtain

$$(x - c_1)(x - \bar{c}_1) = x^2 - (c_1 + \bar{c}_1)x + c_1\bar{c}_1$$

which is a quadratic polynomial that has real coefficients because both $c_1 + \bar{c}_1$ and $c_1\bar{c}_1 = |c_1|^2$ are real numbers by (6) of Section 7.2 and (3) of Section 7.3. The same analysis applies to each of the factors $x - c_2, x - c_3, \ldots, x - c_n$. Thus each of the factors in (2) is a real number or a linear polynomial with real coefficients or combines with a second factor in (2) to form a quadratic polynomial with real coefficients. This proves the theorem. ∎

Example 3. Let $f(x) = -\frac{1}{2}x^3 + 3x^2 - 2x + \sqrt{3}$. Show that $f(x)$ has a real root.

Solution. Since $f(x)$ has degree 3, Theorem 8.8 implies that $f(x)$ can be expressed either as the product of three linear polynomials with real coefficients or as the product of one linear and one quadratic polynomial, each with real coefficients. In either case, $f(x)$ has a linear factor $x - c$, where c is a real number. It then follows from the Factor Theorem that $f(c) = 0$, so c is a real root of $f(x)$. □

By similar reasoning it is possible to prove that any polynomial having real coefficients and odd degree has at least one real root.

EXERCISES 8.4

In Exercises 1–12, find a polynomial $f(x)$ with real coefficients that has the given roots and prescribed value.

1. root: $1 - 2i$; $f(0) = 10$
2. root: $-3 + i$; $f(i) = 3 + 2i$
3. root: $1 + i$; $f(0) = -3$
4. roots: $0, -2, -i$; $f(0) = 0$
5. roots: $0, -2, -i$; $f(1) = 8$
6. roots: $i, -1, 1$; $f(2) = -3$
7. roots: i, i^2, i^3, i^4; $f(-2) = \sqrt{2}$
8. roots: $3, 4 - \sqrt{3}\,i$; $f(0) = 57$
9. roots: $2, 2 + \sqrt{3}\,i$; $f(1) = -28$
10. roots: $1, 2i, 3i$; $f(-1) = 100$
11. roots: $2 - i, 4 - 3i$; $f(2) = \frac{13}{4}$
12. roots: $2, -5, i, \frac{1}{3}i$; $f(0) = -10$

13. Find a polynomial with real coefficients for which -1 is a root of multiplicity 2 and i is a root of multiplicity 3. Express your answer as a product of linear and quadratic polynomials.

14. Find a polynomial with real coefficients for which $\frac{1}{2}$ is a root of multiplicity 3, $1 + 2i$ is a root of multiplicity 2, and i is a root of multiplicity 1. Express your answer as a product of linear and quadratic polynomials.

15. Given that $1 + i$ is a root of the polynomial $2x^3 - 5x^2 + 6x - 2$, factor the polynomial completely.

16. Given that i is a root of the polynomial $2x^4 - x^3 + x^2 - x - 1$, factor the polynomial completely.

17. Given that $-2 + 3i$ is a root of the polynomial $x^3 + 3x^2 + 9x - 13$, factor the polynomial completely.

18. Given that i is a root of multiplicity 2 of the polynomial $2x^5 - x^4 + 4x^3 - 2x^2 + 2x - 1$, factor the polynomial completely.

19. Given that $2 - 3i$ is a root of the polynomial $2x^4 - 8x^3 + 27x^2 - 4x + 13$, factor the polynomial completely.

20. a. Factor $x^4 + 1$ into quadratic factors with real coefficients. (*Hint:* Use the fact that $x^4 + 1 = (x^2 + 1)^2 - (\sqrt{2}x)^2$, and then factor the right side.)
    b. Factor $x^4 + 1$ into linear factors. (*Hint:* Use part (a) and the quadratic formula.)

21. Given that $2 + i$ and $1 - i$ are roots of the polynomial $2x^4 - 12x^3 + 30x^2 - 36x + 20$, factor the polynomial
    a. into a constant and linear factors
    b. into a constant and quadratic factors with real coefficients.

22. Show that the polynomial $-7x^5 + \frac{1}{2}x^3 + \pi x - \sqrt{2}$ has at least one real root.

23. Show that the polynomial $4x^{19} + 2x^{12} - 1$ has at least one real root.

24. Let $f(x) = x^2 - 2x + 1 + 2i$. Show that $2 - i$ is a root of $f(x)$, but its conjugate $2 + i$ is not. Does this contradict Theorem 8.7? Explain.

25. Let $a_6x^6 + a_5x^5 + a_4x^4 + a_3x^3 + a_2x^2 + a_1x + a_0$ be a polynomial with real coefficients and six distinct roots. Show that if one of the roots is real, then at least two of the roots are real.

8.5

ROOTS OF POLYNOMIALS WITH INTEGER COEFFICIENTS

From the Fundamental Theorem of Algebra we know that every nonconstant polynomial has roots, but the theorem gives no hint for finding the roots. Some information about the roots is gained if the coefficients of the polynomial are real: Complex roots come in pairs, that is, the conjugate of a complex root is itself a root. As we saw in the preceding section, this result can be a help in determining the remaining roots of a polynomial one of whose nonreal roots is known, but it does not tell us anything about how to find an initial root of such a polynomial.

In the present section we will address the question of determining integer and, more generally, rational roots of a polynomial whose coefficients are integers.

Integer Roots

Our discussion begins with integer roots. We say that an integer c *divides* a second integer a, or is a *divisor* of a, if there is an integer b such that $a = bc$. Thus the divisors of 12 are

$$1, \quad -1, \quad 2, \quad -2, \quad 3, \quad -3, \quad 4, \quad -4, \quad 6, \quad -6, \quad 12, \quad \text{and} \quad -12$$

whereas the divisors of -7 are

$$1, \quad -1, \quad 7, \quad \text{and} \quad -7$$

Now we are ready to state the following theorem.

THEOREM 8.9

Let $f(x) = a_n x^n + a_{n-1} x^{n-1} + \cdots + a_1 x + a_0$ be a polynomial with integer coefficients. If c is an integer root of $f(x)$, then c is a divisor of a_0.

Since there are only finitely many divisors of a given nonzero integer, Theorem 8.9 provides us with an effective way of determining any integer roots of a polynomial whose coefficients are integers.

Example 1. Let $f(x) = x^3 + 4x^2 + x - 6$. Find the integer roots of $f(x)$.

Solution. In the framework of Theorem 8.9, $a_0 = -6$. Since the divisors of -6 are $1, -1, 2, -2, 3, -3, 6$, and -6, Theorem 8.9 tells us that any integer roots of $f(x)$ must be one of these numbers. Now by substituting one by one the divisors of -6 for x, we find that

$$\begin{aligned}
f(1) &= 0 & f(3) &= 60 \\
f(-1) &= -4 & f(-3) &= 0 \\
f(2) &= 20 & f(6) &= 360 \\
f(-2) &= 0 & f(-6) &= -84
\end{aligned}$$

Consequently the only integer roots of $f(x)$ are $1, -2$, and -3. (From Corollary 8.5 to the Fundamental Theorem of Algebra, there can be only three roots of $f(x)$, because $f(x)$ is a third-degree polynomial.) □

Example 2. Factor $4x^3 - 4x^2 - x + 1$ completely.

Solution. Let

$$f(x) = 4x^3 - 4x^2 - x + 1$$

To start the factorization, let us look for a factor $x - c$ of $f(x)$ with c an integer. Since the constant term is 1, Theorem 8.9 implies that the only possible integer values of c are 1 and -1. But

$$f(1) = 0 \quad \text{and} \quad f(-1) = -6$$

so it follows that $x - 1$ is a factor and $x + 1$ is not. Next we divide $f(x)$ by $x - 1$ by means of synthetic division:

$$\begin{array}{r|rrrr} 1 & 4 & -4 & -1 & 1 \\ & & 4 & 0 & -1 \\ \hline & 4 & 0 & -1 & \lfloor 0 \end{array}$$

As a result,

$$4x^3 - 4x^2 - x + 1 = (x - 1)(4x^2 - 1)$$

Since

$$4x^2 - 1 = (2x + 1)(2x - 1)$$

it follows that the required factorization is given by

$$4x^3 - 4x^2 - x + 1 = (x - 1)(2x + 1)(2x - 1) \quad \square$$

Caution: It is possible, of course, for the method just used of seeking integer roots of a given polynomial to turn up no roots at all. For example, consider

$$f(x) = 8x^3 - 4x^2 - 2x + 1 \tag{1}$$

The coefficient of the constant term is 1, so by Theorem 8.9 the only possible integer roots of $f(x)$ are 1 and -1. But

$$f(1) = 3 \quad \text{and} \quad f(-1) = -9$$

so that there are no integer roots of $f(x)$. In Example 3 we will show that the only roots are $\frac{1}{2}$ and $-\frac{1}{2}$.

Rational Roots

Our next theorem gives us a method of searching out rational roots of polynomials.

THEOREM 8.10

Let $f(x) = a_n x^n + a_{n-1} x^{n-1} + \cdots + a_1 x + a_0$ be a polynomial with integer coefficients. Let $c = p/q$, where p and q are integers without a common factor other than 1 and -1. If c is a root of $f(x)$, then p divides a_0 and q divides a_n.

Notice that if c is an integer in Theorem 8.10, so that $p = c$ and $q = 1$, then the theorem says that c divides a_0, which is the result of Theorem 8.9. Thus Theorem 8.10 generalizes Theorem 8.9.

We will now use Theorem 8.10 to determine the roots of the polynomial given in (1).

Example 3. Let $f(x) = 8x^3 - 4x^2 - 2x + 1$. Find the roots of $f(x)$, and then factor $f(x)$ completely.

Solution. We look for roots of $f(x)$ of the form $c = p/q$. By Theorem 8.10, p must divide 1 and q must divide 8. Thus p must be 1 or -1, and q must be 1, -1, 2, -2, 4, -4, 8, or -8. Therefore c can be 1, -1, $\frac{1}{2}$, $-\frac{1}{2}$, $\frac{1}{4}$, $-\frac{1}{4}$, $\frac{1}{8}$, or $-\frac{1}{8}$. Since

$$f(1) = \quad 3 \qquad f\left(\frac{1}{4}\right) = \frac{3}{8}$$

$$f(-1) = -9 \qquad f\left(-\frac{1}{4}\right) = \frac{9}{8}$$

$$f\left(\frac{1}{2}\right) = \quad 0 \qquad f\left(\frac{1}{8}\right) = \frac{45}{64}$$

$$f\left(-\frac{1}{2}\right) = \quad 0 \qquad f\left(-\frac{1}{8}\right) = \frac{75}{64}$$

it follows that $\frac{1}{2}$ and $-\frac{1}{2}$ are the only rational roots of $f(x)$. In order to factor $f(x)$ completely, we use synthetic division to divide $f(x)$ by $x - \frac{1}{2}$:

$$
\begin{array}{r|rrrr}
\frac{1}{2} & 8 & -4 & -2 & 1 \\
           &   & 4 & 0 & -1 \\
\hline
           & 8 & 0 & -2 & \underline{|0} \\
\end{array}
$$

Therefore

$$f(x) = (x - \tfrac{1}{2})(8x^2 - 2)$$

Next we notice that

$$8x^2 - 2 = 2(4x^2 - 1) = 2(2x - 1)(2x + 1)$$

so the complete factorization of $f(x)$ is given by

$$f(x) = 2\left(x - \frac{1}{2}\right)(2x - 1)(2x + 1) = 8\left(x - \frac{1}{2}\right)^2\left(x + \frac{1}{2}\right)$$

From the factorization we see that the rational roots $\frac{1}{2}$ and $-\frac{1}{2}$ are the only roots of $f(x)$. $\square$

When we employ Theorem 8.10, we may well obtain a large collection of candidates for a root—even larger than the collection of eight candidates we had in Example 3. The following result, which concerns intervals in which roots of a given polynomial must lie, can allow us to ignore many of the candidates and focus on a select few among which we hope to find a root.

THEOREM 8.11 *Intermediate Value Theorem for Polynomials.* Let $f(x)$ be a polynomial with real coefficients, and let a and b be any two distinct

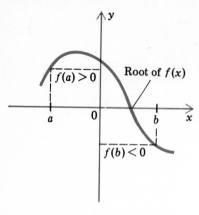

f(a) > 0   Root of f(x)

f(b) < 0

FIGURE 8.1

numbers. If $f(a) > 0$ and $f(b) < 0$, then there is a root of $f(x)$ between a and b.

Figure 8.1 gives a pictorial idea of Theorem 8.11.

Example 4. Let $f(x) = 12x^3 - 20x^2 - x + 6$. Factor $f(x)$ completely.

Solution. If $x - c$ is to be a factor of $f(x)$, where $c = p/q$, then by Theorem 8.10 we know that p must be a factor of 6 and q a factor of 12. Thus p must be

$$1, \quad -1, \quad 2, \quad -2, \quad 3, \quad -3, \quad 6, \quad \text{or} \quad -6$$

and q must be

$$1, \quad -1, \quad 2, \quad -2, \quad 3, \quad -3, \quad 4, \quad -4,$$
$$6, \quad -6, \quad 12, \quad \text{or} \quad -12$$

Consequently the only possible values of c are

$$1, \quad -1, \quad \frac{1}{2}, \quad -\frac{1}{2}, \quad \frac{1}{3}, \quad -\frac{1}{3}, \quad \frac{1}{4}, \quad -\frac{1}{4}, \quad \frac{1}{6}, \quad -\frac{1}{6}, \quad \frac{1}{12}, \quad -\frac{1}{12},$$
$$2, \quad -2, \quad \frac{2}{3}, \quad -\frac{2}{3}, \quad 3, \quad -3, \quad \frac{3}{2}, \quad -\frac{3}{2}, \quad \frac{3}{4}, \quad -\frac{3}{4}, \quad 6, \quad \text{or} \quad -6$$

It seems an enormous task to test all these 24 values of c. However, if we look at, say, $c = -1$ and $c = 0$, we find that

$$f(-1) = -25 < 0 \quad \text{and} \quad f(0) = 6 > 0$$

Therefore by the Intermediate Value Theorem for Polynomials it follows that there is a root of $f(x)$ between -1 and 0, which cuts from 24 to 7 the collection from which we hope to find a rational root:

$$c = -\frac{1}{2}, \quad -\frac{1}{3}, \quad -\frac{1}{4}, \quad -\frac{1}{6}, \quad -\frac{1}{12}, \quad -\frac{2}{3}, \quad \text{or} \quad -\frac{3}{4}$$

When we arrive at $-\frac{1}{2}$ and use synthetic division to divide $f(x)$ by $x + \frac{1}{2}$, we find that

$$
\begin{array}{r|rrrr}
-\frac{1}{2} & 12 & -20 & -1 & 6 \\
 & & -6 & 13 & -6 \\
\hline
 & 12 & -26 & 12 & \underline{|0} \\
\end{array}
$$

Thus

$$12x^3 - 20x^2 - x + 6 = \left(x + \frac{1}{2}\right)(12x^2 - 26x + 12) \tag{2}$$

Next we find that

$$12x^2 - 26x + 12 = 2(6x^2 - 13x + 6)$$

By the quadratic formula,

$$6x^2 - 13x + 6 = 0$$

if and only if

$$x = \frac{-(-13) \pm \sqrt{(-13)^2 - 4(6)(6)}}{2(6)}$$

$$= \frac{13 \pm \sqrt{25}}{12} = \frac{13 \pm 5}{12} = \frac{3}{2} \quad \text{or} \quad \frac{2}{3}$$

It follows that $x - \frac{3}{2}$ and $x - \frac{2}{3}$ are factors of $6x^2 - 13x + 6$, so that

$$6x^2 - 13x + 6 = 6\left(x - \frac{3}{2}\right)\left(x - \frac{2}{3}\right)$$

Consequently by (2) the desired factorization of the given polynomial is

$$12x^3 - 20x^2 - x + 6 = 12\left(x + \frac{1}{2}\right)\left(x - \frac{3}{2}\right)\left(x - \frac{2}{3}\right) \quad \square$$

Irrational Roots Is there any method of finding irrational roots of polynomials? First of all, think about the difficulty of expressing irrational numbers. We already know that irrational numbers do not have repeating decimal expansions. Thus unless we have a special symbol (such as π, e, or $\sqrt{2}$) for an irrational number, about the best we could hope to do is to find the first several digits in its decimal expansion or to find a rational number that is "close" to it. Consequently irrational roots can usually only be approximated. Since most approximation methods depend on more advanced topics, we will not dwell on irrational roots.

EXERCISES 8.5

In Exercises 1–8, determine the divisors of the given number.

1. 2	2. 4	3. -5	4. -1
5. 8	6. 11	7. -24	8. 36

In Exercises 9–16, find all integer roots of the given polynomial.

9. $x^3 + 4x^2 - 7x - 10$ 10. $2x^3 - 3x^2 - 11x + 6$

11. $2x^3 + 3x^2 - 17x + 12$ 12. $x^3 - 2x^2 - 5x + 6$

13. $x^4 + 3x^3 - 3x - 1$

14. $x^4 + x^3 - x^2 - x - 6$

15. $x^4 - 5x^3 - 10x^2 + 20x + 24$

16. $x^4 + 2x^3 + 2x^2 + x$

In Exercises 17–30, use Theorems 8.9 and 8.10 to factor the given polynomial completely.

17. $2x^2 + 5x - 3$

18. $3x^3 + 6x^2 - 4x - 8$

19. $x^3 - 5x^2 + x - 5$

20. $x^3 + x - 2$

21. $x^3 - 9x^2 + 26x - 24$

22. $2x^3 + x^2 - 5x + 2$

23. $4x^3 + 8x^2 - 11x - 15$

24. $12x^3 + 23x^2 - 3x - 2$

25. $2x^4 - 2x^3 + 4x^2 + 2x - 6$

26. $4x^4 + 8x^3 + 7x^2 + 6x + 3$

27. $x^5 + 3x^4 + x^3 - x^2 - 4$

28. $12x^3 + 28x^2 - 7x - 5$

29. $24x^4 + 2x^3 - 5x^2 - x$

30. $16x^4 - 20x^3 - 56x^2 + 15x$

In Exercises 31–40, show that the given polynomial has no rational roots.

31. $x^3 + x + 3$

32. $2x^3 - 17x^2 + 12x - 2$

33. $x^5 - x - 1$

34. $2x^6 - 1$

35. $2x^7 - 3$

36. $3x^4 + 3x^2 + 4x + 4$

37. $x^5 - x^4 + x^3 - x^2 - x - 1$

38. $2x^6 + 4x^5 - 3x^3 + 5x - 1$

39. $-2x^7 - 4x^4 + x - 1$

40. $x^8 + x^7 + 4$

In Exercises 41–44, show that the given number is not rational.

41. $\sqrt{2}$ (*Hint:*

42. $\sqrt{3}$

    Let $f(x) = x^2 - 2$

43. $2^{1/3}$

    and use Theorem 8.10.)

44. $8^{1/4}$

In Exercises 45–48, show that the polynomial has a root in the given interval.

45. $x^4 - x^3 + x^2 - x - 1$; $[-1, 1]$

46. $3x^4 + 8x^3 - 4x + 5$; $[-2, 0]$

47. $x^5 + x^3 - 8$; $[-1, 2]$

48. $x^3 + 3x^2 - 1$; $[-\frac{1}{2}, 1]$

*49. Let $f(x) = 4x^4 - 9x^2 + 1$.

    a. Use the Intermediate Value Theorem for Polynomials to determine in which of the unit intervals determined by -3, -2, -1, 0, 1, 2, 3 a root of $f(x)$ must occur.

    b. Using the result of part (a) and Theorem 8.6, show that $f(x) > 0$ for $x > 3$.

50. Determine the solutions of the equation $4x^3 - 3x + 1 = 0$. (This equation is related to the problem of trisecting an angle.)

51. The problem of showing that a 60° angle cannot be trisected with a straightedge and compass can be reduced to showing that the equation $8x^3 - 6x - 1 = 0$ has no rational solutions. Show that indeed the equation has no rational solutions.

*52. Occasionally one wishes to know whether a polynomial with integer coefficients can be factored into two other polynomials with positive degree that also have integer coefficients. The following result, due to Eisenstein, a student of Gauss, gives a condition that guarantees that the polynomial cannot be so factored.

    Eisenstein's Criterion: Let $f(x) = a_n x^n + a_{n-1} x^{n-1} + \cdots +$

$a_1x + a_0$ be a polynomial with integer coefficients. If there is a prime number p such that

    i. p divides each of the numbers $a_0, a_1, a_2, \ldots, a_{n-1}$

    ii. p does not divide a_n

    iii. p^2 does not divide a_0

then $f(x)$ cannot be written as the product of two polynomials each with integer coefficients and positive degree.

Use Eisenstein's Criterion to show that none of the following polynomials can be factored into two polynomials with integer coefficients and positive degree.

  a. $x^3 + 3x^2 + 9x + 6$ (*Hint:* Take $p = 3$.)

  b. $4x^4 - 15x^3 + 20x^2 + 1000x - 10$

  c. $5x^3 + 21x^2 - 14x + 7$

  d. $7x^5 + 543x^4 - 219x^3 + 72x^2 - 63x + 240$

KEY TERMS

factor

synthetic division

complete factorization

multiplicity of a root

integer root

rational root

KEY THEOREMS

Division Algorithm

Remainder Theorem

Factor Theorem

Fundamental Theorem of Algebra

Intermediate Value Theorem for Polynomials

REVIEW EXERCISES

In Exercises 1–2, perform the indicated long division.

1. $x^2 - x - 1 \overline{\smash{\big)}\, x^4 - 2x + 1}$

2. $x^3 + 1 \overline{\smash{\big)}\, x^7 + x^6 - x^4 - x^3 + 2}$

In Exercises 3–8, use synthetic division to find the quotient and remainder when the first polynomial is divided by the second.

3. $-x^2 - 5x - 10; \; x + 3$

4. $2x^3 + 5x^2 - 7x + 7; \; x - \frac{1}{2}$

5. $x^3 + 2x + 3i; \; x + i$

6. $x^4 + 2x^3 - 2x^2 - 3x + 2; \; x + 2$

7. $x^6 - 3x^5 - (1 + i)x^4 + (1 - i)x^3 + 6ix^2 + 4x + 3; \; x - 3$

8. $x^8 + 1; \; x + 1$

In Exercises 9–12, use synthetic division to show that the first polynomial is a factor of the second.

9. $x + 3$; $2x^3 + 5x^2 - 2x + 3$

10. $x + \frac{1}{2}$; $2x^4 + 5x^3 + 2x^2 - x - \frac{1}{2}$

11. $x - 2i$; $3x^5 + (-4 - 6i)x^4 + 7ix^3 + (-4 + i)x^2 + (3 + 4i)x - 2i$

12. $x + i$; $x^7 - i$

In Exercises 13–16, use the Remainder Theorem to find the remainder when the first polynomial is divided by the second.

13. $-x^4 + 5x^3 + 3x^2 - 4$; $x + 1$

14. $3x^6 - x^3 - 2x + 1$; $x - \sqrt{3}$

15. $ix^4 - 2x^3 - 3x - 2i$; $x + i$

16. $x^8 + 1$; $x - i$

In Exercises 17–22, verify that the linear polynomial is a factor of the other polynomial, and then factor the other polynomial into linear and constant factors.

17. $x^3 - 2x^2 - 5x + 6$; $x + 2$

18. $x^4 - 4x^3 - 2x^2 + 8x$; $x - 4$

19. $3x^4 - 8x^3 + 6x^2 - 1$; $x + \frac{1}{3}$

20. $x^3 + ix^2 + (i - 3)x + 2 - 2i$; $x - 1 + i$

21. $x^4 - x^2 - \sqrt{2}x$; $x - \sqrt{2}$

22. $x^3 - 3x^2 + x - 3$; $x + i$

In Exercises 23–26, factor the given polynomial into linear factors.

23. $x^3 - 4x^2 + 8x$

24. $x^3 + x^2 - x - 1$

25. $x^4 - x^2 - 12$

26. $x^4 + 2x^2 - 63$

In Exercises 27–30, show that the given number is a root of the polynomial. Then factor the polynomial completely.

27. $x^3 + 7x^2 + 19x + 21$; -3

28. $2x^3 - x^2 - x - 3$; $\frac{3}{2}$

*29. $x^3 - 2ix^2 + (i - 2)x + i + 1$; i

30. $x^4 + 4$; $1 + i$

In Exercises 31–34, find the roots of the polynomial, along with their multiplicities.

31. $(x - 2i)^2(x^2 + 49)^4(x - 3)^7$

32. $(3 - 4x)^2(2x^2 + 4x + 3)^5$

33. $x^3(\sqrt{6}x - \sqrt{2})^4(9x^2 + 1)^2$

34. $(x - 1)(x + 2)(x^3 - x^2 + x - 1)$

In Exercises 35–36, write a polynomial with the given roots and multiplicities. Leave the polynomial in factored form.

35. $3 - 5i$ with multiplicity 3, $2i$ with multiplicity 4, 0 with multiplicity 1.

36. $\sqrt{3} - 2i$ with multiplicity 10, -9 with multiplicity 7, $3 - \sqrt{3} + 4i$ with multiplicity 6.

37. Show that 3 is a root of multiplicity 3 of the polynomial $x^4 - 11x^3 + 45x^2 - 81x + 54$, and then find the other root.

38. Show that i is a root of multiplicity 4 of the polynomial $x^6 - 4ix^5 + 10x^4 - 60ix^3 - 95x^2 + 64ix + 16$, and then find the other roots.

In Exercises 39–40, find a polynomial $f(x)$ with real coefficients that has the given roots and prescribed value.

39. roots: $2, -1 - 3i; f(1) = -26$

40. roots: $i - 1, 2i - 4; f(2) = 80$

41. Given that $2 - i$ is a root of multiplicity 2 of the polynomial $x^4 - 8x^3 + 26x^2 - 40x + 25$, factor the polynomial completely.

42. Given that -1 is a root of multiplicity 2 and i is a root of multiplicity 1 of the polynomial $x^5 - 2x^4 - 6x^3 - 6x^2 - 7x - 4$, factor the polynomial completely.

43. Given that $1 + i$ and $-1 - 2i$ are roots of the polynomial $x^4 + 3x^2 - 6x + 10$, factor the polynomial into quadratic factors with real coefficients.

44. Show that the polynomial $3x^{11} - 4x^8 + 3x^5 - 1$ has at least one real root.

In Exercises 45–48, find all integer roots, if any, of the given polynomial.

45. $x^5 + 4x^3 + x^2 + 4$
46. $2x^4 - x^3 + x^2 + x$
47. $x^4 - x^3 - 6x^2 - 6x - 72$
48. $12x^3 + 20x^2 - x - 6$

In Exercises 49–54, use Theorems 8.9 and 8.10 to factor the given polynomial completely.

49. $2x^3 - 4x^2 - x + 2$
50. $8x^3 - 12x^2 + 12x - 18$
51. $x^5 - x^4 - x^3 - x^2 - 2x$
52. $6x^3 + 7x^2 - x - 2$
53. $16x^3 + 28x^2 - 28x + 5$
54. $18x^4 + 9x^3 - 8x^2 - 4x$

In Exercises 55–58, show that the given polynomial has no rational roots.

55. $2x^4 + 3x^3 - x^2 - 1$
56. $x^5 - 3x + 4$
57. $3x^6 - 2$
58. $2x^8 - 1$

In Exercises 59–60, show that the polynomial has a root in the given interval.

59. $x^3 - 4x^2 + 4; [-1, 0]$
60. $x^4 + 2x^3 - 4x^2 - 4x + \frac{1}{2}; [-3, -1]$

61. Show that $x - 1$ is a factor of the polynomial $529x^{1345} - 284x^{612} - 245$.

62. Determine the relationship that must hold between a and b in order that $x + 1$ be a factor of the polynomial $x^3 + ax^2 - x - b$.

DISCRETE
ALGEBRA

The title of this final chapter refers to the fact that each of the topics to be discussed is related to the discrete collection of natural numbers. We open the chapter with a presentation of the axiom of mathematical induction, which makes it possible to prove statements concerning the natural numbers whose proofs are otherwise unmanageable. Section 9.2 introduces sequences, that is, functions whose domains are sets of integers. Most notable among sequences are arithmetic sequences and geometric sequences, whose properties we discuss in Sections 9.3 and 9.4. In Section 9.5 we turn to permutations and combinations, which are related to counting the number of elements in finite sets, and we use these ideas in the final section as we discuss the expansion of $(x + y)^n$ for an arbitrary positive integer n.

9.1
MATHEMATICAL INDUCTION

To set the stage for this axiom, let us consider sums of consecutive odd positive integers:

$$1 = 1$$
$$1 + 3 = 4$$
$$1 + 3 + 5 = 9 \qquad (1)$$
$$1 + 3 + 5 + 7 = 16$$
$$1 + 3 + 5 + 7 + 9 = 25$$

From the equations above, can you guess a formula for the sum

$$1 + 3 + 5 + \cdots + (2n - 1)$$

of the smallest n odd positive integers? If we notice that the right sides of the respective equations in (1) are 1^2, 2^2, 3^2, 4^2, and 5^2, then it becomes apparent that each of those equations has the form

$$1 + 3 + 5 + \cdots + (2n - 1) = n^2 \qquad (2)$$

This suggests that (2) should be valid for *any* positive integer n. But how can we prove that a formula such as (2) is valid for every positive integer n? For a specific value, such as $n = 11$, we could compute the value of each side of (2) and check that the two values are the same. But it would be impossible to check (2) in this way for every positive integer n, since that would require infinitely many calculations. This is where the new axiom plays its role.

Informally, the axiom of mathematical induction states that we can prove a formula such as (2) for every positive integer n by verifying two statements:

    i. The formula is true for $n = 1$.
    ii. For any given positive integer k, if the formula is true for $n = k$, then it is also true for $n = k + 1$.

Let us verify (i) and (ii) for the formula in (2). In the first place, for $n = 1$ we have $2n - 1 = 1$, so that (2) becomes $1 = 1^2$, which is true. This verifies (i). Now to verify (ii), we suppose that any positive integer k is given and that the formula is true for $n = k$, that is,

$$1 + 3 + 5 + \cdots + (2k - 1) = k^2 \qquad (3)$$

From this we will prove that the formula is true for $n = k + 1$; that is, we will prove that

$$1 + 3 + 5 + \cdots + (2k - 1) + [2(k + 1) - 1] = (k + 1)^2$$

which can be rewritten as

$$1 + 3 + 5 + \cdots + (2k - 1) + (2k + 1) = (k + 1)^2 \qquad (4)$$

We accomplish this by adding $2k + 1$ to both sides of (3) and then combining terms to obtain

$$1 + 3 + 5 + \cdots + (2k - 1) + (2k + 1) = k^2 + (2k + 1)$$
$$= (k + 1)^2$$

which is (4). Thus (ii) is verified.

We have proved (i) and (ii) for the formula in (2), so our version of mathematical induction above would imply that (2) is true for every positive integer n.

Now let us state the axiom of mathematical induction formally.

> **THE AXIOM OF MATHEMATICAL INDUCTION**   For each positive integer n, let a statement (or formula or equation) $S(n)$ be given. Suppose that
>
> i. $S(1)$ is true.
> ii. For any positive integer k, if $S(k)$ is true, then $S(k + 1)$ is true.
>
> Then $S(n)$ is true for every positive integer n.

Step (ii) is frequently called the *inductive step*.

The ideas involved in mathematical induction might be illuminated by an analogy utilizing dominoes. Suppose an infinite string of dominoes is lined up (Figure 9.1a), and identify $S(n)$ as the statement "the nth domino falls over." In order for us to conclude

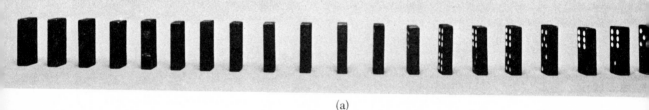

(a)

(b)

FIGURE 9.1

that all dominoes will fall over, we need only see the initial domino fall over, and be assured that once an arbitrary kth domino falls, the neighboring $(k + 1)$st domino also falls (Figure 9.1b).

Mathematical induction is often used to prove that formulas are true for all positive integers. We will illustrate its use now.

Example 1. Use mathematical induction to prove that for any positive integer n,

$$1 + 2 + 3 + \cdots + n = \frac{1}{2} n(n + 1) \qquad (5)$$

Solution. Let $S(n)$ be the equation in (5). We must verify (i) and (ii) in the axiom of mathematical induction. First, $S(1)$ is the equation

$$1 = \frac{1}{2} (1)(1 + 1)$$

which is true. Therefore (i) is verified. To verify (ii), we assume that $S(k)$ is true, that is,

$$1 + 2 + 3 + \cdots + k = \frac{1}{2} k(k + 1) \qquad (6)$$

From $S(k)$ we will prove that $S(k + 1)$ is true, that is,

$$1 + 2 + 3 + \cdots + k + (k + 1) = \frac{1}{2} (k + 1)(k + 2)$$

To accomplish this, we add $k + 1$ to both sides of (6) and rearrange the right side:

$$1 + 2 + 3 + \cdots + k + (k + 1) = \frac{1}{2} k(k + 1) + (k + 1)$$

$$= (k + 1) \left(\frac{1}{2} k + 1 \right)$$

$$= \frac{1}{2} (k + 1)(k + 2)$$

Thus from $S(k)$ we have obtained $S(k + 1)$, thereby verifying (ii). Consequently mathematical induction implies that (5) is valid for any positive integer n. □

Caution: Let us emphasize that in applying mathematical induction correctly, we must verify that *both* conditions (i) and (ii) hold. If we neglect one or the other of the conditions, we can be led to false statements (see Exercises 17 and 18). Moreover, in the verification of (ii), we do not prove that $S(k + 1)$ is true. We prove only that *if* $S(k)$ is true, *then* $S(k + 1)$ is also true.

Example 2. Use mathematical induction to prove that for any positive integer n,

$$x^n - 1 = (x - 1)(1 + x + x^2 + \cdots + x^{n-1}) \tag{7}$$

Solution. In order to verify (i) and (ii) of the mathematical induction axiom, we let the equation in (7) be $S(n)$. Then $S(1)$ is the equation

$$x^1 - 1 = (x - 1)(1)$$

which is obviously valid. Thus (i) is verified. For (ii) we let k be any given positive integer and assume that $S(k)$ is valid, that is,

$$x^k - 1 = (x - 1)(1 + x + x^2 + \cdots + x^{k-1}) \tag{8}$$

From $S(k)$ we will prove that $S(k + 1)$ is valid, that is,

$$x^{k+1} - 1 = (x - 1)(1 + x + x^2 + \cdots + x^{k-1} + x^k) \tag{9}$$

Applying (8), we alter the right side of (9) as follows:

$(x - 1)(1 + x + x^2 + \cdots + x^{k-1} + x^k)$

$$= (x - 1)[(1 + x + x^2 + \cdots + x^{k-1}) + x^k]$$

$$= \underbrace{(x - 1)(1 + x + x^2 + \cdots + x^{k-1})}_{\Downarrow (8) \atop (x^k - 1)} + (x - 1)x^k$$

$$= \qquad\qquad (x^k - 1) \qquad\qquad + (x - 1)x^k$$

$$= x^k - 1 + x^{k+1} - x^k$$

$$= x^{k+1} - 1$$

Thus from $S(k)$ we have obtained $S(k + 1)$, so (ii) is verified. Consequently mathematical induction implies that (7) is valid for all positive integers n. $\square$

Mathematical induction can also be used in order to prove that inequalities are valid.

Example 3. Use mathematical induction to prove that for any positive integer n,

$$\text{if } |x| < 1, \quad \text{then} \quad |x^n| < 1 \tag{10}$$

Solution. Here we let $S(n)$ be the statement in (10). To verify (i) in the axiom, we notice first that $S(1)$ is the statement

$$\text{if } |x| < 1, \quad \text{then} \quad |x^1| < 1$$

Since $x^1 = x$, $S(1)$ is true, so (i) is verified. To verify (ii), we let k be any given positive integer and assume that $S(k)$ is valid, which means that

$$\text{if } |x| < 1, \quad \text{then} \quad |x^k| < 1 \tag{11}$$

From $S(k)$ we will prove that $S(k + 1)$ is valid, which means proving that

$$\text{if } |x| < 1, \quad \text{then} \quad |x^{k+1}| < 1 \tag{12}$$

Using properties of the absolute value and then (11), we find that

$$\text{if } |x| < 1, \quad \text{then} \quad |x^{k+1}| = |x^k x| = |x^k||x| < 1 \cdot 1 = 1 \tag{13}$$

Thus from $S(k)$ we have obtained $S(k + 1)$, thereby verifying (ii). By mathematical induction, (10) is true for all positive integers n. $\square$

Extended Mathematical Induction

Occasionally we wish to prove that a statement or formula $S(n)$ is true for all integer values of n greater than or equal to some integer m, where $m \neq 1$. There is an extended version of mathematical induction that can be used to accomplish this. The inductive step (ii) remains essentially unchanged, and in step (i) we simply replace $S(1)$ by $S(m)$.

> **AXIOM OF EXTENDED MATHEMATICAL INDUCTION** Let m be an integer. For each integer $n \geq m$, let a statement (or formula or equation) $S(n)$ be given. Suppose that
>
>   i. $S(m)$ is true.
>   ii. For each integer $k \geq m$, if $S(k)$ is true, then $S(k + 1)$ is true.
>
> Then $S(n)$ is true for every integer $n \geq m$.

To illustrate extended mathematical induction, we prove one of the Laws of Logarithms that was presented without proof in Chapter 5.

Example 4. Let a be a positive number. Prove that

$$\ln a^n = n \ln a \quad \text{for any integer } n \geq 0 \tag{14}$$

Solution. Let $S(n)$ be the statement in (14). Here we have $m = 0$, so $S(m)$ is the statement

$$\ln a^0 = 0 \ln a \tag{15}$$

Since $a^0 = 1$ and $\ln 1 = 0$, (15) is true, and hence $S(m)$ is true. This verifies (i). Now we let k be an integer greater than or equal to 0 and suppose that $S(k)$ is valid, that is,

$$\ln a^k = k \ln a \qquad (16)$$

From this we will prove that $S(k + 1)$ is valid, that is,

$$\ln a^{k+1} = (k + 1) \ln a$$

Using the formula $\ln (bc) = \ln b + \ln c$ with $b = a^k$ and $c = a$, along with (16), we have

$$\ln a^{k+1} = \ln (a^k \cdot a) = \ln a^k + \ln a$$
$$= k \ln a + \ln a = (k + 1) \ln a$$

which is $S(k + 1)$. Thus from $S(k)$ we have obtained $S(k + 1)$, thereby verifying (ii). By extended mathematical induction, (14) is true for all nonnegative integers n. ☐

EXERCISES 9.1

In Exercises 1–16, use mathematical induction to prove the given formula for every positive integer n.

1. $2 + 4 + 6 + \cdots + 2n = n(n + 1)$
2. $1 + 4 + 7 + \cdots + (3n - 2) = \frac{1}{2}n(3n - 1)$
3. $1 + 5 + 9 + \cdots + (4n - 3) = n(2n - 1)$
4. $1 + 2 + 2^2 + \cdots + 2^{n-1} = 2^n - 1$
5. $1 + 3 + 3^2 + \cdots + 3^{n-1} = \frac{1}{2}(3^n - 1)$
6. $1 + 4 + 4^2 + \cdots + 4^{n-1} = \frac{1}{3}(4^n - 1)$
7. $\dfrac{1}{2} + \dfrac{1}{2^2} + \dfrac{1}{2^3} + \cdots + \dfrac{1}{2^n} = 1 - \dfrac{1}{2^n}$
8. $\dfrac{1}{3} + \dfrac{1}{3^2} + \dfrac{1}{3^3} + \cdots + \dfrac{1}{3^n} = \dfrac{1}{2}\left(1 - \dfrac{1}{3^n}\right)$
9. $\dfrac{1}{1 \cdot 2} + \dfrac{1}{2 \cdot 3} + \dfrac{1}{3 \cdot 4} + \cdots + \dfrac{1}{n(n + 1)} = \dfrac{n}{n + 1}$
10. $\dfrac{1}{1 \cdot 3} + \dfrac{1}{3 \cdot 5} + \dfrac{1}{5 \cdot 7} + \cdots + \dfrac{1}{(2n - 1)(2n + 1)} = \dfrac{n}{2n + 1}$
*11. $1^2 + 2^2 + 3^2 + \cdots + n^2 = \frac{1}{6}n(n + 1)(2n + 1)$
*12. $1^3 + 2^3 + 3^3 + \cdots + n^3 = \frac{1}{4}n^2(n + 1)^2$
*13. $1^2 + 3^2 + 5^2 + \cdots + (2n - 1)^2 = \frac{1}{3}n(2n - 1)(2n + 1)$
*14. $1^3 + 3^3 + 5^3 + \cdots + (2n - 1)^3 = n^2(2n^2 - 1)$
15. $1 \cdot 2 + 3 \cdot 4 + 5 \cdot 6 + \cdots + (2n - 1)(2n) = \frac{1}{3}n(n + 1)(4n - 1)$
16. $1 \cdot 3 + 2 \cdot 4 + 3 \cdot 5 + \cdots + n(n + 2) = \frac{1}{6}n(n + 1)(2n + 7)$

17. Show that $n^2 - n + 11$ is a prime number for $n = 1, 2, \ldots, 10$ but not for $n = 11$.

18. Consider the proposed equation

$$2 + 4 + 6 + \cdots + 2n \stackrel{?}{=} n^2 + n + 2$$

    a. Under the assumption that the equation is valid for a given positive integer k, prove that it is valid for $k + 1$.

    b. Show that the equation is not valid for $n = 1, 2$, or 3.

   *c. Show that the equation is not valid if n is a multiple of 4, and use parts (a) and (b) to conclude that the equation is false for each positive integer n. (*Hint:* Show that if n is a multiple of 4, then the left side is divisible by 4 but the right side is not.)

In Exercises 19–38, use induction or extended mathematical induction to prove the given inequality or statement for the given values of n.

19. $n \le 2^{n-1}$ for $n \ge 1$

20. $4^n > 2^n + 3^n$ for $n \ge 2$

21. $n^3 > (n + 1)^2$ for $n \ge 3$

*22. $\left(\dfrac{3}{2}\right)^{n-1} \ge n$ for $n \ge 5$

23. 2 is a factor of $n^2 + n$ for each positive integer n.

24. 3 is a factor of $n^3 + 2n$ for each positive integer n.

25. 4 is a factor of $5^n - 1$ for each positive integer n. (*Hint:* $5^{k+1} - 1 = 5(5^k - 1) + 4$.)

26. If $x \ne y$, then $x - y$ is a factor of $x^n - y^n$ for any positive integer n. (*Hint:* $x^{k+1} - y^{k+1} = x(x^k - y^k) + y^k(x - y)$.)

27. If $x \ne -y$, then $x + y$ is a factor of $x^{2n+1} + y^{2n+1}$ for any nonnegative integer n. (*Hint:* $x^{2k+3} + y^{2k+3} = (x^{2k+2} + y^{2k+2})(x + y) - xy(x^{2k+1} + y^{2k+1})$.)

28. If $x > 1$, then $x^n > 1$ for any positive integer n.

29. The product of an even number of negative numbers is positive.

30. The product of an odd number of negative numbers is negative.

31. If x is any number and m and n are positive integers, then $(x^m)^n = x^{mn}$. (*Hint:* Let m be any fixed positive integer and use mathematical induction on n.)

32. For any positive number x and any integer $n \ge 2$, $(1 + x)^n > 1 + nx$.

33. If x is any real number and n any positive integer, then

$$(1 - x)(1 + x)(1 + x^2)(1 + x^4) \cdots (1 + x^{2^n}) = 1 - x^{2^{n+1}}$$

34. For any integer $n \ge 2$ and any real numbers $a_1, a_2, \ldots a_n$,

$$e^{a_1 + a_2 + \cdots + a_n} = e^{a_1} e^{a_2} \cdots e^{a_n}$$

35. For any integer $n \ge 2$ and any real numbers $a_1, a_2, \ldots, a_n$,

$$\ln(a_1 a_2 \cdots a_n) = \ln a_1 + \ln a_2 + \cdots + \ln a_n$$

*36. Let $n \ge 3$. Suppose n distinct points are given, no three of which lie on a single line. Then the total number of lines joining any two of the given points is $\frac{1}{2}n(n - 1)$.

*37. The number of diagonals of a convex polygon of n sides is $\frac{1}{2}n(n-3)$.

38. For any positive integer n, $\frac{1}{2}(n^4 + 5n)$ is an integer.

*39. A game called the ***Tower of Hanoi*** consists of three pegs and an arbitrary number of rings of distinct diameters. At the start of the game all the rings are on one peg, arranged according to size, with the smallest on top and the largest on bottom (Figure 9.2). The object of the game is to transfer all the rings to a different peg so that the rings have the same order on the new peg as on the old. Under the rules of the game, only one ring may be moved at a time, and no ring may be placed on top of a smaller ring. All three pegs may be used. Use mathematical induction to show that if there are n rings, then it is possible to complete the game in $2^n - 1$ moves.

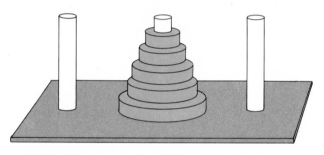

FIGURE 9.2

9.2

SEQUENCES

Earlier we studied functions whose domains were intervals or combinations of intervals. Now we will introduce functions whose domains consist of sets of integers.

DEFINITION 9.1

A ***sequence*** is a function whose domain is the collection of all integers greater than or equal to a given integer m (usually 0 or 1).

For example, the function f defined by

$$f(n) = n^2 \quad \text{for } n \geq 0 \tag{1}$$

is a sequence, since its domain consists of all nonnegative integers. A few specific values of f are

$$f(0) = 0^2 = 0, \quad f(4) = 4^2 = 16, \quad \text{and} \quad f(19) = 19^2 = 361$$

In this particular case, the range consists of all integers that are squares of integers.

Another example of a sequence is given by

$$g(n) = (-1)^n \quad \text{for } n \geq 5 \qquad (2)$$

The range of g consists of -1 and 1, because $(-1)^n = 1$ or -1, depending on whether n is even or odd. A sequence whose range does not consist solely of integers is given by

$$h(n) = \frac{1}{n} \quad \text{for } n \geq 1 \qquad (3)$$

The range of h consists of the numbers $1, \frac{1}{2}, \frac{1}{3}, \ldots$.

Let f represent an arbitrary sequence, and let the value assigned to any integer n in the domain be written a_n (read as "a sub n"). Then a formula for f is

$$f(n) = a_n \quad \text{for } n \geq m \qquad (4)$$

The numbers $a_m, a_{m+1}, a_{m+2}, \ldots$, completely determine the sequence f, and because the values of the sequence can be listed in this way, we normally suppress the expression $f(n)$ appearing in (4) and write $\{a_n\}_{n=m}^{\infty}$ to represent the sequence in (4). Thus the sequences given in (1)–(3) would be given in the form

$$\{n^2\}_{n=0}^{\infty}, \quad \{(-1)^n\}_{n=5}^{\infty}, \quad \text{and} \quad \left\{\frac{1}{n}\right\}_{n=1}^{\infty}$$

In order to get a feeling for the numbers in a sequence, we frequently write out the first few terms of the sequence. For example, we would write

$$0, 1, 4, 9, 16, \ldots \quad \text{for} \quad \{n^2\}_{n=0}^{\infty}$$

$$-1, 1, -1, 1, -1, 1, \ldots \quad \text{for} \quad \{(-1)^n\}_{n=5}^{\infty}$$

$$1, \frac{1}{2}, \frac{1}{3}, \frac{1}{4}, \ldots \quad \text{for} \quad \left\{\frac{1}{n}\right\}_{n=1}^{\infty}$$

The numbers $a_m, a_{m+1}, a_{m+2}, \ldots$ are the **terms** of the sequence $\{a_n\}_{n=m}^{\infty}$, with a_m the first term, a_{m+1} the second term, and so on. The numbers $m, m+1, m+2, \ldots$ are the **indexes** of the sequence, and m is the **initial index**; after all, a_m is the first term one sees when listing them in order. For $\{1/n\}_{n=1}^{\infty}$ we have

1st term, 2nd term, 3rd term, 4th term

$$1, \frac{1}{2}, \frac{1}{3}, \frac{1}{4}, \ldots$$

Example 1. Find the numerical values of the first, second, third, and seventh terms of the sequence $\{a_n\}_{n=m}^{\infty}$, where

a. $a_n = \dfrac{1}{2^n}$ for $n \geq 1$ b. $a_n = 2n - 4$ for $n \geq 0$

c. $a_n = 2n - 4$ for $n \geq 3$

Solution.

a. For this sequence, $m = 1$, so the terms we seek are a_1, a_2, a_3, and a_7:

$$a_1 = \frac{1}{2^1} = \frac{1}{2} \qquad a_3 = \frac{1}{2^3} = \frac{1}{8}$$

$$a_2 = \frac{1}{2^2} = \frac{1}{4} \qquad a_7 = \frac{1}{2^7} = \frac{1}{128}$$

b. Here $m = 0$, so the terms we seek are a_0, a_1, a_2, and a_6:

$$a_0 = 2 \cdot 0 - 4 = -4 \qquad a_2 = 2 \cdot 2 - 4 = 0$$

$$a_1 = 2 \cdot 1 - 4 = -2 \qquad a_6 = 2 \cdot 6 - 4 = 8$$

c. In this case $m = 3$, so the terms we seek are a_3, a_4, a_5, and a_9:

$$a_3 = 2 \cdot 3 - 4 = 2 \qquad a_5 = 2 \cdot 5 - 4 = 6$$

$$a_4 = 2 \cdot 4 - 4 = 4 \qquad a_9 = 2 \cdot 9 - 4 = 14 \quad \square$$

Example 2. Find the specified terms of the following sequences.

a. the tenth term, where $a_n = \dfrac{1}{(n + 1)(n + 2)}$ for $n \geq 1$

b. the eighty-fourth term, where $a_n = \pi$ for $n \geq 3$

Solution.

a. Since $m = 1$, the tenth term is a_{10}. We find that

$$a_{10} = \frac{1}{(10 + 1)(10 + 2)} = \frac{1}{11 \cdot 12} = \frac{1}{132}$$

b. Since $n = 3$, the eighty-fourth term is a_{86}, and since each term is π, it follows that

$$a_{86} = \pi \quad \square$$

The sequence given in part (b) of Example 2 has only one value, namely π. We say that $\{a_n\}_{n=m}^{\infty}$ is a **constant sequence** if it has but one value, that is, if there is a unique number c such that

$$a_n = c \quad \text{for } n \geq m$$

Thus the sequence in part (b) of Example 2 is a constant sequence.

Not all sequences have terms all of which are as easy to calculate as those in Examples 1 and 2 are. A sequence $\{a_n\}_{n=m}^{\infty}$ is given *recursively* (or inductively) if the initial term a_m is given, and if we can determine a_{n+1} by knowing $a_m, a_{m+1}, \ldots, a_n$.

Example 3. Find the first four terms of the sequences described below.

a. Initial term is a_1, $a_1 = 3$, and $a_{n+1} = a_n - 2$ for $n \geq 1$

b. Initial term is a_0, $a_0 = 1$, and $a_{n+1} = \frac{1}{3}a_n$ for $n \geq 0$

Solution.

a. $a_1 = 3$, $a_2 = a_1 - 2 = 3 - 2 = 1$, $a_3 = a_2 - 2 = 1 - 2 = -1$, $a_4 = a_3 - 2 = -1 - 2 = -3$

b. $a_0 = 1$, $a_1 = \frac{1}{3}a_0 = \frac{1}{3} \cdot 1 = \frac{1}{3}$, $a_2 = \frac{1}{3}a_1 = \frac{1}{3} \cdot \frac{1}{3} = \frac{1}{9}$, $a_3 = \frac{1}{3}a_2 = \frac{1}{3} \cdot \frac{1}{9} = \frac{1}{27}$ □

An important sequence that can be defined recursively is given by

$$a_0 = 1, \quad \text{and} \quad a_{n+1} = (n + 1)a_n \quad \text{for } n \geq 0 \tag{5}$$

The value of a_n is called *n factorial* and is denoted by $n!$. The relation $a_{n+1} = (n + 1)a_n$ can be written in factorial form as

$$(n + 1)! = (n + 1)n!$$

The first few terms of $n!$ are given by

$$0! = 1$$

$$1! = 1 \cdot 0! = 1$$

$$2! = 2 \cdot 1! = 2 \cdot 1 = 2$$

$$3! = 3 \cdot 2! = 3 \cdot 2 \cdot 1 = 6$$

$$4! = 4 \cdot 3! = 4 \cdot 3 \cdot 2 \cdot 1 = 24$$

$$5! = 5 \cdot 4! = 5 \cdot 4 \cdot 3 \cdot 2 \cdot 1 = 120$$

$$6! = 6 \cdot 5! = 6 \cdot 5 \cdot 4 \cdot 3 \cdot 2 \cdot 1 = 720$$

$$7! = 7 \cdot 6! = 7 \cdot 6 \cdot 5 \cdot 4 \cdot 3 \cdot 2 \cdot 1 = 5040$$

In general, if $n \geq 1$, then $n!$ is the product of the first n positive integers:

$$n! = n(n - 1)(n - 2) \cdots 3 \cdot 2 \cdot 1 \tag{6}$$

Caution: Formula (6) does not give a value for $0!$; the right side does not even make sense for $n = 0$. However, we defined $0!$ by the formula $0! = 1$ above.

When the terms in a sequence are not easily computed, we sometimes approximate their values.

Example 4. Calculate the first four terms of the sequence $\left\{\left(1 + \dfrac{1}{n}\right)^n\right\}_{n=1}^{\infty}$.

Solution. If $a_n = \left(1 + \dfrac{1}{n}\right)^n$ for $n \geq 1$, then

$$a_1 = \left(1 + \frac{1}{1}\right)^1 = 2$$

$$a_2 = \left(1 + \frac{1}{2}\right)^2 = \left(\frac{3}{2}\right)^2 = \frac{9}{4} = 2.25$$

$$a_3 = \left(1 + \frac{1}{3}\right)^3 = \left(\frac{4}{3}\right)^3 = \frac{64}{27} \approx 2.3703704$$

$$a_4 = \left(1 + \frac{1}{4}\right)^4 = \left(\frac{5}{4}\right)^4 = \frac{625}{256} \approx 2.4414063 \quad \square$$

To calculate higher terms of the sequence in Example 4, say a_{10} or a_{20}, requires ever more patience, or help from a calculator. For instance, by calculator we find that

$$a_{10} = \left(1 + \frac{1}{10}\right)^{10} \approx 2.5937425$$

and

$$a_{20} = \left(1 + \frac{1}{20}\right)^{20} \approx 2.6532977$$

It turns out that as n increases, the value of $(1 + 1/n)^n$ gets nearer and nearer to e. In fact, one can define e to be the value to which the numbers $(1 + 1/n)^n$ tend as n increases without bound.

Not all sequences can be represented by neat numerical formulas. Indeed, if we call a nontrivial factor of an integer n any factor other than 1, -1, n, and $-n$ and if for $n \geq 1$

$$a_n = \text{the number of nontrivial factors of } n$$

then the sequence begins as follows:

$$0, 0, 0, 2, 0, 4, \ldots$$

But there is no simple formula that tells immediately the numerical value of a_n for each positive integer n.

It is time to mention one of the most famous sequences in the mathematical literature, the Fibonacci sequence, named after the thirteenth-century mathematician Fibonacci (whose real name was Leonardo de Pisa). He posed and solved the following problem.

Suppose a rabbit colony starts out with 1 pair of adult rabbits. Assume that each pair of adult rabbits produces a pair of offspring (one male and one female) every month and that rabbits produce their first offspring at the age of 2 months and continue to live forever. How many pairs of adult rabbits are there at the end of n months, for any positive integer n?

To help in the solution of this problem, let us call rabbits less than 1 month old babies, those between 1 and 2 months old adolescents, and all older rabbits adults. From one month to the next, babies become adolescents, adolescents become adults, adults remain adults, and two new babies are born to each pair of adults. At the end of successive months we have:

Month	Pairs of Adults	Pairs of Adolescents	Pairs of Babies
1	1	0	1
2	1	1	1
3	2	1	2
4	3	2	3
5	5	3	5
6	8	5	8
7	13	8	13
8	21	13	21

If for $n \geq 1$

a_n = the number of pairs of adults at the end of n months

then

$$a_1 = 1, \qquad a_2 = 1, \quad \text{and} \quad a_{n+1} = a_n + a_{n-1} \quad \text{for } n \geq 2$$

The sequence $\{a_n\}_{n=1}^{\infty}$ is called the **Fibonacci sequence**.

Graphs of Sequences

The graph of a sequence is not as revealing as graphs of the kinds of functions we described in earlier chapters, because the graph of a sequence is just a collection of isolated points, one on each vertical line $x = n$ for $n \geq m$. The following example illustrates one such graph.

Example 5. Sketch the graph of the sequence

$$\left\{ 2 + (-1)^n \left(\frac{1}{n}\right) \right\}_{n=1}^{\infty}$$

Solution. If $a_n = 2 + (-1)^n \left(\frac{1}{n}\right)$ for $n \geq 1$, then

$$a_1 = 1, \quad a_2 = \frac{5}{2}, \quad a_3 = \frac{5}{3}, \quad a_4 = \frac{9}{4}, \quad a_5 = \frac{9}{5}, \quad a_6 = \frac{13}{6}$$

The graph, which is shown in Figure 9.3, suggests that a_n gets closer and closer to 2 as n increases. $\square$

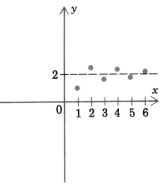

$a_n = 2 + (-1)^n \left(\frac{1}{n}\right)$ for $n \geq 1$

FIGURE 9.3

**Partial Sums
of Sequences**

Upon occasion it is important to find the sum of the first n terms of a sequence. For any sequence $\{a_n\}_{n=1}^{\infty}$ the expression $\sum_{j=1}^{n} a_j$ represents the sum of the first n terms of the sequence and is called the **nth partial sum** of the sequence. Thus

$$\sum_{j=1}^{n} a_j = a_1 + a_2 + a_3 + \cdots + a_n$$

The symbol Σ is a capital Greek sigma, which we use to represent a sum.

Example 6. Find the following partial sums.

a. $\displaystyle\sum_{j=1}^{5} j$ 　　　　 b. $\displaystyle\sum_{j=1}^{6} j^2$ 　　　　 c. $\displaystyle\sum_{j=1}^{4} \frac{1}{j!}$

Solution.

a. $\displaystyle\sum_{j=1}^{5} j = 1 + 2 + 3 + 4 + 5 = 15$

b. $\displaystyle\sum_{j=1}^{6} j^2 = 1^2 + 2^2 + 3^2 + 4^2 + 5^2 + 6^2 = 91$

c. $\displaystyle\sum_{j=1}^{4} \frac{1}{j!} = \frac{1}{1!} + \frac{1}{2!} + \frac{1}{3!} + \frac{1}{4!} = \frac{41}{24}$ 　$\square$

Example 7. Find the nth partial sum $\sum_{j=1}^{n} j$ of the sequence $\{n\}_{n=1}^{\infty}$.

Solution. By Example 1 of Section 9.1,

$$\sum_{j=1}^{n} j = 1 + 2 + 3 + \cdots + n = \frac{1}{2} n(n + 1) \quad \square$$

If the initial index of a sequence is not 1, then the nth partial sum is changed accordingly. For example, the nth partial sum of the sequence $\{a_n\}_{n=0}^{\infty}$, which has initial index 0, is $\sum_{j=0}^{n-1} a_j$. Thus the fourth partial sum of $\{1/2^n\}_{n=0}^{\infty}$ is

$$\sum_{j=0}^{3} \frac{1}{2^j} = \frac{1}{2^0} + \frac{1}{2^1} + \frac{1}{2^2} + \frac{1}{2^3} = 1 + \frac{1}{2} + \frac{1}{4} + \frac{1}{8} = \frac{15}{8}$$

EXERCISES 9.2

In Exercises 1–6, write the given sequence in the form $\{a_n\}_{n=1}^{\infty}$.

1. $\frac{1}{2}, \frac{1}{3}, \frac{1}{4}, \frac{1}{5}, \ldots$

2. $\frac{1}{2}, \frac{2}{3}, \frac{3}{4}, \frac{4}{5}, \ldots$

3. $-1, 1, -1, 1, \ldots$

4. $0, \dfrac{\ln 2}{2}, \dfrac{\ln 3}{3}, \dfrac{\ln 4}{4}, \ldots$

5. $\frac{8}{9}, \frac{16}{27}, \frac{32}{81}, \frac{64}{243}, \ldots$

6. $1, \frac{1}{2}, \frac{1}{6}, \frac{1}{24}, \frac{1}{120}, \ldots$

In Exercises 7–16, find the first four terms of the sequence $\{a_n\}_{n=1}^{\infty}$ with the given nth term.

7. $a_n = \frac{1}{2}\sqrt{3}$

8. $a_n = 3n - 1$

9. $a_n = 2^n$

10. $a_n = \dfrac{1}{3^{n+1}}$

11. $a_n = (-1)^{n+1}$

12. $a_n = (-1)^{n+1} \cdot \dfrac{1}{2^n}$

13. $a_n = \dfrac{1}{n(n+1)}$

14. $a_n = n^n$

15. $a_n = \ln n$

16. $a_n = \dfrac{n-1}{n!}$

In Exercises 17–26, find the first five terms of the sequence defined recursively.

17. $a_1 = 2;\ a_n = 2a_{n-1}$

18. $a_1 = -1;\ a_n = 3a_{n-1}$

19. $a_1 = 3;\ a_n = 4 + a_{n-1}$

20. $a_1 = 3;\ a_n = 4 - a_{n-1}$

21. $a_1 = 32;\ a_n = \frac{1}{2}a_{n-1}$

22. $a_1 = -243;\ a_n = -\frac{1}{3}a_{n-1}$

23. $a_1 = 1;\ a_n = na_{n-1}$

24. $a_1 = 28;\ a_n = \dfrac{1}{a_{n-1}}$

25. $a_1 = 1;\ a_n = \dfrac{na_{n-1}}{(n-1)!}$

26. $a_1 = 2^{1/12};\ a_n = a_{n-1}^n$

In Exercises 27–32, sketch the graph of the sequence with the given nth term.

27. $a_n = \dfrac{2}{n}$

28. $a_n = 1 + \dfrac{1}{2^n}$

29. $a_n = (-1)^n$

30. $a_n = n(-1)^n$

31. $a_n = 2(-1)^{n-1}$

32. $a_n = \dfrac{n}{n+1}$

In Exercises 33–42, find the indicated partial sum.

33. $\displaystyle\sum_{j=1}^{10} 1$

34. $\displaystyle\sum_{j=1}^{6} (-3)$

35. $\displaystyle\sum_{j=1}^{9} (-1)^j$

36. $\displaystyle\sum_{j=1}^{100} (-1)^j$

37. $\displaystyle\sum_{j=1}^{5} [2 + (-1)^j]$

38. $\displaystyle\sum_{j=1}^{4} \dfrac{1}{j}$

39. $\displaystyle\sum_{j=1}^{3} \dfrac{1}{2^j}$

40. $\displaystyle\sum_{j=1}^{6} \dfrac{1}{2^j}$

41. $\displaystyle\sum_{j=1}^{10} \left(\dfrac{1}{j+1} - \dfrac{1}{j}\right)$

42. $\displaystyle\sum_{j=1}^{100} \left(\dfrac{1}{j+1} - \dfrac{1}{j}\right)$

In Exercises 43–48, express the given sum in sigma notation.

43. $1 + 2 + 3 + 4 + 5 + 6 + 7$

44. $1^2 + 2^2 + 3^2 + 4^2 + 5^2 + 6^2$

45. $1^1 + 2^2 + 3^3 + 4^4 + 5^5 + 6^6 + 7^7 + 8^8 + 9^9$

46. $\frac{1}{3} + \frac{2}{4} + \frac{3}{5} + \frac{4}{6} + \frac{5}{7}$

47. $1 + 3 + 5 + 7 + 9 + 11 + 13 + 15$

48. $1 + \frac{1}{1} + 2 + \frac{1}{2} + 3 + \frac{1}{3} + 4 + \frac{1}{4}$

In Exercises 49–56, use a calculator to approximate the indicated values of a_n.

49. $a_n = \dfrac{\ln n}{n^2}$; a_{10}, a_{100}

50. $a_n = \dfrac{2n - 19}{n^2 + 17}$; a_{39}, a_{500}

51. $a_n = \dfrac{n!}{n^n}$; a_5, a_{15}

52. $a_n = e^{-n^2}$; a_2, a_4

53. $a_n = ne^{-n}$; a_{10}, a_{20}

54. $a_n = \sqrt[n]{100}$; a_{10}, a_{200}

55. $a_n = \sqrt[n]{n}$; a_{100}, a_{1000}

56. $a_n = \sqrt[n]{n!}$; a_{10}, a_{20}

In Exercises 57–60, use a calculator to approximate the value of the given partial sum.

57. $\displaystyle\sum_{j=1}^{10} \frac{1}{j}$

58. $\displaystyle\sum_{j=1}^{20} \frac{1}{j}$

59. $\displaystyle\sum_{j=1}^{15} \frac{1}{j^2}$

60. $\displaystyle\sum_{j=1}^{12} \frac{1}{j!}$

61. Find the eighth partial sum of the sequence of prime numbers, whose first five terms are 2, 3, 5, 7, 11 . . .

62. Use mathematical induction to prove that

$$\sum_{j=1}^{n} \frac{1}{2^j} = 1 - \frac{1}{2^n} \quad \text{for } n \geq 1$$

63. Use mathematical induction to prove that the following equations are valid for $n \geq 1$.

a. $\displaystyle\sum_{j=1}^{2n} (-1)^j = 0$

b. $\displaystyle\sum_{j=1}^{2n-1} (-1)^j = -1$

64. Use mathematical induction to prove that if $\{a_n\}_{n=1}^{\infty}$ is defined recursively by $a_1 = 1$ and $a_n = na_{n-1}$ for $n \geq 2$, then $a_n = n!$ for $n \geq 1$.

65. Let $a_n =$ the nth digit after the decimal point in the decimal expansion of π. Write a_5 and a_8.

66. Let $a_n =$ the nth digit after the decimal point in the decimal expansion of $\frac{1}{7}$. Write a_4, a_{10}, and a_{100}.

67. Let $a_n = \dfrac{(1 + \sqrt{5})^n - (1 - \sqrt{5})^n}{2^n\sqrt{5}}$ for $n \geq 1$.

a. Show that $a_1 = 1$ and $a_2 = 1$.

*b. Show that if $n \geq 1$, then $a_{n+2} = a_{n+1} + a_n$, so that $\{a_n\}_{n=1}^{\infty}$ is the Fibonacci sequence. (*Hint:* Evaluate $a_{n+1} + a_n$ by adding separately the parts containing powers of $(1 + \sqrt{5})$ and the parts containing powers of $(1 - \sqrt{5})$, factoring out common factors in the process. Use the fact that $6 + 2\sqrt{5} = (1 + \sqrt{5})^2$ and $6 - 2\sqrt{5} = (1 - \sqrt{5})^2$.)

c. Use parts (a) and (b) to conclude that a_n is an integer for each positive integer n.

68. Let $\{a_n\}_{n=1}^{\infty}$ be the Fibonacci sequence, defined by $a_1 = 1, a_2 = 1$, and $a_{n+1} = a_n + a_{n-1}$ for $n \geq 2$.

a. Show that $a_n = a_{n+2} - a_{n+1}$ for $n \geq 1$.

b. Using (a) and mathematical induction, show that $\sum_{j=1}^{n} a_j = a_{n+2} - 1$ for $n \geq 1$. (*Hint:* In (a) replace n by $n + 1$.)

69. Express the following in terms of factorials:

    a. $2 \cdot 4 \cdot 6 \cdots (2n)$ b. $\dfrac{1 \cdot 3 \cdot 5 \cdots (2n-1)}{2 \cdot 4 \cdot 6 \cdots (2n)}$

9.3

ARITHMETIC SEQUENCES

Suppose we deposit $100 in a savings account paying 6% simple interest. This means that after the first year the balance is $106, and after each succeeding year the balance increases by $6. If a_n denotes the number of dollars in the account after n years, we have

$$a_1 = 106, \qquad a_2 = 112, \qquad a_3 = 118, \qquad a_4 = 124$$

and in general,

$$a_n = 106 + \overbrace{6 + \cdots + 6}^{n-1}$$

which can be written more simply as

$$a_n = 106 + (n-1)6 \tag{1}$$

The sequence $\{a_n\}_{n=1}^{\infty}$ given in (1) has the property that each term after the initial term is obtained from the preceding one by adding the same fixed number, namely, 6. Such a sequence is called an arithmetic sequence (or arithmetic progression).

DEFINITION 9.2

An *arithmetic sequence* is a sequence such that the difference between any term and the succeeding term is constant.

By definition, a sequence $\{a_n\}_{n=1}^{\infty}$ is an arithmetic sequence if and only if there is a number d such that

$$a_{n+1} - a_n = d \quad \text{for } n \geq 1$$

or equivalently,

$$a_{n+1} = a_n + d \quad \text{for } n \geq 1 \tag{2}$$

The number d is called the *common difference* of the sequence. It follows from the discussion before Definition 9.2 that for the sequence given by (1), the common difference $d = 6$.

Example 1. Show that the sequences listed below are arithmetic sequences, and find their common differences.

    a. $\{2n + 3\}_{n=1}^{\infty}$ b. $3, 2, 1, 0, \ldots,$

Solution.

a. If $a_n = 2n + 3$ for $n \geq 1$, then

$$a_{n+1} - a_n = [2(n + 1) + 3] - (2n + 3)$$
$$= 2n + 5 - 2n - 3$$
$$= 2$$

Therefore the common difference is 2, so the sequence is an arithmetic sequence.

b. In this case, notice that $a_n = 4 - n$ for $n \geq 1$. Then for $n \geq 1$ we have

$$a_{n+1} - a_n = [4 - (n + 1)] - (4 - n)$$
$$= 3 - n - 4 + n$$
$$= -1$$

Consequently the common difference is -1, and this sequence is also an arithmetic sequence. □

If one knows the initial term of an arithmetic sequence, which we normally designate by a_1, and the common difference d, then the value of a_n can be calculated by the formula

$$a_n = a_1 + (n - 1)d \quad \text{for } n \geq 1 \tag{3}$$

The formula can be proved by induction.

Example 2. Find a formula for the nth term of the arithmetic sequence whose initial term a_1 and common difference d are given below.

a. $a_1 = 2, d = 5$ b. $a_1 = \frac{2}{3}, d = -\frac{4}{3}$ c. $a_1 = 2, d = 0$

Solution. In each case we use (3):

a. $a_n = 2 + (n - 1)5 = -3 + 5n$
b. $a_n = \frac{2}{3} + (n - 1)(-\frac{4}{3}) = 2 - \frac{4}{3}n$
c. $a_n = 2 + (n - 1)0 = 2$ □

Partial Sums of Arithmetic Sequences

Theorem 9.3, which we present next, includes two convenient formulas for computing partial sums of arithmetic sequences.

THEOREM 9.3

Let $\{a_n\}_{n=1}^{\infty}$ be an arithmetic sequence. The nth partial sum of the sequence is given by either of the following two formulas:

$$\sum_{j=1}^{n} a_j = a_1 + (a_1 + d) + (a_1 + 2d) + \cdots + (a_1 + (n - 1)d)$$

$$= na_1 + \frac{n(n - 1)}{2} \cdot d \tag{4}$$

$$\sum_{j=1}^{n} a_j = a_1 + (a_1 + d) + (a_1 + 2d) + \cdots + (a_1 + (n - 1)d) \tag{5}$$

$$= \frac{n}{2}(a_1 + a_n)$$

Example 3. Find the sum of the first eight terms of the arithmetic sequence that begins 3, 7, ...

Solution. For the given sequence, $a_1 = 3$ and $a_2 = 7$, so that

$$d = a_2 - a_1 = 7 - 3 = 4$$

Thus by (4) with $n = 8$, the sum S is given by

$$S = 8(3) + \frac{8 \cdot 7}{2} 4 = 24 + 112 = 136 \quad \square$$

EXERCISES 9.3

In Exercises 1–6, show that the sequence is arithmetic, and find the common difference.

1. $\{2 + 3n\}_{n=1}^{\infty}$

2. $\{4 - \pi n\}_{n=1}^{\infty}$

3. $\{5 + \frac{1}{2}n\}_{n=1}^{\infty}$

4. $\left\{\dfrac{3 + 2n}{4}\right\}_{n=1}^{\infty}$

5. $\{\ln 2^n\}_{n=1}^{\infty}$

6. $\{e^{1+\ln 3n}\}_{n=1}^{\infty}$

In Exercises 7–14, determine whether the given sequence is arithmetic.

7. 1, 3, 6, 10, . . .

8. 1, −1, 1, −1, . . .

9. 3, 6, 9, 12, . . .

10. −2, 3, 8, 13, . . .

11. $\{2n + 1\}_{n=1}^{\infty}$

12. $\{3 - 2n^2\}_{n=1}^{\infty}$

13. $\{2 - 3 \ln n\}_{n=1}^{\infty}$

14. $\{\ln (3^n)\}_{n=1}^{\infty}$

In Exercises 15–22, find the fourth term and the nth term of the arithmetic sequence whose initial term a_1 and common difference d are given.

15. $a_1 = 2, d = 3$

16. $a_1 = \frac{3}{4}, d = -1$

17. $a_1 = 2\pi, d = 0$

18. $a_1 = \sqrt{2}, d = 2\sqrt{2}$

19. $a_1 = -3, d = \frac{2}{3}$

20. $a_1 = -3, d = -\frac{2}{3}$

21. $a_1 = e, d = 2$

22. $a_1 = 0, d = -\frac{1}{2}$

In Exercises 23–28, find the seventh term and the nth term of the arithmetic sequence whose first two terms are given.

23. 4, 7, . . .

24. 3, 2, . . .

25. 0, $-\frac{3}{2}$, . . .

26. $\sqrt{3}, 3\sqrt{3}$, . . .

27. ln 2, ln 4, . . .

28. $\log_2 3, \log_2 1$, . . .

In Exercises 29–34, find a formula for the nth term of the arithmetic sequence $\{a_n\}_{n=1}^{\infty}$ that satisfies the given conditions.

29. $a_2 = 5, a_4 = 3$

30. $a_4 = 3, d = 5$

31. $a_3 = 6$, $a_6 = 1$

32. $a_9 = 16$, $d = 2$

33. $a_4 = 3$, $a_6 - a_1 = 5$

34. $a_6 = -5$, $a_{10} - a_2 = -12$

In Exercises 35–42, find $\sum_{j=1}^{n} a_j$ for the arithmetic sequence that satisfies the given conditions.

35. $a_1 = 3$, $d = 2$, $n = 7$

36. $a_1 = 1$, $d = -3$, $n = 8$

37. $a_1 = -2$, $d = 1$, $n = 12$

38. $a_3 = -\sqrt{2}$, $d = \sqrt{2}$, $n = 18$

39. $a_2 = \ln 4$, $d = \ln 2$, $n = 20$

40. $a_1 = 2$, $a_6 = -18$, $n = 6$

41. $a_1 = -3\pi$, $a_{10} = 21\pi/2$, $n = 10$

42. $a_2 = 1$, $a_6 = -7$, $n = 7$

43. In the arithmetic sequence that begins -3, 4, . . . , is 277 a term? Explain why or why not.

44. In the arithmetic sequence that begins $2, 5,$ . . . , is 1000 a term? Explain why or why not.

45. The sum of the second and the sixth terms of a given arithmetic sequence is 2, and the sum of the fifth and the ninth terms is -10. Find a formula for the nth term of the sequence.

46. Find the sum of all positive even integers less than 100.

47. Find the sum of all positive odd integers less than 100.

48. The first three terms of an arithmetic sequence have the form x, $2x + 1$, and $4x - 1$ for a suitable real number x. Determine x and the nth term a_n of the sequence.

49. The sum of the first five terms of an arithmetic sequence is 55, and the fifth term is 15. Determine the nth term a_n of the sequence.

50. The sum of the first six terms of an arithmetic sequence is 24, and the common difference is 24. Determine the nth term a_n of the sequence.

51. Let a and b be real numbers. Show that $(a^2 - 2ab - b^2)^2$, $(a^2 + b^2)^2$, and $(a^2 + 2ab - b^2)^2$ form the first three terms of an arithmetic sequence, and find the common difference.

52. The number of times each hour that a chime clock strikes is equal to the number of the hour. How many times does the clock strike in a 24-hour day?

53. During a 12-week summer vacation a student saves $10 the first week, $15 the second week, $20 the third week, and so on. How much money does the student save during the summer?

54. A boy wishes to save for 8 weeks in order to buy a bicycle for $66. If he saves $3 the first week, then by how much must he increase his savings each week in order to have $66 after 8 weeks?

55. A bricklayer plans to build a brick structure on which to mount a sign. If the first layer of bricks is to contain 30 bricks, the second 29, and third 28, and so on with the top layer containing 9 bricks, how many bricks will the bricklayer need?

*56. The Great Pyramid of Cheops was originally about 480 feet tall and had a square base approximately 754 feet on a side. If each layer was 3 feet tall and succeeding layers were indented 2.35 feet (see Figure 9.4), approximate the area of the stone exposed on the four sides of the pyramid.

The Great Pyramid of Cheops

FIGURE 9.4

57. According to legend, during the last weeks of his life the French mathematician Abraham de Moivre noticed that he needed a quarter-hour more sleep each night than on the preceding night, and he predicted that as soon as he needed a 24-hour sleep he would die. If all this were true, how much sleep did he need after the 12-hour sleep and before death?

9.4

GEOMETRIC SEQUENCES

If instead of successive terms of a sequence having the same difference, the successive terms have the same ratio, the sequence is called a geometric sequence (or geometric progression).

DEFINITION 9.4

A *geometric sequence* is a sequence such that the quotient of any term with its predecessor is the same nonzero number, called the *common ratio* of the geometric sequence.

For example, the sequences

$$1, 2, 4, 8, 16, \ldots \quad \text{and} \quad \{3^{-n}\}_{n=0}^{\infty}$$

are geometric. For the first of these sequences the quotient of any term with its predecessor is 2, and for the second it is $\frac{1}{3}$.

If $\{a_n\}_{n=1}^{\infty}$ is a geometric sequence and r its common ratio, then for any $n \geq 1$,

$$\frac{a_{n+1}}{a_n} = r \qquad (1)$$

This means that

$$a_2 = a_1 r$$
$$a_3 = a_2 r = (a_1 r)r = a_1 r^2$$
$$a_4 = a_3 r = (a_1 r^2)r = a_1 r^3$$

and in general,

$$a_n = a_1 r^{n-1} \qquad (2)$$

a formula that can be proved inductively (see Exercise 70).

Example 1. Find the sixth term and the common ratio of the geometric sequence $\{a_n\}_{n=1}^{\infty}$ that begins 2, 6, ...

Solution. We are given that $a_1 = 2$ and $a_2 = 6$. Therefore

$$r = \frac{a_2}{a_1} = \frac{6}{2} = 3$$

so by (2),

$$a_n = a_1 r^{n-1} = 2 \cdot 3^{n-1}$$

Taking $n = 6$, we find that

$$a_6 = 2(3^5) = 2(243) = 486 \quad \square$$

Example 2. Find a formula for the nth term of the geometric sequence $\{a_n\}_{n=1}^{\infty}$ such that
a. $a_1 = 4$, $r = 2$ b. $a_1 = 1$, $r = \frac{1}{5}$
c. $a_1 = -1$, $r = -\frac{1}{3}$

Solution. In each case we use (2):
a. $a_n = 4(2)^{n-1} = 2^{n+1}$

b. $a_n = 1\left(\frac{1}{5}\right)^{n-1} = \frac{1}{5^{n-1}}$

c. $a_n = (-1)\left(-\frac{1}{3}\right)^{n-1} = \frac{-1}{(-3)^{n-1}} = \frac{(-1)^n}{3^{n-1}} \quad \square$

Example 3. Suppose that the third and sixth terms of a geometric sequence $\{a_n\}_{n=1}^{\infty}$ are $\frac{9}{2}$ and $-\frac{243}{16}$, respectively. Find a formula for the nth term of the sequence.

Solution. Once we know a_1 and r, the nth term is given by (2). For r we use (2) and find that

$$\frac{a_6}{a_3} = \frac{a_1 r^5}{a_1 r^2} = r^3$$

so by hypothesis,

$$r^3 = \frac{a_6}{a_3} = \frac{-243/16}{9/2} = -\frac{27}{8}$$

This means that $r = -\frac{3}{2}$. A second application of (2) tells us that

$$a_1 = \frac{1}{r^2} \cdot a_3 = \frac{1}{(-3/2)^2} \cdot \frac{9}{2} = \frac{4}{9} \cdot \frac{9}{2} = 2$$

It follows from this information and from (2) that

$$a_n = 2\left(-\frac{3}{2}\right)^{n-1} \quad \text{for } n \geq 1 \quad \square$$

Partial sums of a geometric sequence can be expressed by a convenient formula. Before we present the formula, we observe that when written out, a partial sum of a geometric sequence has the form

$$\sum_{j=1}^{n} ar^{j-1} = ar^0 + ar^1 + ar^2 + \cdots + ar^{n-1}$$

or more simply,

$$\sum_{j=1}^{n} ar^{j-1} = a + ar + ar^2 + \cdots + ar^{n-1} \qquad (3)$$

THEOREM 9.5   Let $\{ar^{n-1}\}_{n=1}^{\infty}$ be a geometric sequence with $r \neq 0$ and $r \neq 1$. Then the nth partial sum of the sequence is given by

$$\sum_{j=1}^{n} ar^{j-1} = a \cdot \frac{1 - r^n}{1 - r} \qquad (4)$$

Proof   By (3), along with Example 2 in Section 9.1 (with x replaced by r), we conclude that

$$\sum_{j=1}^{n} ar^{j-1} = a + ar + ar^2 + \cdots + ar^{n-1}$$

$$= a(1 + r + r^2 + \cdots + r^{n-1})$$

$$= a \cdot \frac{1 - r^n}{1 - r} \quad \blacksquare$$

Example 4. Find the sum S of the first 10 terms of the geometric sequence $\frac{1}{4}, \frac{1}{8}, \ldots$.

Solution. Applying (1) with $n = 1$, we find that

$$r = \frac{a_2}{a_1} = \frac{\frac{1}{8}}{\frac{1}{4}} = \frac{1}{2}$$

By (4) with $a = \frac{1}{4}$, $r = \frac{1}{2}$, and $n = 10$, we have

$$S = \frac{1}{4} \cdot \frac{1 - \left(\frac{1}{2}\right)^{10}}{1 - \frac{1}{2}} = \frac{1}{2}\left(1 - \frac{1}{1024}\right) = \frac{1023}{2048} \quad \square$$

Next we consider the sequence $\{(\frac{1}{2})^n\}_{n=1}^{\infty}$. The terms start out

$$\frac{1}{2}, \frac{1}{4}, \frac{1}{8}, \frac{1}{16}, \frac{1}{32}, \frac{1}{64}, \cdots$$

Since each term in the sequence is half its predecessor, the value $(\frac{1}{2})^n$ of the nth term can be made as small as we like by taking n large enough. We express this by saying that $(\frac{1}{2})^n$ approaches 0 as n increases without bound. More generally if $|r| < 1$, then r^n approaches 0 as n increases without bound. It follows that

$$\text{if } |r| < 1, \quad \text{then} \quad a \cdot \frac{1 - r^n}{1 - r} \text{ approaches } \frac{a}{1 - r}$$

as n increases without bound. Thus Theorem 9.5 implies that as n increases without bound, the nth partial sum of the geometric sequence $\{ar^{n-1}\}_{n=1}^{\infty}$ approaches the number $a/(1 - r)$. It is therefore natural to define the sum of all the numbers in the geometric sequence $\{ar^{n-1}\}_{n=1}^{\infty}$ to be $a/(1 - r)$. We denote this sum by $\sum_{j=1}^{\infty} ar^{j-1}$, an expression we call a *geometric series*. Thus

$$\sum_{j=1}^{\infty} ar^{j-1} = a + ar + ar^2 + ar^3 + \ldots = \frac{a}{1 - r} \text{ for } |r| < 1 \quad (5)$$

Example 5. Find the numerical value of the following geometric series.

a. $\displaystyle\sum_{j=1}^{\infty} \left(\frac{1}{2}\right)^{j-1}$ b. $\displaystyle\sum_{j=1}^{\infty} 4\left(-\frac{2}{3}\right)^{j-1}$ c. $\displaystyle\sum_{j=1}^{\infty} (-2)\left(\frac{3}{5}\right)^{j}$

Solution.

a. By (5) with $a = 1$ and $r = \frac{1}{2}$, we have

$$\sum_{j=1}^{\infty} \left(\frac{1}{2}\right)^{j-1} = \frac{1}{1 - \frac{1}{2}} = 2$$

b. By (5) with $a = 4$ and $r = -\frac{2}{3}$, we have

$$\sum_{j=1}^{\infty} 4 \left(-\frac{2}{3}\right)^{j-1} = \frac{4}{1-(-\frac{2}{3})} = \frac{12}{5}$$

c. Notice that the geometric series is not in the form $\sum_{j=1}^{\infty} ar^{j-1}$, so in order to apply (5), we alter the series as follows:

$$\sum_{j=1}^{\infty} (-2) \left(\frac{3}{5}\right)^{j} = \sum_{j=1}^{\infty} (-2) \left(\frac{3}{5}\right)\left(\frac{3}{5}\right)^{j-1} = \sum_{j=1}^{\infty} \left(-\frac{6}{5}\right)\left(\frac{3}{5}\right)^{j-1}$$

Now (5) applies, with $a = -\frac{6}{5}$ and $r = \frac{3}{5}$, and we obtain

$$\sum_{j=1}^{\infty} (-2) \left(\frac{3}{5}\right)^{j} = \sum_{j=1}^{\infty} \left(-\frac{6}{5}\right)\left(\frac{3}{5}\right)^{j-1} = \frac{-\frac{6}{5}}{1-\frac{3}{5}} = -3 \quad \square$$

Repeating Decimals

It is well known that $\frac{1}{3}$ can be represented as a repeating decimal:

$$\frac{1}{3} = 0.3333 \ldots \tag{6}$$

The way we arrive at this formula is as follows. First, by $0.3333 \ldots$ we mean the sum

$$0.3 + 0.03 + 0.003 + 0.0003 + \cdots$$

that is,

$$\frac{3}{10} + \frac{3}{100} + \frac{3}{1000} + \frac{3}{10,000} + \cdots$$

or equivalently,

$$\frac{3}{10} + \frac{3}{10} \left(\frac{1}{10}\right) + \frac{3}{10} \left(\frac{1}{10}\right)^{2} + \cdots$$

This is the geometric series whose 1st term is $\frac{3}{10}$ and whose common ratio is $\frac{1}{10}$. In sum notation the series is

$$\sum_{j=1}^{\infty} \frac{3}{10} \left(\frac{1}{10}\right)^{j-1}$$

By (5) with $a = \frac{3}{10}$ and $r = \frac{1}{10}$, we find that

$$\sum_{j=1}^{\infty} \frac{3}{10} \left(\frac{1}{10}\right)^{j-1} = \frac{\frac{3}{10}}{1-\frac{1}{10}} = \frac{3}{10} \cdot \frac{10}{9} = \frac{3}{9} = \frac{1}{3}$$

This completes our proof of (6).

In a similar way we can identify any repeating decimal with a suitable rational number.

Example 6. Write 0.535353 . . . as a fraction.

Solution. Notice first that

$$0.535353 \ldots = \frac{53}{100} + \frac{53}{100}\left(\frac{1}{100}\right) + \frac{53}{100}\left(\frac{1}{100}\right)^2 + \cdots$$

which is the geometric series whose first term is $\frac{53}{100}$ and whose common ratio is $\frac{1}{100}$. In sum notation the series is

$$\sum_{j=1}^{\infty} \frac{53}{100}\left(\frac{1}{100}\right)^{j-1}$$

By (5) with $a = \frac{53}{100}$ and $r = \frac{1}{100}$, we have

$$\sum_{j=1}^{\infty} \frac{53}{100}\left(\frac{1}{100}\right)^{j-1} = \frac{\frac{53}{100}}{1 - \frac{1}{100}} = \frac{53}{100} \cdot \frac{100}{99} = \frac{53}{99}$$

Therefore

$$0.535353 \ldots = \frac{53}{99} \quad \square$$

Example 7. Write 0.4298298298 . . . as a fraction.

Solution. First we separate the repeating part from the rest:

$$0.4298298298 = 0.4 + 0.0298298298 \ldots$$

Next we observe that

0.0298298298 . . .

$$= \frac{298}{10,000} + \frac{298}{10,000}\left(\frac{1}{1000}\right) + \frac{298}{10,000}\left(\frac{1}{1000}\right)^2 + \cdots$$

which is the geometric series whose first term is $\frac{298}{10,000}$ and whose common ratio is $\frac{1}{1000}$. In sum notation the series is

$$\sum_{j=1}^{\infty} \frac{298}{10,000}\left(\frac{1}{1000}\right)^{j-1}$$

By (5) with $a = \frac{298}{10,000}$ and $r = \frac{1}{1000}$, we have

$$\sum_{j=1}^{\infty} \frac{298}{10,000}\left(\frac{1}{1000}\right)^{j-1} = \frac{\frac{298}{10,000}}{1 - \frac{1}{1000}} = \frac{298}{10,000} \cdot \frac{1000}{999} = \frac{298}{9990}$$

Consequently

0.4298298298 . . .

$$= 0.4 + 0.0298298298 \ldots$$

$$= \frac{4}{10} + \frac{298}{9990} = \frac{3996}{9990} + \frac{298}{9990} = \frac{4294}{9990} = \frac{2147}{4995} \quad \square$$

EXERCISES 9.4

In Exercises 1–6, show that the sequence is geometric, and find the common ratio.

1. $\{4^n\}_{n=1}^{\infty}$

2. $\left\{\left(\frac{5}{2}\right)^n\right\}_{n=1}^{\infty}$

3. $\{(-0.3)^{n-1}\}_{n=1}^{\infty}$

4. $\{(\sqrt{2})^{n+1}\}_{n=1}^{\infty}$

5. $\{4^{2n}\}_{n=1}^{\infty}$

6. $\{3^{n/2}\}_{n=1}^{\infty}$

In Exercises 7–14, determine whether the given sequence is geometric.

7. $-1, -2, -4, -8, \ldots$

8. $-1, 2, -4, 8, \ldots$

9. $1, 3, 5, 7, \ldots$

10. $0, 3, 6, 9, \ldots$

11. $\{-3(-2)^{n+1}\}_{n=1}^{\infty}$

12. $\{e^n\}_{n=1}^{\infty}$

13. $\{5n^n\}_{n=1}^{\infty}$

14. $\{15 + 2^n\}_{n=1}^{\infty}$

In Exercises 15–22, find the fifth term and the nth term of the geometric sequence whose initial term a_1 and common ratio r are given.

15. $a_1 = 4, r = 2$

16. $a_1 = 6, r = -\frac{1}{2}$

17. $a_1 = 3, r = \frac{1}{6}$

18. $a_1 = \frac{1}{3}, r = \frac{3}{2}$

19. $a_1 = 10, r = 3$

20. $a_1 = 0.5, r = \frac{1}{10}$

21. $a_1 = c, r = 5$

22. $a_1 = c, r = c^2$

In Exercises 23–30, find the sixth term and the nth term of the geometric sequence $\{a_n\}_{n=1}^{\infty}$ whose first two terms are given.

23. $\frac{1}{3}, 1, \ldots$

24. $5, 20, \ldots$

25. $6, 3, \ldots$

26. $1, -2, \ldots$

27. $256, -64, \ldots$

28. $2, 2^2, \ldots$

29. $\frac{1}{6}, -\frac{1}{18}, \ldots$

30. $\ln 10, \ln 100, \ldots$

In Exercises 31–36, find a formula for the nth term of the geometric sequence $\{a_n\}_{n=1}^{\infty}$ that satisfies the given conditions.

31. $a_2 = 6, a_3 = 9$

32. $a_1 = 4, a_4 = -\frac{1}{16}$

33. $a_3 = 250, a_4 = 1250$

34. $a_2 = 192, a_7 = 6$

35. $a_3 = -\frac{4}{5}, r = \frac{2}{5}$

36. $a_8 = 1024, r = 2$

In Exercises 37–40, find the nth partial sum (for the designated value of n) of the geometric sequence $\{a_n\}_{n=1}^{\infty}$ that satisfies the given conditions.

37. $a_1 = 9, r = 3, n = 3$

38. $a_1 = 5, r = 2, n = 5$

39. $a_2 = 8, r = 2, n = 4$

40. $a_3 = -8, r = -\frac{1}{2}, n = 5$

In Exercises 41–48, find the given partial sum.

41. $\sum_{j=1}^{5} 3^{j-1}$

42. $\sum_{j=1}^{4} \left(\frac{3}{2}\right)^{j-1}$

43. $\sum_{j=1}^{7} \left(-\frac{1}{2}\right)^{j-1}$

44. $\sum_{j=1}^{7} \left(-\frac{1}{2}\right)^{j}$

45. $\sum_{j=1}^{5} \frac{4}{3^j}$

46. $\sum_{j=1}^{5} 8\left(\frac{3}{2}\right)^{j}$

47. $\sum_{j=1}^{4} 2\left(-\frac{3}{4}\right)^{j}$

48. $\sum_{j=1}^{3} 3\left(\frac{4}{5}\right)^{j+1}$

In Exercises 49–56, find the numerical value of the geometric series.

49. $\displaystyle\sum_{j=1}^{\infty} \left(\frac{1}{3}\right)^{j-1}$ 50. $\displaystyle\sum_{j=1}^{\infty} \left(-\frac{1}{2}\right)^{j-1}$ 51. $\displaystyle\sum_{j=1}^{\infty} 2\left(\frac{1}{4}\right)^{j-1}$

52. $\displaystyle\sum_{j=1}^{\infty} \frac{1}{3}\left(-\frac{3}{4}\right)^{j-1}$ 53. $\displaystyle\sum_{j=1}^{\infty} 6\left(\frac{2}{3}\right)^{j}$ 54. $\displaystyle\sum_{j=1}^{\infty} \frac{1}{2}\left(-\frac{4}{5}\right)^{j+1}$

55. $\displaystyle\sum_{j=1}^{\infty} 2(0.3)^{j}$ 56. $\displaystyle\sum_{j=1}^{\infty} \frac{1}{3}(0.62)^{j-1}$

In Exercises 57–64, write the given repeating decimal as a fraction.

57. 0.777 . . .   58. 0.141414 . . .

59. 0.959595 . . .   60. 0.0959595 . . .

61. 5.432432432 . . .   62. 1.697697697 . . .

63. 3.1474747 . . .   64. 2.6858585 . . .

65. If $\{a_n\}_{n=1}^{\infty}$ is a geometric sequence such that $a_5/a_3 = 16$ and $a_2 = 1$, find a formula for a_n.

66. If the product of the second and sixth terms of a geometric sequence is 1 and the common ratio is 3, find a formula for a_n.

67. If 3 and 108 are two terms of a geometric sequence separated by exactly one term, find its value.

68. The first two terms of a geometric sequence are $\sqrt[3]{2}$ and $\sqrt{2}$. Find the third and fourth terms.

69. Show that there are two distinct geometric sequences having second term 12 and fourth term 27. Find a formula for the nth term of each sequence.

70. Prove (2) by using (1) and mathematical induction.

71. Let $\{a_n\}_{n=1}^{\infty}$ be a geometric sequence with positive terms and common ratio r. Prove that $\{\ln a_n\}_{n=1}^{\infty}$ is an arithmetic sequence, and find the common difference d. (*Hint:* Use (2).)

72. Starting with your parents, how many ancestors do you have going back 8 generations?

73. The enrollment at a certain college doubled every 5 years from 1960 to 1980. If the enrollment was 8000 in 1980, what was the enrollment in 1960?

C∗74. When a tennis ball is dropped, it is supposed to bounce back to approximately 55% of its original height. Use a calculator to determine the distance the ball travels during its initial ten bounces (down and up) if it is dropped from a height of 6 feet.

9.5

PERMUTATIONS AND COMBINATIONS

Suppose that a 7-member committee sets out to pick 3 of its members to be officers. The first member picked will be chairman, the next vice-chairman, and the third secretary. The following two questions typify the kind of question we will discuss in this section.

Question A: In how many different ways can the set of officers be formed from the committee?

Question B: How many different sets of officers can be formed if we disregard which officer has what title?

As we set out to answer these questions, let us list the 7 committee members as a, b, c, d, e, f, and g. Question A is equivalent to asking in how many ways we can write down an ordered triplet consisting of 3 of the letters a through g. For example,

$$(a, b, c), \quad (d, f, e), \quad (g, e, b), \quad \text{and} \quad (b, e, g)$$

are four such triplets. Here we let the first letter of each triplet denote the chairman, the second letter the vice chairman, and the third letter the secretary. Each of the triplets listed above is considered to be distinct from the others with respect to Question A (even though the last two contain exactly the same letters), because the order of the letters in the triplet is taken into account.

By contrast, Question B is equivalent to asking in how many ways we can write down a triplet of 3 of the letters a through g without regard to the order of the letters. In this case,

$$(b, e, g), \quad (b, g, e), \quad (e, b, g), \quad (e, g, b),$$
$$(g, b, e), \quad \text{and} \quad (g, e, b)$$

are identified, because they contain exactly the same letters and because order is irrelevant for Question B. In this case our sole concern is which 3 people are selected.

It follows from the discussion above that the answers to Questions A and B are different. Before we set out to find the numerical answers to the questions, let us put the questions into a more general setting:

    i. How many distinct ordered sets of k objects can be chosen from a collection of n objects?

    ii. How many distinct sets of k objects may be chosen from a collection of n objects?

We apply the terms "permutations" and "combinations" to the sets in (i) and (ii), respectively. These terms are defined formally now.

DEFINITION 9.6

Let k and n be integers with n positive and $0 \le k \le n$. A *permutation* of n objects taken k at a time is an ordered set of k objects from the set of n objects. A *combination* of n objects taken k at a time is a set of k objects from the set of n objects (without regard to order).

The difference between a permutation and a combination is that order is all-important with respect to a permutation but is irrelevant with respect to a combination.

A Counting Principle

Before proceeding further in analyzing permutations and combinations, we discuss a method of counting sets of objects. To illustrate the method, let us count the number of full names that can be formed by using "Barbara" or "Sharon" as the first name and "Lee," "Lopez," or "Smith" as the last name. We may regard the selection of such a full name as a 2-stage process, and we display the possibilities in the following diagram, called a *tree diagram*:

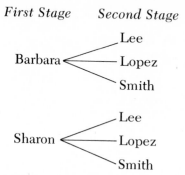

We see that there are 6 possible full names. Indeed, there are 2 possible first names and 3 possible last names, making a total of $2 \times 3 = 6$ possible full names. Thus the total number of possible names is the product of the number of choices at the first stage and at the second stage.

Now let us count the number of full names that can be formed by using "David" or "Mark" as the first name, "Charles" or "Christopher" as the middle name, and "Green," "Jansen," or "Mantle" as the last name. Here we regard the selection of such a full name as a 3-stage process, displayed in the following tree diagram:

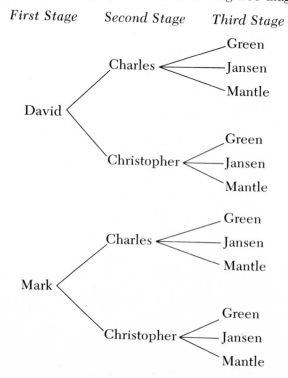

This time we see that there are 12 possible full names. After all, there are 2 possible first names, 2 possible middle names, and 3 possible last names, making a total of $2 \times 2 \times 3 = 12$ possible full names. Thus the total number of possible full names is the product of the number of choices at each of the 3 stages.

In general, if the items to be counted may be selected in k stages, and if we know the number of choices at each of the k stages, then the total number of items is the product of the number of choices at each stage. We now state this result formally.

> **COUNTING PRINCIPLE** Suppose that k objects are to be selected. Assume that the first object can be selected in n_1 distinct ways, and after the first object has been selected the second object can be selected in n_2 distinct ways, and so on, with the kth object selectable in any of n_k distinct ways after the first $k - 1$ objects have been selected. Then the total number of distinct ways of selecting the k objects is the product
>
> $$n_1 n_2 n_3 \cdots n_k$$

In the following example we illustrate the use of the Counting Principle.

Example 1. How many 7-digit telephone numbers are there if neither 0 nor 1 can be used as the first or second digit?

Solution. For each of the first and second digits of the telephone number, there are 8 possible choices: 2, 3, 4, 5, 6, 7, 8, and 9. Therefore

$$n_1 = n_2 = 8$$

For each of the remaining digits of the telephone number there are 10 possibilities: 0, 1, 2, 3, 4, 5, 6, 7, 8, and 9. Therefore

$$n_3 = n_4 = n_5 = n_6 = n_7 = 10$$

By the Counting Principle, the total number N of permissible 7-digit telephone numbers is given by

$$N = n_1 n_2 n_3 n_4 n_5 n_6 n_7 = 8 \cdot 8 \cdot 10 \cdot 10 \cdot 10 \cdot 10 \cdot 10$$

$$= 64 \times 10^5 = 6,400,000 \quad \square$$

Permutations

The number of permutations of n objects taken k at a time is often written symbolically as $_nP_k$ (or sometimes P_k^n or $P(n, k)$). In this notation the number of permutations of 7 objects taken 3 at a time is $_7P_3$.

Suppose now that we wish to determine the numerical value of $_nP_k$, where n is any positive integer and k is any integer with

$0 \le k \le n$. We define $_nP_0 = 1$. For $k \ge 1$ we consider the process of forming a permutation of n objects taken k at a time as a k-stage process. In the first stage we select the first object, with n possible choices for this object. In the second stage we select the second object, with only $n - 1$ choices for this object, since the object selected first is no longer available. Similarly, there are only $n - 2$ choices for the third object, since the objects selected first and second are no longer available. Thus we have the following setup:

Object	*Number of choices*
First	n
Second	$n - 1$
Third	$n - 2$
$\vdots$	$\vdots$
kth	$n - (k - 1)$

Therefore by the Counting Principle, it follows that the number $_nP_k$ of ways of choosing k objects from the total collection of n objects is given by the product of the possible choices just listed. That is,

$$_nP_k = n(n - 1)(n - 2) \cdots [n - (k - 2)][n - (k - 1)] \qquad (1)$$

We are now in a position to answer question A. Indeed, by (1)

$$_7P_3 = 7 \cdot 6 \cdot 5 = 210$$

This tells us that there are 210 ways of choosing three officers from the 7-member committee.

Example 2. Suppose 9 horses are set to run the Kentucky Derby. How many ways are there of the horses occupying the first three places in the order of finish?

Solution. Since we are interested in determining the number of ordered sets of 3 horses selected from 9 horses (corresponding to the order in which the first 3 horses cross the finish line), we seek $_9P_3$. By (1) with $n = 9$ and $k = 3$, we have

$$_9P_3 = 9 \cdot 8 \cdot 7 = 504 \quad \square$$

If we let $k = n$, then (1) becomes

$$_nP_n = n(n - 1)(n - 2) \cdots 2 \cdot 1 = n!$$

so that the number $_nP_n$ of permutations of n objects taken n at a time is $n!$. But notice that a permutation of n objects taken n at a time corresponds to an ordering of the n objects. We conclude that there are $n!$ ways of ordering a set of n objects.

We can obtain a simpler version of (1) by noticing that

$$n(n-1)(n-2) \cdots [n-(k-1)]$$

$$= n(n-1)(n-2) \cdots [n-(k-1)] \frac{(n-k)[n-(k+1)] \cdots 2 \cdot 1}{(n-k)[n-(k+1)] \cdots 2 \cdot 1}$$

$$= \frac{n(n-1)(n-2) \cdots [n-(k-1)](n-k)[n-(k+1)] \cdots 2 \cdot 1}{(n-k)[n-(k+1)] \cdots 2 \cdot 1}$$

$$= \frac{n!}{(n-k)!}$$

Consequently

$$_nP_k = \frac{n!}{(n-k)!} \tag{2}$$

Example 3. Calculate $_6P_4$ by means of (2).

Solution. By (2),

$$_6P_4 = \frac{6!}{(6-4)!} = \frac{6!}{2!} = \frac{720}{2} = 360 \quad \square$$

Next, suppose that we have 5 marbles, including 3 indistinguishable black ones and 2 indistinguishable white ones. Let us denote the black marbles by B_1, B_2, and B_3, and the white marbles by W_1 and W_2. Then the two ordered sets

$$B_1B_2B_3W_1W_2 \quad \text{and} \quad B_2B_3B_1W_2W_1$$

would be indistinguishable, since both have 3 black marbles followed by 2 white marbles. Thus of the 5! ordered sets of the 5 marbles, many are indistinguishable from one another. It can be proved that there are only

$$\frac{5!}{3!2!}$$

or 10, distinguishable ordered sets of the 5 marbles, rather than the 5!, or 120, that would exist if all 5 marbles were different.

More generally, suppose that there are n objects, but n_1 of them are of one type and are indistinguishable, n_2 are of a second type and are indistinguishable, and so on, with a final kth type of n_k indistinguishable objects. Then it can be proved that the number of distinguishable ordered sets that can be formed from the n objects is

$$\frac{n!}{n_1! \, n_2! \cdots n_k!} \tag{3}$$

Example 4. Suppose 6 indistinguishable nickels, 4 indistinguishable dimes, and 2 indistinguishable quarters are drawn out of a bowl, one by one. How many distinguishable ordered sets can be formed from the coins?

Solution. There are 12 coins in all, so using (3) with $n = 12$, $n_1 = 6$, $n_2 = 4$, and $n_3 = 2$, we find that there are

$$\frac{12!}{6!4!2!} = 13,860$$

distinguishable ordered sets of the 12 coins. $\square$

Combinations

With permutations we were able to determine in how many ways 3 officers could be selected from 7 committee members. Now we will study the number of sets of 3 officers that can be chosen from the 7 committee members, without regard to which officer has what office. This will involve combinations.

The number of combinations of n objects taken k at a time is written symbolically as $_nC_k$. We will derive a formula for $_nC_k$ from formula (2) for $_nP_k$.

To that end we observe that the number $_nP_k$ of permutations of n objects taken k at a time takes into account the order in which the k objects are selected. By contrast, for the number $_nC_k$ of combinations the order in which the k objects are selected is irrelevant, so that each distinct combination counted in $_nC_k$ corresponds to $k!$ distinct permutations in $_nP_k$—the $k!$ ways the k elements can be ordered. Therefore

$$_nC_k = \frac{_nP_k}{k!}$$

Together with (2), this means that

$$_nC_k = \frac{n!}{k!(n-k)!} \qquad (4)$$

From (4) we conclude that

$$_7C_3 = \frac{7!}{3!(7-3)!} = \frac{7!}{3!4!} = \frac{7 \cdot 6 \cdot 5 \cdot 4!}{3 \cdot 2 \cdot 1 \cdot 4!} = \frac{7 \cdot 6 \cdot 5}{3 \cdot 2} = 35$$

Since the number of sets of 3 officers that can be chosen from a 7-member committee is $_7C_3$, it follows that there are 35 such sets of officers possible. This then answers Question B posed at the beginning of the section.

Example 5. Determine the number of 5-card hands that can be dealt from a 52-card deck.

Solution. Since the order in which cards are dealt into the hand is immaterial to the composition of the hand, what we seek is $_{52}C_5$, and by (4) with $n = 52$ and $k = 5$, we have

$$_{52}C_5 = \frac{52!}{5!(52 - 5)!} = \frac{52!}{5!47!}$$

$$= \frac{52 \cdot 51 \cdot 50 \cdot 49 \cdot 48(47!)}{5 \cdot 4 \cdot 3 \cdot 2 \cdot 1(47!)} = \frac{52 \cdot 51 \cdot 50 \cdot 49 \cdot 48}{5 \cdot 4 \cdot 3 \cdot 2}$$

$$= 2{,}598{,}960 \quad \square$$

As we will see in the following section, when the nth power $(x + y)^n$ of the binomial $x + y$ is written out, the coefficients of the terms that appear in the formula are equal to $_nC_k$ for various values of k. But in that context they are normally written in the alternative notation $\binom{n}{k}$ and are called **binomial coefficients**. From (4) we find that

$$\binom{n}{k} = \frac{n!}{k!(n - k)!} \tag{5}$$

Example 6. Find the numerical value of $\binom{8}{5}$.

Solution. By (5),

$$\binom{8}{5} = \frac{8!}{5!(8 - 5)!} = \frac{8 \cdot 7 \cdot 6(5!)}{(5!)3 \cdot 2 \cdot 1} = \frac{8 \cdot 7 \cdot 6}{3 \cdot 2} = 56 \quad \square$$

Since $0! = 1$ by definition, it follows from (5) that

$$\binom{n}{0} = \frac{n!}{0!(n - 0)!} = \frac{n!}{0!n!} = \frac{n!}{n!} = 1$$

and

$$\binom{n}{n} = \frac{n!}{n!(n - n)!} = \frac{n!}{n!0!} = \frac{n!}{n!} = 1$$

Similarly, since $1! = 1$, we have

$$\binom{n}{1} = \frac{n!}{1!(n - 1)!} = \frac{n(n - 1)!}{(n - 1)!} = n$$

and likewise,

$$\binom{n}{n - 1} = n$$

EXERCISES 9.5

In Exercises 1–24, calculate the given number.

1. The number of permutations of 5 objects taken 3 at a time.
2. The number of permutations of 6 objects taken 2 at a time.
3. The number of permutations of 7 objects taken 5 at a time.
4. The number of combinations of 5 objects taken 3 at a time.
5. The number of combinations of 6 objects taken 2 at a time.
6. The number of combinations of 7 objects taken 5 at a time.

7. $_4P_1$ 8. $_4P_3$ 9. $_5P_5$
10. $_7P_3$ 11. $_{12}P_3$ 12. $_{100}P_2$
13. $_4C_2$ 14. $_6C_3$ 15. $_6C_6$
16. $_5C_3$ 17. $_5C_2$ 18. $_8C_5$
19. $\binom{2}{1}$ 20. $\binom{4}{3}$ 21. $\binom{6}{4}$
22. $\binom{7}{4}$ 23. $\binom{17}{15}$ 24. $\binom{40}{3}$

25. Let n be an integer greater than 1.

 a. Find a formula for $\binom{n}{n-2}$.

 b. Show that $\sum_{j=1}^{n} j = \binom{n+1}{n-1}$.

26. Show that $_{2n+1}C_n = {}_{2n+1}C_{n+1}$ for any positive integer n.

27. Let n and j be integers with $0 \le j \le n$. Show that

$$\binom{n}{j} = \binom{n}{n-j}$$

28. Show that $(_{12}C_5)(_7C_4) = (_{12}C_4)(_8C_5)$.

29. A student is eligible for 3 mathematics courses and 4 physics courses. In how many ways can the student choose one mathematics course and one physics course?

30. A department store has 4 entrances on the first floor, 3 escalators from the first floor to the second floor, and 2 escalators from the second floor to the third floor. In how many ways can a customer enter the store and ascend to the third floor by escalator?

31. a. How many distinct license plate numbers are there that have 2 letters (neither of which may be I or O) followed by 3 digits?
 b. If in part (a) the letters must be distinct and so must the digits, how many distinct license plate numbers are there?

32. At a sandwich bar there are 3 kinds of bread, 4 kinds of cold cuts, 2 kinds of cheese, and 3 kinds of dressing. How many different kinds of sandwiches can be prepared with one kind each of bread, cold cuts, cheese, and dressing?

33. A corporation president must choose a vice-president and a treasurer from among 10 young executives.
 a. In how many ways can the president choose two executives to

fill the positions (without regard to which executive gets which position)?

    b. In how many ways can the president choose two executives to fill the positions (considering which executive gets which position)?

34. Find the number of permutations of 4 letters selected from the word "rainbow."

35. A classroom contains 20 seats; in how many ways can a teacher assign seats to 15 students? Leave your answer in factorial form.

36. How many sets of three volunteers are possible from a group of 10 people?

37. In how many ways can 5 patients be given appointments with a doctor if there are 7 appointments available?

38. A doctor must reschedule 4 of 7 appointments. In how many ways can the 4 appointments be chosen?

39. On Halloween a trick-or-treater is offered any 3 of 5 different kinds of candy bars. In how many ways can the choices be made?

40. Find an expression for the number of distinct 13-card bridge hands that can be dealt from a deck of 52 cards. Do not carry out the arithmetic operations.

41. A basketball coach has 2 centers, 5 guards, and 4 forwards. How many different teams consisting of 1 center, 2 guards, and 2 forwards can the coach choose? (*Hint:* Consider the selection a 3-stage process.)

42. A person receives a list of five names from a computerized dating service. In how many different orders can the person date the five people listed?

43. A basketball team has 12 members.
    a. In how many ways can the 12 members be introduced one by one before a game?
    b. In how many ways can they be introduced if the five preselected starting players are to be introduced first?

44. A person has 5 quarters, 3 dimes, and 4 nickels. In how many ways can the person choose 2 quarters, 1 dime, and 2 nickels to pay for a 70¢ purchase? (Assume that the coins of each denomination are distinguishable.)

45. A baker has 5 vanilla, 3 chocolate, and 4 lemon layers of cake. In how many distinguishable ways can the baker stack the layers to make a cake with 12 layers?

46. Find the number of (possibly meaningless) 5-letter words with exactly 1 vowel. (*Hint:* There are 5 vowels and 21 consonants in the alphabet. Consider the formation of the word as a six-stage process, the first stage of which involves selecting the position in the word that the vowel is to occupy.)

47. One student claims to another, "I'll bet there are at least 10 times as many distinguishable ordered sets that can be formed from the letters in "Mississippi" as there are distinguishable ordered sets that can be formed from the letters in "Tennessee." Is the student correct?

48. How many 9-digit Social Security numbers are there containing four 5's, two 6's, and three 9's?

9.6

THE BINOMIAL THEOREM

In Section 1.6 we wrote out the following formulas for $(x + y)^2$ and $(x + y)^3$:

$$(x + y)^2 = x^2 + 2xy + y^2 \qquad (1)$$

$$(x + y)^3 = x^3 + 3x^2y + 3xy^2 + y^3 \qquad (2)$$

Now we will derive a formula for $(x + y)^n$ for any positive integer n.

In preparation for the formula of $(x + y)^n$, let us analyze the right sides of the equations in (1) and (2). Notice that each of their terms has the form (coefficient) x^iy^j for suitable integers i and j. For the terms in (1), $i + j = 2$, whereas for the terms in (2), $i + j = 3$.

Now let us consider the expansion of $(x + y)^4$. Notice that

$$(x + y)^4 = (x + y)(x + y)(x + y)(x + y)$$

$$= (xx + xy + yx + yy)(xx + xy + yx + yy)$$

$$= xxxx + xxxy + xxyx + xxyy + xyxx$$

$$\quad + xyxy + xyyx + xyyy + yxxx + yxxy$$

$$\quad + yxyx + yxyy + yyxx + yyxy + yyyx + yyyy$$

Because $xy = yx$, this can be condensed to

$$(x + y)^4 = x^4y^0 + x^3y^1 + x^3y^1 + x^2y^2 + x^3y^1$$

$$\quad + x^2y^2 + x^2y^2 + x^1y^3 + x^3y^1 + x^2y^2 \qquad (3)$$

$$\quad + x^2y^2 + x^1y^3 + x^2y^2 + x^1y^3 + x^1y^3 + x^0y^4$$

Observe that the term x^4y^0 appears only once on the right side of (3) because it arises by choosing x from each of the four expressions $x + y$ occurring in $(x + y)^4$ and never choosing y. Analogously, x^3y appears four times in (3) because we can pick exactly one y from any of the four expressions $x + y$ occurring in $(x + y)^4$. In general, the coefficient of x^iy^j (with $i + j = 4$) is equal to the number of ways of choosing j y's from the four expressions $x + y$ (the remaining choices being x's). Thus the coefficient of x^iy^j is $\binom{4}{j}$, and consequently (3) can be rewritten as

$$(x + y)^4 = \binom{4}{0} x^4y^0 + \binom{4}{1} x^3y^1 + \binom{4}{2} x^2y^2$$

$$\quad + \binom{4}{3} x^1y^3 + \binom{4}{4} x^0y^4$$

$$= x^4 + 4x^3y + 6x^2y^2 + 4xy^3 + y^4$$

In general, if n is a positive integer and we expand $(x + y)^n$ by writing

$$
(x + y)^n = \overbrace{(x + y)(x + y) \cdots (x + y)}^{n \text{ expressions } (x + y)}
\tag{4}
$$

and then multiplying out, the result is a sum of terms of the form (coefficient)$x^i y^j$, where $i + j = n$. The coefficient of $x^i y^j$ is equal to the number of ways of choosing j y's from the n expressions $x + y$ (the remaining $n - j$ choices being x's) in the parentheses on the right side of (4). Thus that coefficient is $\binom{n}{j}$. The resulting formula for $(x + y)^n$, which we state below, is known as the Binomial Theorem. It was one of the first theorems of Isaac Newton, proved in 1665, yet it was discovered nearly 600 years earlier by the great Persian poet and mathematician Omar Khayyám (1044?–1123?).

THEOREM 9.7 *The Binomial Theorem.* Let x and y be any real numbers and let n be any positive integer. Then

$$
\begin{aligned}
(x + y)^n = &\binom{n}{0} x^n + \binom{n}{1} x^{n-1}y + \binom{n}{2} x^{n-2}y^2 + \cdots \\
&+ \binom{n}{n-1} xy^{n-1} + \binom{n}{n} y^n
\end{aligned}
\tag{5}
$$

or more succinctly,

$$
(x + y)^n = \sum_{j=0}^{n} \binom{n}{j} x^{n-j}y^j
\tag{6}
$$

The right side of (5) is known as the **binomial expansion** of the binomial $(x + y)^n$. The Binomial Theorem is also valid if x or y is complex.

Example 1. Find the binomial expansion of $(x + y)^6$.

Solution. By (5) with $n = 6$, we have

$$
\begin{aligned}
(x + y)^6 = &\binom{6}{0} x^6 + \binom{6}{1} x^5y + \binom{6}{2} x^4y^2 + \binom{6}{3} x^3y^3 \\
&+ \binom{6}{4} x^2y^4 + \binom{6}{5} xy^5 + \binom{6}{6} y^6 \\
= &\, x^6 + 6x^5y + 15x^4y^2 + 20x^3y^3 + 15x^2y^4 + 6xy^5 + y^6 \quad \square
\end{aligned}
$$

Example 2. Find the binomial expansion of $(2r^2 - 3s)^5$.

Solution. By (5) with $n = 5$, and with x replaced by $2r^2$ and y replaced by $-3s$, we have

$$(2r^2 - 3s)^5 = \binom{5}{0}(2r^2)^5 + \binom{5}{1}(2r^2)^4(-3s) + \binom{5}{2}(2r^2)^3(-3s)^2$$

$$+ \binom{5}{3}(2r^2)^2(-3s)^3 + \binom{5}{4}(2r^2)(-3s)^4 + \binom{5}{5}(-3s)^5$$

$$= 32r^{10} + 5(16r^8)(-3s) + 10(8r^6)(9s^2)$$

$$+ 10(4r^4)(-27s^3) + 5(2r^2)(81s^4) + (-243s^5)$$

$$= 32r^{10} - 240r^8s + 720r^6s^2 - 1080r^4s^3$$

$$+ 810r^2s^4 - 243s^5 \quad \square$$

Sometimes it is necessary to isolate the kth term in the binomial expansion of $(x + y)^n$. By (6) the kth term is the one with the coefficient $\binom{n}{k-1}$.

Example 3. Write the sixth term in the binomial expansion of $\left(\dfrac{1}{x} + y\right)^{12}$.

Solution. By (6) with $n = 12$ and $j = 5$, the sixth term is

$$\binom{12}{5}\left(\frac{1}{x}\right)^{12-5} y^5$$

Since

$$\binom{12}{5} = \frac{12!}{7!5!} = \frac{12 \cdot 11 \cdot 10 \cdot 9 \cdot 8(7!)}{(7!)5 \cdot 4 \cdot 3 \cdot 2 \cdot 1}$$

$$= \frac{12 \cdot 11 \cdot 10 \cdot 9 \cdot 8}{5 \cdot 4 \cdot 3 \cdot 2} = 792$$

and

$$\left(\frac{1}{x}\right)^{12-5} = \left(\frac{1}{x}\right)^7 = x^{-7}$$

the sixth term can be written more simply as $792x^{-7}y^5$. $\square$

Example 4. Let a be a positive number and n a positive integer. Use the Binomial Theorem to prove that $(1 + a)^n \geq 1 + na$.

Solution. By (5) in the Binomial Theorem, with x replaced by 1 and y by a, we have

$$(1 + a)^n = \binom{n}{0}1^n + \binom{n}{1}1^{n-1}a + \binom{n}{2}1^{n-2}a^2 + \cdots$$

$$+ \binom{n}{n-1}a^{n-1} + \binom{n}{n}a^n$$

$$= 1 + na + \binom{n}{2} a^2 + \cdots + \binom{n}{n-1} a^{n-1} + \binom{n}{n} a^n$$

Since a is positive by hypothesis, all terms on the right side are positive, so it follows that $(1 + a)^n$ is greater than the sum, $1 + na$, of the first two terms. □

Pascal's Triangle

If we write down the values of the binomial coefficients of $(x + y)^n$ for $n = 0, 1, 2, 3, 4$, and 5 in the following fashion, we obtain a triangular array:

	Coefficients					Binomial
		1				$(x + y)^0$
	1		1			$(x + y)^1$
	1	2		1		$(x + y)^2$
1		3	3		1	$(x + y)^3$
1	4		6	4	1	$(x + y)^4$
1	5	10	10	5	1	$(x + y)^5$

The triangular array that arises from writing out the binomial coefficients of $(x + y)^n$ for $n = 0, 1, 2, 3, \ldots$ is called *Pascal's triangle*, after the French mathematician Blaise Pascal (1623–1662), who wrote a treatise on properties of the triangular array.* Notice that the borders of the array consist of 1's and any number in the array not on the border is the sum of the two closest numbers in the preceding row.

EXERCISES 9.6

In Exercises 1–10, use the Binomial Theorem to expand the given power of the given binomial.

1. $(x + y)^5$
2. $(x - y)^3$
3. $(x + 2)^4$
4. $(2x - 1)^6$
5. $(x + y)^7$
6. $(2x - \frac{1}{2})^5$
7. $\left(a - \frac{1}{a}\right)^5$
8. $(z^2 - 1)^7$
9. $\left(\sqrt{x} + \frac{1}{\sqrt{x}}\right)^8$
10. $(x^3 - x^{-3})^6$

* The triangular array is not actually due to Pascal, and in fact the array appeared even as early as 1303 in a work by a Chinese mathematician, Chu Shi Kei. Nevertheless, Pascal's name is attached to it because of his treatise on the triangular array.

In Exercises 11–14, use the Binomial Theorem to calculate the value of the given expression.

11. $(\sqrt{2} + 1)^4$ 12. $(\sqrt{2} + 1)^4 + (\sqrt{2} - 1)^4$

13. $(2 - i)^5$ 14. $(2 - i)^5 - (2 + i)^5$

15. Find the fourth term of $(x + y)^8$.

16. Find the seventh term of $(2x - 3)^{10}$.

17. Find the coefficient of $x^6 y^9$ in the binomial expansion of $(2x^2 - y^3)^6$.

18. Find the coefficient of x^4 in the binomial expansion of $(1 - 2x)^7$.

19. Find the constant term in the binomial expansion of $\left(x - \dfrac{1}{x}\right)^{10}$.

20. a. Fill in the seventh row of Pascal's Triangle. (*Hint:* See Example 1.)
    b. Fill in the eighth row of Pascal's Triangle.

21. Use the first three terms of the expansion of $(1 - 0.02)^7$ to approximate $(0.98)^7$.

22. Without using a calculator, prove that $(1.012)^{30} \geq 1.36$.

23. In the binomial expansion of a certain binomial, the fifth term appears before simplification as

$$\frac{12 \cdot 11 \cdot 10 \cdot 9}{4 \cdot 3 \cdot 2 \cdot 1} \left(-\frac{1}{3}x\right)^8 (y^3)^4$$

   a. Write the original binomial.
   b. Determine the seventh term in the binomial expansion.

24. The coefficients of the sixth and eleventh terms in the binomial expansion of $(x + y)^n$ are the same. Determine the value of n.

25. a. Show that the sum of the coefficients in the binomial expansion of $(x + y)^n$ is 2^n. (*Hint:* Let $x = y = 1$.)
    b. Show that the sum of the coefficients in the binomial expansion of $(x - y)^n$ is 0.

26. Show that the sum of the numbers in the nth row of Pascal's Triangle is twice the sum of the preceding line. (*Hint:* Use Exercise 25(a).)

KEY TERMS

mathematical induction
sequence
term of a sequence
factorial
partial sum
arithmetic sequence
common difference
geometric sequence

common ratio
geometric series
repeating decimal
permutation
combination
binomial expansion
binomial coefficient
Pascal's triangle

KEY FORMULAS

$$\sum_{j=1}^{n} a_j = na_1 + \frac{n(n-1)}{2} \cdot d = \frac{n}{2}(a_1 + a_n)$$

$$\sum_{j=1}^{n} ar^{j-1} = a\frac{1 - r^n}{1 - r}$$

$$\sum_{j=1}^{\infty} ar^{j-1} = \frac{a}{1 - r} \quad \text{for } |r| < 1$$

$$n! = n(n-1)(n-2) \cdots 2 \cdot 1$$

$$_nP_k = \frac{n!}{(n-k)!}$$

$$_nC_k = \frac{n!}{k!(n-k)!}$$

$$(x + y)^n = \sum_{j=0}^{n} \binom{n}{j} x^{n-j}y^j$$

KEY THEOREMS AND LAWS

Axiom of Mathematical Induction
Axiom of Extended Mathematical Induction
Binomial Theorem

REVIEW EXERCISES

In Exercises 1–4, use mathematical induction to prove the given formula for all positive integers n.

1. $1 + 3 + 5 + \cdots + (2n - 1) = n^2$
2. $2 + 5 + 8 + \cdots + (3n - 1) = \frac{1}{2}n(3n + 1)$
3. $1 + 5 + 5^2 + \cdots + 5^{n-1} = \frac{1}{4}(5^n - 1)$
4. $\dfrac{1}{1 \cdot 4} + \dfrac{1}{4 \cdot 7} + \dfrac{1}{7 \cdot 10} + \cdots + \dfrac{1}{(3n - 2)(3n + 1)} = \dfrac{n}{3n + 1}$

5. Prove by induction that 3 is a factor of $n^3 + 2n$ for every integer $n \geq 0$.

*6. Let

$$a_n = \sqrt{2}^{\left(\sqrt{2}^{\left(\sqrt{2}^{\cdot^{\cdot^{\cdot(\sqrt{2})}}}\right)}\right)}$$

Show that $a_n < 2$ for each $n \geq 1$. (*Hint:* Let $a_1 = \sqrt{2}$ and $a_{n+1} = (\sqrt{2})^{a_n}$ for $n \geq 1$. Use mathematical induction to prove that $a_n < 2$ for $n \geq 1$.)

7. For $n \geq 1$ let

$$a_n = \left(1 + \frac{1}{1}\right)\left(1 + \frac{1}{2}\right)\left(1 + \frac{1}{3}\right) \cdots \left(1 + \frac{1}{n}\right)$$

a. Use the fact that for each positive integer k,

$$1 + \frac{1}{k} = \frac{k+1}{k}$$

to rewrite the formula for a_n. Then cancel where possible to discover a formula that shows that a_n is an integer.

b. Use mathematical induction to prove the formula you obtained in part (a).

8. Use mathematical induction to prove that $4^n > n^2$ for every integer $n \geq 0$.

In Exercises 9–12, find the first five terms of the sequence $\{a_n\}_{n=1}^{\infty}$.

9. $a_n = 4n + (-1)^n$

10. $a_n = 2^{n-3} + 2^{3-n}$

11. $a_1 = 3; \ a_{n+1} = -2a_n + 4$

12. $a_1 = 0; \ a_{n+1} = 2^{a_n}$

In Exercises 13–20, find the indicated partial sum.

13. $\sum\limits_{j=1}^{5} (-2)$

14. $\sum\limits_{j=1}^{8} (-1)^{j+1}$

15. $\sum\limits_{j=1}^{4} \left(-\frac{4}{j^2} \right)$

16. $\sum\limits_{j=1}^{10} j$

17. $\sum\limits_{j=1}^{16} (-5 + 7j)$

18. $\sum\limits_{j=2}^{20} \left(\frac{3}{4} + \frac{j}{2} \right)$

19. $\sum\limits_{j=1}^{4} 2 \left(\frac{1}{3} \right)^{j-1}$

20. $\sum\limits_{j=1}^{7} (-2)^j$

21. Find the nth term a_n of the arithmetic sequence whose first term is 5 and whose common difference is $\frac{3}{4}$.

22. Find the nth term a_n of the arithmetic sequence whose second term is 7 and whose common difference is -3.

23. Find the nth term a_n of the arithmetic sequence whose fifth term is 13 and whose ninth term is 5.

24. Find the nth term a_n of the geometric sequence whose first term is 2 and whose common ratio is $\frac{1}{3}$.

25. Find the nth term a_n of the geometric sequence whose fifth term is $\frac{1}{4}$ and whose common ratio is $-\frac{1}{2}$.

26. Find the nth term a_n of the geometric sequence whose third term is -5 and whose sixth term is 625.

In Exercises 27–30, find the numerical value of the geometric series.

27. $\sum\limits_{j=1}^{\infty} 3 \left(-\frac{4}{7} \right)^{j-1}$

28. $\sum\limits_{j=1}^{\infty} - \left(\frac{99}{100} \right)^{j-1}$

29. $\sum\limits_{j=1}^{\infty} -6 \left(\frac{11}{23} \right)^j$

30. $\sum\limits_{j=1}^{\infty} (0.8)^{j+1}$

31. Express $0.151515 \ldots$ as a fraction.

32. Express $3.062062062 \ldots$ as a fraction.

In Exercises 33–42, calculate the given number.

33. The number of permutations of 6 objects taken 3 at a time.

34. The number of permutations of 8 objects taken 6 at a time.

35. The number of combinations of 4 objects taken 2 at a time.

36. The number of combinations of 8 objects taken 5 at a time.

37. $_5P_2$ 38. $_{50}P_3$ 39. $_7C_5$

40. $_{10}C_7$ 41. $\binom{8}{4}$ 42. $\binom{15}{12}$

In Exercises 43–46, use the Binomial Theorem to expand the given power of the binomial.

43. $(x + y)^8$ 44. $(x - y)^6$

45. $(- 3x + 4)^3$ 46. $(2x^2 - y)^5$

47. Use the Binomial Theorem to calculate
   a. $(\sqrt{2} - \sqrt{3})^4 + (\sqrt{2} + \sqrt{3})^4$
   b. $(1 - 4i)^3 - (1 + 4i)^3$

48. Find the fifth term of $(\pi x - y)^6$.

49. Find the coefficient of the third term in the binomial expansion of $(x + 2y)^{16}$.

50. Suppose a university receives $1,000,000 in contributions in 1986, and each year thereafter the amount of contribution increases by $100,000. How much will the university have received in contributions by the end of 1995?

51. Which is more valuable—a gift consisting of one dollar each day for 100 days, or a gift consisting of 10¢ the first day, 20¢ the second day, 40¢ the third day, and so on for 10 days?

C 52. Suppose a king grants a peasant one grain of wheat the first day, 2 grains of wheat the second day, 4 grains of wheat the third day, and so on for 30 days. Use a calculator to compute the number of grains of wheat the peasant would receive.

53. How many 7-digit telephone numbers are there that neither begin nor end in 0?

54. A student has 5 mathematics books, 3 science books, and 2 history books. In how many ways can the books be placed on a shelf in a bookcase if all the books are distinct and the mathematics books are to be together, the science books together, and the history books together?

55. a. How many 2-card Blackjack hands can be dealt from a 52-card deck?
    b. How many Blackjack hands are there consisting of an ace and another card that is either a king, queen, jack, or ten? (*Hint:* There are 4 each of aces, kings, queens, jacks, and tens in a 52-card deck.)

56. A club having 15 members needs 4 officers: President, Vice-President, Secretary, and Treasurer.
    a. In how many ways can 4 members be selected as officers, without regard to who has which office?
    b. In how many ways can 4 members be selected to fill the offices?
    c. If the President can be Treasuer as well but the other officers must be distinct, in how many ways can the officers be selected?

57. A swimming club has 4 blue, 4 red, and 2 green sun umbrellas. In how many distinguishable ways can the sun umbrellas be distributed to 10 tables arranged in a straight line?

TABLES

TABLE A   Powers and Roots (from 1 to 100)

n	n^2	$\sqrt{n}$	n^3	$\sqrt[3]{n}$	n	n^2	$\sqrt{n}$	n^3	$\sqrt[3]{n}$
1	1	1.000	1	1.000	51	2,601	7.141	132,651	3.708
2	4	1.414	8	1.260	52	2,704	7.211	140,608	3.733
3	9	1.732	27	1.442	53	2,809	7.280	148,877	3.756
4	16	2.000	64	1.587	54	2,916	7.348	157,464	3.780
5	25	2.236	125	1.710	55	3,025	7.416	166,375	3.803
6	36	2.449	216	1.817	56	3,136	7.483	175,616	3.826
7	49	2.646	343	1.913	57	3,249	7.550	185,193	3.849
8	64	2.828	512	2.000	58	3,364	7.616	195,112	3.871
9	81	3.000	729	2.080	59	3,481	7.681	205,379	3.893
10	100	3.162	1,000	2.154	60	3,600	7.746	216,000	3.915
11	121	3.317	1,331	2.224	61	3,721	7.810	226,981	3.936
12	144	3.464	1,728	2.289	62	3,844	7.874	238,328	3.958
13	169	3.606	2,197	2.351	63	3,969	7.937	250,047	3.979
14	196	3.742	2,744	2.410	64	4,096	8.000	262,144	4.000
15	225	3.873	3,375	2.466	65	4,225	8.062	274,625	4.021
16	256	4.000	4,096	2.520	66	4,356	8.124	287,496	4.041
17	289	4.123	4,913	2.571	67	4,489	8.185	300,763	4.062
18	324	4.243	5,832	2.621	68	4,624	8.246	314,432	4.082
19	361	4.359	6,859	2.668	69	4,761	8.307	328,509	4.102
20	400	4.472	8,000	2.714	70	4,900	8.367	343,000	4.121
21	441	4.583	9,261	2.759	71	5,041	8.426	357,911	4.141
22	484	4.690	10,648	2.802	72	5,184	8.485	373,248	4.160
23	529	4.796	12,167	2.844	73	5,329	8.544	389,017	4.179
24	576	4.899	13,824	2.884	74	5,476	8.602	405,224	4.198
25	625	5.000	15,625	2.924	75	5,625	8.660	421,875	4.217
26	676	5.099	17,576	2.962	76	5,776	8.718	438,976	4.236
27	729	5.196	19,683	3.000	77	5,929	8.775	456,533	4.254
28	784	5.292	21,952	3.037	78	6,084	8.832	474,552	4.273
29	841	5.385	24,389	3.072	79	6,241	8.888	493,039	4.291
30	900	5.477	27,000	3.107	80	6,400	8.944	512,000	4.309
31	961	5.568	29,791	3.141	81	6,561	9.000	531,441	4.327
32	1,024	5.657	32,768	3.175	82	6,724	9.055	551,368	4.344
33	1,089	5.745	35,937	3.208	83	6,889	9.110	571,787	4.362
34	1,156	5.831	39,304	3.240	84	7,056	9.165	592,704	4.380
35	1,225	5.916	42,875	3.271	85	7,225	9.220	614,125	4.397
36	1,296	6.000	46,656	3.302	86	7,396	9.274	636,056	4.414
37	1,369	6.083	50,653	3.332	87	7,569	9.327	658,503	4.431
38	1,444	6.164	54,872	3.362	88	7,744	9.381	681,472	4.448
39	1,521	6.245	59,319	3.391	89	7,921	9.434	704,969	4.465
40	1,600	6.325	64,000	3.420	90	8,100	9.487	729,000	4.481
41	1,681	6.403	68,921	3.448	91	8,281	9.539	753,571	4.498
42	1,764	6.481	74,088	3.476	92	8,464	9.592	778,688	4.514
43	1,849	6.557	79,507	3.503	93	8,649	9.644	804,357	4.531
44	1,936	6.633	85,184	3.530	94	8,836	9.695	830,584	4.547
45	2,025	6.708	91,125	3.557	95	9,025	9.747	857,375	4.563
46	2,116	6.782	97,336	3.583	96	9,216	9.798	884,736	4.579
47	2,209	6.856	103,823	3.609	97	9,409	9.849	912,673	4.595
48	2,304	6.928	110,592	3.634	98	9,604	9.899	941,192	4.610
49	2,401	7.000	117,649	3.659	99	9,801	9.950	970,299	4.626
50	2,500	7.071	125,000	3.684	100	10,000	10.000	1,000,000	4.642

TABLE B **Exponential Functions e^x and e^{-x}**

x	e^x	e^{-x}	x	e^x	e^{-x}
0.00	1.0000	1.0000	1.5	4.4817	0.2231
0.01	1.0101	0.9901	1.6	4.9530	0.2019
0.02	1.0202	0.9802	1.7	5.4739	0.1827
0.03	1.0305	0.9702	1.8	6.0496	0.1653
0.04	1.0408	0.9608	1.9	6.6859	0.1496
0.05	1.0513	0.9512	2.0	7.3891	0.1353
0.06	1.0618	0.9418	2.1	8.1662	0.1225
0.07	1.0725	0.9324	2.2	9.0250	0.1108
0.08	1.0833	0.9331	2.3	9.9742	0.1003
0.09	1.0942	0.9139	2.4	11.023	0.0907
0.10	1.1052	0.9048	2.5	12.182	0.0821
0.11	1.1163	0.8958	2.6	13.464	0.0743
0.12	1.1275	0.8869	2.7	14.880	0.0672
0.13	1.1388	0.8781	2.8	16.445	0.0608
0.14	1.1503	0.8694	2.9	18.174	0.0550
0.15	1.1618	0.8607	3.0	20.086	0.0498
0.16	1.1735	0.8521	3.1	22.198	0.0450
0.17	1.1853	0.8437	3.2	24.533	0.0408
0.18	1.1972	0.8353	3.3	27.113	0.0369
0.19	1.2092	0.8270	3.4	29.964	0.0334
0.20	1.2214	0.8187	3.5	33.115	0.0302
0.21	1.2337	0.8106	3.6	36.598	0.0273
0.22	1.2461	0.8025	3.7	40.447	0.0247
0.23	1.2586	0.7945	3.8	44.701	0.0224
0.24	1.2712	0.7866	3.9	49.402	0.0202
0.25	1.2840	0.7788	4.0	54.598	0.0183
0.30	1.3499	0.7408	4.1	60.340	0.0166
0.35	1.4191	0.7047	4.2	66.686	0.0150
0.40	1.4918	0.6703	4.3	73.700	0.0136
0.45	1.5683	0.6376	4.4	81.451	0.0123
0.50	1.6487	0.6065	4.5	90.017	0.0111
0.55	1.7333	0.5769	4.6	99.484	0.0101
0.60	1.8221	0.5488	4.7	109.95	0.0091
0.65	1.9155	0.5220	4.8	121.51	0.0082
0.70	2.0138	0.4966	4.9	134.29	0.0074
0.75	2.1170	0.4724	5.0	148.41	0.0067
0.80	2.2255	0.4493	5.5	244.69	0.0041
0.85	2.3396	0.4274	6.0	403.43	0.0025
0.90	2.4596	0.4066	6.5	665.14	0.0015
0.95	2.5857	0.3867	7.0	1096.6	0.0009
1.0	2.7183	0.3679	7.5	1808.0	0.0006
1.1	3.0042	0.3329	8.0	2981.0	0.0003
1.2	3.3201	0.3012	8.5	4914.8	0.0002
1.3	3.6693	0.2725	9.0	8103.1	0.0001
1.4	4.0552	0.2466	10.0	22026	0.00005

TABLE C Common Logarithms (base 10)

x	0	1	2	3	4	5	6	7	8	9
1.0	.0000	.0043	.0086	.0128	.0170	.0212	.0253	.0294	.0334	.0374
1.1	.0414	.0453	.0492	.0531	.0569	.0607	.0645	.0682	.0719	.0755
1.2	.0792	.0828	.0864	.0899	.0934	.0969	.1004	.1038	.1072	.1106
1.3	.1139	.1173	.1206	.1239	.1271	.1303	.1335	.1367	.1399	.1430
1.4	.1461	.1492	.1523	.1553	.1584	.1614	.1644	.1673	.1703	.1732
1.5	.1761	.1790	.1818	.1847	.1875	.1903	.1931	.1959	.1987	.2014
1.6	.2041	.2068	.2095	.2122	.2148	.2175	.2201	.2227	.2253	.2279
1.7	.2304	.2330	.2355	.2380	.2405	.2430	.2455	.2480	.2504	.2529
1.8	.2553	.2577	.2601	.2625	.2648	.2672	.2695	.2718	.2742	.2765
1.9	.2788	.2810	.2833	.2856	.2878	.2900	.2923	.2945	.2967	.2989
2.0	.3010	.3032	.3054	.3075	.3096	.3118	.3139	.3160	.3181	.3201
2.1	.3222	.3243	.3263	.3284	.3304	.3324	.3345	.3365	.3385	.3404
2.2	.3424	.3444	.3464	.3483	.3502	.3522	.3541	.3560	.3579	.3598
2.3	.3617	.3636	.3655	.3674	.3692	.3711	.3729	.3747	.3766	.3784
2.4	.3802	.3820	.3838	.3856	.3874	.3892	.3909	.3927	.3945	.3962
2.5	.3979	.3997	.4014	.4031	.4048	.4065	.4082	.4099	.4116	.4133
2.6	.4150	.4166	.4183	.4200	.4216	.4232	.4249	.4265	.4281	.4298
2.7	.4314	.4330	.4346	.4362	.4378	.4393	.4409	.4425	.4440	.4456
2.8	.4472	.4487	.4502	.4518	.4533	.4548	.4564	.4579	.4594	.4609
2.9	.4624	.4639	.4654	.4669	.4683	.4698	.4713	.4728	.4742	.4757
3.0	.4771	.4786	.4800	.4814	.4829	.4843	.4857	.4871	.4886	.4900
3.1	.4914	.4928	.4942	.4955	.4969	.4983	.4997	.5011	.5024	.5038
3.2	.5051	.5065	.5079	.5092	.5105	.5119	.5132	.5145	.5159	.5172
3.3	.5185	.5198	.5211	.5224	.5237	.5250	.5263	.5276	.5289	.5302
3.4	.5315	.5328	.5340	.5353	.5366	.5378	.5391	.5403	.5416	.5428
3.5	.5441	.5453	.5465	.5478	.5490	.5502	.5514	.5527	.5539	.5551
3.6	.5563	.5575	.5587	.5599	.5611	.5623	.5635	.5647	.5658	.5670
3.7	.5682	.5694	.5705	.5717	.5729	.5740	.5752	.5763	.5775	.5786
3.8	.5798	.5809	.5821	.5832	.5843	.5855	.5866	.5877	.5888	.5899
3.9	.5911	.5922	.5933	.5944	.5955	.5966	.5977	.5988	.5999	.6010
4.0	.6021	.6031	.6042	.6053	.6064	.6075	.6085	.6096	.6107	.6117
4.1	.6128	.6138	.6149	.6160	.6170	.6180	.6191	.6201	.6212	.6222
4.2	.6232	.6243	.6253	.6263	.6274	.6284	.6294	.6304	.6314	.6325
4.3	.6335	.6345	.6355	.6365	.6375	.6385	.6395	.6405	.6415	.6425
4.4	.6435	.6444	.6454	.6464	.6474	.6484	.6493	.6503	.6513	.6522
4.5	.6532	.6542	.6551	.6561	.6571	.6580	.6590	.6599	.6609	.6618
4.6	.6628	.6637	.6646	.6656	.6665	.6675	.6684	.6693	.6702	.6712
4.7	.6721	.6730	.6739	.6749	.6758	.6767	.6776	.6785	.6794	.6803
4.8	.6812	.6821	.6830	.6839	.6848	.6857	.6866	.6875	.6884	.6893
4.9	.6902	.6911	.6920	.6928	.6937	.6946	.6955	.6964	.6972	.6981
5.0	.6990	.6998	.7007	.7016	.7024	.7033	.7042	.7050	.7059	.7067
5.1	.7076	.7084	.7093	.7101	.7110	.7118	.7126	.7135	.7143	.7152
5.2	.7160	.7168	.7177	.7185	.7193	.7202	.7210	.7218	.7226	.7235
5.3	.7243	.7251	.7259	.7267	.7275	.7284	.7292	.7300	.7308	.7316
5.4	.7324	.7332	.7340	.7348	.7356	.7364	.7372	.7380	.7388	.7396
x	0	1	2	3	4	5	6	7	8	9

TABLE C   Common Logarithms (cont.)

x	0	1	2	3	4	5	6	7	8	9
5.5	.7404	.7412	.7419	.7427	.7435	.7443	.7451	.7459	.7466	.7474
5.6	.7482	.7490	.7497	.7505	.7513	.7520	.7528	.7536	.7543	.7551
5.7	.7559	.7566	.7574	.7582	.7589	.7597	.7604	.7612	.7619	.7627
5.8	.7634	.7642	.7649	.7657	.7664	.7672	.7679	.7686	.7694	.7701
5.9	.7709	.7716	.7723	.7731	.7738	.7745	.7752	.7760	.7767	.7774
6.0	.7782	.7789	.7796	.7803	.7810	.7818	.7825	.7832	.7839	.7846
6.1	.7853	.7860	.7868	.7875	.7882	.7889	.7896	.7903	.7910	.7917
6.2	.7924	.7931	.7938	.7945	.7952	.7959	.7966	.7973	.7980	.7987
6.3	.7993	.8000	.8007	.8014	.8021	.8028	.8035	.8041	.8048	.8055
6.4	.8062	.8069	.8075	.8082	.8089	.8096	.8102	.8109	.8116	.8122
6.5	.8129	.8136	.8142	.8149	.8156	.8162	.8169	.8176	.8182	.8189
6.6	.8195	.8202	.8209	.8215	.8222	.8228	.8235	.8241	.8248	.8254
6.7	.8261	.8267	.8274	.8280	.8287	.8293	.8299	.8306	.8312	.8319
6.8	.8325	.8331	.8338	.8344	.8351	.8357	.8363	.8370	.8376	.8382
6.9	.8388	.8395	.8401	.8407	.8414	.8420	.8426	.8432	.8439	.8445
7.0	.8451	.8457	.8463	.8470	.8476	.8482	.8488	.8494	.8500	.8506
7.1	.8513	.8519	.8525	.8531	.8537	.8543	.8549	.8555	.8561	.8567
7.2	.8573	.8579	.8585	.8591	.8597	.8603	.8609	.8615	.8621	.8627
7.3	.8633	.8639	.8645	.8651	.8657	.8663	.8669	.8675	.8681	.8686
7.4	.8692	.8698	.8704	.8710	.8716	.8722	.8727	.8733	.8739	.8745
7.5	.8751	.8756	.8762	.8768	.8774	.8779	.8785	.8791	.8797	.8802
7.6	.8808	.8814	.8820	.8825	.8831	.8837	.8842	.8848	.8854	.8859
7.7	.8865	.8871	.8876	.8882	.8887	.8893	.8899	.8904	.8910	.8915
7.8	.8921	.8927	.8932	.8938	.8943	.8949	.8954	.8960	.8965	.8971
7.9	.8976	.8982	.8987	.8993	.8998	.9004	.9009	.9015	.9020	.9025
8.0	.9031	.9036	.9042	.9047	.9053	.9058	.9063	.9069	.9074	.9079
8.1	.9085	.9090	.9096	.9101	.9106	.9112	.9117	.9122	.9128	.9133
8.2	.9138	.9143	.9149	.9154	.9159	.9165	.9170	.9175	.9180	.9186
8.3	.9191	.9196	.9201	.9206	.9212	.9217	.9222	.9227	.9232	.9238
8.4	.9243	.9248	.9253	.9258	.9263	.9269	.9274	.9279	.9284	.9289
8.5	.9294	.9299	.9304	.9309	.9315	.9320	.9325	.9330	.9335	.9340
8.6	.9345	.9350	.9355	.9360	.9365	.9370	.9375	.9380	.9385	.9390
8.7	.9395	.9400	.9405	.9410	.9415	.9420	.9425	.9430	.9435	.9440
8.8	.9445	.9450	.9455	.9460	.9465	.9469	.9474	.9479	.9484	.9489
8.9	.9494	.9499	.9504	.9509	.9513	.9518	.9523	.9528	.9533	.9538
9.0	.9542	.9547	.9552	.9557	.9562	.9566	.9571	.9576	.9581	.9586
9.1	.9590	.9595	.9600	.9605	.9609	.9614	.9619	.9624	.9628	.9633
9.2	.9638	.9643	.9647	.9652	.9657	.9661	.9666	.9671	.9675	.9680
9.3	.9685	.9689	.9694	.9699	.9703	.9708	.9713	.9717	.9722	.9727
9.4	.9731	.9736	.9741	.9745	.9750	.9754	.9759	.9763	.9768	.9773
9.5	.9777	.9782	.9786	.9791	.9795	.9800	.9805	.9809	.9814	.9818
9.6	.9823	.9827	.9832	.9836	.9841	.9845	.9850	.9854	.9859	.9863
9.7	.9868	.9872	.9877	.9881	.9886	.9890	.9894	.9899	.9903	.9908
9.8	.9912	.9917	.9921	.9926	.9930	.9934	.9939	.9943	.9948	.9952
9.9	.9956	.9961	.9965	.9969	.9974	.9978	.9983	.9987	.9991	.9996
x	0	1	2	3	4	5	6	7	8	9

TABLE D   Natural Logarithms of Numbers (base e)

n	$\log_e n$	n	$\log_e n$	n	$\log_e n$
		4.5	1.5041	9.0	2.1972
0.1	−2.3026	4.6	1.5261	9.1	2.2083
0.2	−1.6094	4.7	1.5476	9.2	2.2192
0.3	−1.2040	4.8	1.5486	9.3	2.2300
0.4	−0.9163	4.9	1.5892	9.4	2.2407
0.5	−0.6931	5.0	1.6094	9.5	2.2513
0.6	−0.5108	5.1	1.6292	9.6	2.2618
0.7	−0.3567	5.2	1.6487	9.7	2.2721
0.8	−0.2231	5.3	1.6677	9.8	2.2824
0.9	−0.1054	5.4	1.6864	9.9	2.2925
1.0	0.0000	5.5	1.7047	10	2.3026
1.1	0.0953	5.6	1.7228	11	2.3979
1.2	0.1823	5.7	1.7405	12	2.4849
1.3	0.2624	5.8	1.7579	13	2.5649
1.4	0.3365	5.9	1.7750	14	2.6391
1.5	0.4055	6.0	1.7918	15	2.7081
1.6	0.4700	6.1	1.8083	16	2.7726
1.7	0.5306	6.2	1.8245	17	2.8332
1.8	0.5878	6.3	1.8405	18	2.8904
1.9	0.6419	6.4	1.8563	19	2.9444
2.0	0.6931	6.5	1.8718	20	2.9957
2.1	0.7419	6.6	1.8871	25	3.2189
2.2	0.7885	6.7	1.9021	30	3.4012
2.3	0.8329	6.8	1.9169	35	3.5553
2.4	0.8755	6.9	1.9315	40	3.6889
2.5	0.9163	7.0	1.9459	45	3.8067
2.6	0.9555	7.1	1.9601	50	3.9120
2.7	0.9933	7.2	1.9741	55	4.0073
2.8	1.0296	7.3	1.9879	60	4.0943
2.9	1.0647	7.4	2.0015	65	4.1744
3.0	1.0986	7.5	2.0149	70	4.2485
3.1	1.1314	7.6	2.0281	75	4.3175
3.2	1.1632	7.7	2.0412	80	4.3820
3.3	1.1939	7.8	2.0541	85	4.4427
3.4	1.2238	7.9	2.0669	90	4.4998
3.5	1.2528	8.0	2.0794	100	4.6052
3.6	1.2809	8.1	2.0919	110	4.7005
3.7	1.3083	8.2	2.1041	120	4.7875
3.8	1.3350	8.3	2.1163	130	4.8676
3.9	1.3610	8.4	2.1282	140	4.9416
4.0	1.3863	8.5	2.1401	150	5.0106
4.1	1.4110	8.6	2.1518	160	5.0752
4.2	1.4351	8.7	2.1633	170	5.1358
4.3	1.4586	8.8	2.1748	180	5.1930
4.4	1.4816	8.9	2.1861	190	5.2470

ANSWERS TO ODD-NUMBERED EXERCISES

SECTION 1.1

1. 6   3. 1   5. 22   7. $a^2 + 3a + 2$ 9. $-a^2 + 1$ 11. -2

13. $27 + 10\sqrt{2}$ 15. $\frac{64}{9}$ 17. $ac + bc + ad + bd - 3a - 3b + 2c + 2d - 6$

19. $\frac{3}{4}$ 21. $\frac{3}{10}$ 23. $-\frac{5}{7}$ 25. $\frac{5}{8}$ 27. $\frac{6}{5}$ 29. $\frac{3}{4}$

31. 0.022024709 33. 0.034176350 35. l.c.d.: 24; $\frac{7}{24}$ 37. l.c.d.: 45; $\frac{46}{45}$

39. l.c.d.: 180; $-\frac{7}{180}$ 41. $-\dfrac{4a}{a^2 - 1}$ 43. $\dfrac{2b - 3a + 4}{ab}$

45. $(a + 1)(b + 1) = ab + a + b + 1$ 47. $a - (b + c) = a - b - c$

49. $(a + b)^3 = a^3 + 3a^2b + 3ab^2 + b^3$ 51. $\dfrac{1}{a + b}$ remains $\dfrac{1}{a + b}$

53. The decimal expansion neither terminates nor repeats.

57. a. $a = 0, b = 1, c = 1$ (other solutions possible)
    b. $a = 2, b = 2, c = 2$ (other solutions possible)

SECTION 1.2

1.

3. $x \geq \sqrt{2}$ 5. $-1 < 6 - r < 1$ 7. $z < 0$ 9. $|x - 2| < 0.01$

11. $c \leq \frac{1}{10}$ 13. $\sqrt{2} > 1$ 15. $(-2)^2 > 3$ 17. $0 < \pi - 3$ 19. $\sqrt{16} = 4$

21. 1   23. 6   25. 1   27. 8.5   29. 5   31. 1   33. -4

35. π 37. 2   39. 16   41. 10   43. x^2 45. $x^2 + 1$ 47. $a - 4$

49. $a - b$ 51. 0

SECTION 1.3

1. 1   3. $\frac{1}{8}$ 5. $\frac{1}{4}$ 7. $\frac{1}{1000}$ 9. 4^5 11. -2^9 13. $-\frac{1}{7}$ 15. 2^8
17. $\frac{1}{100}$ 19. 4^7 21. 4^{30} 23. $\frac{8}{9}$ 25. $\frac{9}{4}$ 27. $\frac{1}{147}$ 29. a^{12}
31. y^{-8} 33. b^5 35. c^8 37. x^3y^6 39. $\frac{9}{4}x^{-8}$ 41. $r^{16}s^{14}$ 43. t^2
45. $\dfrac{1}{ab}$ 47. $\dfrac{a^3b^3}{(a+b)^3}$ 49. $\dfrac{a^2b^2}{a+b}$ 51. Positive   53. Negative
55. 4.832×10^2 57. 1.009×10^0, or 1.009 59. 9.999×10^{-1}
61. 2.7168876×10^7 63. 1.7659446×10^{-1} 65. 3×10^{-7}
67. 1.817×10^9 69. 8.7445702×10^{-2} 71. 7.8066378×10^{-3}
73. $C = 2\pi r$ 75. $A = s^2$ 77. $A = \frac{4}{3}\pi r^3$ 79. $V = \frac{1}{3}hr^2$
81. $0.000000000000000000000000000167$
85. a. $a = 0$ and either $m \le 0$ or $n \le 0$
    b. $n \le 0$ and either $a = 0$ or $b = 0$
87. 9.2529×10^7

SECTION 1.4

1. 4   3. $\frac{1}{5}$ 5. 1.3 7. 0.2 9. $4\sqrt{3}$ 11. $\frac{1}{12}$ 13. $2\sqrt{10} \times 10^5$
15. $6\sqrt{2}$ 17. 10^7 19. $\dfrac{\sqrt{6}}{6}$ 21. $\frac{5}{7}$ 23. $ab\sqrt{b}$ 25. $4xy^2z\sqrt{2xy}$
27. $\dfrac{p^2}{3q^4}\sqrt{\dfrac{p}{3}}$ 29. $\dfrac{8}{r}\sqrt{\dfrac{2}{r}}$ 31. $\dfrac{1}{c^{12}d^4\sqrt{d}}$ 33. $x + y$ 35. $(4a - 7b)^2$
37. $2ab^2$ 39. $\sqrt{a} + \sqrt{b}$ 41. 3   43. 0.5 45. 4   47. 3   49. $2\sqrt[3]{2}$
51. $2\sqrt[4]{8}$ 53. ab^2 55. $-\dfrac{1}{5xy^2\sqrt[3]{z}}$ 57. $\dfrac{x+y}{3x^3y^2}$ 59. t
61. $(4x + 2y)^2$ 63. $\dfrac{\sqrt{3}}{3}$ 65. $\sqrt{2} - 1$ 67. $\sqrt{6} - 2$ 69. $\dfrac{19 + 8\sqrt{3}}{13}$
71. $\dfrac{a + b + 2\sqrt{ab}}{a - b}$ 73. $s = \sqrt[3]{V}$ 75. $r = \sqrt[3]{\dfrac{3V}{4\pi}}$

SECTION 1.5

1. 32   3. $\frac{1}{25}$ 5. $\frac{1}{256}$ 7. $\frac{1}{32}$ 9. $\frac{125}{27}$ 11. 2   13. $250\sqrt{2}$ 15. $\frac{8}{27}$
17. $\dfrac{1}{2\sqrt{2}}$ 19. x 21. $216a^{9/2}b^6$ 23. $x^2y^4z^{16/3}$ 25. $\dfrac{z^{9/4}}{8}$
27. $x^{2/5}y^{24/5}z^{14/5}$ 29. $4a$ 31. $\left(\dfrac{p}{q}\right)^{2/3}$ 33. $\sqrt{p^2 + q^2}$ 35. $\dfrac{z^2}{64(x-y)^3}$
37. $b^{13/6}$ 39. 1.7411011 41. 1.4645919 43. 0.89921804

SECTION 1.6

1. 3, 1, 10   3. $\sqrt{5}, -4 + \sqrt{5}, -4 + \sqrt{5}$ 5. $7x^2 - 6$
7. $-x^3 + x^2 + 2x + 4$ 9. $8x^2 - 22x + 15$ 11. $\frac{1}{8}x^2 + 3x + 12$
13. $7x^2 - 2x - 3$ 15. $12x - 15$ 17. $\sqrt{2}\, y^3 - 2y$ 19. $2r^2 - 2r + 25$
21. $81x^4 - 162x^3 + 117x^2 - 36x + 4$ 23. $2y^6 + 16$

25. $4x^{25} + 7x^{20} - 23x^{15} + 10x^{10}$ 27. $4x^2 + xy + 2y^2$ 29. $x^2 + xy - y^2$

31. $-2x^2y^2 + 6xy - 12y^2$ 33. $\frac{1}{4}x^2 + \frac{1}{4}xy + \frac{1}{16}y^2$

35. $64x^3 - 48x^2y^2 + 12xy^4 - y^6$ 37. $2xh + h^2$ 39. $x^3 - y^3$ 41. $r^5s - rs^5$

43. $x^2 + 4y^2 + 9z^2 + 4xy - 12yz - 6xz$ 45. $u + 2u^{1/2}v^{1/2} + v$ 47. $\dfrac{1}{r^2} - \dfrac{2}{rs} + \dfrac{1}{s^2}$

SECTION 1.7

1. $(x + 2)(x + 6)$ 3. $(x - 1)(x - 6)$ 5. $(t + 1)(t - 4)$ 7. $(y + 4)(y + 9)$

9. $(x - 4)(x - 15)$ 11. $(7 - b)(3 - b)$ 13. $(x + 2)(x - 2)$ 15. $(a - 7)^2$

17. $(x + \frac{11}{2})^2$ 19. $(4 + 3z)(4 - 3z)$ 21. $(x - \sqrt{2})^2$ 23. $(2x + 1)(x + 3)$

25. $(x - 3)(-x + 2)$ or $(x - 2)(-x + 3)$ 27. $(7x - 4)(x + 3)$

29. $(3t - 1)(2t + 6)$ 31. $(x + y)^2$ 33. $(x + 2y)(x - 2y)$

35. $(12a + b)(12a - b)$ 37. $(x - y)^2(x + y)^2$ 39. $(5x + y)(x - 3y)$

41. $x(x + 2)$ 43. $x^6(x - 1)$ 45. $x(3x + 5)(x + 1)$ 47. $(x - 3)^2(x + 3)^2$

49. $(x - 2)(x + 1)(x - 1)$ 51. $(x + 1)^2(x - 1)$ 53. $(3x + 2y + 4)(3x + 2y - 3)$

55. $(x - 5 + 2y)(x - 5 - 2y)$ 57. $(x + 1)(x^2 - x + 1)$

59. $(x - 2)(x^2 + 2x + 4)$ 61. $(2x - 1)(4x^2 + 2x + 1)$

63. $(2x - 1)(16x^4 + 8x^3 + 4x^2 + 2x + 1)$

65. $(x + y)(x^2 - xy + y^2)(x - y)(x^2 + xy + y^2)$

67. $(x + 1)(x^4 - x^3 + x^2 - x + 1)(x - 1)(x^4 + x^3 + x^2 + x + 1)$

69. $[(x - 2y)^2 + 1](x - 2y + 1)(x - 2y - 1)$

SECTION 1.8

1. All real numbers except -2 and -3; $-\frac{2}{3}$ 3. All real numbers except 1; $-\frac{10}{7}$

5. $\dfrac{1}{x + 4}$ 7. $\dfrac{y - 8}{y - 3}$ 9. $\dfrac{b - 1}{4b - 5}$ 11. $\dfrac{(s + 3)(s - 2)(s + 2)}{s - 4}$

13. $\dfrac{x + y}{x^2 + xy + y^2}$ 15. $(x + 1)(x - 2)$ 17. $y(y - 2)$

19. $\dfrac{(z + 3)^3}{(z + 2)(z - 2)(z - 1)}$ 21. $\dfrac{(x + 3)^2}{(x - 1)^2}$ 23. $\dfrac{(b + 1)(b - 1)}{b - 9}$ 25. $\dfrac{(x - y)^2}{x + y}$

27. $\dfrac{3x - 1}{x(x - 1)}$ 29. $\dfrac{3y - 4}{(y + 3)(y - 3)}$ 31. $-\dfrac{8}{(y - 3)(y + 8)}$ 33. $\dfrac{20z^2 + 2z - 13}{(6z + 1)(z + 4)}$

35. $\dfrac{5}{t - 3}$ 37. $\dfrac{7u(u - 2)}{(3u - 1)(2u - 3)}$ 39. $\dfrac{v - 1}{v + 2}$ 41. $\dfrac{a^2 + b^2}{ab}$ 43. $\dfrac{b^2}{a^2 + b^2}$

45. $\dfrac{ab^2 - 2b + 4}{ab^3}$ 47. $\dfrac{qr + pr + pq}{pqr}$ 49. $-\dfrac{1}{x(x + h)}$ 51. $\dfrac{x - y}{x + y}$

53. $\dfrac{1 + x^2}{2 + x}$ 55. x 57. $\dfrac{(5z - 3)(z + 4)}{12(z + 5)}$ 59. $x - y$ 61. $\dfrac{\sqrt{y} - \sqrt{x}}{\sqrt{xy}}$

63. $\dfrac{\sqrt{x} + \sqrt{3}}{x - 3}$ 65. $\dfrac{\sqrt{x} + 9}{x - 81}$ 67. $\sqrt{x} - \sqrt{3}$ 69. $\dfrac{x - 2\sqrt{xy} + y}{x - y}$

CHAPTER 1 REVIEW

1. $\frac{5}{12}$ 3. $\frac{1}{27}$ 5. 16 7. 0.3 9. 9 11. $-\frac{2}{3}$ 13. $a - \sqrt{5} \geq 0$

15. 4.3   17. 1.59×10^5 19. 6.4×10^{-4} 21. $\dfrac{1}{8a^2}$ 23. $6\sqrt{2}$

25. $\frac{1}{15}$ 27. $(x - 5)|x - 5|$ 29. $8x^3$ 31. $\dfrac{2a}{\sqrt[3]{b}}$ 33. $\dfrac{2ab}{a^2 + b^2}$ 35. 1

37. $10a^2 - 22a + 4$ 39. $34x^2 - 137x + 24$ 41. $x - 3x^{1/3} + 3x^{-1/3} - x^{-1}$
43. $8x^6 + 60x^4y + 150x^2y^2 + 125y^3$ 45. $\sqrt{5} + 2$ 47. Negative
49. $(x + 3)(x - 9)$ 51. $(t - 10)^2$ 53. $(4x - 1)(3x - 2)$
55. $(y - 1)^3(y + 1)(y - 3)$ 57. $(z^2 + 2)(z + 1)(z - 1)$ 59. $(x + 1)(x^2 + 1)$
61. $(x + 7y)(x - 5y)$ 63. $(4x^2 + 1)(2x + 1)(2x - 1)$

65. All real numbers except -4 and -5 67. -14 69. $\frac{28}{97}$ 71. $\dfrac{x - 1}{x + 2}$

73. $\dfrac{(2x - 1)(x - 2)}{(x + 4)(3x - 1)}$ 75. 1   77. $\dfrac{3x^2 - x - 6}{(x - 2)(x + 2)}$ 79. $-\dfrac{2(x^2 + 5x - 3)}{x(x + 1)(x - 1)}$

81. $\dfrac{y}{y - x}$ 83. $\dfrac{2y + 3x}{y - 2x}$ 85. $\dfrac{x - 1}{x + 1}$ 87. $S = 2s^2 + 4sh$

89. $A = \dfrac{\sqrt{3}}{4}s^2$ 91. Jill, by \$484

SECTION 2.1

1. 0   3. $-\frac{1}{3}$ 5. 1   7. 7   9. -3 11. $-\frac{2}{7}$ 13. $\frac{1}{2}$ 15. -5
17. $\frac{1}{12}$ 19. $-\frac{7}{13}$ 21. -1 23. $\frac{18}{7}$ 25. $-\frac{7}{3}$ 27. $-\frac{1}{7}$ 29. $\frac{3}{2}$
31. 1, 7   33. $-\frac{7}{2}, \frac{3}{2}$ 35. $-9, -3$ 37. 0, 36   39. 1, 3   41. $x = y$
43. $x = 3y - \frac{1}{2}$ 45. $x = -\frac{1}{2}y + \frac{5}{4}$ 47. $x = 2y + 12$ 49. $x = -\frac{7}{3}y + \frac{1}{15}$
51. $y = \frac{1}{2}x - \frac{7}{2}$ 53. $y = \frac{2}{3}x - 4$ 57. -2 59. 8
61. a. 30.48   b. 25   c. 39.370079   63. 212 degrees Fahrenheit

65. $C = \frac{5}{9}(F - 32)$ 67. $p = \dfrac{fq}{q - f}$

SECTION 2.2

1. 83   3. 96   5. Karen is 8, and her father is 32.   7. 17 and 12
9. 4700 student tickets, 1800 nonstudent tickets
11. 100 dollar bills, 20 quarters, 50 dimes, 40 nickels   13. 24   15. \$25.50
17. 96   19. \$2200 in the 6% account, \$800 in the 8% account
21. 7.5% on the \$4200 investment, 6.5% on the \$2400 investment
23. Length: 30 inches; width: 20 inches   25. 30 meters by 50 meters   27. 9
29. 55,160,000 square meters
31. 30 pounds of 40% silver alloy, 20 pounds of 30% silver alloy   33. 1.6 gallons
35. $\frac{1}{5}$ hour (12 minutes)   37. 2400 miles
39. a. 6.5 miles   b. $2\frac{1}{4}$ hours   41. 12 minutes
43. $\frac{36}{23}$ hour (approximately 1 hour and 34 minutes)
45. $\frac{36}{5}$ hour (7 hours and 12 minutes)   47. \$50   49. b. 48 miles per hour

SECTION 2.3

1. $4, -4$ 3. $3, -3$ 5. -2 7. 25   9. $\frac{2}{3}$ 11. $\sqrt{2}$ 13. $-\frac{1}{4}$
15. $-4, -1$ 17. $2, -5$ 19. 7, 3   21. $\frac{1}{5}\sqrt{15}, -\frac{1}{5}\sqrt{15}$ 23. $1, \frac{1}{2}$

25. $1 + \sqrt{7}, 1 - \sqrt{7}$ 27. $5 + 3\sqrt{3}, 5 - 3\sqrt{3}$ 29. $-1 + \frac{1}{2}\sqrt{10}, -1 - \frac{1}{2}\sqrt{10}$
31. $\frac{3}{2} + \frac{1}{2}\sqrt{5}, \frac{3}{2} - \frac{1}{2}\sqrt{5}$ 33. None   35. $1 + \sqrt{5}, 1 - \sqrt{5}$
37. $\frac{1}{2} + \frac{1}{2}\sqrt{7}, \frac{1}{2} - \frac{1}{2}\sqrt{7}$ 39. $1 + \frac{1}{3}\sqrt{6}, 1 - \frac{1}{3}\sqrt{6}$ 41. $-\frac{1}{2}, -\frac{1}{3}$
43. $\frac{2}{3}(\sqrt{3} + \sqrt{6}), \frac{2}{3}(\sqrt{3} - \sqrt{6})$ 45. $-\frac{1}{2}\sqrt{2} + 1, -\frac{1}{2}\sqrt{2} - 1$ 47. None
49. $\frac{3}{2} + \frac{1}{2}\sqrt{29}, \frac{3}{2} - \frac{1}{2}\sqrt{29}$ 51. $-2 + 2\sqrt{3}, -2 - 2\sqrt{3}$ 53. $3 + \sqrt{14}, 3 - \sqrt{14}$
55. $6, -6$ 57. 0 59. $2, -2$ 61. $6.7949548, -1.1282881$

63. $1.4336420, -1.0196293$ 65. $r = \sqrt[4]{\dfrac{A}{\pi}}$ 67. $r = \sqrt{\dfrac{V}{\pi h}}$ 69. $-\frac{1}{2} + \frac{1}{2}\sqrt{5}$

SECTION 2.4

1. $-11, -4$ 3. $7, -4$ 5. $2, -2, \sqrt{2}, -\sqrt{2}$ 7. $3, -3$ 9. None
11. $\sqrt[3]{28}, -1$ 13. 16 15. 36 17. 81 19. 16 21. $8, 1$
23. 16 25. $\sqrt{5}, -\sqrt{5}$ 27. $7, -2$ 29. $\frac{1}{3}$ 31. $\sqrt{3}, -\sqrt{3}$ 33. 5

35. 1 37. 1 39. $\frac{3}{2}$ 41. $\dfrac{\sqrt[3]{10}}{3}$ 43. $2^{1/n}$ (and $-2^{1/n}$ if n is even)

45. -1 47. $\sqrt{3}, -\sqrt{3}$ 49. $\sqrt[3]{23}$ 51. $0, -1$ 53. $0, 1, -1$
55. $2, -2, 4$ 57. $3, -3$ 59. 1 61. $0, 2, -2, 3$ 63. a. -1 b. -1
65. $a = \sqrt{\left(\dfrac{Q}{E}\right)^{2/3} - 1}$ 67. $r = \sqrt[4]{\dfrac{8\eta l V}{\pi p}}$ 69. $c \approx 2.5118864, d \approx 15.848932$

SECTION 2.5

1. 2 seconds   3. $\frac{3}{2}\sqrt{5}$ seconds (approximately 3.35 seconds)   5. 1 second
7. $\sqrt{61}$ inches   9. $-\frac{1}{2} + \frac{1}{2}\sqrt{199}$ meters, $\frac{1}{2} + \frac{1}{2}\sqrt{199}$ meters   11. 3000 feet
13. 40 miles per hour   15. 300 and 400 miles per hour
17. 170 miles per hour
19. 40 miles per hour in midday, 24 miles per hour in rush hour   21. 32, 33
23. $(0, 2\sqrt{5})$ and $(0, -2\sqrt{5})$ 25. 9 inches by 7 inches   27. A 10-foot strip
29. 28 inches and 16 inches   31. 4 inches
33. a. \$3 and \$10; 70 umbrellas and 0 umbrellas, respectively
    b. \$5; 50 umbrellas
35. 8 feet

SECTION 2.6

1. Open, bounded   3. Closed, unbounded   5. Half-open, bounded
7. Closed, bounded   9. Open, bounded   11. Half-open, bounded
13. Open, unbounded   15. $(-4, 3]$ 17. $(-1.1, -0.9)$ 19. $(-8, \infty)$
21. $[-1, 1)$ 23. $(-\infty, 3]$ 25. $(-\infty, \frac{1}{4}]$ 27. $(-\infty, -\frac{2}{7}]$ 29. $(-\infty, \frac{5}{2})$
31. $(6, \infty)$ 33. $(-\infty, 7]$ 35. $(-\infty, -\frac{3}{2}]$ 37. $(-5, -2)$ 39. $[-3, -1)$
41. $[-23, -15]$ 43. $(1.99, 2.01)$ 45. $(a, a + d)$
49. Between 20 and 120, inclusive   51. From \$0 to \$5.50 inclusive
53. From \$0 to \$2 inclusive

SECTION 2.7

1. $(-\infty, 1], [2, \infty)$ 3. $[-3, -1]$ 5. $(-\infty, 3), (3, \infty)$ 7. $(-2, 2)$

9. $[1, 3]$ 11. $(-3, -1), (1, 3)$ 13. $(2, 3)$ 15. $(-\infty, -3), (5, \infty)$
17. $[-9, 0]$ 19. $(-\frac{1}{2}, 4)$ 21. $(-\infty, -5), (4, \infty)$ 23. $(-2, \frac{3}{2}]$ 25. $(\frac{1}{2}, \frac{5}{9}]$
27. $(0, \frac{3}{20})$ 29. $(-1, 0), (1, \infty)$ 31. $[0, \infty)$ 33. $(2, \infty)$
35. $(-\infty, -5], [-3, 0]$ 37. $[-3, -1], [1, 3]$ 39. $(0, 1)$
41. $(-\frac{4}{3}, -\frac{2}{7}), (\frac{1}{6}, \infty)$ 43. $(-\infty, 0), (1, \infty)$ 45. $(-\frac{1}{2} - \frac{1}{2}\sqrt{5}, -\frac{1}{2} + \frac{1}{2}\sqrt{5})$
47. $(-1 - \frac{1}{2}\sqrt{2}, -1 + \frac{1}{2}\sqrt{2})$ 49. $(-\infty, -\sqrt{2} - 4), (-\sqrt{2} + 4, \infty)$
51. $(-\infty, -\frac{1}{2}\sqrt{2}), (0, \frac{1}{2}\sqrt{2}), (\frac{3}{2}, \infty)$ 53. $(-\infty, -2), [-1, \infty)$ 59. $(-1, 0), (1, \infty)$
61. Between 1 and 2 seconds   63. Between 20 and 50   67. Between 0 and 2 inches

SECTION 2.8

1. $(-4, 4)$ 3. $[-\frac{1}{5}, \frac{1}{5}]$ 5. $(-\infty, -\frac{9}{2}], [\frac{9}{2}, \infty)$ 7. $(-\infty, -0.01), (0.01, \infty)$
9. $(2, 8)$ 11. $[-6, 0]$ 13. $(-\infty, 6), (8, \infty)$ 15. $(-\infty, \frac{5}{3}], [\frac{7}{3}, \infty)$
17. $(\frac{7}{4}, \frac{13}{4})$ 19. $(-1, 2)$ 21. $(-\infty, \frac{2}{3}), (\frac{2}{3}, \infty)$ 23. $(-\infty, \infty)$ 25. $[-\frac{1}{8}, \frac{5}{8}]$
27. $(-\infty, -\frac{10}{3}), (6, \infty)$ 29. $(-\infty, 1], [\frac{5}{2}, \infty)$ 31. $(-18, 6)$ 33. $(-6, 6)$
35. $(3, 4], [6, 7)$ 37. $[-\frac{7}{6}, -\frac{5}{6}], [-\frac{1}{2}, -\frac{1}{6}]$
39. a. $(-2\sqrt{5}, 2\sqrt{5})$ b. $(-\infty, -2\sqrt{5}), (2\sqrt{5}, \infty)$

CHAPTER 2 REVIEW

1. 26 3. 3 5. $-\frac{2}{19}$ 7. $\frac{1}{2}, -\frac{5}{3}$ 9. $1, \frac{7}{3}$ 11. $\sqrt{3}, -\sqrt{3}$
13. None   15. $\frac{1}{4} + \frac{1}{4}\sqrt{17}, \frac{1}{4} - \frac{1}{4}\sqrt{17}$ 17. $3, -3$ 19. $-3 + \sqrt{6}, -3 - \sqrt{6}$
21. $\sqrt{2}, -\sqrt{2}$ 23. $\frac{1}{2}, 1$ 25. $-4 - 2\sqrt{5}$ 27. $0, -3 + 3\sqrt{2}, -3 - 3\sqrt{2}$
29. $0, \frac{1}{2}, 1, -1$ 31. $2, -2, 2\sqrt{2}, -2\sqrt{2}$ 33. $2\sqrt{3}, -2\sqrt{3}, 2\sqrt{5}, -2\sqrt{5}$
35. $(-\infty, \frac{5}{2}]$ 37. $(-\infty, \frac{9}{2})$ 39. $[-\frac{1}{2}, \infty)$ 41. $(-\infty, 4), (5, \infty)$
43. $(-2, -\frac{1}{3})$ 45. $[-\frac{1}{2}, \frac{2}{5})$ 47. $(-\infty, -\frac{7}{6}], (3, \infty)$ 49. $(-\infty, -\sqrt{2}], [\sqrt{2}, \infty)$
51. $[-4, 9]$ 53. -1 55. $(3, 8)$ 57. $58°C$ and approximately $-88.3°C$
59. 13 miles   61. $\frac{9}{2}$ hours   63. Between 1 and 2 seconds   65. 4
67. \$10,000   69. 54   71. $r = \sqrt{\dfrac{3V}{\pi h}}$ 73. $r = \dfrac{A_n - P}{Pn}$

75. $x = \dfrac{Qy}{17{,}860y - 1.798Q}$

SECTION 3.1

1.   3.   5.

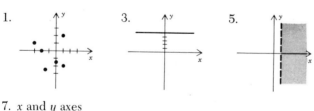

7. x and y axes

9.   11.   13.

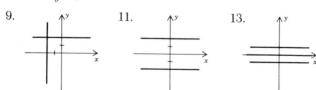

15.   17.   19.

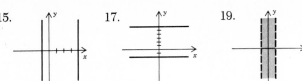

21. Second and fourth quadrants, not including axes   23. 2   25. 13
27. $2\sqrt{10}$ 29. 13   31. $\sqrt{29 - 2\sqrt{2}}$ 33. $5a$ 35. $(6, 6)$
37. $(0, 7)$ 39. $(0.7, 2.85)$ 41. $(0, a)$ 43. $(4 + \sqrt{3}, 0)$ and $(4 - \sqrt{3}, 0)$
45. $(\frac{1}{3}a + \frac{2}{3}c, \frac{1}{3}b + \frac{2}{3}d)$

SECTION 3.2

1. $-14; -14; -14$ 3. $4; 8; t^4 - 3t^2 + 4$ 5. $2; \dfrac{4a^2 - 2a}{a^2 - 0.09}$ 7. $2; \sqrt{7}; |x|$

9. $-504; 2z^2 - z^9$ 11. $5; 0$ 13. $11; 3$ 15. $\frac{4}{5}; \frac{5}{6}$ 17. $-2; -6$

19. $\dfrac{2}{3}; \dfrac{3\sqrt{2}}{3\sqrt{2} - 1}$ 21. All real numbers   23. All real number except $0, \frac{1}{2}$

25. All real numbers   27. All real numbers except $1, -6$
29. All real numbers   31. All negative numbers
33. All numbers in $(-\infty, -2]$ and $[2, \infty)$ 35. All numbers in $(-\infty, 0]$ and $(1, \infty)$
37. All real numbers   39. All numbers in $[1, \infty)$
41. All real numbers except 0 43. c, d, e   45. a, c, d   47. Yes
53. a. 9.1206685 b. 236.13394 c. 161.43322 d. -18.327447
55. a. $C = 2\pi r$ for $r \geq 0$ b. $C = \pi d$ for $d \geq 0$

57. a. $C = 2.54I$ for $I \geq 0$ b. $I = \dfrac{1}{2.54} C$ for $C \geq 0$

SECTION 3.3

1. $p = 4s$ 3. $S = 6s^2$ 5. a. $V = s^3$ b. 37 cubic inches

7. a. $l_f = \dfrac{1250}{381} l_m$ b. $l_y = cl_m$, where $c = \dfrac{1250}{1143}$ 9. $p = ch$

11. Approximately 2.498×10^{19} ergs   13. $P = ci^2$ 15. $R = cv^2$ 17. $\lambda = \dfrac{c}{\nu}$

19. $F = \dfrac{cq_1q_2}{r^2}$

SECTION 3.4

1. y intercept: 3;   3. y intercept: -2;   5. y intercept: 5;
   x intercept: -3 x intercept: 1 x intercepts: 5, -5

7. y intercept: 2;
    x intercept: 2

9. y intercept: 3;
    x intercepts: $\sqrt{3}, -\sqrt{3}$

11. no y intercept;
     x intercept: -1

13. y intercept: -1;
     x intercept: 1

15. y intercept: 0;
     x intercept: 0

17. y intercept: 0;
     x intercept: 0

19. y intercept: 1; x intercept: -1 21. y intercept: 0; x intercepts: 0,4
23. y intercept: 9; x intercept: 3   25. y intercept: -1; no x intercept
27. y intercept: $\frac{1}{2}$; x intercept: -1 29. y intercept: $\sqrt{2}$; x intercept: 2
31. y axis   33. Neither   35. y axis   37. y axis   39. Neither
41. Odd   43. Neither   45. Neither   47. Even   49. a, c, g, h
51. The graph of f is one unit to the right of the graph of g.
53.

SECTION 3.5

1. 3   3. $-\frac{5}{2}$ 5. $\frac{2}{3}$ 7. $y - (-4) = 0(x - 3)$ (or $y + 4 = 0$)
9. $y = 0 = \frac{2}{3}(x - 3)$ 11. $y - \frac{3}{2} = -1(x - 0)$ 13. $y - 1 = 2x$
15. $y + 2 = -4(x + 1)$ 17. $x = 2$ 19. $y = 0x - 2$(or $y = -2$)
21. $y = -\frac{4}{5}x - 1$ 23. $y = -\sqrt{3}x + \frac{1}{2}$ 25. $m = 0; b = 3$
27. $m = 3; b = -\frac{1}{2}$ 29. $m = -\frac{2}{3}; b = 2$ 31. $m = 3; b = 0$
33. Perpendicular   35. Perpendicular   37. Neither   39. Parallel
41. Neither   43. Perpendicular
45.   47.   49.   51.

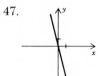

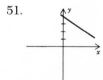

53. b. $y - y_1 = \dfrac{y_2 - y_1}{x_2 - x_1}(x - x_1)$ 55. a. $\frac{1}{3}$ b. -5 c. $-\frac{3}{4}$ 57. a. $\dfrac{x}{-1} + \dfrac{y}{2} = 1$

b. $\dfrac{x}{3} + \dfrac{y}{\frac{1}{2}} = 1$ c. $\dfrac{x}{0.7} + \dfrac{y}{0.3} = 1$ d. $\dfrac{x}{-2} + \dfrac{y}{-3} = 1$ 59. a, c, f   61. $-\frac{4}{3}$

63. $f(x) = -2x - 6$ 65. $f(x) = 4x + 4$ 67. $y - 5 = -\frac{4}{3}(x + 2)$ 69. $(-1, 2)$

SECTION 3.6

1. $f(x) = x - 1, g(x) = 2x$ 3. $f(x) = x^4, g(x) = \dfrac{3}{x}$ 5. $f(t) = \sqrt{t}, g(t) = \dfrac{1}{2t}$

7. Domain: all real numbers; rule: $(g \circ f)(x) = 2x^2 - 3x + 2$

9. Domain: $(0, \infty)$; rule: $(g \circ f)(x) = \sqrt{\dfrac{1}{2x}}$

11. Domain: $[1, \infty)$; rule: $(g \circ f)(x) = \sqrt{x - 1} + 1$

13. Domain: all real numbers except $0, \frac{1}{2}, -\frac{1}{2}$; rule: $(g \circ f)(x) = \dfrac{4x^2}{1 - 4x^2}$

15. Domain: all real numbers except 0; rule: $(g \circ f)(x) = x$

17. Domain: all real numbers except -3 and -1; rule: $(g \circ f)(x) = -\dfrac{4x + 2}{x + 3}$

19. Domain: $(-\infty, -\sqrt{3}]$ and $[\sqrt{3}, \infty)$; rule: $(g \circ f)(x) = \sqrt{x^2 - 3}$

25. $(9, 10]$ 27. m 29. $V(r(t)) = \frac{9}{2}\pi t^6$

SECTION 3.7

1. $f^{-1}(x) = \frac{1}{2}x$ 3. $f^{-1}(x) = 3 - 2x$ 5. $f^{-1}(x) = x^{1/3}$ 7. $f^{-1}(x) = [\frac{1}{3}(x + 5)]^{1/3}$

9. $f^{-1}(x) = (1 - x)^{1/5}$ 11. $f^{-1}(x) = x^2$ for $x \geq 0$ 13. $f^{-1}(x) = x^2 + 3$ for $x \geq 0$

15. $f^{-1}(x) = \dfrac{1}{x^2}$ for $x > 0$ 17. $f^{-1}(x) = 2 - \dfrac{1}{2x^3}$ 19. $f^{-1}(x) = \dfrac{x + 1}{1 - x}$

21. $f^{-1}(x) = \left(\dfrac{3x + 1}{2 - x}\right)^{1/3}$ 23.   25.

27.   29. $f = f^{-1}$ 31.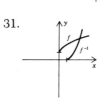

SECTION 3.8

1. $x^2 + y^2 = 9$ 3. $(x + 1)^2 + (y - 4)^2 = 4$ 5. $x^2 + y^2 = 17$

7. $(x + 2)^2 + (y - 3)^2 = 25$

9. y intercept: 3; symmetry with respect to y axis

11. y intercept: 0; x intercept: 0; symmetry with respect to x axis

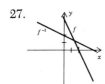

13. x intercept: 1;
    symmetry with respect to x axis

15. y intercept: 0; x intercept: 0;
    symmetry with respect to origin

17. Circle: y intercepts and x intercepts: 2 and -2; all symmetries
19. y intercept: 0; x intercept: 0;
    no symmetry

21. y intercepts: $\frac{1}{2}$ and $-\frac{1}{2}$; x intercepts: 1 and -1 23. x intercept: 3
25. x intercepts: 1 and -1 27. y intercepts: $\sqrt{5}$ and $-\sqrt{5}$; x intercept: 5
29. y intercepts: 3 and -2 31. y intercepts: 1 and -3; x intercepts: 1 and 3

SECTION 3.9

1.

3.

5.

7.

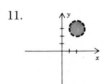

9.

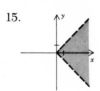

11.

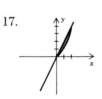

13.

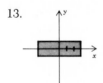

15.

17.

CHAPTER 3 REVIEW

1. a. $3\sqrt{5}$ b. $\frac{1}{2}\sqrt{82}$ 3. All real numbers except -3
5. All real numbers except $-2 + \sqrt{11}$ and $-2 - \sqrt{11}$ 7. All numbers in $[\frac{2}{3}, \infty)$
9. All numbers in $(-\infty, -5]$ and $[5, \infty)$ 13. $-\frac{1}{3}$

15. a. All numbers in $(-\infty, -2]$ and $[2, \infty)$

17. y intercept: -5;
    x intercept: $\frac{5}{2}$; no symmetry

19. y intercept: -6; x intercepts: $\sqrt{6}$ and $-\sqrt{6}$;
    symmetry with respect to y axis

21. y intercept: 2; x intercepts: 2 and -2;
    symmetry with respect to y axis

23. y intercept: 0; x intercept: 0;
    no symmetry

25. No intercepts;
    symmetry with respect to origin

27. y intercept: $\sqrt{3}$; x intercept: $-\sqrt{2}$;
    no symmetry

29. y intercepts: 3 and -5;
    x intercepts: 3 and 5; no symmetry

31. y intercepts: $-3 + \sqrt{3}$ and $-3 - \sqrt{3}$;
    no symmetry

33. y intercepts: $\sqrt{3}$ and $-\sqrt{3}$; x intercept: -3;
    symmetry with respect to x axis

35.

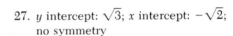
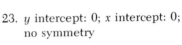

37.

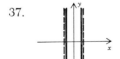

39.

41. $(f \circ g)(x) = 1/x^2$ for all x except 0
$\quad (g \circ f)(x) = 125/x^2$ for all x except 0

43. $(f \circ g)(x) = (2x + 3)^{3/2}$ for all $x \geq -\frac{3}{2}$
$\quad (g \circ f)(x) = \sqrt{2x^3 + 3}$ for all $x \geq -(\frac{3}{2})^{1/3}$

45. $(f \circ g)(x) = \sqrt{x^2 - 4}$ for all x in $(-\infty, -2]$ and $[2, \infty)$
$\quad (g \circ f)(x) = \sqrt{x^2 - 4}$ for all x in $(-\infty, -2]$ and $[2, \infty)$

49. No   51. $f^{-1}(x) = \dfrac{1}{x} - 3$ 53. $f^{-1}(x) = \dfrac{2x + 1}{4 - 3x}$ 55. No inverse

57. $f^{-1}(x) = x^2 + 2$ for $x \geq 0$ 59. $2\sqrt{a + b}$

61. $(7 + 2\sqrt{5}, 2 + \sqrt{5})$ and $(7 - 2\sqrt{5}, 2 - \sqrt{5})$ 63. $(x - \frac{1}{2})^2 + (y + \frac{1}{4})^2 = \frac{67}{16}$

65. $y + 3 = -\frac{1}{2}(x + 1)$ 67. $6y - 8x = 7$ 69. No

75. a. $L = \dfrac{cI}{D^2}$ b. Multiplied by 9

SECTION 4.1

1. Vertex: $(0, 0)$; y intercept: 0;
x intercept: 0

3. Vertex: $(0, 1)$;
y intercept: 1

5. Vertex: $(0, 8)$; y intercept: 8;
x intercepts: 2 and -2

7. Vertex: $(1, 0)$; y intercept: $-\frac{3}{2}$;
x intercept: 1

9. Vertex: $\left(-\dfrac{1}{3}, -\dfrac{1}{3}\right)$; y intercept: $-\dfrac{1}{6}$; x intercepts: $-\dfrac{1}{3} + \dfrac{\sqrt{2}}{3}$ and $-\dfrac{1}{3} - \dfrac{\sqrt{2}}{3}$

11. Vertex: $(-1, -1)$; y intercept: 0;
x intercepts: 0 and -2

13. Vertex: $(1, 1)$;
y intercept: 2

15. Vertex: $(3, 4)$; y intercept: -5;
    x intercepts: 5 and 1

17. Vertex: $(-\frac{1}{3}, \frac{4}{3})$; y intercept: 1;
    x intercepts: $\frac{1}{3}$ and -1

19. Vertex: $(-\frac{1}{4}, -\frac{9}{8})$; y intercept: -1; x intercepts: $\frac{1}{2}$ and -1

21. $y = \frac{1}{2}(x - 1)^2$ 23. $y = -(x - 1)^2 + 1$ 25. a. $g(x) = -2x^2 + 4$
    b. $g(x) = -2x^2$ c. $g(x) = -2(x - \frac{1}{2})^2 + 3$ d. $g(x) = -2(x + 4)^2 + 3$ 27. 3
29. 8   33. 12 feet

SECTION 4.2

1. Minimum value: 5   3. Minimum value: -20 5. Maximum value: 1
7. Minimum value: -28 9. Minimum value: $-\frac{9}{4}$
11. 1 mile by $\frac{1}{2}$ mile   13. In half   15. 3 inches each   17. 4 by 4
19. -8 and -8 21. 8 and 24   23. $(2, -2)$ 25. $(\frac{1}{2}\sqrt{2}, \frac{1}{2})$ and $(-\frac{1}{2}\sqrt{2}, \frac{1}{2})$
27. \$2.50   29. 100 feet; 1.5 seconds

SECTION 4.3

1. y intercept: 1;
   x intercept: -1

3. y intercept: $\frac{1}{4}$;
   x intercept: $-\frac{1}{2}$

5. y intercept: -2;
   x intercept: -1

7. y intercept: 1;
   x intercept: -1

9. y intercept: 2;
   x intercepts: -1, 1, and 2

11. y intercept: 0;
    x intercepts: -2, -1, and 0

13. y intercept: 0;
    x intercepts: 0 and 1

15. y intercept: -1;
    x intercepts: 1 and -1

17. y intercept: 0;
    x intercepts: -1, 0, and 2

19. y intercept: 0;
    x intercepts: -1, 0, and 1

21. y intercept: 0;
    x intercepts: 0 and 1

23. y intercept: 4;
    x intercepts: -2, -1, 1, and 2

25. 2   27. -9

SECTION 4.4

1. No intercepts;
   vertical asymptote: $x = 0$;
   horizontal asymptote: $y = 0$

3. y intercept: 2; no x intercepts;
   vertical asymptote: $x = 2$;
   horizontal asymptote: $y = 0$

5. y intercept: -1; no x intercepts;
   vertical asymptote: $x = -1$;
   horizontal asymptote: $y = 0$

7. y intercept: -4; no x intercepts;
   vertical asymptote: $x = -\frac{1}{2}$;
   horizontal asymptote: $y = 0$

9. y intercept: 0; x intercept: 0;
   vertical asymptote: $x = 2$;
   horizontal asymptote: $y = 1$

11. y intercept: 0; x intercept: 0;
    vertical asymptotes: $x = 2$ and $x = -2$;
    horizontal asymptote: $y = 0$

13. y intercept: -1; no x intercepts;
    vertical asymptotes: $x = 1$ and $x = -2$;
    horizontal asymptote: $y = 0$

15. y intercept: 0; x intercept: 0;
    vertical asymptotes: $x = -3$ and $x = -1$;
    horizontal asymptote: $y = 0$

17. y intercept: 0; x intercept: 0;
    vertical asymptotes: $x = 2$ and $x = -\frac{1}{2}$;
    horizontal asymptote: $y = \frac{1}{2}$

19. No y intercept; x intercept: $-\frac{1}{2}$;
    vertical asymptote: $x = 0$;
    horizontal asymptote: $y = 4$

21. y intercept: $-\frac{1}{4}$; t intercept: $-\frac{1}{3}$;
    no vertical asymptotes:
    horizontal asymptote: $y = 0$

23. y intercept: $\frac{3}{4}$; x intercept: -1;
    vertical asymptote: $x = -4$;
    horizontal asymptote: $y = 0$

SECTION 4.5

1. Vertex: $(0, 0)$;
   axis: $y = 0$

3. Vertex: $(-5, 1)$;
   axis: $y = 1$

5. Vertex: $\left(-\frac{1}{4}, -\frac{1}{2}\right)$;
   axis: $y = -\frac{1}{2}$

7. Vertex: $\left(\frac{1}{2}, \frac{1}{2}\right)$;
   axis: $y = \frac{1}{2}$

9. Vertex: $(3, 2)$;
   axis: $x = 3$

11. Vertex: $(-1, 3)$;
    axis: $y = 3$

13. Vertex: $(-2, -3)$;
    axis: $x = -2$

15. $y = \frac{5}{4}x^2$ 17. $y - 5 = \frac{1}{2}(x + 2)^2$ 19. $y - 4 = (x - 3)^2$

21. The lines $x - 3y + 4 = 0$ and $x + 3y - 2 = 0$ 23. The point $(-\frac{1}{2}, \frac{1}{3})$

25. -15 27. $(3, -\frac{1}{2})$ and $(3, 1)$ 29. $y = \dfrac{316}{(1750)^2}\, x^2$

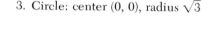

SECTION 4.6

1. Center: $(0, 0)$;
   major axis between $(-4, 0)$ and $(4, 0)$;
   minor axis between $(0, -2)$ and $(0, 2)$;
   vertices: $(-4, 0)$, $(4, 0)$

3. Circle: center $(0, 0)$, radius $\sqrt{3}$

5. Center: $(0, 0)$;
   major axis between $(0, -3)$ and $(0, 3)$;
   minor axis between $(-2, 0)$ and $(2, 0)$;
   vertices: $(0, -3)$, $(0, 3)$

7. Center: $(4, -3)$;
   major axis between $(4, -7)$ and $(4, 1)$;
   minor axis between $(1, -3)$ and $(7, -3)$;
   vertices: $(4, -7)$, $(4, 1)$

9. Center: $(-2, -1)$;
   major axis between $(-6, -1)$ and $(2, -1)$;
   minor axis between $(-2, -3)$ and $(-2, 1)$;
   vertices: $(-6, -1)$, $(2, -1)$

11. Center: $(1, -2)$; major axis between
    $(1, -6)$ and $(1, 2)$; minor axis between
    $(-1, -2)$ and $(3, -2)$;
    vertices: $(1, -6)$, $(1, 2)$

13. Center: $(-3, 3)$;
    major axis between $(-3, -2)$ and $(-3, 8)$;
    minor axis between $(-7, 3)$ and $(1, 3)$;
    vertices: $(-3, -2)$, $(-3, 8)$

15. $\dfrac{x^2}{4} + \dfrac{y^2}{9} = 1$ 17. $\dfrac{x^2}{4} + \dfrac{y^2}{16} = 1$ 19. $\dfrac{(x-3)^2}{49} + \dfrac{(y-2)^2}{25} = 1$

21. $\dfrac{(x+2)^2}{1} + \dfrac{(y+5)^2}{4} = 1$ 23. $4(x + \frac{3}{2})^2 + \dfrac{4(y - \frac{1}{2})^2}{25} = 1$

25. $\dfrac{4x^2}{9} + \dfrac{(y-4)^2}{1} = 1$ and $\dfrac{x^2}{1} + \dfrac{4(y-4)^2}{9} = 1$ 29. $4 + \frac{3}{2}\sqrt{3}$ and $4 - \frac{3}{2}\sqrt{3}$

31. b. $(\frac{24}{5}, -\frac{12}{5})$ 33. a. Approximately 1.473×10^8 (kilometers)

b. $\dfrac{x^2}{(1.495)^2 \times 10^{16}} + \dfrac{y^2}{(1.473)^2 \times 10^{16}} = 1$

SECTION 4.7

1. Center: $(0, 0)$;
   vertices: $(-1, 0)$, $(1, 0)$;
   asymptotes: $y = x$, $y = -x$

3. Center: $(0, 0)$;
   vertices: $(-3, 0)$, $(3, 0)$;
   asymptotes: $y = \frac{2}{3}x$, $y = -\frac{2}{3}x$

5. Center: $(0, 0)$;
   vertices: $(0, -2)$, $(0, 2)$;
   asymptotes: $y = \frac{2}{5}x$, $y = -\frac{2}{5}x$

7. Center: $(-2, -1)$;
   vertices: $(-6, -1)$, $(2, -1)$;
   asymptotes: $y + 1 = \frac{5}{4}(x + 2)$,
   $y + 1 = -\frac{5}{4}(x + 2)$

9. Center: $(0, -1)$;
   vertices: $(0, -2)$, $(0, 0)$;
   asymptotes $y + 1 = \frac{1}{2}x$, $y + 1 = -\frac{1}{2}x$

11. Center: $(3, 0)$;
    vertices: $(3, -3)$, $(3, 3)$;
    asymptotes: $y = \frac{3}{4}(x - 3)$, $y = -\frac{3}{4}(x - 3)$

13. Center: $(1, -3)$;
   vertices: $(-1, -3), (3, -3)$;
   asymptotes: $y + 3 = \frac{3}{2}(x - 1), y + 3 = -\frac{3}{2}(x - 1)$

15. $\dfrac{x^2}{4} - \dfrac{y^2}{36} = 1$ 17. $\dfrac{y^2}{16} - \dfrac{11x^2}{256} = 1$ 19. $\dfrac{(x - 2)^2}{4} - \dfrac{(y + 3)^2}{4} = 1$
21. $\frac{1}{2}\sqrt{3}$ and $-\frac{1}{2}\sqrt{3}$ 23. b. $\left(\frac{13}{6}, -\frac{5}{4}\right)$ 25. Parabola; vertex: $(1, \frac{1}{3})$
27. Hyperbola; center: $(3, 4)$ 29. Parabola; vertex: $(-3, 2)$ 31. Hyperbola; center:
$(0, 1)$ 33. Parabola; vertex: $(-3, -1)$ 35. Ellipse; center: $(-\frac{1}{3}, \frac{4}{3})$

CHAPTER 4 REVIEW

1. Vertex: $(0, -4)$; axis: $x = 0$;
   y intercept: -4; x intercepts: 4 and -4

3. Vertex: $(-2, -3)$; axis: $x = -2$;
   y intercept: -7; no x intercepts

5. y intercept: 2; x intercept: $2^{1/3}$

7. y intercept: 2; no x intercepts

9. y intercept: 0; x intercepts: 0 and 2

11. y intercept: 9; x intercepts: $-3, -1, 1,$ and 3

13. y intercept: -1; x intercept: -1;
    vertical asymptote: $x = 1$;
    horizontal asymptote: $y = 1$

15. No y intercept; x intercept: -2;
    vertical asymptotes: $x = 0$ and $x = -4$;
    horizontal asymptote: $y = 0$

17. Circle: center $(-1, 2)$, radius $\sqrt{3}$

19. Center: $(\frac{1}{2}, -\frac{1}{2})$;
    vertices: $(-\frac{3}{2}, -\frac{1}{2})$ and $(\frac{5}{2}, -\frac{1}{2})$;
    asymptotes: $y + \frac{1}{2} = x - \frac{1}{2}$ and
    $y + \frac{1}{2} = -(x - \frac{1}{2})$

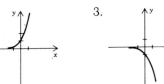

21. Center: $(-3, -4)$; vertices:
    $(-3, -4 - 2\sqrt{3})$, $(-3, -4 + 2\sqrt{3})$;
    asymptotes: $y + 4 = \dfrac{2\sqrt{3}}{3}(x + 3)$
    and $y + 4 = -\dfrac{2\sqrt{3}}{3}(x + 3)$

23. $-\frac{1}{3}$ 25. $-9, -1$ 27. 350, and 350 people   29. 4096 feet   31. For
the square: $\dfrac{8}{\pi + 4}$ feet; for the circle: $\dfrac{2\pi}{\pi + 4}$ feet   33. Length: $1 - \frac{1}{2}\sqrt{2}$; width: $\frac{1}{4}\sqrt{2} + \frac{1}{2}$

SECTION 5.1

1.

3.

5. The line $y = 1$

7.

9.

11.

13.

15. π^2 17. $5^{\sqrt{6}}$ 19. $\frac{1}{4}$ 21. 3^{π} 23. 2 25. 7 27. $\frac{1}{3}$

29. $\frac{1}{4}$ 31. $\frac{1}{64}$ 33. 1.4049476 35. 0.30119421 37. 1.6487213

39. 11 41. 22 43. 0 and $\frac{12}{5}$ 45. $-2\sqrt{2} - 1$ 47. $\sqrt{5}$ and $-\sqrt{5}$

49.

53. $e^{(e^x)}$ 55. b. $e^{1.5} \approx 4.48168907$, sum ≈ 4.481686598

59. a. Approximately 6.76×10^{-6} atmospheres   b. Approximately 3.07×10^{-3} atmospheres
    c. Approximately 5.99×10^{-1} atmospheres   d. Approximately 1.12 atmospheres

SECTION 5.2

1. $\log_{10} 10 = 1$ 3. $\log_{1/2} 32 = -5$ 5. $\log_x 16 = -4$ 7. $10^3 = 1000$

9. $4^{2.5} = 32$ 11. $e^2 = x$ 13. 3   15. 4   17. -4 19. 0   21. $\frac{3}{2}$
23. -1 25. $\frac{1}{3}$ 27. -2 29. -1 31. -4 33. $\frac{5}{2}$ 35. $\frac{4}{3}$
37. 6   39. 4   41. 5   43. 1   45. 16   47. 2   49. $\frac{1}{16}$ 51. 1
53. 12   55. 67   57. 4   59. $\frac{2}{3}\sqrt{3}$ 61. 11, -11 63. 8   65. 4
67. 5   69. 2, -2 71. 2   73. 9   75. $\frac{1}{3}$ 77. $\frac{1}{4}$
79.   81.   83.   85.

87. No, since $\log_{10} \frac{1}{6} < 0$

SECTION 5.3

1. 0.7781   3. 1.0791   5. -0.4771 7. 0.5229   9. -0.2498
11. -0.6990 13. 0.73855   15. 2.9957   17. -1.9459 19. 3.8918
21. 1.6095   23. 3.1987   25. 0.8797   27. 1.7590   29. 15   31. $\frac{7}{12}$
33. 1.6021   35. -0.6021 37. 5.7318   39. -6.2682 41. 27.318
43. $\log_2 5$ 45. $\log_2 \dfrac{x^6}{y}$ 47. $\log_{10} \dfrac{x^6}{y}$ 49. $\log_a x + \log_a y - \log_a z$
51. $3 \log_a z + \frac{1}{2}\log_a x + \frac{1}{2}\log_a y$ 53. $\frac{1}{2}\log_a x + \frac{5}{2}\log_a y - \frac{3}{2}\log_a z$ 55. 4
57. $\frac{1}{2}$ 59. 3   61. 2.8073   63. 1.6309   65. 0.6309

SECTION 5.4

1. 0.0043   3. 2.7160   5. $-2 + 0.9917$ 7. 4.7185   9. 0.0746
11. $-1 + 0.1593$ 13. 0.5160   15. 7.1858   17. $-3 + 0.1575$
19. 4.1016   21. 7.51   23. 99,600   25. 631,000,000   27. 0.00515
29. 9.445   31. 155.4   33. 0.02083   35. 0.4043   37. 70.40
39. 3.870   41. 1.673×10^9 43. 0.002243   45. a. 89.78 kilograms
b. 10.22 kilograms   c. 0.7471 meters   d. 0.6194 meters

SECTION 5.5

1. $\dfrac{\log 5}{\log 2}$ 3. $\dfrac{1}{2}\left(\dfrac{\log 51}{\log 5}\right)$ 5. $\dfrac{\log 5}{\log 4} - 2$ 7. $\dfrac{3 \log 5 + 2 \log 3}{5 \log 3 - 2 \log 5}$
9. $0, \dfrac{\log 3}{\log 2}$ 11. $-2, \dfrac{\log 4}{\log 5} + 2$ 15. 10^{-7} watts per square meter
17. 20 decibels   19. $10^{0.06} \approx 1.1481536$ 21. $10^{8.9} \approx 794{,}328{,}230$ 23. a. 0
b. 3   c. 4   25. $\log 2 \approx 0.3$ 27. $10^{-1.4} \approx 0.039810717$ 29. Highly acidic
31. Wheat

SECTION 5.6

1. $\frac{1}{3}\ln 4$ 3. $\dfrac{1}{1.2}\ln 2$ 5. 3   7. 0, $\ln 2$ 9. 0

13. a. $f(t) = 1000 \, e^{(\ln 1.537)t}$ b. $\dfrac{\ln 2}{\ln 1.537} \approx 1.61$ weeks   15. No   17. At 32.3°C

19. 4.687 billion   23. Approximately 20.35%   25. 28,987 years old

27. Approximately 79%   29. Approximately 1.2%

31. Approximately 65.25 million years

33. a. 29.92 inches of mercury b. Approximately 24.50 inches of mercury

c. Approximately 17.28 inches of mercury   35. No

SECTION 5.7

1. a. $1276.28 b. $1283.36 c. $1284.00

3. a. $23,673.64 b. $24,513.57 c. $24,593.30

5. $759.41

7. The $1500 deposit

9. Approximately 11.3%

11. Approximately 67.71%

13. a. Approximately 11.55 years b. Approximately 7.70 years

15. a. Approximately $1,019,433,514 b. Approximately $1,575,839,259

17. b. Approximately 7.76% c. $r = k[(E + 1)^{1/k} - 1]$

19. a. $3286.39 b. $65,727.80

21. $32,686.13

23. Approximately 15.83%

CHAPTER 5 REVIEW

1.   3.

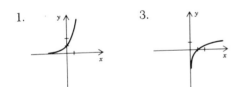

5. 2   7. $6 + \ln 6$ 9. $-\frac{1}{2} + \frac{1}{2}\sqrt{5}, -\frac{1}{2} - \frac{1}{2}\sqrt{5}$ 11. $\frac{1}{8}$ 13. $\frac{7}{5}$

15. $-\dfrac{\log 7}{\log 2}$ 17. $0, \left(\dfrac{\log 2}{\log 3}\right)^{1/3}$ 19. $\frac{1}{5}\ln 4.37$ 21. -4 23. 5

25. 2.6233   27. $-3 + 0.8865$ 29. 0.6160   31. 8.8051   33. 3.56

35. 4.733   37. 2.079   39. 0.4343   41. $x = e^{(y-a)/b}$ 45. $a = \frac{1}{5}, c = -2$

47. 8.1   49. a. Approximately 23.1 years b. Approximately 5.78 years earlier

51. Approximately 26.58 minutes   53. Approximately 110 A.D.   55. $4451.08

SECTION 6.1

5. $(1, -2)$ 7. $(3, 1)$ 9. $\left(a, \dfrac{7a}{5} + \dfrac{6}{5}\right)$ for any number a 11. $(-1, 4)$

13. $(1.7, 2.1)$ 15. No solution   17. No solution   19. $(1, 3), (-1, -3)$
21. $(4, 2)$ 23. $(3\sqrt{5}/5, -6\sqrt{5}/5), (-3\sqrt{5}/5, 6\sqrt{5}/5)$ 25. $(\frac{1}{2}, \frac{1}{4}), (-\frac{1}{3}, \frac{1}{9})$
27. $(\log_2 3, 3)$ 29. $(a, -b)$

31. a. $\left(\dfrac{s + \sqrt{s^2 - 4}}{2}, \dfrac{2}{s + \sqrt{s^2 - 4}} \right), \left(\dfrac{s - \sqrt{s^2 - 4}}{2}, \dfrac{2}{s - \sqrt{s^2 - 4}} \right)$ 33. a. $a > \frac{3}{4}$
  b. $a = \frac{3}{4}$ c. $a < \frac{3}{4}$ 35. $a = -\frac{3}{7}, b = \frac{5}{7}$ 37. 12 and 18, or -12 and -18
39. $2\sqrt{10}, 2\sqrt{10}$ 41. 17 feet and 24 feet
43. $\frac{9}{8}$ quarts tap water, $\frac{15}{8}$ quarts stove water
45. Girl is 12 years old; father is 40 years old.   47. 2 centimeters and 4 centimeters

SECTION 6.2

1. $(3, -1)$ 3. $(2, 4)$ 5. $(-2, -3)$ 7. $\left(a, \dfrac{a}{10} - 1\right)$ for any number a

9. No solution   11. $\left(a, \dfrac{3a}{2} - \dfrac{7}{4}\right)$ for any number a

13. $(4a, 2b)$ 17. $a = 3, b = -4$ 19. \$20 and \$30
21. 2 ounces of 24-carat, 3 ounces of 14-carat   23. 60

SECTION 6.3

5. $(4, 0, -2)$ 7. $(1, -1, 3)$ 9. $(a, 0, a)$ for any number a 11. No solution
13. $(0, 0, 0)$ 15. $(4, 1, -1)$ 17. $(3, -1, -2)$ 19. $(-\frac{1}{2}, \frac{1}{2}, -1)$
21. $(\frac{1}{2}, -\frac{2}{3}, \frac{1}{6})$ 23. $(1.5, -1, 0.5)$ 25. $a = 3, b = -2, c = 5$
27. 14, 16, 28   29. 8   31. 320

SECTION 6.4

1. $\begin{bmatrix} 2 & -4 \\ -5 & 1 \end{bmatrix}, \left[\begin{array}{cc|c} 2 & -4 & 7 \\ -5 & 1 & 6 \end{array}\right]$

3. $\begin{bmatrix} 1 & 1 \\ 0 & 1 \end{bmatrix}, \left[\begin{array}{cc|c} 1 & 1 & 2 \\ 0 & 1 & 1 \end{array}\right]$

5. $\begin{bmatrix} \frac{1}{3} & -\frac{1}{4} & \frac{1}{2} \\ \frac{2}{5} & 0 & 1 \\ 1 & -\frac{1}{2} & 0 \end{bmatrix}, \left[\begin{array}{ccc|c} \frac{1}{3} & -\frac{1}{4} & \frac{1}{2} & 1 \\ \frac{2}{5} & 0 & 1 & 3 \\ 1 & -\frac{1}{2} & 0 & 4 \end{array}\right]$

13. $(1, -1)$ 15. $(\frac{7}{2}, \frac{1}{2})$ 17. $(0.6, 0.4)$ 19. $(-2, -1, 3)$ 21. $(2, 5, 1)$
23. $(-c + 2, 5c - 2, c)$ for any number c

SECTION 6.5

1. $M_{21} = 10, M_{22} = -32, M_{23} = 20, A_{21} = -10, A_{22} = -32, A_{23} = -20$

3. 8   5. 0   7. -55 9. -28 11. -14 13. 4   15. 97

17. 1   23. 3, -2 29. a. 5 b. 40

SECTION 6.6

1. $(1, -2)$ 3. $(3, 1)$ 5. $(3, -1)$ 7. $(2, 4)$ 9. $(-2, -3)$

11. $(4, 1, -2)$ 13. $(3, -2, 8)$ 15. $(1, -2, 3, 0)$ 17. $(0, -1, 5, 3)$

SECTION 6.7

1. $\begin{bmatrix} 5 & -2 & 3 \\ -3 & 5 & 7 \end{bmatrix}$ 3. $\begin{bmatrix} -6 & 12 \\ 15 & -36 \end{bmatrix}$ 5. $\begin{bmatrix} -2 & 5 \\ -8 & 5 \end{bmatrix}$

7. $\begin{bmatrix} 3 & \frac{3}{2} \\ -4 & 1 \end{bmatrix}$ 9. $\begin{bmatrix} -5 \\ -3 \end{bmatrix}$ 11. $\begin{bmatrix} 6 & -8 \\ -3 & 4 \end{bmatrix}$

13. $\begin{bmatrix} 4 & -2 & 5 \\ 18 & -2 & -14 \\ -9 & 18 & 9 \end{bmatrix}$ 15. $\begin{bmatrix} 1 & 0 & 0 \\ 0 & 1 & 0 \\ 0 & 0 & 1 \end{bmatrix}$ 17. $[-7]$

19. $AB = \begin{bmatrix} -8 & -6 \\ -7 & -3 \end{bmatrix}$, $BA = \begin{bmatrix} -11 & 9 \\ 2 & 0 \end{bmatrix}$

21. $AB = \begin{bmatrix} -5 & 5 & 7 \\ 27 & 3 & 1 \\ 2 & 2 & -3 \end{bmatrix}$, $BA = \begin{bmatrix} 10 & -3 & 9 \\ 1 & -6 & 1 \\ 6 & 13 & -9 \end{bmatrix}$

23. $(AB)C = A(BC) = \begin{bmatrix} -8 & 0 \\ 35 & 9 \end{bmatrix}$

25. Inverse: $\begin{bmatrix} 0 & 1 \\ 1 & 0 \end{bmatrix}$ 27. No inverse

29. Inverse: $\dfrac{1}{15} \begin{bmatrix} 2 & -1 \\ 3 & 6 \end{bmatrix}$

31. $\dfrac{1}{ab} \begin{bmatrix} b & 0 \\ -c & a \end{bmatrix}$

33. Inverse: $\begin{bmatrix} 1 & 0 & 0 \\ 0 & 1 & 0 \\ 0 & 0 & 1 \end{bmatrix}$

35. No inverse

37. Inverse: $\begin{bmatrix} 2 & -3 & 1 \\ -6 & 9 & -2 \\ -3 & 5 & -1 \end{bmatrix}$

39. Inverse: $\frac{1}{10} \begin{bmatrix} 11 & 18 & 2 \\ 6 & 8 & 2 \\ -7 & -16 & -4 \end{bmatrix}$

41. $(-1, -2)$ 43. $(-6, 1)$ 45. $(1, -1, 2)$ 47. $(\frac{1}{2}, -\frac{1}{4}, \frac{3}{4})$

SECTION 6.8

1. Yes   3. No   5. No
7.   9.

11. $(0, 0), (3, 0), (3, 2), (0, 5)$ 13. $(0, 0), (4, 0), (18, 7), (0, 7)$

15. $(-\frac{4}{3}, \frac{1}{3}), (4, 3), (2, 7)$ 17. $(0, 0), (10, 0), (8, 4), (4, 8), (0, 10)$

19. $(0, -3), (-3, 0)$ 21. $(\frac{1}{2}\sqrt{2}, \frac{1}{2}), (-\frac{1}{2}\sqrt{2}, \frac{1}{2})$

SECTION 6.9

1. Maximum: 5 at $(-1, 1)$; minimum: -14 at $(3, -2)$
3. Maximum: 4 at $(4, 6)$; minimum: 0 at $(0, 0)$
5. Maximum: 28 at $(-4, 2)$; minimum: -2 at $(2, 2)$
7. Maximum: 19 at $(6, 21)$; minimum: 3 at $(0, 9)$
9. Maximum: 42 at $(5, 8)$; minimum: 0 at $(0, 0)$
11. \$70   13. \$4000
15. 180 bushels of peaches and \$1260 of profit   17. \$1200

CHAPTER 6 REVIEW

1. $(9, -2)$ 3. No solution   5. $(2b, b)$ for any number b
7. $(\frac{1}{2}, -1), (\frac{1}{4}, -2)$ 9. $(1, 2)$ 11. $(1, -2, 4)$
13. $(3c, 2c - 1, c)$ for any number c 15. $(-3, 5, 7)$ 17. $(0, \frac{2}{5}, -\frac{3}{5})$ 19. 22
21. 132   23. -51 25. $(5, 7)$ 27. $(-3, 4, 5)$

29. $\begin{bmatrix} 0 & -13 & 11 \\ 2 & -14 & 28 \\ 14 & 11 & 0 \end{bmatrix}$

31. Inverse: $\dfrac{1}{8}\begin{bmatrix} 4 & 1 \\ 0 & 2 \end{bmatrix}$ 33. Inverse: $\dfrac{1}{2}\begin{bmatrix} 1 & 1 \\ -1 & 1 \end{bmatrix}$

35. Inverse: $\dfrac{1}{3}\begin{bmatrix} 11 & 1 & -5 \\ -4 & 3 & 3 \\ -6 & 0 & 3 \end{bmatrix}$

37. $(-7, -10)$ 39. $(2, 1, 0)$
41. $(0, 0), (\frac{3}{4}, 0), (\frac{1}{2}, \frac{1}{3}), (0, \frac{2}{3})$ 43. $(0, 0), (5, 0), (4, 4), (2, 8), (0, 10)$

45. Maximum: $\frac{8}{3}$ at $(0, \frac{2}{3})$; minimum: 0 at $(0, 0)$
47. Maximum: 16 at $(4, 4)$; minimum: 0 at $(0, 0)$
51. $9, -9$ 53. \$0.15   55. \$4000 at 8%, \$16,000 at 10%
57. 300 doors, 400 windows

SECTION 7.1

1. $6 + 9i$ 3. $\frac{3}{4} + \frac{1}{3}i$ 5. $4 + 3i$ 7. i 9. -2 11. $-12 - \frac{3}{2}i$
13. $-\frac{3}{2}i$ 15. $-23 + 2i$ 17. $3i$ 19. i 21. -1 23. $26 - 18i$
25. i 27. $-4i$ 29. $-25 + 34i$ 31. 0 33. $4i$ 35. $\sqrt{15}i$
37. i 39. $12 + 5i$ 41. $5i, -5i$ 43. $4\sqrt{3}i, -4\sqrt{3}i$
45. $5 + 5i, 5 - 5i$ 47. $2 + 3i, 2 - 3i$ 49. $1 + i, 1 - i$
51. $2, -1 + \sqrt{3}i, -1 - \sqrt{3}i$ 53. $3 - 3i$ 55. $1 + \frac{1}{2}i, -2 - i$
57. $\dfrac{\sqrt{2}}{2} + \dfrac{\sqrt{2}}{2}i, -\dfrac{\sqrt{2}}{2} - \dfrac{\sqrt{2}}{2}i$

SECTION 7.2

1. $7 - 5i$ 3. $6 + 19i$ 5. 23 7. $6 + 2i$ 9. $5 + \frac{1}{2}i$ 11. $\frac{1}{2} - \frac{1}{2}i$
13. $-i$ 15. $\frac{1}{15} - \frac{2}{15}i$ 17. i 19. $\frac{3}{5} + \frac{2}{5}i$ 21. $-\frac{24}{25} + \frac{7}{25}i$ 23. 169

25. $\frac{12}{5} + \frac{4}{5}i$ 29. b. $3 + 4i$

SECTION 7.3

1. 1 3. 1 5. 5 7. $\sqrt{2}$ 9. $5\sqrt{2}$ 11. $\sqrt{26}/\sqrt{20}$ 13. $\sqrt{2}/4$
15. Circle: center 0, radius 1   17. Circle: center 0, radius $\sqrt{3}$
19. Circle: center 1, radius 1   21. Circle: center $-2 + 4i$, radius $\frac{3}{2}$

CHAPTER 7 REVIEW

1. $5 + 3i$ 3. $\frac{65}{4}$ 5. $120 + 119i$ 7. $\dfrac{\sqrt{2}}{512} - \dfrac{\sqrt{6}}{512}\, i$ 9. $-3 - 4i$

11. $-4 - 4i$ 13. $-\frac{2}{13} + \frac{3}{13}i$ 15. 1 17. 10 19. $\sqrt{5}/2$

21. $7i, -7i$ 23. $\dfrac{1}{2} + \dfrac{\sqrt{3}}{2}\, i, \dfrac{1}{2} - \dfrac{\sqrt{3}}{2}\, i$ 25. $1 + \sqrt{5}i, 1 - \sqrt{5}i$

27. $\dfrac{1}{2}(\sqrt{3} - i), \dfrac{1}{2}(-\sqrt{3} - i)$ 29. No solution

31. Circle: center 0, radius $\frac{2}{3}$

33. Circle: center $1 + i$, radius 3

35. All points on the line $y = x$

SECTION 8.1

1. Quotient: $2x^2 + 3x - 2$; remainder: $-7x + 9$
3. Quotient: $x^2 - 1$; remainder: 2
5. Quotient: $3x - 2$; remainder: 1
7. Quotient: $3x - 5$; remainder: 12
9. Quotient: $2x^2 + 4x + 2$; remainder: 6
11. Quotient: $6x^2 - 3x + 3$; remainder: 0
13. Quotient: $5x^4 - 4x^3 + x^2 - 3x + 31$; remainder: -124
15. Quotient: $x^6 + x^5 + x^4 + x^3 + x^2 + x + 1$; remainder: 0
17. Quotient: $x^4 + ix^3 - x^2 - ix + 1 - i$; remainder: $2 + i$
31. $1, \frac{1}{3}$

SECTION 8.2

1. 1 3. 21 5. $3\sqrt{2} - 2$ 7. $a^4 + a^2$ 9. $2 - 6i$ 11. $(x - 2)^3$
13. $(x + 2)^2(x - 1)$ 15. $2(x - \frac{1}{2})(x + 3)(x - 1)$ 17. $x(x - 2)(x + 3i)(x - 3i)$
19. $2x(x - \frac{1}{2}i)(x - 2i)(x - i)$ 29. $x^2 - x - 6$ 31. $x^3 - 12x^2 - 13x$
33. $x^3 - 3x^2 - 6x + 8$ 35. $x^4 - 5x^2 + 4$ 37. $x^2 + 25$
39. $x^2 - 2\sqrt{2}x + 6$ 41. $(x + 2i)(x - 2i)$ 43. $(x + \sqrt{7}i)(x - \sqrt{7}i)$
45. $2(x + i)(x - i)$ 47. $x(x + i)(x - i)$ 49. $-\frac{3}{2}$

SECTION 8.3

1. 1 (multiplicity 2), $-\frac{1}{2}$ (multiplicity 3)
3. 0 (multiplicity 2), $\sqrt{3}i$ (multiplicity 2), $-\sqrt{3}i$ (multiplicity 2)
5. 3 (multiplicity 1), -3 (multiplicity 1), $\sqrt{3}i$ (multiplicity 1), $-\sqrt{3}i$ (multiplicity 1)
7. 0 (multiplicity 2), $-\frac{1}{2}$ (multiplicity 2), $\dfrac{\sqrt{3}}{3}\, i$ (multiplicity 3), $-\dfrac{\sqrt{3}}{3}\, i$ (multiplicity 3)
9. $x(x - 1)(x - 2)$ 11. $(x - 2)(x + 2)(x - \sqrt{6})(x + \sqrt{6})$ 13. $(x + 3i)^2(x - 3i)^2$
15. $-(x - 2 + 3i)(x - 2 - 3i)$ 19. 0 21. $(x - 2)^3(x - i)(x + i)$
23. $(x + i)^4(x - 2i)$ 25. $(x - 3)^4(x - 1 + i)$ 27. $(x - 1 - i)^2(x - 1 + i)^2(x + 6)^3$
29. 2 31. 2

SECTION 8.4

1. $f(x) = 2x^2 - 4x + 10$
3. $f(x) = -\frac{3}{2}x^2 + 3x - 3$
5. $f(x) = \frac{4}{3}(x^4 + 2x^3 + x^2 + 2x)$
7. $f(x) = \dfrac{\sqrt{2}}{15}(x^4 - 1)$
9. $f(x) = 7(x^3 - 6x^2 + 15x - 14)$
11. $f(x) = \frac{1}{4}(x^4 - 12x^3 + 62x^2 - 140x + 125)$
13. $(x + 1)^2(x^2 + 1)^3$
15. $(x - 1 - i)(x - 1 + i)(2x - 1)$
17. $(x - 1)(x + 2 - 3i)(x + 2 + 3i)$
19. $(x - 2 - 3i)(x - 2 + 3i)(\sqrt{2}x + i)(\sqrt{2}x - i)$
21. a. $2(x - 2 - i)(x - 2 + i)(x - 1 + i)(x - 1 - i)$
    b. $2(x^2 - 4x + 5)(x^2 - 2x + 2)$

SECTION 8.5

1. $1, -1, 2, -2$ 3. $1, -1, 5, -5$ 5. $1, -1, 2, -2, 4, -4, 8, -8$
7. $1, -1, 2, -2, 3, -3, 4, -4, 6, -6, 8, -8, 12, -12, 24, -24$
9. $-1, 2, -5$ 11. $1, -4$ 13. $1, -1$ 15. $-1, 2, -2, 6$
17. $(x + 3)(2x - 1)$ 19. $(x - 5)(x - i)(x + i)$ 21. $(x - 2)(x - 3)(x - 4)$
23. $(x + 1)(2x - 3)(2x + 5)$ 25. $2(x - 1)(x + 1)\left(x - \dfrac{1}{2} + \dfrac{\sqrt{11}}{2}i\right)\left(x - \dfrac{1}{2} - \dfrac{\sqrt{11}}{2}i\right)$
27. $(x - 1)(x + 2)^2(x - i)(x + i)$ 29. $x(2x - 1)(3x + 1)(4x + 1)$
49. a. $(-2, -1), (-1, 0), (0, 1), (1, 2)$

CHAPTER 8 REVIEW

1. Quotient: $x^2 + x + 2$; remainder: $x + 3$
3. Quotient: $-x - 2$; remainder: -4
5. Quotient: $x^2 - ix + 1$; remainder: $2i$
7. Quotient: $x^5 + (-1 - i)x^3 + (-2 - 4i)x^2 + (-6 - 6i)x - 14 - 18i$; remainder: $-39 - 54i$
13. -7 15. 0 17. $(x + 2)(x - 1)(x - 3)$ 19. $3(x + \frac{1}{3})(x - 1)^3$
21. $x(x - \sqrt{2})\left(x + \dfrac{\sqrt{2}}{2} + \dfrac{\sqrt{2}}{2}i\right)\left(x + \dfrac{\sqrt{2}}{2} - \dfrac{\sqrt{2}}{2}i\right)$ 23. $x(x - 2 + 2i)(x - 2 - 2i)$
25. $(x - 2)(x + 2)(x - \sqrt{3}i)(x + \sqrt{3}i)$ 27. $(x + 3)(x + 2 + \sqrt{3}i)(x + 2 - \sqrt{3}i)$
29. $(x - i)(x - 1)(x + 1 - i)$
31. $2i$ (multiplicity 2), $7i$ (multiplicity 4), $-7i$ (multiplicity 4), 3 (multiplicity 7)
33. 0 (multiplicity 3), $\dfrac{\sqrt{3}}{3}$ (multiplicity 4), $\frac{1}{3}i$ (multiplicity 2), $-\frac{1}{3}i$ (multiplicity 2)
35. $x(x - 3 + 5i)^3(x - 2i)^4$ 37. 2 39. $2x^3 + 12x - 40$
41. $(x - 2 + i)^2(x - 2 - i)^2$ 43. $(x^2 - 2x + 2)(x^2 + 2x + 5)$ 45. -1
47. $-3, 4$ 49. $(x - 2)(\sqrt{2}x - 1)(\sqrt{2}x + 1)$ 51. $x(x + 1)(x - 2)(x - i)(x + i)$
53. $(2x + 5)(4x - 1)(2x - 1)$

SECTION 9.2

1. $\left\{\dfrac{1}{n+1}\right\}_{n=1}^{\infty}$ 3. $\left\{(-1)^n\right\}_{n=1}^{\infty}$ 5. $\left\{\dfrac{2^{n+2}}{3^{n+1}}\right\}_{n=1}^{\infty}$ 7. $\dfrac{\sqrt{3}}{2}, \dfrac{\sqrt{3}}{2}, \dfrac{\sqrt{3}}{2}, \dfrac{\sqrt{3}}{2}$

9. 2, 4, 8, 16   11. 1, −1, 1, −1   13. $\dfrac{1}{2}, \dfrac{1}{6}, \dfrac{1}{12}, \dfrac{1}{20}$

15. 0, ln 2, ln 3, ln 4   17. 2, 4, 8, 16, 32   19. 3, 7, 11, 15, 19

21. 32, 16, 8, 4, 2   23. 1, 2, 6, 24, 120   25. 1, 2, 3, 2, $\frac{5}{12}$

27.   29.   31.

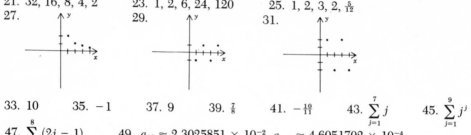

33. 10   35. −1   37. 9   39. $\frac{7}{8}$ 41. $-\frac{10}{11}$ 43. $\displaystyle\sum_{j=1}^{7} j$ 45. $\displaystyle\sum_{j=1}^{9} j^j$

47. $\displaystyle\sum_{j=1}^{8} (2j-1)$ 49. $a_{10} \approx 2.3025851 \times 10^{-2}$, $a_{100} \approx 4.6051702 \times 10^{-4}$

51. $a_5 = 3.84 \times 10^{-2}$, $a_{15} \approx 2.9862814 \times 10^{-6}$

53. $a_{10} \approx 4.5399930 \times 10^{-4}$; $a_{20} \approx 4.1223072 \times 10^{-8}$

55. $a_{100} \approx 1.0471285$, $a_{1000} \approx 1.0069317$

57. 2.9289683   59. 1.5804403   61. 77   65. $a_5 = 9$, $a_8 = 5$

69. a. $2^n n!$ b. $\dfrac{(2n)!}{2^{2n}(n!)^2}$

SECTION 9.3

1. 3   3. $\frac{1}{2}$ 5. ln 2   7. No   9. Yes   11. Yes   13. No

15. $a_4 = 11$, $a_n = 2 + 3(n-1) = -1 + 3n$ 17. $a_4 = 2\pi$, $a_n = 2\pi$

19. $a_4 = -1$, $a_n = -3 + \dfrac{2}{3}(n-1) = -\dfrac{11}{3} + \dfrac{2n}{3}$

21. $a_4 = e + 6$, $a_n = e + 2(n-1) = e - 2 + 2n$

23. $a_7 = 22$, $a_n = 4 + 3(n-1) = 1 + 3n$

25. $a_7 = -9$, $a_n = -\dfrac{3}{2}(n-1) = \dfrac{3}{2} - \dfrac{3n}{2}$

27. $a_7 = 7 \ln 2$, $a_n = n \ln 2$ 29. $a_n = 6 - (n-1) = 7 - n$

31. $a_n = \dfrac{28}{3} - \dfrac{5}{3}(n-1) = 11 - \dfrac{5n}{3}$ 33. $a_n = n - 1$ 35. 63   37. 42

39. 210 ln 2   41. $\dfrac{75\pi}{2}$ 43. Yes   45. $a_n = 9 - 2n$ 47. 2500

49. $a_n = 7 + 2(n-1) = 5 + 2n$ 51. Common difference: $4a^3b - 4ab^3$

53. $450   55. 429   57. 846 hours

SECTION 9.4

1. 4   3. −0.3   5. 16   7. Yes   9. No   11. Yes   13. No

15. $a_5 = 64$, $a_n = 2^{n+1}$ 17. $a_5 = \frac{1}{432}$, $a_n = 3(\frac{1}{6})^{n-1}$ 19. $a_5 = 810$, $a_n = 10(3^{n-1})$

21. $a_5 = 625c$, $a_n = c(5^{n-1})$ 23. $a_6 = 81$, $a_n = 3^{n-2}$ 25. $a_6 = \frac{3}{16}$, $a_n = 6(\frac{1}{2})^{n-1}$

27. $a_6 = -\frac{1}{4}$, $a_n = 256(-\frac{1}{4})^{n-1}$ 29. $a_6 = -\frac{1}{1458}$, $a_n = \frac{1}{6}(-\frac{1}{3})^{n-1}$ 31. $a_n = 4(\frac{3}{2})^{n-1}$

33. $a_n = 10(5)^{n-1}$ 35. $a_n = -5(\frac{2}{5})^{n-1}$ 37. 117   39. 60   41. 121

43. $\frac{43}{64}$ 45. $\frac{484}{243}$ 47. $-\frac{75}{128}$ 49. $\frac{3}{2}$ 51. $\frac{8}{3}$ 53. 12   55. $\frac{6}{7}$

57. $\frac{7}{9}$ 59. $\frac{95}{99}$ 61. $\frac{201}{37}$ 63. $\frac{3116}{990}$ 65. $a_n = 4^{n-2}$ 67. 18

69. $8(\frac{3}{2})^{n-1}$ and $-8(-\frac{3}{2})^{n-1}$ 71. $d = \ln r$ 73. 500

SECTION 9.5

1. 60   3. 2520   5. 15   7. 4   9. 120   11. 1320   13. 6

15. 1   17. 10   19. 2   21. 15   23. 136   25. a. $\frac{1}{2} n(n-1)$

29. 12   31. a. 576,000   b. 397,440   33. a. 45   b. 90   35. $\frac{20!}{5!}$ ($\approx 2 \times 10^{16}$)

37. 2520   39. 10   41. 120   43. a. 12! (= 479,001,600)
b. 5! × 7! (= 604,800)   45. 27,720   47. No

SECTION 9.6

1. $x^5 + 5x^4y + 10x^3y^2 + 10x^2y^3 + 5xy^4 + y^5$ 3. $x^4 + 8x^3 + 24x^2 + 32x + 16$
5. $x^7 + 7x^6y + 21x^5y^2 + 35x^4y^3 + 35x^3y^4 + 21x^2y^5 + 7xy^6 + y^7$
7. $a^5 - 5a^3 + 10a - \frac{10}{a} + \frac{5}{a^3} - \frac{1}{a^5}$ 9. $x^4 + 8x^3 + 28x^2 + 56x + 70 + \frac{56}{x} + \frac{28}{x^2} + \frac{8}{x^3} + \frac{1}{x^4}$
11. $17 + 12\sqrt{2}$ 13. $-38 - 41i$ 15. $56x^5y^3$ 17. -160 19. -252 21. 0.8684
23. a. $\left(-\frac{1}{3}x + y^3\right)^{12}$ b. $\frac{308}{243} x^6 y^{18}$

CHAPTER 9 REVIEW

9. 3, 9, 11, 17, 19   11. 3, -2, 8, -12, 28   13. -10 15. $-\frac{205}{36}$

17. 872   19. $\frac{89}{27}$ 21. $a_n = 5 + \frac{3}{4}(n-1) = \frac{17 + 3n}{4}$

23. $a_n = 21 - 2(n-1) = 23 - 2n$ 25. $a_n = 4(-\frac{1}{2})^{n-1}$ 27. $\frac{21}{11}$ 29. $-\frac{11}{2}$
31. $\frac{5}{33}$ 33. 120   35. 6   37. 20   39. 21   41. 70
43. $x^8 + 8x^7y + 28x^6y^2 + 56x^5y^3 + 70x^4y^4 + 56x^3y^5 + 28x^2y^6 + 8xy^7 + y^8$
45. $-27x^3 + 108x^2 - 144x + 64$ 47. a. 98   b. 104i   49. 480
51. The ten-day gift   53. 8,100,000   55. a. 1326   b. 64   57. 3150

INDEX

ALGEBRAIC FORMULAS

$(a + b)^2 = a^2 + 2ab + b^2$

$(a - b)^2 = a^2 - 2ab + b^2$

$x^2 - a^2 = (x + a)(x - a)$

$x^3 - a^3 = (x - a)(x^2 + ax + a^2)$

$x^3 + a^3 = (x + a)(x^2 - ax + a^2)$

$\dfrac{a}{b} + \dfrac{c}{d} = \dfrac{ad + bc}{bd}$

$\dfrac{a}{b} \cdot \dfrac{c}{d} = \dfrac{ac}{bd}$

$\dfrac{a}{b} \div \dfrac{c}{d} = \dfrac{ad}{bc}$

LINES

Slope: $m = \dfrac{y_2 - y_1}{x_2 - x_1}$

Point–slope equation: $y - y_1 = m(x - x_1)$

Slope–intercept equation: $y = mx + b$

LAWS OF INEQUALITIES

If $a < b$, then $a + c < b + c$.

If $a < b$ and $c > 0$, then $ac < bc$.

If $a < b$ and $c < 0$, then $ac > bc$.

$ac > 0$ if $a > 0$ and $c > 0$, or if $a < 0$ and $c < 0$.

$ac < 0$ if $a < 0$ and $c > 0$, or if $a > 0$ and $c < 0$.

ABSOLUTE VALUE

$|a| = \begin{cases} a & \text{if } a \geq 0 \\ -a & \text{if } a < 0 \end{cases}$

$|a + b| \leq |a| + |b|$

$|ab| = |a||b|$

$|x| < c$ if and only if $-c < x < c$

QUADRATIC FORMULA

The solutions of $ax^2 + bx + c = 0$ are given by

$$x = \frac{-b \pm \sqrt{b^2 - 4ac}}{2a}$$